Handbook of Pesticides

Handbook of Pesticides

Edited by Edwin Tan

SYRAWOOD
PUBLISHING HOUSE

New York

Published by Syrawood Publishing House,
750 Third Avenue, 9th Floor,
New York, NY 10017, USA
www.syrawoodpublishinghouse.com

Handbook of Pesticides
Edited by Edwin Tan

© 2017 Syrawood Publishing House

International Standard Book Number: 978-1-68286-391-6 (Hardback)

Cataloging-in-publication Data

Handbook of pesticides / edited by Edwin Tan.
     p. cm.
Includes bibliographical references and index.
ISBN 978-1-68286-391-6
1. Pesticides. 2. Agricultural chemicals. 3. Pests--Control--Equipment and supplies. I. Tan, Edwin.
SB951 .H36 2017
632.95--dc23

Printed in the United States of America.

TABLE OF CONTENTS

Preface..IX

Chapter 1   Reconciling Pesticide Reduction with Economic and Environmental
Sustainability in Arable Farming..1
Martin Lechenet, Vincent Bretagnolle, Christian Bockstaller, François Boissinot,
Marie-Sophie Petit, Sandrine Petit, Nicolas M. Munier-Jolain

Chapter 2   Increasing Minimum Daily Temperatures are Associated with Enhanced Pesticide
use in Cultivated Soybean along a Latitudinal Gradient in the Mid-Western
United States...11
Lewis H. Ziska

Chapter 3   Enantioselective Metabolism of Quizalofop-Ethyl in Rat...19
Yiran Liang, Peng Wang, Donghui Liu, Zhigang Shen, Hui Liu, Zhixin Jia,
Zhiqiang Zhou

Chapter 4   Swimming with Predators and Pesticides: How Environmental Stressors Affect
the Thermal Physiology of Tadpoles..27
Marco Katzenberger, John Hammond, Helder Duarte, Miguel Tejedo,
Cecilia Calabuig, Rick A. Relyea

Chapter 5   Insecticide Resistance Status of United States Populations of Aedes albopictus
and Mechanisms Involved..38
Sébastien Marcombe, Ary Farajollahi, Sean P. Healy, Gary G. Clark, Dina M. Fonseca

Chapter 6   Development of Composite Indices to Measure the Adoption of
Pro-Environmental Behaviours across Canadian Provinces..48
Magalie Canuel, Belkacem Abdous, Diane Bélanger, Pierre Gosselin

Chapter 7   Environmental Risk Factors and Amyotrophic Lateral Sclerosis (ALS).......................62
Yu Yu, Feng-Chiao Su, Brian C. Callaghan, Stephen A. Goutman,
Stuart A. Batterman, Eva L. Feldman

Chapter 8   Molecular Mechanisms of Reduced Nerve Toxicity by Titanium Dioxide
Nanoparticles in the Phoxim-Exposed Brain of Bombyx mori......................................71
Yi Xie, Binbin Wang, Fanchi Li, Lie Ma, Min Ni, Weide Shen, Fashui Hong, Bing Li

Chapter 9   Imidacloprid Alters Foraging and Decreases Bee Avoidance of Predators....................81
Ken Tan, Weiwen Chen, Shihao Dong, Xiwen Liu, Yuchong Wang, James C. Nieh

Chapter 10   Prenatal and Postnatal Exposure to DDT by Breast Milk Analysis in Canary Islands..........89
Oriol Vall, Mario Gomez-Culebras, Carme Puig, Ernesto Rodriguez-Carrasco,
Arelis GomezBaltazar, Lizzeth Canchucaja, Xavier Joya, Oscar Garcia-Algar

Chapter 11 Persistence and Dissipation of Chlorpyrifos in Brassica Chinensis, Lettuce, Celery, Asparagus Lettuce, Eggplant, and Pepper in a Greenhouse.........................96
Meng-Xiao Lu, Wayne W. Jiang, Jia-Lei Wang, Qiu Jian, Yan Shen, Xian-Jin Liu, Xiang-Yang Yu

Chapter 12 Trypsin-Catalyzed Deltamethrin Degradation.........................104
Chunrong Xiong, Fujin Fang, Lin Chen, Qinggui Yang, Ji He, Dan Zhou, Bo Shen, Lei Ma, Yan Sun, Donghui Zhang, Changliang Zhu

Chapter 13 Four Common Pesticides, their Mixtures and a Formulation Solvent in the Hive Environment have High Oral Toxicity to Honey Bee Larvae.........................109
Wanyi Zhu, Daniel R. Schmehl, Christopher A. Mullin, James L. Frazier

Chapter 14 Field Efficacy of Vectobac GR as a Mosquito Larvicide for the Control of Anopheline and Culicine Mosquitoes in Natural Habitats in Benin, West Africa.........................120
Armel Djénontin, Cédric Pennetier, Barnabas Zogo, Koffi Bhonna Soukou, Marina Ole-Sangba, Martin Akogbéto, Fabrice Chandre, Rajpal Yadav, Vincent Corbel

Chapter 15 Prenatal Exposure to Organophosphate Pesticides and Neurobehavioral Development of Neonates.........................127
Ying Zhang, Song Han, Duohong Liang, Xinzhu Shi, Fengzhi Wang, Wei Liu, Li Zhang, Lixin Chen, Yingzi Gu, Ying Tian

Chapter 16 Habitat Availability is a More Plausible Explanation than Insecticide Acute Toxicity for U.S. Grassland Bird Species Declines.........................137
Jason M. Hill, J. Franklin Egan, Glenn E. Stauffer, Duane R. Diefenbach

Chapter 17 Self-Reported Parental Exposure to Pesticide during Pregnancy and Birth Outcomes: The MecoExpo Cohort Study.........................145
Flora Mayhoub, Thierry Berton, Véronique Bach, Karine Tack, Caroline Deguines, Adeline Floch-Barneaud, Sophie Desmots, Erwan Stéphan-Blanchard, Karen Chardon

Chapter 18 Identification of Genomic Features in Environmentally Induced Epigenetic Transgenerational Inherited Sperm Epimutations.........................159
Carlos Guerrero-Bosagna, Shelby Weeks, Michael K. Skinner

Chapter 19 Paraoxonase Enzyme Protects Retinal Pigment Epithelium from Chlorpyrifos Insult.........................173
Jagan Mohan Jasna, Kannadasan Anandbabu, Subramaniam Rajesh Bharathi, Narayanasamy Angayarkanni

Chapter 20 Does Occupational Exposure to Solvents and Pesticides in Association with Glutathione S-Transferase A1, M1, P1, and T1 Polymorphisms Increase the Risk of Bladder Cancer? The Belgrade Case-Control Study.........................183
Marija G. Matic, Vesna M. Coric, Ana R. Savic-Radojevic, Petar V. Bulat, Marija S. Pljesa- Ercegovac, Dejan P. Dragicevic, Tatjana I. Djukic, Tatjana P. Simic, Tatjana D. Pekmezovic

Chapter 21 **Selection and Evaluation of Potential Reference Genes for Gene Expression
Analysis in the Brown Planthopper,** *Nilaparvata lugens* **(Hemiptera: Delphacidae)
using Reverse-Transcription Quantitative PCR**...**191**
Miao Yuan, Yanhui Lu, Xun Zhu, Hu Wan, Muhammad Shakeel, Sha Zhan,
Byung-Rae Jin, Jianhong Li

Permissions

List of Contributors

Index

PREFACE

Pesticides are a form of biocide used for protecting plants and crops from pests. Pesticides play a significant role in crop yield. This book will offer information about a wide variety of pesticides and also about its sub-fields such as herbicides, fungicides, insect growth regulators, etc. It elucidates new techniques and their applications in a multidisciplinary approach to understand the need and use of pesticides. This book provides significant information of this discipline to help develop a good understanding of pest control and its processes. It will help the readers in keeping pace with the rapid changes in this field.

This book has been a concerted effort by a group of academicians, researchers and scientists, who have contributed their research works for the realization of the book. This book has materialized in the wake of emerging advancements and innovations in this field. Therefore, the need of the hour was to compile all the required researches and disseminate the knowledge to a broad spectrum of people comprising of students, researchers and specialists of the field.

At the end of the preface, I would like to thank the authors for their brilliant chapters and the publisher for guiding us all-through the making of the book till its final stage. Also, I would like to thank my family for providing the support and encouragement throughout my academic career and research projects.

Editor

Reconciling Pesticide Reduction with Economic and Environmental Sustainability in Arable Farming

Martin Lechenet[1], Vincent Bretagnolle[2], Christian Bockstaller[3,4], François Boissinot[5], Marie-Sophie Petit[6], Sandrine Petit[1], Nicolas M. Munier-Jolain[1]*

1 Institut National de la Recherche Agronomique, Unité Mixte de Recherche 1347 Agroécologie, Dijon, Côte d'Or, France, 2 Centre d'Etudes Biologiques de Chizé - Centre National de Recherche Scientifique, Beauvoir sur Niort, Deux-Sèvres, France, 3 Institut National de la Recherche Agronomique, Unité de Recherche 1121 Agronomie et Environnement, Colmar, Haut-Rhin, France, 4 Université de Lorraine, Vandœuvre-lès-Nancy, Meurthe-et-Moselle, France, 5 Chambre d'Agriculture des Pays de la Loire, Angers, Maine-et-Loire, France, 6 Chambre Régionale d'Agriculture de Bourgogne, Quetigny, Côte d'Or, France

Abstract

Reducing pesticide use is one of the high-priority targets in the quest for a sustainable agriculture. Until now, most studies dealing with pesticide use reduction have compared a limited number of experimental prototypes. Here we assessed the sustainability of 48 arable cropping systems from two major agricultural regions of France, including conventional, integrated and organic systems, with a wide range of pesticide use intensities and management (crop rotation, soil tillage, cultivars, fertilization, etc.). We assessed cropping system sustainability using a set of economic, environmental and social indicators. We failed to detect any positive correlation between pesticide use intensity and both productivity (when organic farms were excluded) and profitability. In addition, there was no relationship between pesticide use and workload. We found that crop rotation diversity was higher in cropping systems with low pesticide use, which would support the important role of crop rotation diversity in integrated and organic strategies. In comparison to conventional systems, integrated strategies showed a decrease in the use of both pesticides and nitrogen fertilizers, they consumed less energy and were frequently more energy efficient. Integrated systems therefore appeared as the best compromise in sustainability trade-offs. Our results could be used to re-design current cropping systems, by promoting diversified crop rotations and the combination of a wide range of available techniques contributing to pest management.

Editor: Raul Narciso Carvalho Guedes, Federal University of Viçosa, Brazil

Funding: Funding for the study was provided by the French National Research Agency ANR (STRA-08-02 Advherb project) and from Région Bourgogne. The Burgundy farms network was developed in the framework of the Réseau Mixte Technologique "Systèmes de Culture Innovants" and the project "Plus d'Agronomie, Moins d'intrants" initiated by Région Bourgogne. The long term experiment at Dijon-Epoisses was partly funded by the European Network of Excellence ENDURE. The funders had no role in study design, data collection and analysis, decision to publish, or preparation of the manuscript.

Competing Interests: The authors have declared that no competing interests exist.

* E-mail: nicolas.munier-jolain@dijon.inra.fr

Introduction

Reconciling agricultural productivity with other components of sustainability remains one of the greatest challenges for agriculture [1]. A key issue will be to achieve substantial reductions in the level of pesticide use for environmental and health reasons [2,3]. Agriculture in temperate climates is widely dominated by conventional intensive farming systems, with highly specialized crop productions and a heavy reliance on pesticides and mineral fertilizers [4]. However, increasing environmental concerns about intensive farming practices has contributed to the emergence of innovative farming systems, such as organic and integrated farming, typically presented as alternative paths to reduce pesticide use as compared to current conventional systems [5,6,7]. Whether these systems better meet sustainability criteria has been a matter of debate [8,9]. Integrated farming, recently promoted in Europe through the 2009/128/EC European directive [10], is defined as a crop protection management based on Integrated Pest Management (IPM) principles, which emphasizes physical and biological regulation strategies to control pests while reducing the reliance on pesticides [11]. It can be regarded as an intermediate between conventional farming, with high levels of inputs, and organic

farming, which prohibits the use of synthetic pesticides and fertilizers. Organic and integrated farming have in common the combined use of management approaches to replace, at least in part, synthetic inputs. However, unlike organic farming which is growing both in Europe (by 40 to 50% between 2003 and 2010 [12]) and in the US (by 270% between 2000 and 2008 [13]), integrated arable crop production is not expanding because it is perceived by farmers as a complex system which is difficult to implement, labour-consuming, and associated with reduced and unpredictable economic profitability [14,15]. As a consequence, the amount of pesticides sprayed has only decreased slightly in Europe (-3.6% from 2000 to 2007 [16]) and in the US (-7.5% from 2000 to 2007 [17]). Moreover, this decrease can be partly attributed to the substitution of older chemistry, applied at high dosage, by new products that are efficient at lower doses, which actually cannot be considered as a reduction of pesticide reliance. In France, the national action plan, ECOPHYTO 2018, which had set a target of a 50% decrease in pesticide use by the year 2018, is currently far from achieving this goal [18].

So far, assessments of cropping system sustainability have compared few – typically two or three – experimental prototypes that represent conventional, organic or integrated strategies

[19,20]. However, this approach fails to capture the diversity within each of these farming strategies. Given the diversity of crop management options within a conventional, an integrated or an organic strategy, which might lead to contrasted performances, the generic value of experimental results ignoring this variability may be argued. We assessed the sustainability of 48 cropping systems located in regions of intensive arable farming and covering a wide range of pesticide use levels and cultivation techniques such as crop rotations, from monoculture to highly diversified crop rotations, soil tillage (e.g. inversion tillage, shallow tillage or direct drilling), fertilization (mineral or organic fertilizers), or weed management (e.g. only based on herbicide use, including mechanical weeding). More details about the cropping system sample are available in the online SI section (Dataset S1). All the studied cropping systems were followed for between three and 12 years, between 1999 and 2012. Eight cropping systems were organic, 30 were based on integrated farming and 10 were conventional (Figure 1). Using eight sustainability indicators to evaluate the performance of the study systems, our aims were: (i) to identify possible conflicts between the reduction of pesticide reliance and other components of sustainability; and, (ii) to assess the potential of organic and integrated strategies for improving agricultural sustainability.

As the performance of a cropping system depends not only on the combination of management options it implements, but also on the local production situation [21] (including biophysical and socio-economic local aspects), we standardized the indicators of performance and pesticide use, using a ratio of the performances of the cropping systems over those of a local reference system. This enabled us to focus solely on the effects of the management strategies on sustainability indicators. The local references were cropping systems selected as representative of the most widespread crops and practices within each production situation. Pesticide use

was measured as the Treatment Frequency Index (TFI), which is a commonly used indicator in Europe to estimate the cropping system dependence on pesticides [22]. In our sample, organic cropping systems did not use any pesticides (synthetic or natural) so their relative TFI, expressed as a ratio of the local reference TFI, was zero. Integrated cropping systems displayed TFI values that were on average half (−47%) of the local references (Table S1).

Results

Table S1 presents the mean and standard deviation for each performance indicator according to the management strategy (organic, integrated and conventional). The second tab of Dataset S1 provides performance details for each cropping system of the sample.

Productivity and energy efficiency

Given the primary role of agriculture remains to produce food and other goods, we used an indicator of productivity, expressed as the total yearly amount of energy produced by a cropping system, whatever the crops cultivated (Figure 2a). The productivity of organic cropping systems was below that of their local reference (Figure 3b), ranging from −22% to −76%. For non-organic cropping system, productivity was uncorrelated to relative TFI (Figure 2a and Table 1), with some cropping systems that had a low reliance on pesticides even exceeding the productivity of the local reference. Cropping system productivity may strongly depend on crop type, especially if the whole above-ground biomass is harvested or not. Crops other than grain crops were frequently grown in integrated farming, as they are typically associated with low pesticide requirements and can contribute to weed control in subsequent crops [23]. They typically consist of

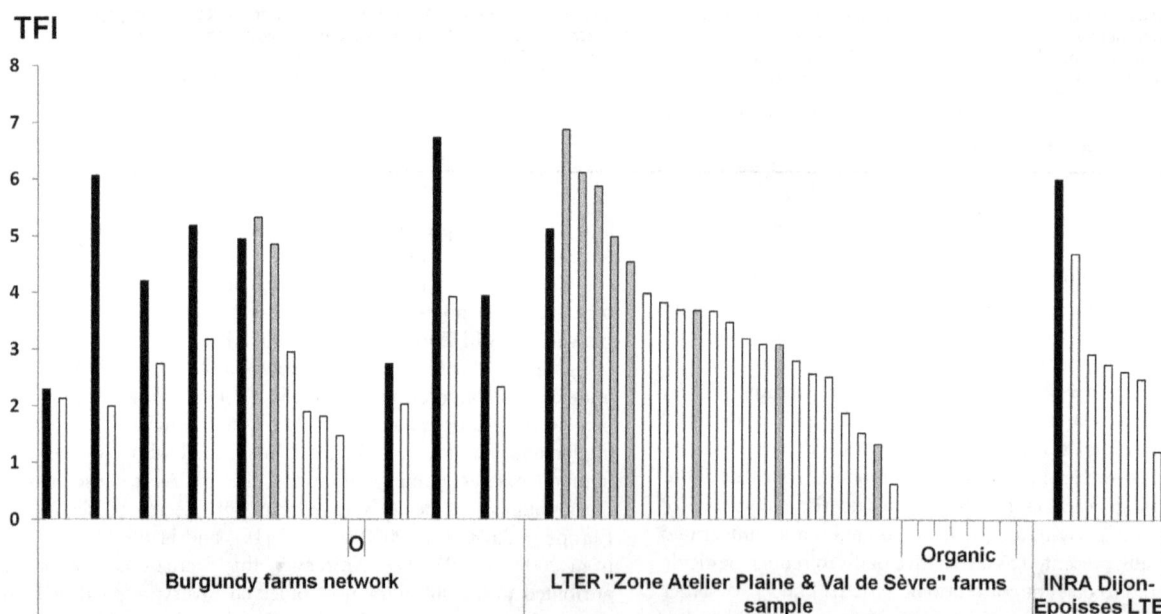

Figure 1. Distribution of the Treatment Frequency Index (TFI) for the studied arable cropping systems. Average TFI for each cropping system composing the study sample. At each site, black bars correspond to the local reference, grey bars to conventional cropping systems and white bars to integrated cropping systems. The sample also includes eight organic cropping systems with TFI = 0 and labelled with "O" or "Organic". Details about the cropping systems are available in Dataset S1.

forage crops, dedicated to livestock feeding with limited energy efficiency, or of crops used for non-food applications. However, distinguishing cropping systems based on grain crops or on crops in which all above-ground biomass is harvested did not change the observed pattern. In systems with grain crops only, productivity was not correlated with relative TFI (Table 1), suggesting that a reduction in pesticide use intensity may not be necessarily translated into a decrease in productivity. The second indicator of energy productivity we used was the energy efficiency of cropping systems, resulting from a ratio between energy output and energy input. It evaluated the ability of a cropping system to convert energy inputs into outputs. Organic cropping systems were significantly less energy efficient than other systems (Figures 2b and 3c, Table 2). Despite their energy consumption being lower (Table 2), notably due to their low reliance on nitrogen fertilizers, it was not sufficient to offset their limited productivity. Energy consumption was negatively correlated with relative TFI in integrated and conventional systems which cultivated only grain crops (Table 1). Energy efficiency was also negatively correlated with relative TFI in these systems, although the relationship was weak and only marginally significant ($r_s = -0.35$, $P = 0.07$). The systems with the highest energy efficiency, whether they included crops with all above-ground biomass harvested or not, were mostly integrated systems (Figures 2b and 3c).

Environmental impact

The environmental impact of cropping systems and their reliance on external inputs were assessed with the indicator I-Pest [24] and with estimates of fuel and nitrogen fertilizer consumption. I-Pest is a predictive indicator that assesses the environmental impacts of pesticide use as the risk of contamination of the air, and surface and ground waters (see Figure S1). As the organic cropping systems composing the sample did not used synthetic or natural pesticide, their cumulated I-Pest was 0. As expected for the rest of the sample, cumulated I-Pest was strongly and positively correlated to relative TFI (Figure 2c and Table 1). Fuel and nitrogen fertilizers together amounted to more than 60% of the total energy inputs for all tested cropping systems. Organic systems consumed more fuel than the rest of the sample (Table 2), with their average consumption exceeding the local references by 17% (Figure 3d). Organic cropping systems had, nonetheless, a lower reliance on N fertilization than the rest of the sample (Figure 3g, Table 2), in line with their lower yield targets and the frequent occurrence of crops with low N requirements used in organic rotations. No relation was detected between fuel consumption and relative TFI in non-organic systems (Figure 2d, Table 1), but a positive correlation was clearly visible between relative TFI and the amount of nitrogen fertilizers applied (Figure 2e).

Economic sustainability and workload

Economic sustainability was assessed by considering (i) the profitability, i.e. the average semi-net margin over a range of ten real price scenarios for agricultural products, fuel and fertilisers, and (ii) the sensitivity of this profitability in a context of price volatility, i.e. the relative standard deviation of the semi-net margin. The range of price scenarios used for the calculations was set to reflect the variability of the economic context over the last decade. Profitability, when averaged over the ten price scenarios was not correlated with relative TFI for integrated and conventional systems (Figure 2f and Table 1), and no significant difference appeared with organic systems (Mann-Whitney test, P>0.9). It suggests that low pesticide use would not necessarily result in lower economic return. The strong variability observed within each class (Figure 3f), most notably for integrated cropping systems,

confirmed that strategies to reduce pesticide use could even lead to an increase in profitability. As integrated cropping systems were, in contrast to organic systems, evaluated with a conventional price reference, the most profitable integrated systems were able to efficiently reduce their production costs. No relation was detected between the sensitivity to price volatility and relative TFI in conventional and integrated systems (Figure 2g, Table 1). Sensitivity to price volatility was significantly lower in organic cropping systems than in other systems (Table 2), most probably because: (i) they were based on more diversified crop rotations, which spread risks and buffered semi-net margin at the farming system scale; and, (ii) their crop rotations typically included crops with low N demand, that had reduced reliance on N inputs, whose price is directly related to the volatile price of fossil fuels.

The issue of social sustainability was addressed using the 'workload' indicator, which gives emphasis to the potential for bottlenecks where available workforce is a limiting factor at the farm scale (Figure 2h). Workload was calculated for each technical operation but excluded time devoted to transport and crop monitoring. Workload was found not correlated with relative TFI in non-organic cropping systems (Table 1), and no significant difference was found with the organic group (Mann-Whitney test, P>0.1), so that reducing pesticide use does not necessarily imply an increased workload. Indeed, in integrated systems, labour requirements ranged from low to high relative values (Fig 2h). The level of workload was, however, related to the type of fertilization, with cropping systems having organic fertilization requiring an average of 13% greater working time, as compared to mineral fertilizer-based cropping systems (Table 2).

Crop diversity

Diversification of crop rotations is often presented as an efficient management tool for controlling pests and to improve agricultural sustainability [25,26]. We used a crop sequence indicator, Isc [27], which estimates the consistency of the crop sequence with regard to the potential of input reduction, by addressing effects of crop rotation on pathogens, pests, weeds, soil structure and nitrogen supply of preceding crops. Even if no significant correlation appeared between Isc and relative TFI (Table 3), organic and integrated cropping systems displayed significantly higher Isc values than conventional systems (Table 2). A negative correlation between Isc and productivity suggests that diversifying crop rotation may reduce cropping system productivity (Table 3), but the Spearman correlation test was no longer significant when organic cropping systems were excluded (P = 0.07). We did not detect any significant relationship between energy efficiency and crop diversification, whether organic cropping systems were included or not (P = 0.44). No correlation was observed between Isc and semi-net margin, but workload appeared to be lower for systems with higher Isc (Table 3). We found the expected negative correlation between Isc and N fertilization rates, and consequently between Isc and energy consumption (Table 3). We focused therefore more particularly on cropping systems including legume crops, which also displayed higher Isc values than the rest of the sample (Table 2). The role of legume in improving energy efficiency at the cropping system scale was clearly demonstrated by the correlation between the frequency of occurrence of legumes in the crop rotation and the energy efficiency ($r_s = 0.37$, $P < 0.05$). The sensitivity to price volatility was negatively correlated with the frequency of occurrence of legumes in the crop rotation ($r_s = -0.33$, $P = 0.02$), but positively correlated with the level of N fertilization ($r_s = 0.49$, $P = 5*10^{-4}$). Fostering exogenous N independence therefore appeared as an efficient way to limit income variability.

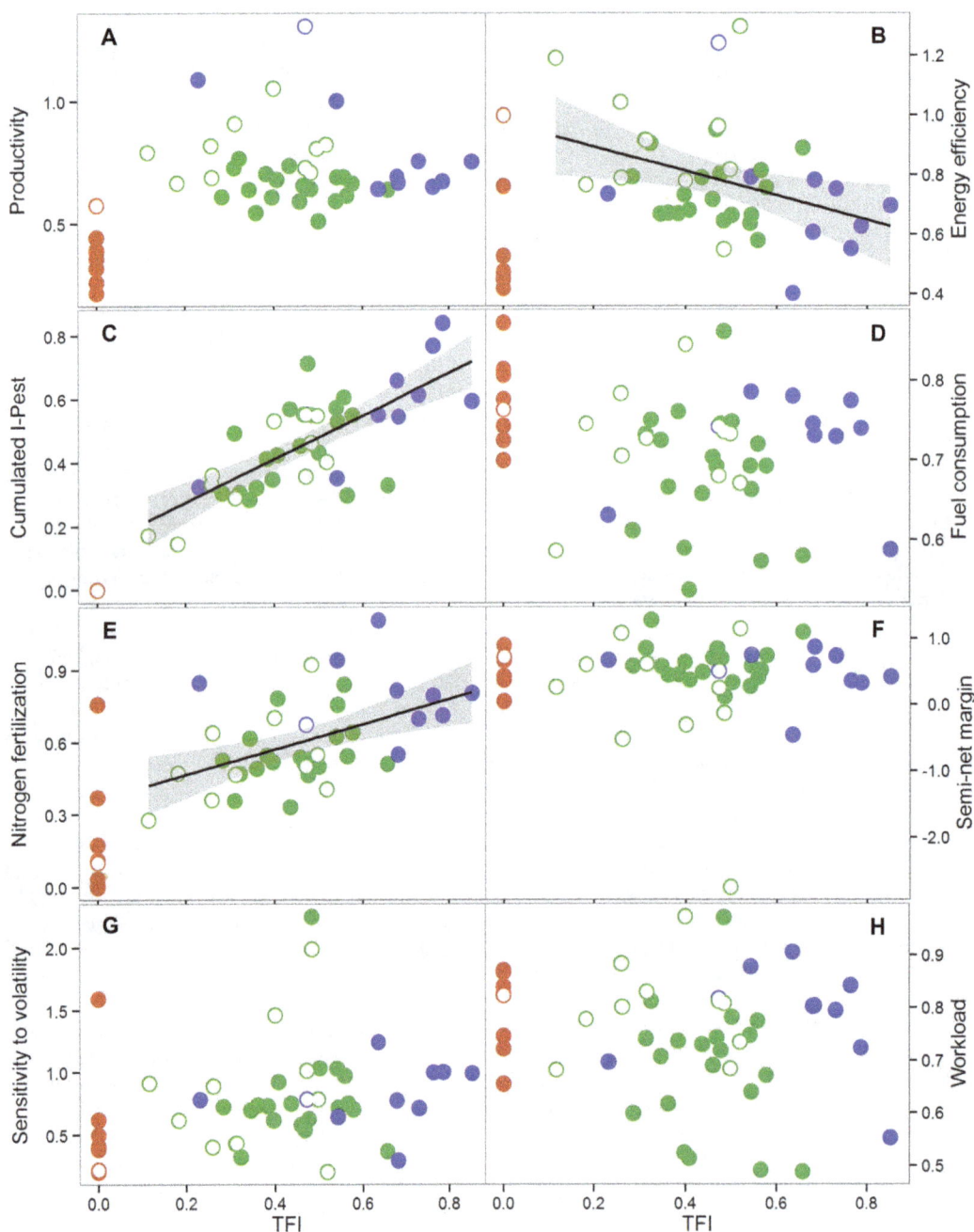

Figure 2. Relationship between sustainability indicators and relative TFI. Cropping system performances according to their relative TFI. Conventional, integrated and organic cropping systems are represented by blue, green and red symbols respectively. Filled symbols correspond to the cropping systems with grain crops only and empty symbols refer to the cropping systems including crops for which the whole above-ground biomass is harvested. Each sustainability indicators is expressed as the natural logarithm of the ratio between the cropping system and the local reference indicators. Linear regressions are represented with their standard error for cumulated I-Pest (Pearson correlation test: $r_p = 0.74$, $P = 5*10^{-8}$), nitrogen fertilization (Pearson correlation test: $r_p = 0.48$, $P = 0.002$), and energy efficiency (Pearson correlation test: $r_p = -0.38$, $P = 0.02$). Performance metric included: a) productivity, b) energy efficiency, c) cumulated I-Pest, d) fuel consumption, e) nitrogen fertilization, f) semi-net margin, g) sensitivity to price volatility, h) workload.

Discussion

This work was aimed at detecting cropping systems able to reconcile low pesticide use and other components of sustainability. Our original multiple dimensions approach, based on a precise description of management practices, was designed to compare and contrast numerous cropping systems from different produc-

tion situations. This approach, applied at the large-scale, was able to provide generic knowledge about potential trade-offs between the different issues of agricultural sustainability.

Sustainability of integrated and organic farming

Our results show that achieving a low level of pesticide use is possible without triggering negative side effect on any of the

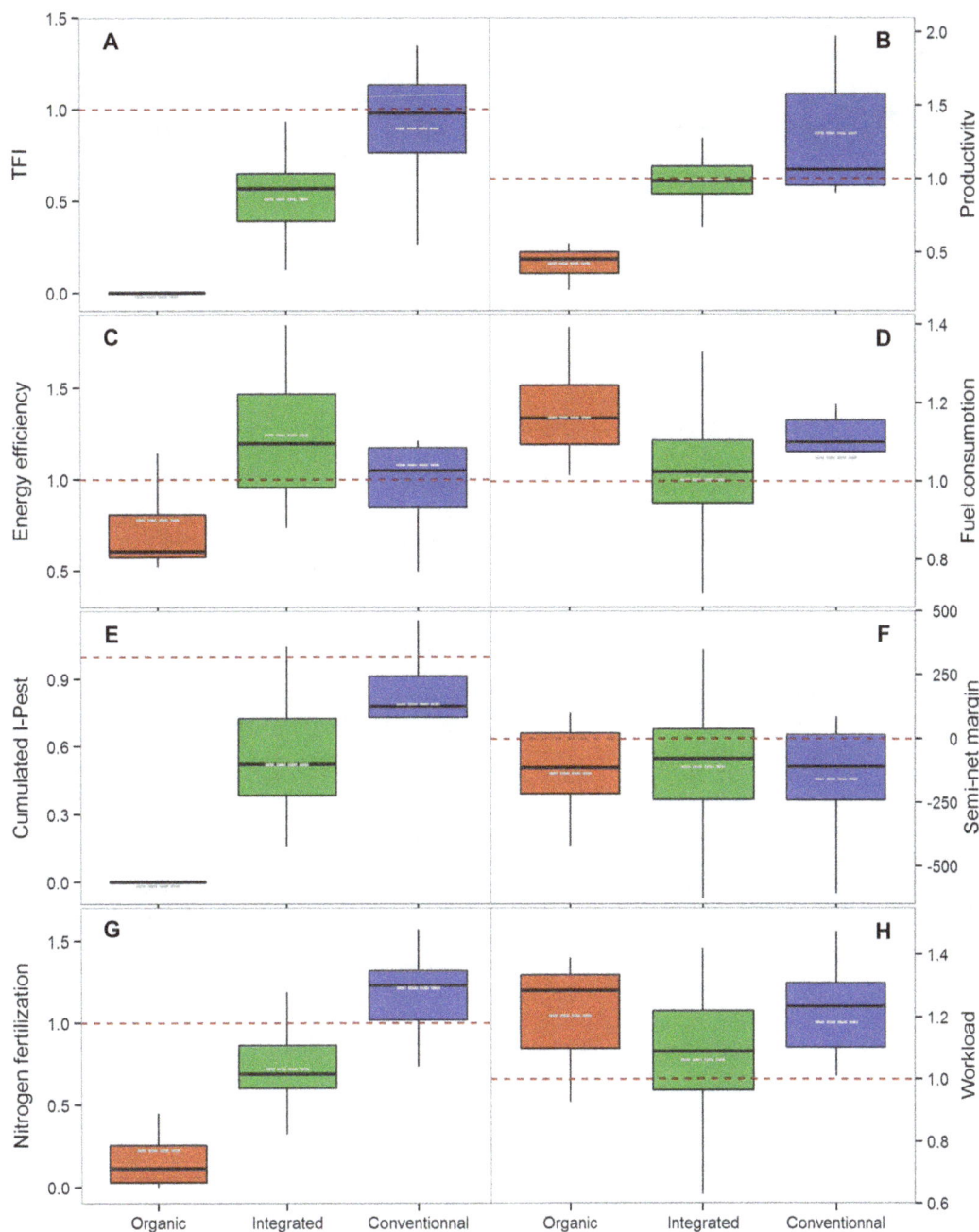

Figure 3. Cropping systems distribution according to sustainability indicators. Performance indicators are expressed as a ratio of the local reference indicator, except for semi-net margin, expressed as a difference with the local reference. Conventional, integrated and organic cropping systems are represented by blue, green and red box plots respectively. The horizontal black bars and grey dashed bars correspond to median and mean values respectively. The horizontal red dashed bar recalls the position of the local references. Outliers are not represented. Performance metrics included: a) Treatment Frequency Index, b) productivity (organic farming: one outlier, $v = 0.78$; integrated farming: two outliers, $v1 = 1.48$ and $v2 = 1.87$), c) energy efficiency (organic farming: one outlier, $v = 1.72$), d) fuel consumption (integrated farming: one outlier, $v = 1.37$; conventional farming: two outliers $v1 = 0.8$ and $v2 = 0.88$), e) cumulated I-Pest (conventional farming: two outliers $v1 = 0.38$ and $v2 = 0.42$), f) semi-net margin, g) nitrogen fertilization (organic farming: one outlier, $v = 1.13$; integrated farming: two outliers, $v1 = 1.32$ and $v2 = 1.52$), h) workload (conventional farming: one outlier $v = 0.74$).

components of cropping system sustainability we assessed in this study. Integrated cropping systems were not only associated with low pesticide use and low risks for contamination of air and water with pesticide residues; they also displayed lower energy consumption than more intensive cropping systems and are likely to improve energy efficiency without impact on productivity and

profitability. Lower pesticide usage in arable cropping systems did not imply a heavier workload, another critical point conditioning strongly the adoption of an innovative strategy.

Organic farming prohibits the use of synthetic pesticides and fertilizers, and this approach is often associated with low nitrogen fertilization, as observed in our sample. In addition to the positive

Table 1. Rank correlation between TFI and sustainability indicators for integrated and conventional cropping systems.

Spearman correlation	Productivity	Productivity (grain crops only)	Energy consumption	Energy consumption (grain crops only)	N fertilization	Pesticides environmental impact
r_s	−0.17 (NS)	0.06 (NS)	0.30 (NS)	0.42	0.51	0.67
P-value	0.3	0.8	0.06	0.03	$9*10^{-4}$	$4*10^{-6}$
Spearman correlation	**Fuel consumption**	**Energy efficiency**	**Semi-net margin**	**Sensitivity to prices volatility**	**Workload**	**Crop Sequence Indicator (Isc)**
r_s	0.06 (NS)	−0.40	−0.05 (NS)	0.22 (NS)	$9*10^{-3}$ (NS)	−0.22
P-value	0.7	0.01	0.8	0.2	0.95	0.2

Spearman rank correlation tests ($\alpha = 0.05$). r_s is the Spearman correlation coefficient. Values of r_s followed by (NS) are not significant.

effects on environmental quality, numerous studies underlined other environmental benefits of organic farming such as effects on pollinator dynamics [28], on landscape floristic composition [29], as well as on soil microbial diversity [5,30]. Here we demonstrated that organic farming does not necessarily affect profitability and workload, and conversely, might strengthen farm financial stability in a variable and unpredictable economic context. Organic cropping systems were however less productive and less energy efficient than integrated systems in our sample. Although highly dependent on crops and production context, productivity in organic farming was already reported as lower than in conventional farming by other comparative studies [31]. The poor land use efficiency associated with organic farming is a key issue in the current land sharing – land sparing debate about the growing competition for land use [32], and notably urban sprawl [33] as well as the necessity to keep natural spaces undisturbed [34,35]. Both aspects – environmental benefits of organic farming and the limited productivity per unit of land – should therefore be considered by decision makers in their incentives for sustainable agriculture.

Crop diversification

Our results support the hypothesis that crop diversification may be an effective means to enhance cropping system performance. At the cropping system scale, crop diversification provides agronomic advantages, such as the regulation of pests, diseases and weeds [36,37,25]. In our sample, the most diversified cropping systems, which displayed the highest values of the crop sequence indicator, Isc, were indeed less dependent on pesticides. Their low environmental impact on water and air quality makes crop diversification an interesting potential pathway for reducing the damage caused by agriculture on natural resources (e.g., biodiversity [38]), as well as on human health (e.g. neurological degenerative disorders [39]). By mitigating the adverse effects of climate variability, crop diversification may also improve system resilience for productivity [40], with the increasing likelihood of extreme weather events requiring farm adaptation [41]. Economic market volatility is an additional source of variation and risk factor for farm economic stability. We found that crop diversification, particularly through the introduction of legumes in the crop rotation, is likely to limit dependence on inputs that have unstable prices. By allowing a decrease in the use of exogenous N fertilizer across a crop rotation, legume cultivation reduces production cost fluctuations and consequently makes the cropping system less sensitive to market volatility. Legumes come with a supplementary advantage [42] in the face of the considerable amount of fossil energy necessary to produce mineral N fertilizers, and we noted a substantial increase in energy efficiency for crop rotations where

Table 2. Significantly different groups for a given performance indicator.

Indicator	Test designation	P-value	Statistic W
Productivity	Difference of productivity between organic cropping systems and the rest of the sample	$2*10^{-8}$	318
Productivity	Difference of productivity between cropping systems including crops with the whole above-ground biomass harvested and the rest of the sample	$5*10^{-4}$	76
Energy efficiency	Difference of energy efficiency between organic cropping systems and the rest of the sample	0.005	258
Energy consumption	Difference of energy consumption between organic cropping systems and the rest of the sample	0.001	272
Fuel consumption	Difference of fuel consumption between organic cropping systems and the rest of the sample	0.02	72
N fertilization rate	Difference of N fertilization rate between organic cropping systems and the rest of the sample	$2*10^{-4}$	285
Sensitivity to price volatility	Difference of sensitivity to price volatility between organic cropping systems and the rest of the sample	0.01	250
Workload	Difference of workload between cropping systems based on organic fertilization and the rest of the sample	0.045	189
Crop sequence indicator	Difference of Isc between organic cropping systems and conventional cropping systems	0.01	69
Crop sequence indicator	Difference of Isc between integrated cropping systems and conventional cropping systems	0.001	255
Crop sequence indicator	Difference of Isc between cropping systems including legumes and the rest of the sample	$6*10^{-6}$	63

Mann-Whitney tests ($\alpha = 0.05$). All P-values are below 0.05, indicating that the differences between means of the sub-samples are significant for the corresponding indicators.

Table 3. Rank correlation between Crop Sequence Indicator Isc and sustainability indicators.

Spearman correlation	Productivity	Energy consumption	N fertilization	Pesticides environmental impact	Fuel consumption
r_s	−0.35	−0.42	−0.41	−0.39	−0.07 (NS)
P-value	0.01	0.003	0.004	0.006	0.7
Spearman correlation	Energy efficiency	Semi-net margin	Sensitivity to price volatility	Workload	
r_s	−0.05 (NS)	0.10 (NS)	−0.23 (NS)	−0.29	
P-value	0.7	0.5	0.1	0.04	

Spearman correlation tests ($\alpha = 0.05$). r_s is the Spearman correlation coefficient. Values of r_s followed by (NS) are not significant.

legume crops are more frequent. The most part of legumes introduced as diversification crops are however forage crops, and livestock production is commonly considered more energy consuming than plant production [43]. Conversely, the use of farmyard manure may contribute to reduce mineral fertilizers reliance for grain and forage crop production. The necessity of (i) integrating these situation-dependent parameters into energy balancing calculations, and, (ii) evaluating other environmental indicators [44] will be critical for the assessment of livestock production as a management option for enhancing agricultural sustainability.

A key agronomical advantage of crop diversification is related to the management of weed resistance to herbicides. Crop diversification is an efficient means to alternate herbicide modes of action and to introduce diversified measures of weed control, allowing changing selection pressure on weed communities and thus maintaining a sensitive weed population (i.e. maintaining high herbicide efficiency) [45].

Our results demonstrate a negative correlation between the Isc value and workload. We can nevertheless assume that diversifying crop rotations increases cropping system complexity and time devoted to field observations. Another aspect is that crop diversification may lead to a more evenly distributed workload over the seasons. Crops diversity implies a greater diversity in sowing and harvest periods, which are both times of peak labour that strongly influence task organisation and farmer decision making [15]. By reducing the amplitude of these peaks in labour, crop diversification could contribute to ensuring greater farmer decisional flexibility at the farm scale.

Beyond technical and organizational issues at the farm level, diversifying crop production as a component of an integrated strategy at regional or national scale would inevitably lead to important changes in production volumes, as well as markedly changing agricultural sectors within each production basin. It would definitely require an adaptation in the organisation of the whole agricultural sector and the development of new local markets. These economic and social lock-ins are rightly highlighted as the main limiting constraints hindering crop diversification [46]. However, by creating a particular economic sub-context, niche markets can be attractive and able to support innovation. Promoting such niche markets, for integrated farming development, would be the first step along an accelerating cycle of improvement based on mutually positive feed-backs between production and outlets.

Materials and Methods

In all cases, the field studies did not involve endangered or protected species.

For future permissions about the private farm network of Burgundy, please contact Marie-Sophie Petit (co-author of the research article, Chambre Régionale d'Agriculture de Bourgogne) and Sandrine Petit (co-author of the research article, INRA).

For future permissions about the private farms survey carried out on the LTER "Zone Atelier Plaine & Val de Sèvre", please contact Nicolas Munier-Jolain (corresponding author, INRA).

Study areas

The main objective of this study was to highlight potential conflicts between pesticide use and a set of sustainability indicators, so the cropping systems we consider were selected to maximize the contrast across the range of possible pesticide use intensities. The sample of cropping systems we used originates from:

(i) A long term experiment conducted since 2000 at the INRA Dijon-Epoisses farm in Bretenière (Burgundy, eastern France; 47°20′N, 5°2′E) in order to assess Integrated Weed Management-based cropping systems [15,26]. Seven cropping systems were tested between 2000 and 2012 including different combinations of technical levers likely to reduce pesticide reliance.

(ii) An experimental network (bringing together 14 cropping systems) monitored (1) by the local agricultural extension services and coordinated by the Chambre Régionale d'Agriculture de Bourgogne, and (2) by the INRA de Dijon. This network involved contrasting private farms of the Burgundy region, and was developed to test feasibility of innovative cropping systems with reduced pesticide use in a realistic context.

(iii) A survey of private farms carried out in 2010 on the LTER "Zone Atelier Plaine & Val de Sèvre" [47] located in the Poitou-Charentes region (450 km^2 study area in western France), and set up to explore a diversity of pesticide reliance, including organic farming, conventional intensive systems, and intermediate, IPM-based systems. Twenty nine varied cropping systems were surveyed in this area.

Cropping systems classification

Details of cropping systems, including crop sequences, performances, and detailed crop management operations are made available in the Dataset S1 and S2. Cropping systems were considered as conventional, integrated or organic according to the following rules. Cropping systems complying with the organic farming specifications were treated as 'organic'. Other systems were considered as 'integrated', when they were based either on diversified crop rotations including unusual alternative crops for the production situation (i.e. not present in the local reference crop rotation), or when crop management included at least one non-chemical management approach that contributed to the control of pests, diseases or weeds. These included for instance biocontrol, mechanical weeding and false seed bed techniques. Systems that were not classified as 'organic' or 'integrated' were classified as 'conventional'.

Local reference definition

For each of the 48 systems, a local cropping system reference was selected to reflect the most widespread crop rotation and associated technical management, as well as the typical agricultural performance in the production situation. Using this local cropping system reference made it possible to distinguish the effects of agronomic strategies from the effects of the production situation (soil, climate, economic and social context) when assessing the various components of sustainability of each cropping system. The Dijon-Epoisses experiment included a reference standard system that follows recommendations of local extension services [26], and which was used as the local reference. For each farm of the network across Burgundy, the local reference was defined as the cropping system implemented within the farm before the set-up of the alternative cropping system, even though crop management sequences were slightly updated according to expert appraisal to match with current standards (e.g. active ingredients allowed). For the Zone Atelier "Plaine et Val de Sèvre", local expert knowledge was used to select one system from the survey, with a standard crop rotation for the area, and a crop management representative of local practices. This system was then used as the local reference for all the remaining surveyed systems of the area.

Assessment of sustainability

The assessment of sustainability at the cropping system scale was based on a range of indicators covering economic, environmental and social issues. The Treatment Frequency Index (TFI) [22] estimates the number of registered doses applied, for each pesticide, per hectare and per crop season. Averaged over the cropping system, this indicator summarizes the level of dependence on pesticides, which should be distinguished from the environmental impact of pesticide use. This indicator is calculated for each pesticide application according to the following formula:

$$TFI = \frac{Application\ rate \times Treated\ surface\ area}{Registered\ dose \times Plot\ surface\ area}$$

The application rate and the registered dose were both expressed for a given commercial product (which possibly contains several active ingredients). The recommended application dose depends obviously on the treated crop and on the targeted pest. Here we defined the registered dose as the lowest application dose which is recommended for a given crop. The TFI for a given crop season was then calculated as the sum of the TFI for each pesticide application performed during this crop season. Productivity was evaluated as the amount of energy harvested yearly. This approach allowed the comparison of different crop rotations that included crops with different yielding potentials and different energy content. For each crop, yields were transformed into the energy metric using their Lower Heating Value (LHV) [48], which corresponds to the amount of energy released per unit of mass by the combustion of the harvested biomass. Energy consumption

was estimated from the conversion of inputs into energy according to the Dia'terre reference database [48]. Dia'terre is an assessment tool developed by the Agency for the Environment and Energy Management (ADEME) in the framework of the French Plan for Energy Performance (PPE) to evaluate a carbon-energy balance at the farm scale. The reference database used to design this assessment tool provides energy values for indirect energy consumption associated with the production of farming inputs. For instance, the calculation of the energy cost associated with the production of nitrogen fertilizers integrates the energy necessary from raw material production (e.g. Haber–Bosch process) through materials processing, manufacture and distribution. We used the reference energy cost provided by the carbon calculator Dia'terre to compute the energy balances of the cropping systems. In this way, the inputs necessary for crop production were converted into energy, using the energy cost of fertilizers, pesticides, seeds, water spread for irrigation, fuel consumed by the equipment and the amount of steel necessary to manufacture this equipment, i.e. energy cost of mechanization (see Dataset S3). The energy requirements for preparing farmyard manure are farm-specific and very difficult to quantify precisely. A simplification was consequently required: following previous studies based on energy balancing methods in crop production [49], the energy equivalent of farmyard manure was equated with that of the mineral fertilizers they substituted (using a substitution value related to the fertilizing efficiency of manure). Energy efficiency was computed from the ratio between productivity and energy consumption. For assessing the economic productivity, the gross product derived from the direct conversion of crops yields into economic values. The 'semi-net' margin was calculated as the gross product per hectare from which we subtracted the input costs (fertilizers, pesticides, seeds, fuel, water and mechanisation). This 'semi-net' margin assessed the system profitability without taking into account subsidies or incentives. The sensitivity to price volatility was defined as the relative standard deviation of the semi-net margin calculated over ten contrasting real price scenarios selected between 2000 and 2010, and thus measured the ability of a cropping system to generate a stable income in a variable economic context. The ten scenarios integrated the prices of crops but also the prices of volatile inputs such as fertilizers or fuel. Each price scenario was defined at a given moment between 2000 and 2010, and it therefore reflected the correlations between the prices of crop products and inputs. This approach notably made it possible to integrate better the effects coming from crop diversity (proportion of cereal crops, oil crops or protein crops) on cropping system profitability and economic stability. Fuel consumption and workload were estimated according to in-field cropping operations only, without considering fuel and time consumed for farm-to-field transports, or extra-workload dedicated to equipment maintenance or field observations. The size, fuel requirements and working output of the various equipment types were standardized for all cropping systems, and defined from a national database [50], consistent with the aim of evaluating management strategies, and of ignoring the potential effects of the equipment specifications (See Dataset S4 for the details of the equipment used for the calculations).

Pesticide environmental impact was expressed as cumulated I-Pest [24]. This indicator measures the risk associated with pesticide application for three compartments of the environment, namely the air, the surface water and the groundwater. This risk indicator, ranging from 0 to 1 (maximum risk), and calculated for each active substance application, is based on: (i) field inherent sensitivity to pesticide transfer toward these three compartments; (ii) characteristics of the active substance (e.g. ecotoxicity, mobility, half-life); and, (iii) information about the conditions of the spraying operation (e.g. amount of active substances employed, canopy cover at the date of treatment) in order to calculate three impact factors, one for each compartment. I-Pest index is obtained using fuzzy decision trees that allow the aggregation of these three impact factors into one synthetic indicator. The diagram presented in Figure S1 illustrates how this indicator of pesticide environmental impact was computed for each pesticide active substance that was sprayed within the field.

The crop sequence indicator Isc [27] is used as an additional indicator to quantify the agronomic effects of crop diversification. Isc ranges on a qualitative scale between 0 and 10 (best value) and is calculated as shown in the following equation:

$$Isc = kp \times kr \times kd$$

Isc is based on the assessment of the effects of the previous crop on the current crop (kp), with respect to the development of pathogens, pests and weeds, to soil structure and nitrogen supply. kp, ranging between 1 and 6, was assessed for 470 couples crop/ previous crop. kp is corrected by two factors taking into account the crop frequency (kr ranging between 0.3 and 1.2) and the crop rotation whole diversity (kd ranging between 1.0 and 1.4). Isc yields respectively 0.5 for wheat monoculture, 3.3 for a rape/ wheat rotation, 5.1 for a rape/wheat/barley rotation, and 7.6 for a maize/wheat/sunflower/spring barley rotation.

Computation of sustainability indicators at the cropping system level

As a first step each indicator was calculated for each cropping operation composing our database (Dataset S2). These values of indicators were summed over the crop season year, and then averaged across years and across plots, each plot being considered as a replicate of a given cropping system. Each indicator was therefore calculated at the cropping system level, integrating (i) the different crops composing the crop sequence, (ii) the variability of crop production related with the inter-annual climatic variability, and (iii) the possible variation in plot properties.

All sustainability indicators were expressed per hectare and per year. For distinguishing specifically the effects of the management strategy on cropping system sustainability from the effects of the production situation, each indicator computed for a given cropping system was then expressed as a ratio (or as a distance in the case of semi-net margin) between the system indicator and the local reference indicator. To increase the quality of the graphs drawing the relationship between sustainability indicators and pesticide use, values of assessment indicators were translated into natural logarithm (Figure 2), which reduced the visual effect of extreme values.

Statistical analyses

Spearman and Pearson correlations were estimated using the 'rcorr' correlation matrix function in the *Hmisc* package of R v2.15.0 [51]. The difference between the means of two sub-samples for a given indicator was tested with a non-parametric Mann-Whitney test ('wilcox.test' function with two samples) in the *stats* package of R v2.15.0.

Supporting Information

Figure S1 Simplified description of the assessment process of pesticide environmental impact in the I-Pest model.

Dataset S1 Cropping systems details. A.xlsx file describing the cropping systems of the studied sample (e.g. crop rotation, tillage and weed management strategies). This file also provides information about the local reference associated with the evaluation of each cropping system. The second tab provides the respective performances of each cropping system described in the first tab.

Dataset S2 Cropping operations database. A.xlsx file which provides the details of all cropping operations carried out in each cropping system: type of cropping operation, date (when recorded), application rates (for pesticides, fertilizers, seeds and irrigation) and proportion of the plot surface targeted.

Dataset S3 Energy balancing database. A.xlsx file with two sheets. The first sheet provides energy cost values for inputs: pesticides active substances, fuel, fertilizers, irrigation water and seeds. The second sheet includes the Lower Heating Values (LHV) for usual crops, that is to say the energy contained in one mass unit of crop harvested.

Dataset S4 Standard equipment characteristics. A.xlsx file describing the technical characteristics of the standard equipment we associated with each cropping operation. Details include the purchase price, the payback period and the maintenance cost to calculate the mechanization costs, but also the equipment size and weight, the working output, the fuel consumption rate and the energy cost value.

Table S1 Means and standard deviations for the range of performance indicators according to the management strategy. A.xlsx file summarizing and comparing the performances of organic, integrated and conventional cropping systems which compose the study sample. Significant difference between groups was tested with a Mann-Whitney test.

Acknowledgments

We thank A. Villard, C. Vivier and M. Geloen for their contribution to the experimental network in Burgundy, and D. Meunier, P. Farcy, P. Chamoy for technical assistance. We particularly want to thank D. Bohan for the precious advice on style and the language corrections.

Author Contributions

Conceived and designed the experiments: NMJ VB CB SP. Performed the experiments: ML FB. Analyzed the data: ML FB. Contributed reagents/ materials/analysis tools: CB MSP. Wrote the paper: NMJ ML. Discussed the results and commented on the manuscript: NMJ ML VB CB SP MSP FB.

References

1. Foley JA, Ramankutty N, Brauman KA, Cassidy ES, Gerber JS, et al. (2011) Solutions for a cultivated planet. Nature 478: 337–342.

2. Pimentel D (1995) Amounts of pesticides reaching target pests: environmental impacts and ethics. Journal of Agricultural and Environmental Ethics 8: 17–29.

3. Richardson M (1998) Pesticides-friend or foe? Water science and technology 37: 19–25.

4. Tilman D, Cassman KG, Matson PA, Naylor R, Polasky S (2002) Agricultural sustainability and intensive production practices. Nature 418: 671–677.

5. Maeder P, Fliessbach A, Dubois D, Gunst L, Fried P, et al. (2002) Soil fertility and biodiversity in organic farming. Science 296: 1694–1697.

6. Holland JM, Frampton GK, Cilgi T, Wratten SD (1994) Arable acronyms analysed - a review of integrated arable farming systems research in Western Europe. Annals of applied biology 125: 399–438.

7. Ferron P, Deguine JP (2005) Crop protection, biological control, habitat management and integrated farming. A review. Agronomy for Sustainable Development 25: 17–24.

8. Trewavas A (2001) Urban myths of organic farming. Nature 410: 409–410.

9. Pimentel D, Hepperly P, Hanson J, Douds D, Seidel R (2005) Environmental, energetic, and economic comparisons of organic and conventional farming systems. Bioscience 55: 573.

10. Directive 2009/128/EC of the European Parliament and of the Council (2009) Official journal of the European Union. Available: http://eur-lex.europa.eu/LexUriServ/LexUriServ.do?uri = OJ:L:2009:309:0071:0086:en:PDF Accessed 2013 Sep 10.

11. Munier-Jolain N, Dongmo A (2010) Evaluation de la faisabilité technique de systèmes de Protection Intégrée en termes de fonctionnement d'exploitation et d'organisation du travail. Comment adapter les solutions aux conditions locales? Innovations Agronomiques 8: 57–67.

12. European commission (2010) Eurostat Agriculture online database. Available: http://epp.eurostat.ec.europa.eu/portal/page/portal/agriculture/farm_structure/database. Accessed 2013 Jul 19.

13. USDA Economic Research Service (2010) U.S. certified organic farmland acreage, livestock number, and farm operations. Available: http://www.ers.usda.gov/data-products/organic-production.aspx "\l ".UiWVOX8QO89. Accessed 2013 Jul 22.

14. Bastiaans L, Paolini R, Baumann DT (2008) Focus on ecological weed management: what is hindering adoption? Weed Research 48: 481–491.

15. Pardo G, Riravololona M, Munier-Jolain N (2010) Using a farming system model to evaluate cropping system prototypes: Are labour constraints and economic performances hampering the adoption of Integrated Weed Management? European Journal of Agronomy 33: 24–32.

16. Food and Agriculture Organization of the United Nations (FAO) (2013) FAOSTAT Resources: Pesticides Use. Available: http://faostat.fao.org/site/424/DesktopDefault.aspx?PageID = 424#ancor. Accessed 2013 Jul 25.

17. U.S. Environmental Protection Agency (EPA) (2011) Pesticides industry sales and usage: 2006 and 2007 market estimates. Washington, D.C.: U.S. Environmental Protection Agency. Available: www.epa.gov/opp00001/pestsales/07pestsales/market_estimates2007.pdf. Accessed 2013 Jul 25.

18. Ministère de l'Agriculture, de l'Agro-alimentaire et de la Forêt (2012) Note de suivi du plan Ecophyto 2018: tendances de 2008 à 2011 du recours aux produits phytopharmaceutiques. Available: http://agriculture.gouv.fr/IMG/pdf/121009_Note_de_suivi_2012_cle0a995a.pdf. Accessed 2013 Jul 26.

19. Reganold JP, Glover JD, Andrews PK, Hinman HR (2001) Sustainability of three apple production systems. Nature 410: 926–930.

20. Deike S, Pallutt B, Christen O (2008) Investigations on the energy efficiency of organic and integrated farming with specific emphasis on pesticide use intensity. European Journal of Agronomy 28: 461–470.

21. Aubertot JN, Robin MH (2013) Injury Profile SIMulator, a qualitative aggregative modelling framework to predict crop injury profile as a function of cropping practices, and the abiotic and biotic environment. I. Conceptual bases. PLoS one 8: e73202.

22. OECD (2001) Environmental Indicators for Agriculture, Volume 3: Methods and Results. Available: www.oecd.org/tad/sustainable-agriculture/40680869.pdf. Accessed 2014 Feb 26.

23. Meiss H, Mediene S, Waldhardt R, Caneill J, Bretagnolle V, et al. (2010) Perennial lucerne affects weed community trajectories in grain crop rotations. Weed Research 50: 331–340.

24. Van der Werf H, Zimmer C (1998) An indicator of pesticide environmental impact based on a fuzzy expert system. Chemosphere 36: 2225–2249.

25. Davis AS, Hill JD, Chase CA, Johanns AM, Liebman M (2012) Increasing cropping system diversity balances productivity, profitability and environmental health. PLoS one 7: e47149.

26. Chikowo R, Faloya V, Petit S, Munier-Jolain NM (2009) Integrated Weed Management systems allow reduced reliance on herbicides and long-term weed control. Agriculture, Ecosystems and Environment 132: 237–242.

27. Bockstaller C, Girardin P (2000) Using a crop sequence indicator to evaluate crop rotations. 3rd International Crop Science Congress 2000 ICSC, Hambourg, 17–22 August 2000, p. 195.

28. Andersson GKS, Rundlöf M, Smith HG (2012) Organic Farming Improves Pollination Success in Strawberries. PLoS one 7: e31599.

29. Aavik T, Liira J (2010) Quantifying the effect of organic farming, field boundary type and landscape structure on the vegetation of field boundaries. Agriculture, Ecosystems and Environment 135: 178–186.

30. Li R, Khafipour E, Krause DO, Entz MH, de Kievit TR, et al. (2012) Pyrosequencing Reveals the Influence of Organic and Conventional Farming Systems on Bacterial Communities. PLoS one 7: e51897.

31. Seufert V, Ramankutty N, Foley JA (2012) Comparing the yields of organic and conventional agriculture. Nature 485: 229–232.

32. Foley JA, Defries R, Asner GP, Barford C, Bonan G, et al. (2005) Global consequences of land use. Science 309: 570–574.

33. Theobald DM (2001) Land-use dynamics beyond the American urban fringe. Geographical Review 91: 544.

34. Phalan B, Onial M, Balmford A, Green R (2011) Reconciling food production and biodiversity conservation: land sharing and land sparing compared. Science 333: 1289–1291.

35. Hulme MF, Vickery JA, Green RE, Phalan B, Chamberlain DE, et al. (2013) Conserving the birds of Uganda's banana-coffee arc: land sparing and land sharing compared. PLoS one 8: e54597.

36. Altieri MA, Nicholls CI, Ponti L (2009) Crop diversification strategies for pest regulation in IPM systems. Integrated pest management Cambridge University Press, Cambridge, UK. pp. 116–130.

37. Krupinsky J, Bailey K, McMullen M, Gossen B, Turkington T (2002) Managing plant disease risk in diversified cropping systems. Agronomy Journal 94: 198–209.

38. Beketov MA, Kefford BJ, Schäfer RB, Liess M (2013) Pesticides reduce regional biodiversity of stream invertebrates. Proceedings of the National Academy of Sciences 110: 11039–11043.

39. Ascherio A, Chen H, Weisskopf MG, O'Reilly E, McCullough ML, et al. (2006) Pesticide exposure and risk for Parkinson's disease. Annals of Neurology 60: 197–203.

40. Di Falco S, Chavas JP (2008) Rainfall shocks, resilience, and the effects of crop biodiversity on agroecosystem productivity. Land Economics 84: 83–96.

41. Reidsma P, Ewert F, Lansink AO, Leemans R (2010) Adaptation to climate change and climate variability in European agriculture: The importance of farm level responses. European Journal of Agronomy 32: 91–102.

42. Nemecek T, von Richthofen JS, Dubois G, Casta P, Charles R, et al. (2008) Environmental impacts of introducing grain legumes into European crop rotations. European Journal of Agronomy 28: 380–393.

43. Pimentel D, Pimentel M (2003) Sustainability of meat-based and plant-based diets and the environment. The American Journal of Clinical Nutrition 78: 660S–663S.

44. Halberg N, van der Werf HMG, Basset-Mens C, Dalgaard R, de Boer IJM (2005) Environmental assessment tools for the evaluation and improvement of European livestock production systems. Livestock Production Science 96: 33–50.

45. Beckie HJ (2009) Herbicide Resistance in Weeds: Influence of Farm Practices. Prairie Soils and Crops 2:3.

46. Meynard JM, Messéan A, Charlier A, Charrier F, Farès M, et al. (2013) Freins et leviers à la diversification des cultures. Etudes au niveau des exploitations agricoles et des filières. Synthèse du rapport d'étude, INRA. Available: http://inra.dam.front.pad.brainsonic.com/ressources/afile/223799-6afe9-resource-etude-diversification-des-cultures-synthese.html. Accessed 2013 Mar 15.

47. Centre d'Etudes Biologiques de Chizé (2009) Zone Atelier «Plaine & Val de Sèvre». Available: http://www.zaplainevaldesevre.fr. Accessed 2013 Jun 5.

48. Agence de l'Environnement et de la Maîtrise de l'Energie (ADEME) (2011) Guide des valeurs Dia'terre. Version du référentiel 1.13. Available: http://www2.ademe.fr/servlet/KBaseShow?sort = -1&cid = 96&m = 3&catid = 24390. Accessed 2013 Jul 12.

49. Hülsbergen KJ, Feil B, Biermann S, Rathke GW, Kalk WD, et al. (2001) A method of energy balancing in crop production and its application in a long-term fertilizer trial. Agriculture, Ecosystems and Environment 86: 303–321.

50. Bureau de Coordination du Machinisme Agricole (BCMA) (2012) Simcoguide online decision tool. Available: http://simcoguide.pardessuslahaie.net/#accueil. Accessed 2013 Apr 12.

51. R Development Core Team (2012). R: A language and environment for statistical computing (R Foundation for Statistical Computing, Vienna, Austria).

2

Increasing Minimum Daily Temperatures Are Associated with Enhanced Pesticide Use in Cultivated Soybean along a Latitudinal Gradient in the Mid-Western United States

Lewis H. Ziska*

Crop Systems and Global Change Laboratory, United States Department of Agriculture, Agricultural Research Service, Beltsville, Maryland, United States of America

Abstract

Assessments of climate change and food security often do not consider changes to crop production as a function of altered pest pressures. Evaluation of potential changes may be difficult, in part, because management practices are routinely utilized *in situ* to minimize pest injury. If so, then such practices, should, in theory, also change with climate, although this has never been quantified. Chemical (pesticide) applications remain the primary means of managing pests in industrialized countries. While a wide range of climate variables can influence chemical use, minimum daily temperature (lowest 24 h recorded temperature in a given year) can be associated with the distribution and thermal survival of many agricultural pests in temperate regions. The current study quantifies average pesticide applications since 1999 for commercial soybean grown over a 2100 km North-South latitudinal transect for seven states that varied in minimum daily temperature (1999–2013) from −28.6°C (Minnesota) to −5.1°C (Louisiana). Although soybean yields (per hectare) did not vary by state, total pesticide applications (kg of active ingredient, ai, per hectare) increased from 4.3 to 6.5 over this temperature range. Significant correlations were observed between minimum daily temperatures and kg of ai for all pesticide classes. This suggested that minimum daily temperature could serve as a proxy for pesticide application. Longer term temperature data (1977–2013) indicated greater relative increases in minimum daily temperatures for northern relative to southern states. Using these longer-term trends to determine short-term projections of pesticide use (to 2023) showed a greater comparative increase in herbicide use for soybean in northern; but a greater increase in insecticide and fungicide use for southern states in a warmer climate. Overall, these data suggest that increases in pesticide application rates may be a means to maintain soybean production in response to rising minimum daily temperatures and potential increases in pest pressures.

Editor: Matthew Germino, US Geological Survey, United States of America

Funding: This author has no support or funding to report.

Competing Interests: The authors have declared that no competing interests exist.

* E-mail: l.ziska@ars.usda.gov

Introduction

Considerable research effort has focused on determining the impact of anthropogenic climate change on global agriculture [1–4]. Of merited interest in this regard are the physical aspects of climate (e.g. carbon dioxide, temperature, precipitation, extreme weather events) that directly alter crop biology (e.g. growth, phenology, sterility, yields) and the resulting consequences for food security [5–9].

However, research efforts related to assessing the agricultural impacts of rising CO_2 and climate change do not always consider trophic interactions. Overall, changes to the biology and competitive abilities of agricultural pests (insects, pathogens, weeds) relative to potential crop yield losses has not been well quantified [10–11]. This is an important omission as the role of pests on constraining crop production is significant and well recognized. For example, weed competition can result in potential crop losses of ~34% globally, with insect pests and pathogens resulting in additional losses of ~18 and 16%, respectively [12].

Such omissions may reflect the complex challenges in relating atmospheric CO_2 and climate variables to potential reductions in crop production related to increased pest pressures. For example, weed growth and fecundity can be directly affected by increasing atmospheric CO_2 as well as rising temperature; insects and pathogens can also be directly affected by temperature, but indirectly by CO_2 and/or climate induced changes to their weed hosts [12–13]. Overall, while a number of pest studies have been conducted, empirical evidence has been eclectic, although it has been suggested that pest pressures will probably increase with climate change (e.g., [14]).

Yet, even if pest pressures regarding crop production were unequivocal and well-characterized in regard to climate change, it would still be difficult to quantify yield reductions *in situ*. This is because there are strong economic incentives at the field level to manage agro-ecosystems to prevent or minimize pest damage. While management methods vary greatly and may include cultural, mechanical, chemical and biological options, among developed countries, such as Australia and the United States, application of chemicals, usually as pesticides, represents the most widely used method for pest control.

But could quantification of pesticide usage in turn, provide an alternative means to gauge changes in pest pressures associated with changes in climate? Among climate variables, it is generally recognized that in temperate regions, the distribution and survival of agricultural pests is often limited by low winter temperatures; i.e., minimum thermal thresholds) [15]. Rising minimum temperatures associated with anthropogenic climate change could extend the potential geographic range of pest species and/or alter their demographics, although long-term changes in species diversity are unclear [16–17]. Climate change assessments have also emphasized that the current and projected increases in global warming are not uniform, and enhanced land-surface temperatures (relative to the global average) are more probable for minimum (Winter) than maximum (Summer) temperatures [18–19].

To ascertain if minimum temperatures reflect pesticide usage, regression analysis of the interrelationship between insecticide, fungicide and herbicide application rates and the minimum daily (24 h) observed temperature was conducted on a commercial crop, soybean, grown over a latitudinal transect of seven mid-western states within a humid region. This analysis was performed using a multi-year (1999–2013), multi-state (Minnesota, Wisconsin, Iowa, Missouri, Arkansas, Mississippi and Louisiana) data series obtained from USDA-NASS pesticide usage surveys in conjunction with state based minimum temperature. If significant correlations were observed between minimum temperature and usage for a given class of pesticide, longer trends (1977–2013) in minimum temperature were analyzed and then used to project near-term (decadal, to 2023) changes in pesticide rates by state.

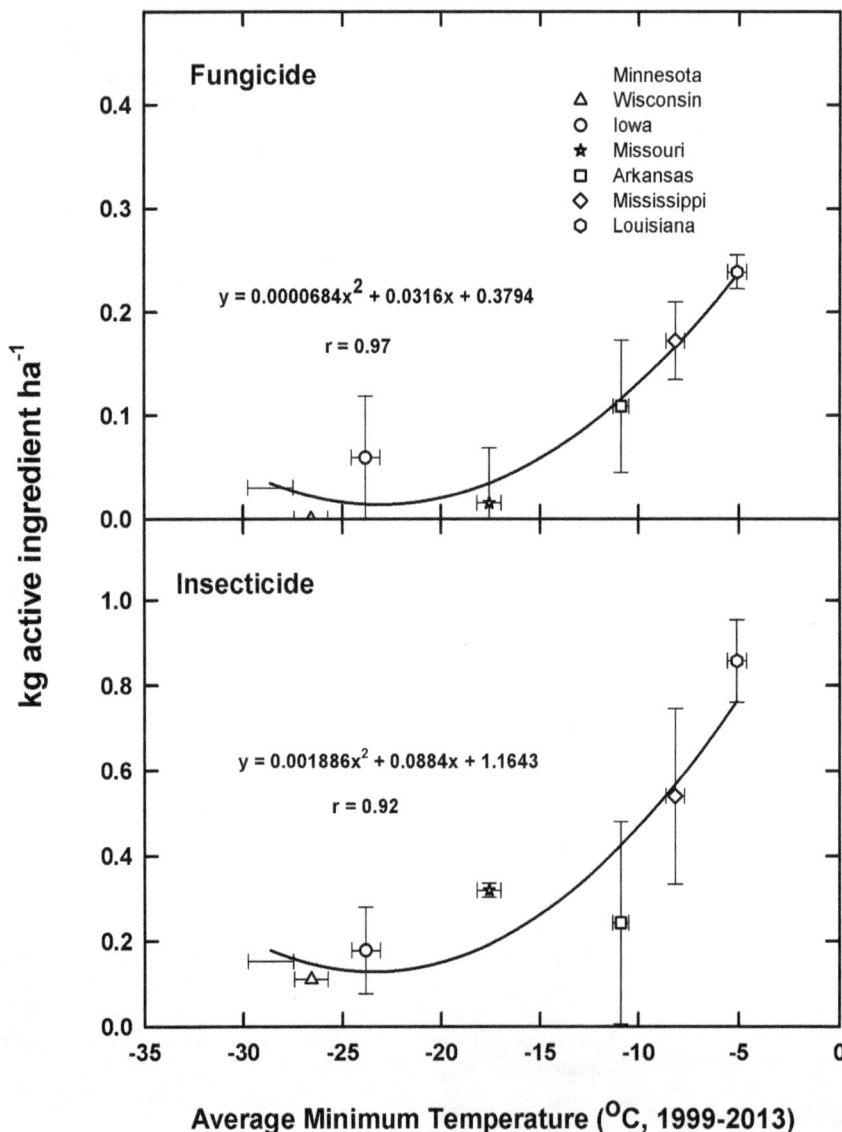

Figure 1. Minimum temperature by state averaged for the period 1999–2013 as a function of insecticide and fungicide usage (kg active ingredient, ai, per hectare) as determined from the National Agricultural Statistical Service survey for the years 1999, 2000, 2001, 2002, 2004, 2006, and 2012) for a north-south transect of seven Midwestern states where commercial soybean is grown. Minimum temperatures were determined as the lowest recorded temperature for a 24 h period during a given calendar year for four locations for a soybean growing area within a state. Lines represent the "best fit" second order polynomial. Bars are ±SE. See Methods for additional details.

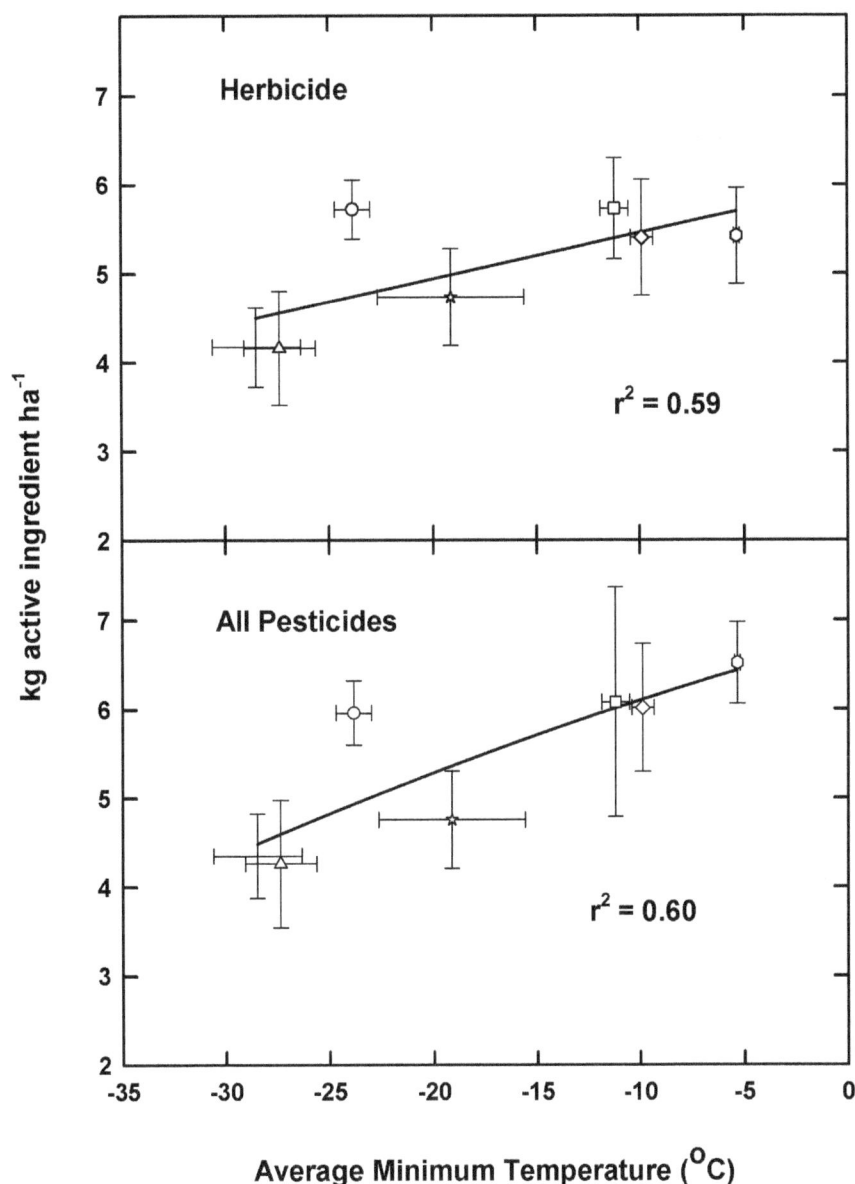

Figure 2. Symbols and regression determined as for figure 1, but for herbicide use and total pesticide (insecticide, fungicide and herbicide) usage for soybean.

Results

For two classes of pesticides, fungicides and insecticides, a second order quadratic function provided the 'best-fit' for minimum temperature and amounts of active ingredient applied for the period 1999–2012. There was a significant correlation (r values of 0.97 and 0.92, **,P<0.01) for fungicide and insecticide soybean applications, respectively over the north-south transect (**Figure 1**). A linear function between minimum daily temperature and herbicide application rates was also observed to be positively correlated and significant (r value of 0.84: *,P<0.05) for herbicide use (**Figure 2**). As most pesticide applications are herbicides, a similar relationship (r = 0.84) was observed for total pesticide use (**Figure 2**).

As the correlation between minimum daily temperature and pesticide usage was significant in all cases, states within the transect were examined to quantify longer term temperature changes (1977 through 2013) in order to assess the increase in minimum winter temperature per decade. Overall, these data indicated that the rise in minimum temperatures was a function of latitude, with states such as Minnesota showing a more rapid increase in minimum temperatures than southern states (e.g., Louisiana) (**Table 1, Figure 3**). However, average soybean yields (2009–2013, MT Ha^{-1}) did not significantly vary as a function of the north-south transect (**Table 1**).

The average decadenal increase in minimum daily temperature was used to estimate potential near term (to 2023) changes in pesticides by category. Because a first order regression was the best-fit for herbicide use, increases in minimum temperature result in a relatively greater increase in herbicide applied for northern compared to southern states. In contrast, because the relationship between insecticide/fungicides and minimum temperatures is best described by a second degree polynomial, temperature changes above a critical threshold, (ca −20°C, **Figure 2**) resulted in

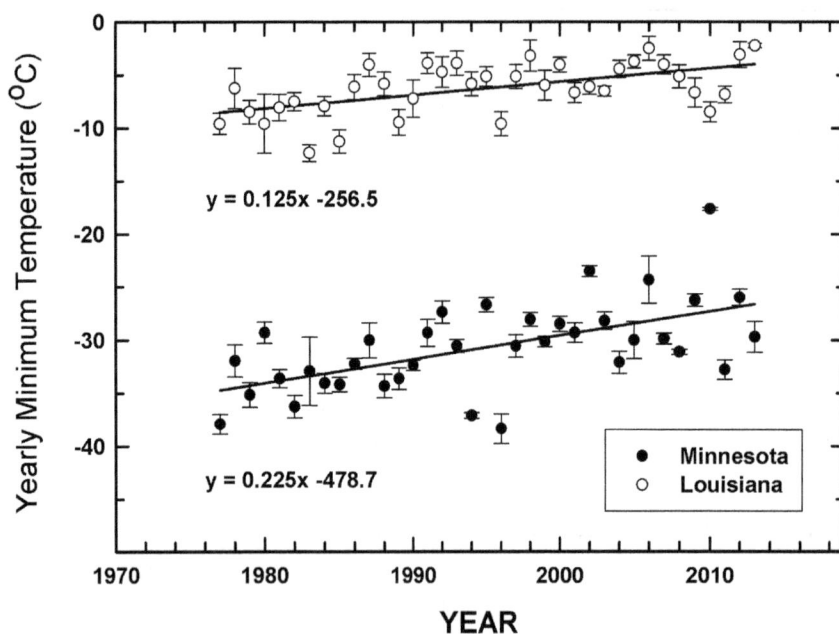

Figure 3. Change in average (±SE) minimum temperature (lowest recorded temperature for a 24 h period during a given calendar year for four locations) for the northernmost and southernmost states (Minnesota, closed circles) and Louisiana (open circles) used in the Midwestern transect. Slope of the regression line indicates the average increase in minimum temperature (°C) per year for each location.

proportionally larger changes in insecticide/fungicide usage. Hence, for those states with minimum temperatures above this threshold, further increases in temperature result in disproportionally larger projected changes in insecticide and fungicide usage (**Table 1**).

Discussion

The need to assess the relative impact of pests on agricultural production in a changing climate has long been recognized [20]; however, few quantitative *in situ* assessments are available. An earlier economic analysis [21] indicated increased pesticide costs as a function of increased rainfall and hotter weather during the 1990s. Application of this analysis to projected climate scenarios to 2090 also indicated general increases in average pesticide cost for a range of crop species [21]. However, this study, while useful, did not distinguish between classes of pesticide, or quantify changes in application amounts as a function of a specific climate variable.

Although agriculture is widespread, there are, in fact, only a handful of crops that are grown over a wide range of climates. For the U.S., soybean, corn and wheat fall into this category; however wheat is grown temporally over a latitudinal gradient as spring (e.g. North Dakota) or winter wheat (e.g. Kansas) with significant differences in pest management. In addition, wheat is not grown widely in the southern states. In contrast, corn is often grown in conjunction with soybean, but data on pesticide use for corn in the southern states (e.g. Mississippi, Arkansas, and Louisiana) is currently not available through NASS. Alternatively, the distribution of soybean in this study encompassed a broad range of minimum daily temperatures (i.e., a ~23°C difference from Minnesota to Louisiana). Such a range would include projections of future surface temperatures associated with anthropogenic climate change, e.g. if minimum daily temperatures in Minnesota became like those of Louisiana [22]; hence soybean was selected for analysis.

There is, as shown here, a significant positive correlation between the range of minimum daily temperatures and pesticide usage for soybean, particularly in regard to fungicides and insecticides. The second order functions for soybean are consistent with insect and fungal biology; i.e., once temperature has reached a critical thermal threshold, it is a significant driver of shifts in insect and pathogen demography [13,23]. For example, in the UK increasing winter temperatures above a biological minimum have been associated with increased northward migration of aphids in Scotland and increased genotype variation among aphid populations [24]. Similar correlations have also been observed between minimum winter temperatures and the southern pine beetle [25].

Patterson et al. [14] provide an extensive list of temperature thresholds in this regard including associated phenological responses and potential shifts in the expansion of insect ranges. These data indicate that for temperate regions, warming would result in increased winter survival as well as an increase in growth and insect fecundity. Similarly, it has been recognized that pathogen development is affected by warm conditions, with mild winters and warmer weather associated with increased outbreaks of powdery mildew; leaf spot disease, leaf rust and rizomania disease [14]. Treharne [26] in turn, has also suggested that increases in minimum daily temperatures would shift the occurrence of plant disease into cooler regions. The relationship between insecticide and fungicide use observed here for soybean in response to rising minimal temperatures is, overall, consistent with the biology of thermal limits and expected shifts in insect and fungal populations in a warmer climate (e.g., [14,23,27]).

For the current study, herbicide applications were a linear function of rising minimum daily temperatures in soybean. In contrast to insect and fungal populations, thermal limits for the occurrence of weeds in soybean were not evident. However, if climate is suitable for crop production, it will also be, de facto, suitable for weed growth.

Although herbicide tolerance in agriculture is increasing [28], for the current study the observed increase in herbicide

Table 1. States included in the latitudinal transect.

State	Soybean (Ha *1000)	Soybean (MT Ha⁻¹)	T_{min}/decade	r value	Insecticide (2023)	Fungicide (2023)	Herbicide (2023)
Minnesota	2,513	2.81	+2.25	0.58,**	0	0	+160.2
Wisconsin	628	2.91	+1.99	0.57,***	0	0	+140.0
Iowa	3,741	3.20	+1.58	0.51,**	+20.3	0	+85.1
Missouri	2,247	2.52	+1.56	0.51,**	+35.6	+12.0	+73.5
Arkansas	1,308	2.62	+1.31	0.54,***	+110.0	+36.5	+70.0
Mississippi	806	2.77	+1.31	0.52,***	+117.7	+39.3	+70.3
Louisiana	449	2.81	+1.25	0.55,***	+170.0	+41.1	+58.6

Soybean area is given as hectares (*1000), average yield was determined from 2009–2013 as metric tons (MT) per hectare, the T_{min} is the increase in minimal daily (24 h) temperature (°C) per decade (determined from 1977–2013). The r value is the correlation coefficient of T_{min} over time, with P<0.05, *; P<0.01, ** and P<.001, ***. The values for insecticide, fungicide and herbicide are determined for 2023 (10 year projection) using the functions for figures 1–3 and are shown as changes in grams of active ingredient (ai) per hectare, with a value of zero indicating no change in application rates over 2013 levels.

applications with minimum temperatures did not appear to be related to a greater number of herbicide resistant weeds along the north-south transect (**Table 2**). Nor was there any apparent relationship with photosynthetic pathway (i.e. the number of C_4 weeds is greater in Minnesota than Louisiana). This suggests no difference in recent CO_2 increases on latitudinal weed selection between C_3 and C_4 weedy species *per se*. However the number of perennial or facultative perennial weed species was significantly greater for the southern soybean locations (e.g. Louisiana, **Table 2**). In weed management, it is generally recognized that perennial weeds can be more competitive and more difficult to control. More competitive because of their ability to quickly regenerate from below-ground structures in the spring and more difficult to control chemically because the elimination of perennial weeds requires the killing of all plant parts (including belowground stems, rhizomes, tubers, etc.) that are capable of producing new shoots [29]. Potential increases in perennial weed establishment with increasing minimum temperatures could, in turn, result in concomitant increases in herbicide usage.

The quantitative differences in decadenal increases in minimum temperature reported here (e.g., 2.25 vs. 1.25°C per decade for Minnesota and Louisiana, respectively since 1977) are consistent with the Intergovernmental Panel on Climate Change (IPCC) projections regarding enhanced warming as a function of latitude [19]. Because of the variation in pesticide category by minimal temperature for soybean, near-term projections (2023) suggest a greater relative increase in herbicide use for the upper Midwest, whereas greater insecticide and fungicide use is projected for the southern states for this same period. This may be due, in part, to the southern states having already surpassed the minimum winter temperatures needed to support insect and pathogen populations for soybean in the southern region. However, if minimum temperatures continue to rise, then similar increases in use for all pesticide categories could occur for the upper Midwest.

At present, soybean in the mid-western United States is grown over a ~2100 km north-south transect that provides a wide range of yearly temperatures. Although such a range of temperatures should, ostensibly, increase pest pressures and subsequent crop loss, no production losses are evident in average yield per hectare (e.g., average yields in Minnesota and Wisconsin do not differ from those of Mississippi and Louisiana). However, insecticide, fungicide and herbicide use increase significantly in conjunction with minimum daily temperatures along the latitudinal transect. Assuming that prophylactic use is not widespread in any one region, this increase should reflect increased pest pressures *per se*.

It should be emphasized however, that the biological basis for rising minimum temperatures and changes in pesticide usage are likely to be complex. While there are empirical correlations between population demographics and temperature, temperature by itself does not reflect the complexity of pest population causations in agriculture. For example, pest-pest interactions, insects as disease vectors, temporal asynchrony between insects and host plants, soil management and herbicide efficacy (e.g. precipitation, windspeed), *inter alia* [10,23,30] will also need to be considered. Rising levels of carbon dioxide are also likely to directly alter secondary plant chemistry and alter plant-arthropod interactions depending on plant species and insect group [31]. In addition, the current analysis assumes future monotonic poleward shifts in minimum temperatures influencing pesticide usage; whereas an increase in extreme climatic events may be the norm [32].

However, the ability to include realistic impacts of agricultural pests in future assessments of climate change and food security is imperative. While there are numerous challenges remaining before

Table 2. Top ten lists of most troublesome weed species in soybean for three states along the North-South transect.

State	Common Name	Scientific Name	Photosynthetic Pathway	Growth Habit
Louisiana	Morning glory spp.	Ipomea spp.	C_3	Annual or Perennial
	Pigweed spp.	Amaranthus spp.	C_4	Annual or Perennial*
	Browntop millet	Urochloa ramose	C_4	Perennial
	Nutsedge spp.	Cyperus spp.	C_4	Perennial
	Redvine	Brunnichia ovate	C_3	Perennial
	Henbit	Lamium amplexicaule	C_3	Annual*
	Dayflower	Commelina spp.	C_3	Annual or Perennial
	Johnson grass	Sorghum halapense	C_4	Perennial
	Prickly sida	Sida spinosa	C_3	Annual
	Red rice	Oryza sativa	C_3	Annual
Missouri	Waterhemp	Amaranthus rudis	C_4	Annual*
	Morning glory	Ipomea spp.	C_3	Annual or Perennial
	Palmer amaranth	Amaranthus palmeri	C_4	Annual*
	Giant ragweed	Ambrosia trifida	C_3	Annual*
	Johnson grass	Sorghum halapense	C_4	Perennial
	Asiatic dayflower	Commelina communis	C_3	Annual
	Horseweed	Conyza canadensis	C_3	Biennial*
	Prickly sida	Sida spinosa	C_3	Annual
	Common ragweed	Ambrosia artemisiifolia	C_3	Annual*
	Eastern black nightshade	Solanum ptycanthum	C_3	Annual or Perennial
Minnesota	Lambsquarters	Chenopodium album	C_3	Annual
	Giant foxtail	Setaria faberii	C_4	Annual
	Waterhemp	Amaranthus rudis	C_4	Annual*
	Wooly cupgrass	Eriochloa villosa	C_4	Annual
	Giant ragweed	Ambrosia trifida	C_3	Annual*
	Yellow foxtail	Setaria viridis	C_4	Annual
	Green foxtail	Setaria lutescens	C_4	Annual
	Quackgrass	Elytrigia repens	C_3	Perennial
	Common ragweed	Ambrosia artemisiifolia	C_3	Annual*
	Wild proso millet	Panicum miliaceum	C_4	Annual

Lists are generated by farmer surveys as reported by the Southern Weed Science Society (SWSS) for Louisiana (LA) and Missouri (MO) [37] and the University of Minnesota extension service for Minnesota (MN) (Frank Forcella, USDA-ARS Personal Communication). C_3/C_4 refers to photosynthetic pathway.
*Indicates herbicide resistant populations within that state.

the impacts of anthropogenic climate change on pest pressures and crop yields can be completely quantified, the current study suggests that farm management, through the increased use of chemical applications, may negate pest pressures on crop production that are incurred as a result of changing climate variables such as temperature. As such, pest-induced reductions in agricultural yield of soybean associated with climate change may be difficult to quantify where pesticides are widely used. At present, while the current analysis suggests that minimum daily temperatures could be used as a proxy to understand climate change impacts on pesticide use in soybean; critical information regarding the environmental and economic consequences of such impacts will require further study.

Materials and Methods

Soybean

For all seven states examined along the north-south transect, soybean is considered a major crop (>300,000 ha planted) [34]. Genetically modified (gmo) soybean (e.g., Round-Up Ready[tm]) was introduced in 1995–1996, mainly as a means for blanket control of post-emergent weeds. Following its introduction it was quickly adopted by growers, with gmo soybean acreage >50% by 1999 [33] (**Table 3**). There is no evidence of different rates of adoption of gmo soybean over the period 1999–2012. For example, by 2001, the amount of soybean area plant to gmo was 63% for both Minnesota and Mississippi (**Table 3**). Since 2005, approximately 90% of all soybeans grown in the U.S. have been herbicide resistant gmo (**Table 3**).

Table 3. Genetically engineered soybean varieties for herbicide tolerance for states included in the north-south transect.

STATE	2000	2001	2002	2003	2004	2005	2006	2007	2008	2009	2010	2011	2012
Minnesota	46	63	71	79	82	83	88	92	91	92	93	95	91
Wisconsin	51	63	78	84	82	84	85	88	90	85	88	91	92
Iowa	59	73	75	84	89	91	91	94	95	94	96	97	93
Missouri	62	69	72	83	87	89	93	91	92	89	94	91	91
Arkansas	43	60	68	84	92	92	92	92	94	94	96	95	94
Mississippi	48	63	80	89	93	96	96	96	97	94	98	98	95
U.S.	54	68	75	81	85	87	89	91	92	91	93	94	93

Values are the percent of all soybeans planted. Data are available at: www.usda.mannlib.cornell.edu/usda/nass/Acre/2010s/2011/Acre-06-30-2011.pdf#page=27.

Pesticide Use Rates

State level pesticide usage rate are available for 1999, 2000, 2001, 2002, 2004, 2006 and 2012 from the USDA National Agricultural Statistical Service Survey (NASS) data online (www.quickstats.nass.usda.gov/#5072A3CD-64E7-384A-8FF0-DDC931164C3B) [34] These data include chemical usage by compound for insecticides, fungicides and herbicides for soybean for the states included in the north-south transect. Although data are also available from 1991 through 1998 from this same source, the rapid rates of gmo adoption for herbicide resistant soybean, and the shift in herbicide usage during this time period was thought to, potentially, obfuscate any potential changes related to climate variables such as temperature. This is because application rates often vary by herbicide mode of action; hence, differential herbicide use would result in different amounts of active ingredient (ai) being applied. However, the large-scale shift to gmo soybean provided greater uniformity in herbicide type over a large regional area (**Table 3**). This is an important consideration as herbicides represent the largest component of total pesticide use (e.g. **Figure 2**). However, the temporal range of yearly averages (1999–2013) was considered insufficiently long to evaluate any potential increase in minimum daily temperatures associated with anthropogenic surface warming and pesticide usage (relative to year to year variability) for any one state.

Temperature Data

A software program developed by Texas A&M University [35] was used to identify at least four weather stations located within soybean growing regions of a given state used in the transect. Information on the lowest recorded (24 h) minimum temperature for a given year was determined and averaged for all weather stations within a state for the 1999–2013 period. These same stations and data sets were used to generate minimum daily temperature over a longer time period, from 1977 (i.e. the start of the rapid increase in global land-ocean temperature index) through 2013 [36].

Statistical Analysis

In examining pesticide use it is important to determine whether there may be fixed or random regional effects within the data. For example, are there state specific factors beyond temperature that would result in systematic shifts in pesticide usage? (e.g. one state using irrigation while another state does not, change in rotation rates, etc.) This analysis has been previously done by Chen and McCarl [21] who found with 99% confidence that a random state effect existed for pesticide use in corn, potatoes, soybeans and wheat [21].

A step-wise regression program (ver. 10.0 Statview, Cary NC, USA) was used to determine the best fit regression line for average minimum daily temperature and pesticide application rate (insecticides, fungicides and herbicides) for all states within the latitudinal transect. If the correlation coefficient was determined to be significant for a given pesticide category (e.g. insecticides); then a longer term minimum daily temperature data set (1977–2013) was generated for each state to quantify increases in minimum temperature by decade. The increase in minimum daily temperature by decade was then used to project potential short-term (2014–2023) pesticide use (by category and state) using the quadratic or linear functions provided in figures 1 and 2.

Author Contributions

Conceived and designed the experiments: LHZ. Performed the experiments: LHZ. Analyzed the data: LHZ. Contributed reagents/materials/analysis tools: LHZ. Wrote the paper: LHZ.

References

1. Adams RM, Rosenzweig C, Peart RM, Ritchie JT, McCarl BA, et al. (1990) Global climate change and U.S. agriculture. Nature 345: 219–224.

2. Reilly J, Tubiello F, McCarl B, Abler D, Darwin R, et al. (2003) U.S. Agriculture and climate change. Clim Change 57: 43–69.

3. Howden SM, Soussana JF, Tubiello FN, Cherti N, Dunlop M, et al. (2007) Adapting agriculture to climate change. Proc Nat Acad Sci (USA) 104: 19691–19696.

4. Lobell DB, Burke MB, Tebaldi C, Mastrandrea MD, Falcon WP, et al. (2008) Prioritizing climate change adaptation needs for food security in 2030. Science 319: 607–610.

5. Tao F, Yokozawa M, Xu Y, Hayashi Y, Zhang Z (2006) Climate changes and trends in phenology and yields of field crops in China, 1981–2000. Ag Forest Met 138: 82–92.

6. Schmidhuber J, Tubiello FN (2007) Global food security under climate change. Proc Nat Acad Sci 104: 19703–19708.

7. Schlenker W, Roberts MJ (2009) Nonlinear temperature effects indicate severe damages to US crop yields under climate change. Proc Nat Acad Sci 106: 15594–15598.

8. Ziska LH, Bunce JA, Shimono H, Gealy DR, Baker JT, et al. (2012) Food security and climate change: On the potential to adapt global crop production by active selection to rising atmospheric carbon dioxide. Proc Royal Soc (B) 279: 4097–4105.

9. Lobell DB, Gourdji SM (2012) The influence of climate change on global crop productivity. Plant Physiol 160: 1686–1697.

10. Scherm H (2004) Climate change: can we predict the impacts on plant pathology and pest management? Can J Plant Pathol 26: 267–273.

11. Gregory PJ, Johnson SN, Newton AC, Ingram JSI (2009) Integrating pests and pathogens into the climate change/food security debate. J of Exp Bot 60: 2827–2838.

12. Oerke E-C (2006) Crop losses to pests. J Ag Sci 144: 31–43.

13. Ziska LH, Runion GB (2007) Future Weed, Pest and Disease Problems for Plants. In: Newton PCD, Carran A, Edwards GR, Niklaus PA, Editors. Agroecosystems in a Changing Climate. CRC Press, Boston, MA, 262–279.

14. Patterson DT, Westbrook JK, Joyce RJV, Lingren PD, Rogasik J (1999) Weeds, insects, and diseases. Climatic Change 43: 711–727.

15. Cammel ME, Knight JD (1992) Effects of climatic change on the population dynamics of crop pests. Adv Ecol Res 22: 117–162.

16. Bradley BA, Blumenthal DM, Wilcove DS, Ziska LH (2010) Predicting plant invasions in an era of global change. TREE 25: 310–318.

17. McDonald AJ, Riha S, Ditomasso A, DeGaetano A (2009) Climate change and the geography of Weed damage: Analysis of US maize systems suggests the potential for significant range transformations. Ag Ecosys Environ 130: 131–140.

18. Vose RS, Easterling DR, Gleason B (2005) Maximum and minimum temperature trends for the globe: An update through 2004. Geophys Res Let 32: 1–5.

19. Intergovernmental Panel on Climate Change (2007) Summary for policymakers. In: Climate Change 2007: The Physical Science Basis. Contribution of Working Group I to the Fourth Assessment Report of the Intergovernmental Panel on Climate Change (Solomon, S. et al., eds), Cambridge University Press.

20. Lough JM, Wigley TML, Palutikof JP (1983) Climate and climate impact scenarios for Europe in a warmer world. J. Clim. Appl. Meteorol. 22: 1673–1684.

21. Chen C-C, McCarl BA (2001) An investigation of the relationship between pesticide usage and Climate change. Climatic Change 50: 475–487.

22. Intergovernmental Panel on Climate Change (2012) Managing the risks of extreme events and disasters to advance climate change adaptation. A special report of Working Groups I and II of the Intergovernmental Panel on Climate Change. Cambridge University Press, Cambridge.

23. Fuhrer J (2003) Agroecosystem responses to combinations of elevated CO_2, ozone, and global climate change. Ag Ecosys Environ 97: 1–20.

24. Malloch G, Highet F, Kasprowicz L, Pickup J, Beilson R, et al. (2006) Microsatellite marker analysis of peach-potato aphids (*Myzus persicae*, Homoptera: Aphididae) from Scottish suction traps. Bull Entomol Res 96: 573–582.

25. Tran JK, Yliqja T, Billings RF, Regniere J, Ayres MP (2007) Impact of minimum winter temperatures on the population dynamics of *Dendroctonus frontalis*. Ecol Appl 17: 882–889.

26. Treharne K (1989) The implications of the 'greenhouse effect' for fertilizers and agrochemicals. In: Bennet RC editor, Ministry of Agriculture, Fisheries and Food, London, UK 67–78.

27. Porter JH, Parry ML, Carter TR (1991) The potential effects of climatic change on agricultural insect pests. Ag For Meteor 57: 221–240.

28. Delye CD, Jasieniuk M, Le Core V (2013) Deciphering the evolution of herbicide resistance in weeds. Trends in Gen 29: 649–655.

29. Zimdahl RL (2007) Weed-crop competition: a review. John Wiley & Sons, New York, NY, 220 pages.

30. Ziska LH (2010) Global Climate Change and Carbon Dioxide: Assessing Weed Biology and Management. In: Rosenzweig C, Hillel D, Ediotrs. Handbook of Climate Change and Agro-ecosystems: Impacts, adaptation and mitigation. World Scientific Publishing, Hackensack, NJ, 191–208.

31. Robinson EA, Ryan GD, Newman JA (2012) A meta-analytical review of the effects of elevated CO_2 on plant-arthropod interactions highlights the importance of interacting environmental and biological variables. New Phytologist 194: 321–336.

32. Hansen J, Sato M, Ruedy R (2012). Perception of climate change. Proc Nat Acad Sci 109: E2415–E2423.

33. Carpenter J, Gianessi L (1999) Herbicide tolerant soybeans: Why growers are adopting Round-up Ready varieties. AgBioForum 2: 65–72.

34. National Agricultural Statistical Service, (NASS) (www.quickstats.nass.usda.gov/#5072A3CD-64E7-384A-8FF0-DDC931164C3B) (Accessed, January, 2014).

35. Yang Y, Wilson LT, Wang J (2010) Development of an Automated Climatic Data Scraping, Filtering and Display System. Comp Elect Ag 71: 77–87.

36. Hansen J, Ruedy R, Glascoe J, Sato M (1999). GISS analysis of surface temperature change. J Geophysical Res: Atmos (1984–2012), 104: 30997–31022.

37. Southern Weed Science Society (SWSS) (2013) SWSS Proceedings. 66: 275–287.

3

Enantioselective Metabolism of Quizalofop-Ethyl in Rat

Yiran Liang[1], Peng Wang[1], Donghui Liu[1], Zhigang Shen[1], Hui Liu[1], Zhixin Jia[2], Zhiqiang Zhou[1]*

1 Department of Applied Chemistry, China Agricultural University, Beijing, PR China, **2** Institute of Materia Medica, Chinese Academy of Medical Sciences and Peking Union Medical College, Beijing, PR China

Abstract

The pharmacokinetic and distribution of the enantiomers of quizalofop-ethyl and its metabolite quizalofop-acid were studied in Sprague-Dawley male rats. The two pairs of enantiomers were determined using a validated chiral high-performance liquid chromatography method. Animals were administered quizalofop-ethyl at 10 mg kg^{-1} orally and intravenously. It was found high concentration of quizalofop-acid in the blood and tissues by both intragastric and intravenous administration, and quizalofop-ethyl could not be detected through the whole study which indicated a quick metabolism of quizalofop-ethyl to quizalofop-acid in vivo. In almost all the samples, the concentrations of (+)-quizalofop-acid exceeded those of (−)-quizalofop-acid. Quizalofop-acid could still be detected in the samples even at 120 h except in brain due to the function of blood-brain barrier. Based on a rough calculation, about 8.77% and 2.16% of quizalofop-acid were excreted through urine and feces after intragastric administration. The oral bioavailability of (+)-quizalofop-acid and (−)-quizalofop-acid were 72.8% and 83.6%.

Editor: Luis Eduardo M. Quintas, Universidade Federal do Rio de Janeiro, Brazil

Funding: Supported by A Foundation for the Author of National Excellent Doctoral Dissertation of PR China, Program for New Century Excellent Talents in University (NCET09-0738), the National Natural Science Foundation of China (21277171, 21307155), the New-Star of Science and Technology supported by Beijing Metropolis, Program for New Century Excellent Talents in University and Program for Changjiang Scholars and Innovative Research Team in University. The funders had no role in study design, data collection and analysis, decision to publish, or preparation of the manuscript.

Competing Interests: The authors have declared that no competing interests exist.

* Email: zqzhou@cau.edu.cn

Introduction

Pesticide is a double edged sword, which plays very important roles in increasing crop production and income, but it also causes some negative effects, such as environmental pollutions [1,2], homicidal and suicidal accident [3], cancer and other diseases [4]. Among the total amount of pesticide in china, more than 40% of them are chiral [5], and this ratio is increasing as more and more complex structures are being developed. Chiral pesticides are composed of two or multiple enantiomers, which have the same physical, chemical properties and affection in achiral environment. However, for the individual enantiomers can interact enantioselectively with enzymes or biological receptors in organisms [6], the biological and physiological properties of enantiomers are often different [7]. For example, (−)-o,p′-DDT is a more active estrogen-mimic in rat and human than (+)-o,p′-DDT [8]. The (R)-form of dichlorprop is active while the other is totally inactive [9], but its inactive form still has oxidative damage to the non-target organisms [10]. Although the enantioselective ecotoxicities of some chiral pesticides to non-target animals, plants and human cancer cell lines have been reported [7], the different properties of the enantiomers are still poorly understood and many chiral pesticides are still used and regulated as if they were achiral.

Quizalofop-ethyl, (2RS)-[(2-(4-((6-chloro-2-quinoxalinyl)oxy)-phenoxy)-ethyl ester] (QE, Fig. 1) is intensively used to control both annual and perennial grass weeds in broadleaf crops, such as alfalfa, bean, cabbage, canola, carrot, lettuce, potato, soybean, sugar beet, tobacco, tomato and turnip [11]. The half-life ($T_{1/2}$) of quizalofop ethyl on onion was about 0.8 day [12]. QE could be rapidly metabolized to its primary metabolite quizalofop-acid (QA)

in soybean, cotton foliage and goat [11,13]. The study of potential effects of QE on the development of rats has been conducted, and the results showed a significant decrease in the number of fetuses alive and a significant increase in the number of rats with retained placenta [14]. QE exits two enantiomeric forms, the (+)- and (−)-form, but the (+)-form has higher herbicidal activities. For the herbicidal mechanism of QE is inhibiting acetyl CoA carboxylase and (+)-form is a more potent inhibitor against acetyl-CoA in chloroplasts [15]. However, the racemate of QE is still widely used owing to the low cost. The inactive enantiomer just causes environmental problems and may have influences on non-target organisms after their use on crops.

A chiral HPLC and a LC-MS/MS method were set up for the separation of the enantiomers and the identification of QE and QA in this work. The stereoselective metabolism of QE in rat in vivo was conducted. The data presented in this study may have some significance for risk assessment.

Materials and Methods

1. Ethics statement

This study and all animal experiments were approved by the local ethics committee (Beijing Association For Laboratory Animal Science), ethical permit number 30749 and carried out with local institutional guidelines.

2. Chemicals and Reagents

Rac-quizalofop-ethyl (98%, technical grade) and rac-quizalofop-acid (99%) were obtained from Institute for the Control of Agrichemicals, Ministry of Agriculture of China. Tween 80 and

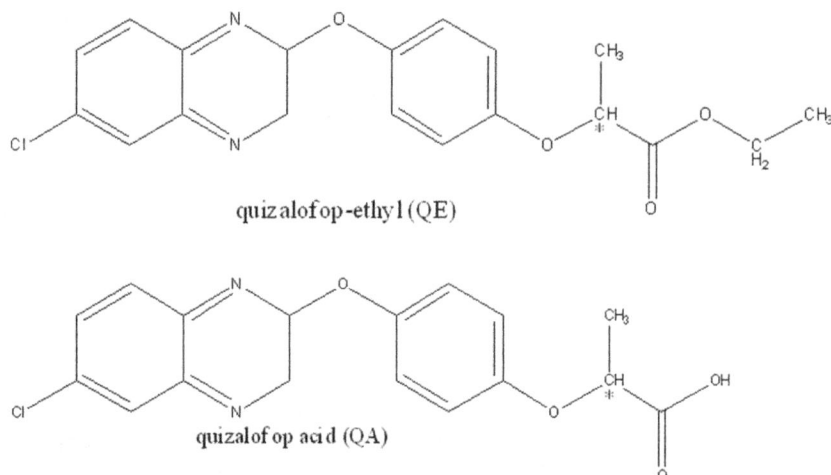

Figure 1. Chemical structures of QE and its primary metabolite QA. Chiral center is denoted by an asterisk (*).

corn oil was obtained from Sigma-Aldrich (St. Louis, MO, USA). Dimethyl sulfoxide, trifluoroacetic acid (TFA), ethyl acetate, n-hexane, acetonitrile, methanol and 2-propanol were purchased from Beijing Chemicals (Beijing, China). Water was purified by Milli-Q water, 18 MΩ·cm. All other chemicals and solvents were of analytical grade and purchased from commercial sources.

3. Animal Experiments

Sprague-Dawley male rats weighing 180–220 g were procured from Experimental Animal Research Institute of China Agriculture University and housed in well-ventilated cages with a 12:12 h light: dark photoperiod. The rats were provided standard pellet diet and water ad libitum throughout the study. The experiments were started only after acclimatization of animals to the laboratory conditions. Before the experiments, the rats were fasted for 12 h, with free access to drinking water at all the times. All the samples were stored immediately at $-20°C$ till the sample processing.

A certain amount of QE dissolved in dimethyl sulfoxide was added in corn oil, after ultrasound and shaking, it turned into a suspension solution and then given to rats by intragastric administration at a dose of 10 mg kg^{-1} b.w. (n = 6) [16]. Blood was sampled from rat tails at 1, 3, 7, 9, 10, 12, 15, 24, 48, 72 and 120 h after the intragastric administration. Control rats received an equal volume of corn oil only. Brain, liver, kidney and lung

were collected at 12 h and 120 h respectively. Urine and feces were gathered throughout the study.

The injection solution for intravenous administration was prepared by dissolving QE in tween 80 and adding with sterile saline (5% tween 80, v/v), which was injected into the caudal vein at 10 mg kg^{-1} body weight. Blood was sampled from rat tails at 1/6, 1/2, 1, 2, 3, 5, 8, 12, 24, 48, 72 and 120 h after the intravenous administration.

4. Sample Preparations

Kidney, lung, liver, brain and feces were homogenized for 3 min to prepare homogenized tissues. The rat blood (0.2 mL), urine (2 mL) and 0.2 g homogenized tissues were transferred to a 15 mL plastic centrifuge tube with the addition of 5 mL of ethyl acetate. To obtain a better extraction, 100 µL HCl (1 mol L^{-1}) was added. The tube was then vortexed for 5 min. After centrifugation at 3500 rpm for 5 min, the upper solution was transformed to a new test tube. Repeat the extraction with another 5 mL of ethyl acetate and combine the upper solution. The extract was dried under a stream of nitrogen gas at 35°C. Then the residue was redissolved in 0.5 mL of 2-propanol or 5 mL of methanol, and finally filtered through a 0.22 µm syringe filter for HPLC and LC -MS/MS analysis.

Table 1. LC-MS/MS conditions: channel mass, apply mode, retention time.

Analyte	Channel mass	mode	RT(min)
QE standard	373+299+271	Positive	2.78
QA standard	343+271+243	Negative	1.69
Blood	343+271+243	Negative	1.69
Kidney	343+271+243	Negative	1.69
Lung	343+271+243	Negative	1.69
Liver	343+271+243	Negative	1.69
Brain	343+271+243	Negative	1.69
Urine	343+271+243	Negative	1.69
Feces	343+271+243	Negative	1.69

Table 2. Calibration data of QE and QA enantiomers in different sample matrixes.

Enantiomers	Matrix	Calibration range (mg kg^{-1})	Standard calibration curve		
			Slope	Intercept	R^2
(+)-QE	blood	0.3–60	63.86	−15.13	0.998
	kidney	0.3–60	72.65	59.04	0.998
	lung	0.3–60	71.55	31.69	0.997
	liver	0.3–60	72.19	29.58	0.998
	brain	0.3–60	64.27	1.39	0.996
	urine	0.3–60	68.44	14.52	0.997
	feces	0.3–60	65.66	21.42	0.998
(−)-QE	blood	0.3–60	61.38	−16.18	0.997
	kidney	0.3–60	70.41	1.17	0.998
	lung	0.3–60	70.78	44.69	0.995
	liver	0.3–60	68.65	−32.81	0.997
	brain	0.3–60	72.32	17.79	0.996
	urine	0.3–60	63.48	14.7	0.998
	feces	0.3–60	66.05	14.56	0.998
(+)-QA	blood	0.3–60	128.15	−43.62	0.998
	kidney	0.3–60	130.19	−32	0.998
	lung	0.3–60	131.65	−26.97	0.995
	liver	0.3–60	130.23	−51.97	0.997
	brain	0.3–60	129.56	−28.1	0.995
	urine	0.3–60	130.48	−43.62	0.998
	feces	0.3–60	132.89	−16	0.998
(−)-QA	blood	0.3–60	119.15	−40.81	0.999
	kidney	0.3–60	128.3	−14.85	0.998
	lung	0.3–60	129.87	−14.99	0.997
	liver	0.3–60	130.33	−43.04	0.996
	brain	0.3–60	128.45	−18.42	0.994
	urine	0.3–60	131.68	−32.61	0.999
	feces	0.3–60	126.73	−19.96	0.998

5. Analytical Procedures

QE and QA were analysed by HPLC using Agilent 1200 series equipped with a G1322A degasser, G1311A quatemary pump, a G1329A automatic liquid sampler, G1314B variable wavelength UV detector and Agilent 1200 Chemstation software. A column attemperator (Tianjin Automatic Science Instrument Co. Ltd,

Table 3. Extraction efficiency of (+)-QE, (−)-QE in blood, kidney, lung, liver, brain, urine and feces.

Tissues	Recovery(%)					
	(+)-QE			(−)-QE		
	0.3 mg kg^{-1}	6 mg kg^{-1}	60 mg kg^{-1}	0.3 mg kg^{-1}	6 mg kg^{-1}	60 mg kg^{-1}
Blood	93.2±5.4	89.4±4.7	88.5±3.7	85.2±5.5	90.3±3.9	92.1±3.3
Kidney	80.1±5.8	78.7±3.8	87.3±4.2	78.0±7.9	80.8±6.3	86.7±4.8
Lung	86.1±9.7	81.1±6.4	85.2±7.8	88.5±4.4	78.3±4.8	84.1±5.1
Liver	86.5±5.5	84.9±7.4	83.9±7.7	80.0±6.9	85.3±4.6	82.7±3.7
Brain	77.6±8.8	80.3±9.4	85.2±7.4	84.2±9.3	79.0±6.2	88.6±6.5
Urine	90.6±4.4	85.4±5.3	88.9±5.2	101.2±4.8	88.6±4.7	90.1±5.5
Feces	86.4±8.4	84.6±8.6	80.9±7.1	78.6±6.5	90.2±5.8	88.5±4.4

Table 4. Extraction efficiency of (+)-QA, (−)-QA in blood, kidney, lung, liver, brain, urine and feces.

Tissues	Recovery(%)					
	(+)-QA			(−)-QA		
	0.3 mg kg^{-1}	6 mg kg^{-1}	60 mg kg^{-1}	0.3 mg kg^{-1}	6 mg kg^{-1}	60 mg kg^{-1}
Blood	101.7±5.1	101.1±3.8	107.7±3.3	102.8±4.3	103.4±4.1	102.4±4.4
Kidney	105.2±6.2	104.0±6.5	103.6±5.5	103.5±9.1	101.0±5.2	102.5±5.6
Lung	104.2±9.3	107.5±5.5	101.9±6.2	106.5±8.2	105.7±4.6	103.5±6.4
Liver	103.5±5.1	105.4±4.3	103.6±5.2	104.4±6.1	103.8±5.4	102.8±7.1
Brain	105.2±8.0	101.5±6.6	100.3±8.4	100.6±10.2	106.1±4.9	103.2±5.8
Urine	98.9±5.7	96.0±4.4	96.3±5.2	100.3±4.8	101.2±4.8	98.6±6.4
Feces	102.6±6.3	100.6±3.5	101.1±6.2	99.8±5.6	101.6±5.3	98.6±8.1

China) was used to control column temperature. The chiral column was chiralpak IC (250×4.6 mm, Daicel Chemical Industries, Tokyo, Japan). A 20 µL sample was injected into the column and eluted with a mobile phase of n-hexane: 2-propanol (92: 8 v/v) at a flow-rate of 0.6 mL min^{-1}. To get a better separating effect, 0.5% TFA was added to the mobile phase. The temperature of the column was adjusted to 15°C. The elution was monitored by UV absorption at 230 nm and quantification was based on direct comparison of the peak-areas with those of standard. Optical rotatory dispersion (ORD) detector was used to determine the elution orders. LC -MS/MS was used to the identification of QE and QA.

Ultrahigh pressure chromatography was performed using Dionex Ultimate 3000 (Dionex, Sunnyvale, CA, USA) with Hypersil GOLD C$_{18}$ column (2.1×100 mm, 3 µm) at 20°C. The mobile phase was methanol-water-formic acid (70:30:0.1%, v/v/v) at a constant flow rate of 0.3 mL/min and the injection volume was 1 µL. A Thermo TSQ Quantum Access Max (Thermo Fisher Scientific, Waltham, MA, USA) with a heated electrospray ionization source (Thermo Fisher Scientific, Waltham, MA, USA) was used to quantitative analysis. MS/MS was operated under the following parameters: spray voltage, 2500 V; vaporizer temperature, 200°C; capillary temperature, 270°C; sheath gas

pressure, 30 arb; aux gas pressure, 15 arb. Identification was performed using selected reaction monitoring (SRM) in positive mode for QE and in negative mode for QA, with a scan time of 0.10 s per transition. Data were acquired in SRM mode as summarized in Table 1.

6. Method validation

Blank tissues obtained from untreated rats were spiked with rac-QE and rac-QA working standard solutions to generate calibration samples ranging from 0.3 to 60 mg L^{-1}. Calibration curves were generated by plotting peak area of each enantiomer versus the concentration of the enantiomer in the spiked samples. The standard deviation (SD) and the relative standard deviation (RSD = SD/mean) were calculated over the entire calibration range. The recoveries were estimated by the peak area ratio of the extracted analytes with an equivalent amount of the standard solution in pure solvents. The limit of detection (LOD) for each enantiomer was considered to be the concentration that produced a signal-to-noise (S/N) ratio of 3. The limit of quantification (LOQ) was defined as the lowest concentration in the calibration curve with acceptable precision and accuracy.

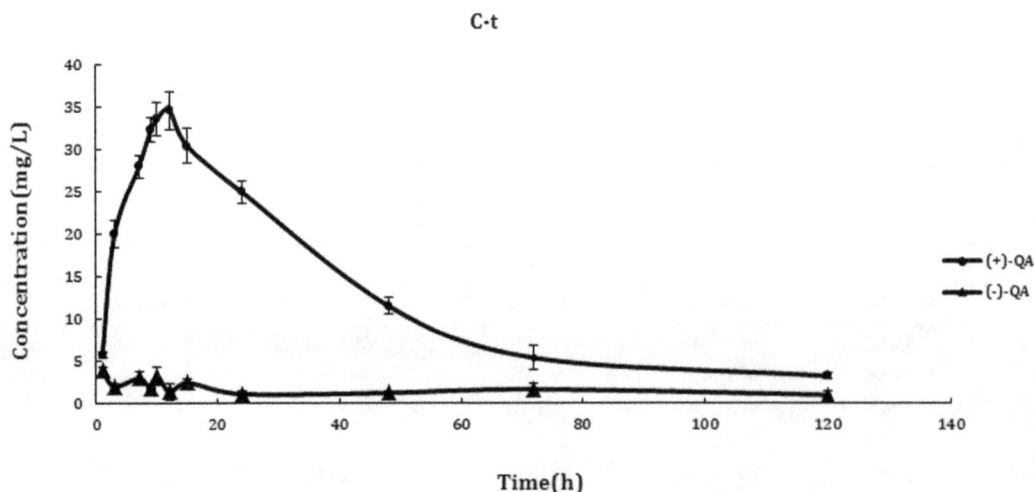

Figure 2. The concentration-time curves of QA enantiomers in blood after intragastric administration. Each point represents the mean ± SD (n = 6). Blood QE level was not detected through the whole study.

C-t

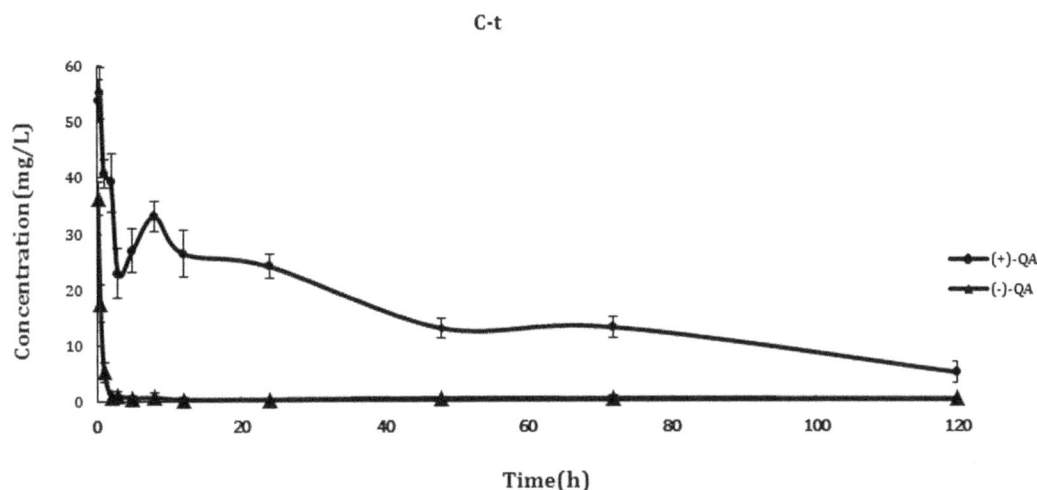

Figure 3. The concentration-time curves of QA enantiomers in blood after intravenous administration. Each point represents the mean ± SD (n = 6). Blood QE level was not detected through the whole study.

7. Data Analysis

Enantiomeric fraction (EF) was used to present the enantioselectivity, defined as: peak areas of $(+)/[(+)+(-)]$. An EF = 0.5 indicates a racemic mixture, whereas preferential degradation of one of the enantiomers made EF under or over 0.5.

The direct excretion rate (ER) of urine and feces was defined as the following exponential: $ER = \frac{C \times m_1}{m_2} \times 100\%$.

Where C is the concentration of QA in urine or feces, mg kg^{-1}; m_1 is the amount of urine or feces, g; m_2 stand for the administered dose of QE. This equation could only reflect the excretion rate approximately base on the assumption that all the QE was metabolized to QA quickly according to the results of this work and the previous studies. The pharmacokinetic parameters such as volume of distribution (Vd) and clearance rate (CL) were generated. The oral bioavailability was calculated as (AUC$_{oral}$/

AUC$_{i.v.}$) × (dose$_{i.v.}$/dose$_{oral}$). The area under the concentration-time curve (AUC) was determined to the last quantifiable concentration using the linear trapezoidal rule and extrapolated to infinity using the terminal phase rate constant. An analysis of variance (ANOVA) was used to determine the statistical differences and $p < 0.05$ was considered to be of statistical significance. Data were presented as the mean ± SD of six parallel experiments.

Results and Discussion

1. Assay Validation

The chromatograms of the control and spiked samples and mass spectrums were shown in Fig. S1 and Fig. S2 in File S1. No endogenous peaks from samples were found to interfere with the

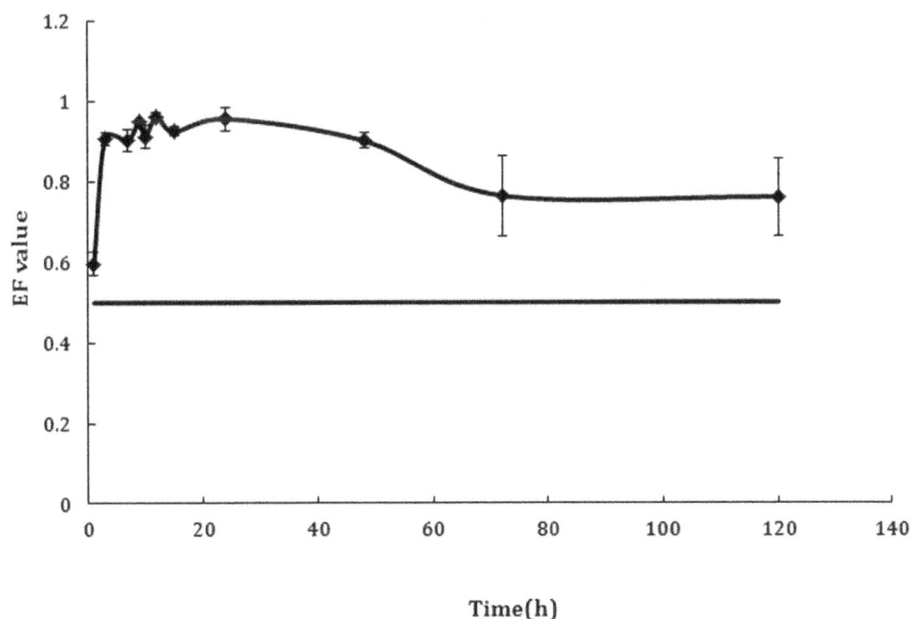

Figure 4. EF-time curve of QA in blood after intragastric administration.

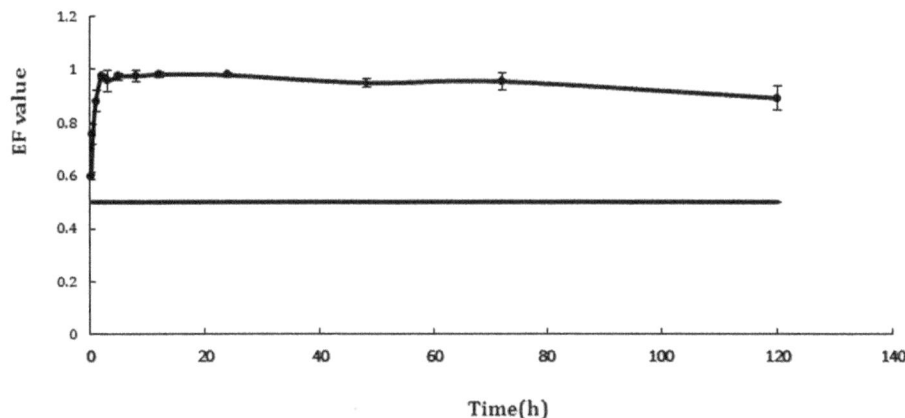

Figure 5. EF-time curve of QA in blood after intravenous administration.

elution of QE and QA. The elution sequence of QE and QA was both (+)/(−).

Linearities of all the tissues were shown in Table 2. Over the concentration range of 0.3–60 mg kg^{-1}, correlation coefficients (R^2) were all higher than 0.994. As shown in Table 3 and Table 4, Extraction efficiency of (+)-QE, (−)-QE, (+)-QA and (−)-QA in samples at the concentrations of 0.3, 6 and 60 mg kg^{-1} (n = 3), ranging from 77% to 108% with RSD of 3%–10%. The LOD and LOQ were 0.1 and 0.3 mg kg^{-1}, respectively.

2. Degradation Kinetics in Rat in vivo

As shown in Fig. 2 and Fig. 3, QE could not be detected in blood after intragastric and intravenous administration of rac-QE, which indicted that QE could be metabolized to QA quickly. However, QA could still be detected even at 120 h in all samples that meant QA could not be easily metabolized by animals. Great difference between the two enantiomers of QA was found in all samples (Fig. 4, Fig. 5). The maximum concentration (Cmax) of (+)-QA in blood was almost ten times higher than that of (−)-QA. Pharmacokinetic parameters and bioavailability of QA after intravenous and oral administration were shown in Table 5. The AUC of (+)-QA and (−)-QA were 1631.202±241.038 mg/L/h and 246.571±70.677 mg/L/h after intragastric administration, and 2239.105±300.554 mg/L/h and 294.751±85.377 mg/L/h after intravenous administration. The oral bioavailability of (+)-QA and (−)-QA were 72.8% and 83.6%. The results revealed a slow clearance of QA from blood.

The reason for not detecting QE in blood after intragastric and intravenous administration could be the rapid deesterification of QE in small intestine and blood. The selective uptake, transport across tissues or protein and elimination of enantiomers may be responsible for the enrichment of (+)-QA [17,18]. The high index of AUC in both intragastric and intravenous administration means that QA was slowly eliminated from plasma and tissues, which may have chronic effects such as reproductive toxicity on rats [15].

QE was also not detected in the tissues. The data of the residue of QA at 12 and 120 h in tissues were shown in Table 6. The EF values in brain, kidney, lung, liver, urine and feces were shown in Fig. 6. Both enantiomers could be detected in brain, kidney, lung and liver at 12 h and 120 h except (−)-QA in brain at 120 h. The concentrations of QA in the tissues were in the order of liver>kidney>lung>brain at 12 h and kidney>liver>lung>brain at 120 h. The relative low concentration of (+) and (−)-QA in brain was mainly due to the function of blood-brain barrier [19]. QA was also found in urine and feces. As shown in Table 7, the rats excreted approximately 8.77% and 2.16% of the administered dose by urine and feces based on the calculation. The relative low amount of QA in urine and feces might be attributed to the fact that QA was degraded to further metabolites or QA was transferred to others tissues.

Conclusions

The stereoselective metabolism of QE and its primary metabolite QA in rats was conducted. QE was rapidly hydrolyzed to QA and could not be detected in all samples. However, QA still could be detected even at 120 h. High index of AUC indicated that QA was more likely to have chronic toxicity to animal and human, especially to the tissues that contained high concentration of QA, such as liver and kidney. (+)-QA occupy a higher proportion than the (−)-isomer in residues and the faster degradation of (−)-QA might contribute to the enantioselectivity. It was also found that urine excretion was not the main pathway of

Table 5. Pharmacokinetic parameters and bioavailability of QA after intravenous and oral administration (n = 6).

Administration routes	Intravenous administration		Oral administration	
	(+)-QA	(−)-QA	(+)-QA	(−)-QA
Vd (ml/kg)	0.279±0.035	9.264±2.519	0.289±0.02	3.302±0.591
CL (ml/min/kg)	0.109±0.014	1.069±0.349	0.147±0.005	0.742±0.271
AUC (mg/h/L)	2239.105±300.554	294.751±85.377	1631.202±241.038	246.571±70.677
Bioavailability (%)	100	100	72.8	83.6

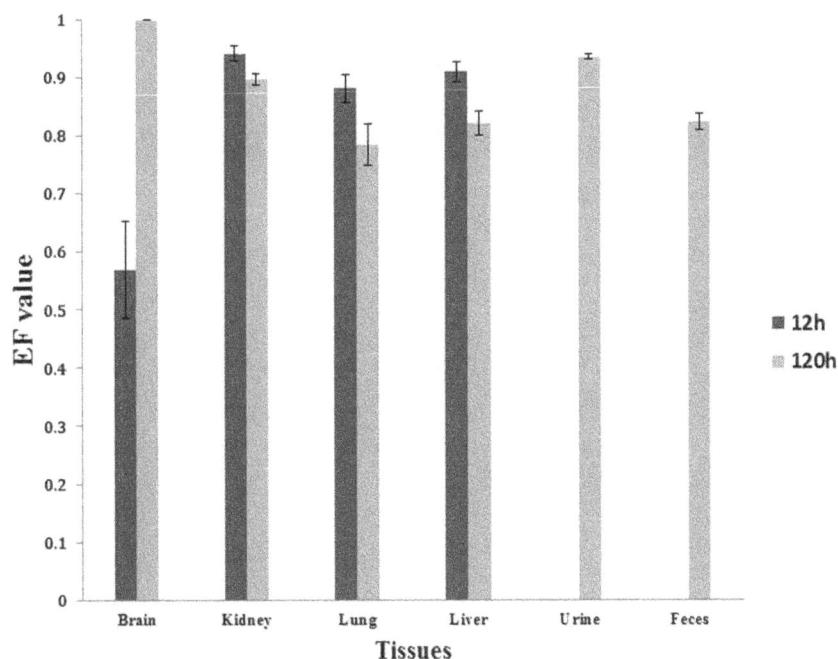

Figure 6. The EF value in brain, kidney, lung, liver, urine and feces.

Table 6. The concentrations of QA in brains, kidneys, lungs and liver at 12 h and 120 h.

Tissues	C(mg kg^{-1})/12 h		C(mg kg^{-1})/120 h	
	(+)-QA	(−)-QA	(+)-QA	(−)-QA
Brain	1.48±0.23	1.12±0.30	0.93±0.88	nd
Kidney	21.77±1.39	1.35±0.25	5.29±0.15	0.60±0.06
Lung	15.19±1.20	2.05±0.38	1.65±0.109	0.45±0.07
Liver	25.58±1.28	2.53±0.62	4.54±0.23	0.98±0.14

Table 7. Excretion rate of (+)-QA (ER$_1$) and (−)-QA (ER$_2$) by urine and feces.

Excreta	ER$_1$	ER$_2$
Urine	8.20%±0.72%	0.57%±0.04%
Feces	1.78%±0.05%	0.38%±0.03%

QA by rat. The data was helpful for full risk assessment of chiral pesticides.

Supporting Information

File S1 Figure S1, Representative HPLC chromatograms of QE and QA extracted from untreated and spiked samples. A1-G1 and A2-G2 represent chromatograms extracted from rat blood, urine, feces, liver, brain, kidney and lung (untreated and spiked with 10 mg L^{-1} of rac-QE and rac-QA respectively). H represents the standard of 10 mg L^{-1} of QA and QE. **Figure S2, Representative MS spectra of QE and QA extracted from untreated and spiked samples.** (A) rat blood; (B) rat urine; (C) rat feces; (D) rat liver; (E) rat brain; (F) rat kidney; (G) rat lung; (1) untreated sample; (2) sample spiked with 1 mg L^{-1} of QE; (3) untreated sample; (4) sample spiked with 1 mg L^{-1} of QA; (H1) standard of 1 mg L^{-1} of QE; (H1) standard of 1 mg L^{-1} of QA.

Author Contributions

Conceived and designed the experiments: YL PW DL ZZ. Performed the experiments: YL ZS HL ZJ. Analyzed the data: YL. Contributed reagents/materials/analysis tools: PW DL ZZ. Wrote the paper: YL PW.

References

1. Ogwok P, Muyonga J, Sserunjogi M (2009) Pesticide residues and heavy metals in Lake Victoria Nile perch, Lates niloticus, belly flap oil. Bulletin of environmental contamination and toxicology 82: 529–533.

2. Lv J, Shi R, Cai Y, Liu Y, Wang Z, et al. (2010) Assessment of 20 organochlorine pesticides (OCPs) pollution in suburban soil in Tianjin, China. Bulletin of environmental contamination and toxicology 85: 137–141.

3. Abhilash P, Singh N (2009) Pesticide use and application: An Indian scenario. Journal of hazardous materials 165: 1–12.

4. Costa LG, Giordano G, Guizzetti M, Vitalone A (2007) Neurotoxicity of pesticides: a brief review. Frontiers in bioscience: a journal and virtual library 13: 1240–1249.

5. Liu W, Gan J, Schlenk D, Jury WA (2005) Enantioselectivity in environmental safety of current chiral insecticides. Proceedings of the National Academy of Sciences of the United States of America 102: 701–706.

6. Liu W, Ye J, Jin M (2009) Enantioselective phytoeffects of chiral pesticides. Journal of agricultural and food chemistry 57: 2087–2095.

7. Marucchini C, Zadra C (2002) Stereoselective degradation of metalaxyl and metalaxyl-M in soil and sunflower plants. Chirality 14: 32–38.

8. Hoekstra PF, Burnison BK, Neheli T, Muir DC (2001) Enantiomer-specific activity of o,p′-DDT with the human estrogen receptor. Toxicology letters 125: 75–81.

9. Williams A (1996) Opportunities for chiral agrochemicals. Pesticide Science 46: 3–9.

10. Wu T, Li X, Huang H, Zhang S (2011) Enantioselective Oxidative Damage of Chiral Pesticide Dichlorprop to Maize. Journal of agricultural and food chemistry 59: 4315–4320.

11. Koeppe MK, Anderson JJ, Shalaby LM (1990) Metabolism of [^{14}C] quizalofop-ethyl in soybean and cotton plants. Journal of agricultural and food chemistry 38: 1085–1091.

12. Sahoo S, Mandal K, Singh G, Kumar R, Chahil G, et al. (2013) Residual behavior of quizalofop ethyl on onion (Allium cepa L.). Environmental monitoring and assessment 185: 1711–1718.

13. Banijamali AR, Strunk RJ, Nag JK, Putterman GJ, Gay MH (1993) identification of [^{14}C] quizalofop-P-tefuryl metabolites in goat urine by nuclear magnetic resonance and mass spectrometry. Journal of agricultural and food chemistry 41: 1122–1128.

14. James d, James M (1999) Evidence on developmental and reproductive toxicity of quizalofop-ethyl. Available: http://oehha.ca.gov/. Accessed 14 November 2013.

15. Kurihara N, Miyamoto J, Paulson G, Zeeh B, Skidmore M, et al. (1997) Chirality in synthetic agrochemicals: bioactivity and safety consideration. Pure Appl Chem 69: 2007–2025.

16. Sun J, Zhu Y, Gu L (2005) Study on the NEL of Quizalofop-P-ethyl. Occupation and Health 11: 006.

17. Covaci A, Gheorghe A, Schepens P (2004) Distribution of organochlorine pesticides, polychlorinated biphenyls and α-HCH enantiomers in pork tissues. Chemosphere 56: 757–766.

18. Ulrich EM, Willett KL, Caperell-Grant A, Bigsby RM, Hites RA (2001) Understanding enantioselective processes: A laboratory rat model for α-hexachlorocyclohexane accumulation. Environmental science & technology 35: 1604–1609.

19. Abbott NJ, Patabendige AA, Dolman DE, Yusof SR, Begley DJ (2010) Structure and function of the blood–brain barrier. Neurobiology of disease 37: 13–25.

4

Swimming with Predators and Pesticides: How Environmental Stressors Affect the Thermal Physiology of Tadpoles

Marco Katzenberger[1]*, **John Hammond**[2], **Helder Duarte**[1], **Miguel Tejedo**[1], **Cecilia Calabuig**[3],
Rick A. Relyea[4]

1 Department of Evolutionary Ecology, Doñana Biological Station - Spanish Council for Scientific Research, Sevilla, Spain, **2** Department of Biology, University of New Mexico, Albuquerque, New Mexico, United States of America, **3** Department of Animal Sciences, Federal Rural University of the Semiarid Region, Mossoró, Rio Grande do Norte, Brazil, **4** Department of Biological Sciences, University of Pittsburgh, Pittsburgh, Pennsylvania, United States of America

Abstract

To forecast biological responses to changing environments, we need to understand how a species's physiology varies through space and time and assess how changes in physiological function due to environmental changes may interact with phenotypic changes caused by other types of environmental variation. Amphibian larvae are well known for expressing environmentally induced phenotypes, but relatively little is known about how these responses might interact with changing temperatures and their thermal physiology. To address this question, we studied the thermal physiology of grey treefrog tadpoles (*Hyla versicolor*) by determining whether exposures to predator cues and an herbicide (Roundup) can alter their critical maximum temperature (CT_{max}) and their swimming speed across a range of temperatures, which provides estimates of optimal temperature (T_{opt}) for swimming speed and the shape of the thermal performance curve (TPC). We discovered that predator cues induced a 0.4°C higher CT_{max} value, whereas the herbicide had no effect. Tadpoles exposed to predator cues or the herbicide swam faster than control tadpoles and the increase in burst speed was higher near T_{opt}. In regard to the shape of the TPC, exposure to predator cues increased T_{opt} by 1.5°C, while exposure to the herbicide marginally lowered T_{opt} by 0.4°C. Combining predator cues and the herbicide produced an intermediate T_{opt} that was 0.5°C higher than the control. To our knowledge this is the first study to demonstrate a predator altering the thermal physiology of amphibian larvae (prey) by increasing CT_{max}, increasing the optimum temperature, and producing changes in the thermal performance curves. Furthermore, these plastic responses of CT_{max} and TPC to different inducing environments should be considered when forecasting biological responses to global warming.

Editor: Michael Sears, Clemson University, United States of America

Funding: This work was funded by a Fundação para a Ciência e Tecnologia (FCT; www.fct.pt) PhD fellowship (SFRH/BD/60271/2009) to MK and a U.S. National Science Foundation (NSF; www.nsf.gov) grant to RAR. The funders had no role in study design, data collection and analysis, decision to publish, or preparation of the manuscript.

Competing Interests: The authors have declared that no competing interests exist.

* E-mail: katzenberger@ebd.csic.es

Introduction

Biological mechanisms underlying a response to environmental changes can be quite complex. To forecast these biological responses, we need to understand how a species' physiology varies through space and time [1,2] and assess how changes in physiological function induced by environmental changes (e.g., increasing environmental temperatures) may interact with phenotypic changes induced by other types of environmental variation [3,4,5,6].

Species can possess the ability to respond to new or altered environments with flexible phenotypes that are environmentally induced and can potentially contribute to adaptive evolution [7]. Stressful environments can induce non-adaptive plasticity, increasing the variance around the mean phenotypic response or distancing it from the favored optimum. Nevertheless, if plasticity is adaptive and promotes establishment and persistence in a new environment, by placing populations close enough to a new phenotypic optimum for directional selection to act, it can predictably enhance fitness and is most likely to facilitate adaptive evolution on ecological timescales [7].

The presence of predators in the environment can induce behavioral and morphological changes in prey that result in the prey being less susceptible to the predator (e.g., [8,9,10,11]). Furthermore, pesticides can also induce behavioral and morphological changes in organisms. Sublethal exposure to pesticides early in life can make the individuals more tolerant of the pesticide later in life [12,13] and they can induce phenotypic changes that resemble predator-induced phenotypes [14,15,16,17]. In other cases, pesticides impede the induction of predator-induced morphology [18,19,20,21].

In the current scenario of climate change, there has been a renewed interest in the thermal physiology of organisms and the estimation of thermal tolerance and sensitivity, using physiological traits such as the critical thermal maximum (CT_{max}; e.g., the temperature at which animals become immobile [22,23]), the optimum temperature (T_{opt}) for performing some function, or the shape of the thermal performance curve (TPC), which describes how an animal's performance changes across a range of temperatures. Although some pesticides are known to affect CT_{max} and burst speed, usually in a negative way (e.g., [24]), there is limited information on how pesticides affect optimum temperature and performance over a range of temperatures (i.e. how pesticides affect TPCs), especially for amphibians. Likewise, much is known about predator-induced changes in organisms, including some interactions with pesticides [17]. Predators also influence thermoregulation and thermal preferences of prey, resulting in behavioral changes and coevolution of thermal optima between species [25]. Other than these behavioral responses that indirectly affect physiology, little is known about whether predator cues can directly affect the thermal physiology of prey.

We addressed these issues by studying the thermal physiology of grey treefrog tadpoles (*Hyla versicolor* LeConte 1825) that were exposed to predator cues and pesticides. Tadpoles are excellent model organisms for this study because they are practically isothermal with their aquatic environment [23] and their thermal physiology traits (CT_{max} and T_{opt}) are not influenced by confounding processes such as dehydration. Tadpoles are also well known for expressing predator-induced changes in behavior and morphology (e.g., [9,26,27]. Furthermore, at least two species of tadpoles can alter their morphology when exposed to the herbicide Roundup and exhibit morphological changes that closely resemble predator-induced changes in tadpoles [17].

Given that pollutants and predators can both affect many aspects of tadpole biology, including development and metamorphosis (e.g., [28,29]), and the interaction of pollutants with other stressors are often negative to the organism (e.g., glyphosate, [30]), we expect the impact of these stressors on the thermal physiology of tadpoles to be mainly negative. Therefore, we hypothesized that tadpoles exposed a sublethal concentration of an herbicide will have reduced tolerance to higher temperatures (CT_{max}) and exhibit a lower optimal temperature (T_{opt}) compared to tadpoles not exposed to the herbicide. Furthermore, because predator cues and the herbicide can induce deeper tails in tadpoles, we hypothesized that tadpoles exposed to either stressor will suffer a vertical shift upward in their TPC across a range of temperatures [31], and have increased swimming performance (e.g., [32]). However, it is also possible that the herbicide will have a negative effect on swimming performance (e.g., [33]) if induced morphological changes are countered by other phenotypic changes that impair swimming ability.

Methods

Inducing the tadpoles

The induction experiment was conducted at the University of Pittsburgh's Pymatuning Laboratory of Ecology in northwest Pennsylvania, USA. The experiment used a completely randomized, 2×2 factorial design comprised of the presence or absence of predator cues crossed with the presence or absence of an herbicide (nominal concentrations of 0 or 2 mg active ingredient per liter (a.e./L). Based on past studies, this herbicide concentration should remain sublethal to gray treefrog tadpoles while inducing morphology changes (e.g., [34,35]).

The four treatment combinations were replicated four times for a total of 16 mesocosms, which consisted of 120-L wading pools, set outdoors (air temperature ranged from 9°C to 28°C), that we filled with 100 L of well water on 11 June 2011. We then added 100 g of dry leaves (*Quercus* spp.) and 5 g of rabbit chow to serve as habitat structure and an initial nutrient source, respectively. We also added an aliquot of zooplankton and phytoplankton that was

a mixture from 5 local ponds. Each mesocosm was equipped with a predator cage constructed of 10×10 cm well pipe covered with window screen at each end. These cages allow the chemical cues emitted during predation to diffuse through the water while preventing the predators from killing the target tadpoles [36,37,38]. Mesocosms were covered with a 60% shade cloth, for the duration of the outdoor experiment.

To obtain tadpoles for the experiment, we collected >20 amplecting pairs of grey treefrogs from a nearby wetland (41° 34′ 9.55″ N, 80° 27′ 22.29″ W) on 18, 21 and 22 May 2011, and allowed them to lay eggs in tubs containing aged well water. Once the eggs hatched, the tadpoles were held in outdoor pools and fed rabbit pellets *ad libitum* until used in the experiment.

On 15 June 2011, which we defined as day 0 of the experiment, we added 40 tadpoles to each mesocosm from a mixture of the clutches with an initial mass ($\pm$SE) of 37.5 ± 2.1 mg per tadpole (subsample, $N = 20$). On day 1, we applied the herbicide treatment. To achieve nominal concentrations of 2 mg a.e./L, we prepared 8 equal mixtures containing 372 µL of stock solution (Roundup Power Max; concentration = 540 g a.e./L) and 250 ml of water. For the eight mesocosms assigned the herbicide treatment, we drizzled one mixture into each mesocosm. For the eight mesocosms assigned the no-herbicide treatment, we drizzled 250 mL of water into each mesocosm. Approximately 1 hr after dosing, we collected water samples from each tank to confirm the concentration of the herbicide. An independent analysis found that the concentrations in the water were 0 and 1.55 mg a.e./L (Mississippi State Chemical Laboratory, Mississippi State, MS). Observing lower actual concentrations is a common phenomenon in mesocosm experiments (reviewed in Brock et al. 2000), likely as the result of binding to surfaces in the mesocosm and degradation of the samples before the testing is conducted. Jones et al. [39] measured little herbicide breakdown for a similar time period, so we assumed there was little change in herbicide concentration during the induction experiment.

After sampling the water, we manipulated the predator environment. For mesocosms assigned the no-predator treatment, the cages remained empty. For mesocosms assigned the predator-cue treatment, we placed a single dragonfly nymph (*Anax junius*) inside the predator cage. Each dragonfly was fed ~300 mg of grey treefrog tadpole biomass every 2 d (see [38]). Prior to each feeding, we observed no tadpoles left in the predator cage, which indicates that the dragonfly nymphs consumed the tadpoles in the cages. The feeding continued until day 10 to allow tadpole growth and induction by the herbicide and predator cues.

Determining the critical thermal maximum of the tadpoles

On day 10, we brought sets of tadpoles into the laboratory to allow them to acclimate at a temperature of 20°C (approximately the average temperature experienced in the mesocosms), with a 12L:12D photoperiod, for 4 to 5 d before testing them for CT_{max} and T_{opt} [22,40]. During acclimation, tadpoles were fed rabbit pellets *ad libitum* and we maintained the predator and herbicide environments to help prevent the loss of any phenotype induction [41]. All tested larvae were below Gosner stage 38 [42]. This is important because tadpoles close to metamorphic climax exhibit a significant decline in thermal tolerance [43].

We obtained upper critical thermal tolerances (CT_{max}) by using a slightly modified version of Hutchison's dynamic method [23]. We exposed tadpoles to a constant heating rate of 0.05°C min^{-1} (3°C h^{-1}), which simulates a natural rate of temperature increase in ponds (H. Duarte, M. Tejedo, J. Hammond, M. Katzenberger, R.A. Relyea, unpublished data from dataloggers; see also [44]) until we observed complete immobility, which signaled the endpoint of the experiment. After reaching CT_{max}, we transferred tadpoles to cooler water (~20°C) to allow recovery. After complete recovery, the tadpoles were weighed and we found that the mass of the tadpoles had increased by 13- to 15-fold since day 0. We tested 3 to 4 tadpoles from each mesocosm, for a total of 56 tadpoles from the 16 mesocosms, as seen in Table 1.

We performed an analysis of variance (ANOVA) that used CT_{max} as the dependent variable, predator cues and herbicide as categorical factors (including the interaction of these factors), and mesocosm nested within the interaction of predator cues and herbicide (i.e. mesocosm nested within treatment). Given that tadpole mass was not correlated with CT_{max} (see results), we did not include it as a covariate. No data transformations were required for this analysis.

Determining the thermal performance curves for tadpole burst speed

Locomotor performance, measured as a TPC, is considered to be a proxy of maximum physiological performance and has been used to estimate optimum temperatures in amphibians [45,46]. We obtained TPCs by measuring each tadpole's maximal burst swimming speed (i.e. burst speed) across a range of temperatures. To determine burst speed, tadpoles were placed individually in a portable thermal bath (patent license ES 2372085), which consists of an opened cross section methacrylate tube (1 m long ×6 cm wide ×3 cm deep) filled with water of a given temperature. We then gently prodded the tadpole with a thin stick to stimulate swimming. Each trial was recorded using a digital camera (30 frames/s) positioned above the tube (JVC Everio GZ-MG505). TPCs were defined using a set of six temperatures (20°, 24°, 28°, 32°, 35° and 38°C). This set includes temperatures tadpoles experienced in the mesocosms (20°-32°C) and two more (35° and 38°C) which they might be exposed to in a scenario of increasing environmental temperatures (but lower than their critical thermal maximum). Temperatures were tested in a random order and, for each temperature, tadpoles from the four treatments were tested in the same session; therefore, all treatments had the same temperature order. Prior to swimming, tadpoles were held individually in 250-ml containers at the test temperature for approximately 1 hr. A different set of tadpoles (total N = 570) was used for each temperature (Table 2) and each wading pool was represented equally in each set.

After the tadpole started to move, we used the software Measurement in Motion [47] to estimate burst speed over three frames (0.1 s) by measuring the distance the center of mass moved between frames [48,49]. After conducting at least three bouts, we used the fastest speed measured for a given tadpole as our measure of that individual's burst speed. Since maximal swimming speed may scale with body size [45] and body size may confound the

effect of speed on escape success [50], we used size-corrected burst speed (using tadpole total length) when constructing TPCs.

To describe the TPCs for burst speed, we used the Template Mode of Variation method (TMV, [51]) which employs a polynomial function to decompose variation among TPCs into three predetermined modes of variation with biological connotation: vertical shift (faster-slower), horizontal shift (hotter-colder), and specialist-generalist trade-offs ([31]; see [51] and supporting information for details on calculations). Since we tested tadpole performance at six temperatures, we assumed that the common template curve was a fourth-degree polynomial, as in previous studies (e.g., [46]). Making this assumption avoids inadequately describing TPCs, which can happen when using a lower-order polynomial [51,52].

In addition to using the TMV method, we also calculated maximum performance (z_{max}) to evaluate changes in maximum swimming speed at the optimum temperature and a more traditional measurement of performance breadth to confirm specialist-generalist trade-offs (using B_{95}, which is the range of temperatures at which performance values exceed 95% of the maximum;[53]). We used B_{95} instead of the traditional B_{80} because the lower limit of B_{80} would fall below 20°C, which is outside the tested range of temperatures. All computations regarding the TPCs, except for B_{95}, were made using the Matlab code by R. Izem (available online in the appendix of [51]). We also confirmed the fit of each treatment's curve and calculated standard error (SE) of each curve's parameters using nlinfit and nlparci functions, respectively, in Matlab [54].

We conducted an ANOVA that used burst speed as the dependent variable, temperature, predator cues and the herbicide (including the interaction of these factors) as categorical factors and, mesocosm nested within the interaction of predator cues and herbicide (i.e. mesocosm nested within treatment). ANOVA analysis was followed by a Tukey post-hoc test.

Assessing the morphology of the tadpoles

After the swimming trials, we determined the mass and developmental stage of each tested tadpole. We then took lateral photos of each tadpole and digitized the images for morphometric measurements. We captured the shape of tadpoles by digitizing 10 landmarks and 15 semi-landmarks (see supporting information; see also [49,55]) on each tadpole using tpsDig2 software [56]. We then extracted partial warps and the uniform component with tpsRelw software [57], which we used as our shape variables in a subsequent analysis. We visualized variation in landmark positions using the thin-plate spline approach (transformation grids, [58] in MorphJ [59]. As an alternative approach to quantify tadpole morphology, we also took the following linear measurements of each tadpole: total tadpole length (TTL, distance between snout and tip of tail fin), body length (BL, distance between snout and point where bottom edge of tail muscle meets body), body depth

Table 1. Critical thermal maximum (CT_{max}), sample size (N) and body mass (Mass) of *Hyla versicolor* tadpoles, in four treatments.

Treatment	N	CT_{max} (°C±SE)	Mass (mg±SE)
Control	13	41.78±0.1	483.7±22.9
Predator	13	42.14±0.1	520.4±29.3
Roundup	15	41.76±0.1	545.4±28.0
Predator + Roundup	15	42.17±0.1	489.8±34.2

Tested tadpoles are representative of the four mesocosms used for each treatment.

Table 2. Experimental temperatures, sample size (N), total tadpole length in mm (TTL±SE), and maximum swimming speeds in cm/s (mean ±SE) for gray treefrog tadpoles that were exposed to predator cues and the herbicide Roundup.

Temp.	Control			Predator			Roundup			Predator + Roundup		
	N	TTL	Speed	N	TTL	Speed	N	TTL	Speed	N	TTL	Speed
20°C	22	37.3±0.6	39.2±1.0	24	38.2±0.6	41.1±1.1	23	38.8±0.6	41.0±1.4	24	37.9±0.7	40.2±1.3
24°C	24	38.1±0.5	41.3±1.1	24	41.0±0.5	46.9±1.5	23	38.8±0.5	44.0±1.3	23	38.0±0.6	44.0±1.3
28°C	24	39.9±0.6	45.4±1.5	24	42.1±0.5	52.3±1.7	24	41.8±0.5	50.7±1.8	24	39.7±0.6	50.3±1.7
32°C	24	39.4±0.6	46.7±1.3	24	39.7±0.5	52.5±1.2	24	40.2±0.6	50.2±1.4	24	39.1±0.6	52.5±1.1
35°C	24	39.6±0.6	45.8±1.7	24	40.8±0.5	51.5±1.6	24	40.6±0.6	47.1±1.7	24	40.4±0.6	50.1±1.8
38°C	24	37.9±0.5	40.2±1.8	24	37.6±0.6	44.6±1.6	24	39.1±0.6	41.5±2.1	23	36.2±0.6	42.3±2.1

(BD, deepest point of the body), tail length (TL, distance between point where bottom edge of tail muscle meets body and tip of tail fin), muscle depth (MD, deepest point of the muscle) and tail depth (TD, maximum depth of the tail fin).

We conducted canonical correlation analysis as a dimension-reducing procedure to obtain two morphological indices (i.e. a linear combination of shape variables); one was for the linear measurements (MI_{lin}) and the other was for the partial warps and uniform component (MI_{geo}). We then examined these two indices for correlations with burst speed (across all treatments; see [55]). To determine if predator cues, herbicide, and their interaction influenced tadpole size (i.e. centroid) or shape (MI_{lin} or MI_{geo}), we performed three ANOVAs followed by Tukey HSD post-hoc tests; mesocosms were nested within the interaction of predator cues and herbicide (i.e. mesocosm nested within treatment). Shape variables (MI_{lin} and MI_{geo}) and tadpole size (centroid) were then used as continuous predictors, along with temperature, predator cues and herbicide as a categorical predictors, in two ANCOVA analysis (testing either MI_{lin} or MI_{geo} separately), to evaluate their effects on burst speed. We performed all analyses using Matlab [54], except when mentioned otherwise, and used a significance level of $\alpha = 0.05$.

All experiments were approved by the University of Pittsburgh's Institutional Animal Care and Use Committee (Protocol #12050451).

Results

Critical thermal maxima of the tadpoles

In our analysis of CT_{max}, there were no differences among mesocosms within a given treatment. We found an effect of predator cues but no effect of the herbicide or the interaction of both (Table 3). Averaged across herbicide treatments, tadpoles exposed to predators had a CT_{max} that was 0.4°C higher than tadpoles not exposed to predators (Table 1). CT_{max} was not correlated with tadpole mass (Pearson's R = -0.17, $p = 0.22$).

Thermal performance curves for tadpole burst speed

When we test tadpole swimming ability across different water temperatures, we found that swimming burst speed varied with temperature (Table 2). When we used the TMV method on size-corrected performance data, we obtained both a common template curve, which provided a good approximation of the common shape of each treatment's curve (Fig. 1), and a three-parameter shape-invariant model (with the use of a fourth-degree polynomial), which explained over 99% of the variation for swimming speed. Decomposition of the total variation into the three pre-determined directions of variation reveals that TPCs for swimming speed vary mostly in the specialist-generalist (53.27%) direction and the vertical (45.98%) direction, but very little in the horizontal (0.59%) direction. This indicates that tadpoles in the control treatment had a wider swimming TPC than tadpoles exposed to predator cues or the herbicide, even when comparing more traditional measures of curve width (B_{95}; Table 4, Fig. 2). Thus, most of the variation in the TPCs is due to specialist-generalist trade-offs and differences in overall performance (faster-slower), rather than changes in T_{opt} (hotter-colder). Indeed, tadpoles raised in the herbicide treatment exhibited only a small decrease in T_{opt} (-0.4°C) while tadpoles raised with predator cues exhibited an increase in T_{opt} (1.5°C). Tadpoles raised with both predators and herbicide exhibited a T_{opt} that was intermediate in magnitude between the latter two treatments but still higher (0.5°C) than tadpoles raised in the control treatment. The only significant difference in T_{opt} was between tadpoles exposed only to

Table 3. ANOVA using CT_{max} as dependent variable, predator cues and Roundup as categorical factors (including the interaction of these factors) and, mesocosm nested within the interaction of predator cues and Roundup, for *Hyla versicolor.*

	SS	d.f.	MS	F	p
Predator	1.993	1	1.993	14.9	<0.001
Roundup	0.006	1	0.006	0.04	0.834
Predator*Roundup	0.009	1	0.009	0.06	0.801
Mesocosm (Predator*Roundup)	1.329	12	0.111	0.83	0.622
Error	5.350	40	0.134		

Univariate tests of significance for CT_{max}. In this model, we used Sigma-restricted parameterization and Type III sum of squares.

herbicide and those exposed only to predator cues (1.8°C; 2-tailed *t*-test, $p<0.05$). Maximal performance (z_{max}) was marginally correlated with performance breadth (Pearson's R = −0.95, $p=0.051$).

Temperature and predator cues both influenced burst speed (Table 5). There was also a significant interaction between predator cues and herbicide. Tadpoles in the control treatment had slower burst speeds accross all temperatures than tadpoles in the other three treatments (all $p<0.05$). Tadpoles raised in the predator treatment were also faster than those from herbicide treatment ($p<0.05$). Furthermore, tadpoles in all treatments containing predator cues or herbicide had higher maximum performance (z_{max}) than tadpoles in the control treatment, so that their burst speed at the optimum temperature was higher than the burst speed of tadpoles raised without any cues. These differences in the parameters of the TPCs can be seen as changes in the overall shape of the curves (Fig. 2). Our analysis of burst speed also revealed a significant effect of mesocosms (nested within treatment), however the magnitude of this effect was much smaller than in other effects, such as the interaction of predator cues and herbicide (Table 5). Nevertheless, we checked for burst speed differences among tanks of the same treatment and temperature

and we found no significant effect of mesocosm on burst speed, in any of the treatment-temperature combinations (all $p>0.05$).

Induced morphology of the tadpoles

We observed size and shape changes in tadpoles exposed to the herbicide and predator cue treatments (Fig. 3). Predator cues and herbicide had no main effects on tadpole centroid size (Table 6a) but they did have a significant interaction; tadpoles exposed to predator cues + herbicide were smaller than those exposed only to the herbicide or only to the predator cues (both $p<0.05$). Similarly, tadpoles in the control treatment were smaller than those exposed only to the herbicide or only to the predator cues (both $p<0.05$). For geometric morphometric measurements, both predator cues and herbicide influenced tadpole shape (Table 6b) and there was a significant interaction between the two factors. Tadpoles raised in the control treatment differed from those raised in the other three treatments (all $p<0.05$), however these did not differ amongst themselves. For linear measurements, only predator cues significantly influenced shape of tadpoles (Table 6c). Tadpoles raised in predator or predator + herbicide treatment differed from those raised in herbicide or control treatments (all $p<0.05$). Mesocosm effect on either centroid or shape (MI_{lin} or MI_{geo}) was non-significant (Table 6). Overall, compared to tadpoles in the control,

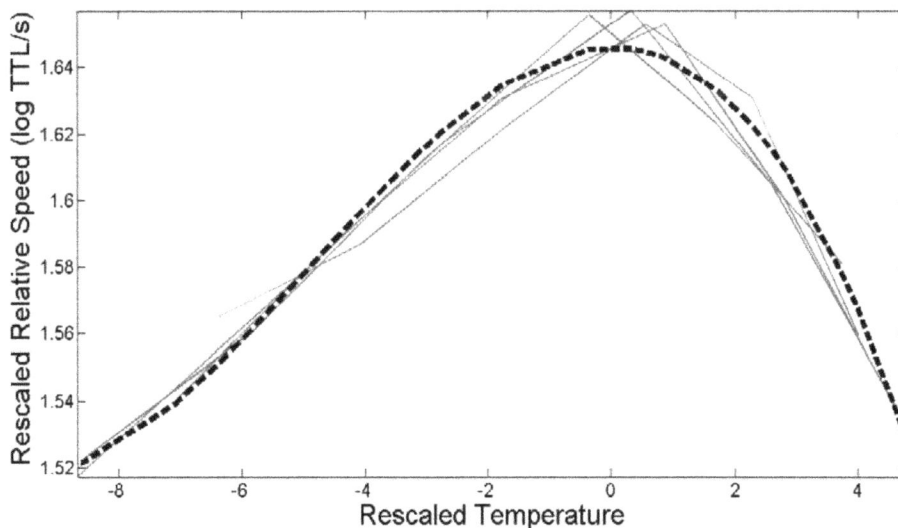

Figure 1. Rescaled thermal performance curves for swimming speed in each treatment with fitted common template shape.
Common template shape z(T) is represented by a dashed line nad the treatments by solid lines. Each thermal performance curve of a treatment (i) and temperature were standardized with respect to the estimates of height (h), location (m; T_{opt}), and width (w) parameters from the fit to model. Rescaled optimum temperature $T_{opt} = 0$. (see [46,51]). Swimming z(T) = $1.6458-0.004T^2-0.00023982T^3+0.000003493T^4$.

Table 4. Parameters of thermal performance curves for maximum swimming speed in four treatments, for *Hyla versicolor*, estimated with TMV method (Izem and Kingsolver, 2005) and nlinfit/nlparci functions in Matlab (Mathworks, 2009).

Treatment	TMV parameters						nlinfit/nlparci		
	h^*	T_{opt}	w	z_{max} 1	z_{max} 2	B_{95}	$h \pm SE$	$T_{opt} \pm SE$	$w \pm SE$
Control	0.12	31.05	1.74	1.07	11.65	18.36	0.12±0.10	31.07±0.76	1.73±0.19
Predator	−0.01	32.52	1.47	1.11	12.96	14.29	−0.01±0.09	32.53±0.52	1.47±0.13
Roundup	0.00	30.70	1.52	1.09	12.27	14.97	0.00±0.09	30.71±0.65	1.52±0.14
Predator + Roundup	−0.11	31.58	1.34	1.12	13.13	12.35	−0.11±0.09	31.58±0.51	1.33±0.11

h^*, height (log TTL/s); T_{opt}, optimal temperature (°C); w, width (dimensionless); z_{max} 1 (TPC), maximum performance (log TTL/s); z_{max} 2 (TTL/s), maximum performance (TTL/s); B_{95}, thermal performance breadth (°C).

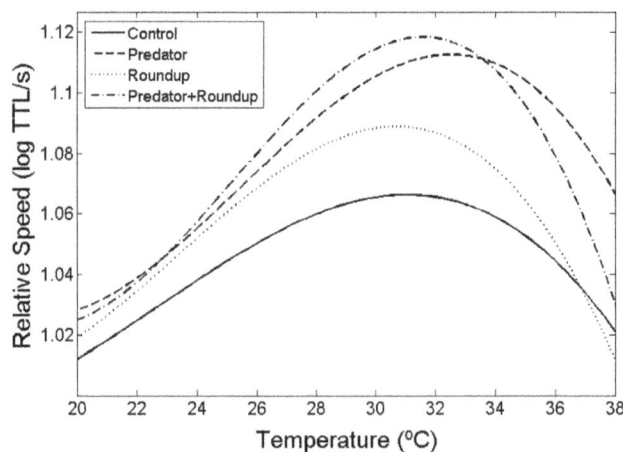

Figure 2. Overall shape of the thermal performance curves for each of the four induction treatments. Each treatment is represented by a thermal performance curve for tadpole swimming speed: control - solid line, predator - dashed line, Roundup - dotted line and predator+Roundup - dash-dot line.

tadpoles in the other three treatments exhibited relatively shorter bodies. Furthermore, in the two treatments containing predator cues, tadpoles exhibited an increase in their relative tail length and tail depth (Fig. 3). Apart from temperature and predator cues, burst speed was also influenced by tadpole's size, either when using morphometric geometric data (Table 7a) or linear measurements (Table 7b). We also found a significant effect of shape on burst speed when using geometric morphometric data (Table 7a).

Discussion

We discovered that predator cues and the herbicide Roundup can affect the thermal physiology of *Hyla versicolor* tadpoles. Predator cues induced tadpoles to have CT_{max} values that were 0.4°C higher whereas the herbicide had no effect. Predator cues and Roundup also influenced the shape of the thermal performance curves, resulting in changes in optimum temperature, performance breadth and maximal performance (Fig. 2). Furthermore, predator cues also induced morphological changes that increased the tadpoles' burst speed.

Roundup, a glyphosate based broad-spectrum systemic herbicide, did not have any effect on CT_{max} estimates of tadpoles. However there have been reports of other contaminants affecting the thermal physiology of vertebrates. Among insecticides, for example, endosulfan (an organochlorine insecticide that affects the central nervous system) and chlorpyrifos (an organophosphate insecticide that inhibits acetylcholinesterase) are known to decrease CT_{max} in fishes [60]. Other environmental contaminants, such as cadmium and copper, can adversely affect the ability of fish to withstand high temperature stress [61,62]. Whether all of these observations in fishes can be extrapolated to other species of aquatic organisms, such as tadpoles, is yet to be determined. Based on these studies and our own results, it seems that the effects of pesticides on CT_{max} may depend on the type of pesticide, the concentration of the pesticide, and how it affects the organism (i.e. its mode of action). There is the possibility that using higher concentrations of the herbicide might induce a decrease in CT_{max}, but higher concentrations will cause tadpole death [63]. Furthermore, the herbicide also did not interfere with the increase in CT_{max} induced by predator cues; tadpoles exposed to predator

Table 5. ANOVA using burst speed as dependent variable, and temperature, mesocosm, predator cues and Roundup as categorical predictors, with mesocosm nested within the interaction of predator cues and Roundup, for *Hyla versicolor*.

	SS	d.f.	MS	F	p
Temperature	0.891	5	0.178	32.17	<0.001
Predator	0.106	1	0.106	19.16	<0.001
Roundup	0.002	1	0.002	0.38	0.537
Predator*Roundup	0.070	1	0.070	12.65	<0.001
Mesocosm (Predator*Roundup)	0.127	12	0.010	1.92	0.03
Predator*Temperature	0.023	5	0.005	0.83	0.528
Roundup*Temperature	0.017	5	0.003	0.62	0.683
Predator*Roundup*Temperature	0.009	5	0.002	0.33	0.903
Error	3.085	546	0.006		

Univariate tests of significance for burst speed. We used Sigma-restricted parameterization and Type III sum of squares.

cues + herbicide had similar CT_{max} values to those exposed only to predator cues.

Different methodological protocols and biological sources can affect estimates of upper thermal tolerances (see [44,64]). For example, the ramping rate used [65,66,67,68], the selection of end-point [23], variations in previous thermal acclimation [40], ontogenetic stage [43], time of day, and photoperiod [69] all may promote shifts in amphibian upper thermal tolerances. We discovered that predatory cues can also affect CT_{max} estimates of prey. An increase in thermal tolerance of predator-induced tadpoles would cause an increase in their warming tolerance,

which is the difference between CT_{max} and maximum temperature of the environment to which an ectotherm is exposed [70,71]. This means that tadpoles exposed to predator cues would be less susceptible to acute thermal stress than tadpoles that were not exposed to predator cues. In contrast, an exposure to the herbicide, at least at the concentration used in our study, would not affect the warming tolerance of tadpoles.

An exposure to predator cues and the herbicide had interactive effects on tadpole burst speed. The interaction occurred because the herbicide alone and predator cues alone each increased burst speed compared to the control, but the combination of the

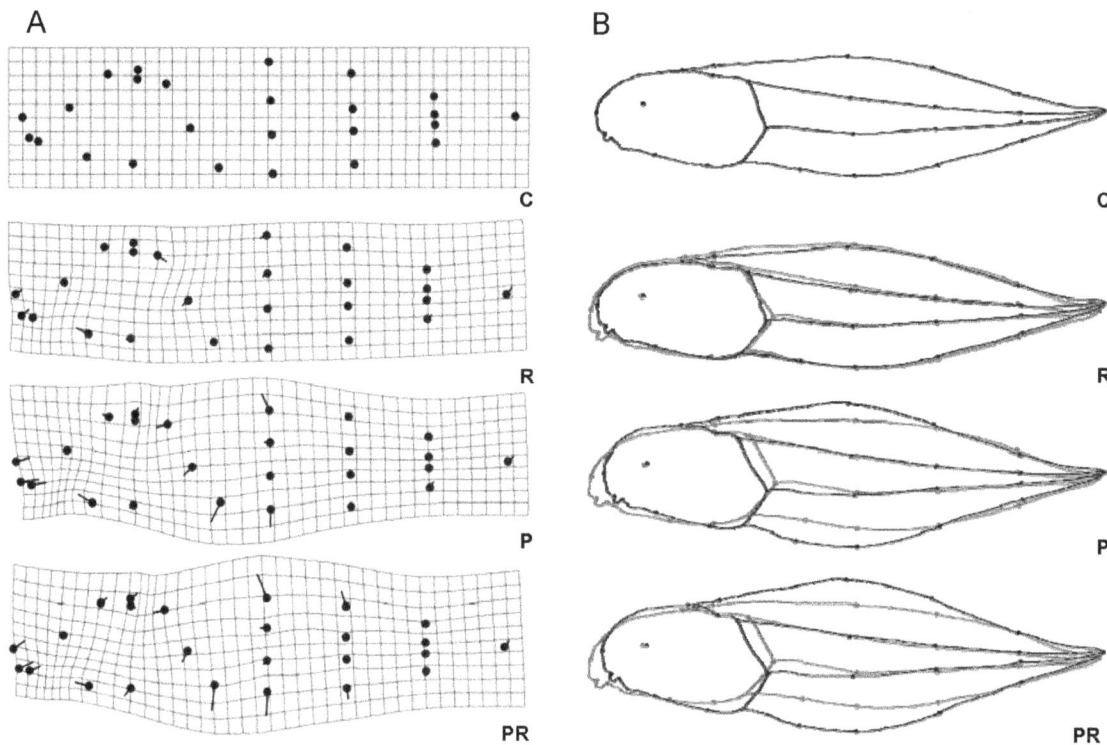

Figure 3. Transformation grids with landmarks and warped outline drawings for each treatment's tadpole shape. a) Transformation grids with landmarks (black dots) and vectors showing direction of variation; b) comparison of warped outline drawings for each treatment shape (black) and control shape (grey). Transformation grids and warped outline drawings were magnified (x5) to better illustrate the differences. C – Control, R – Roundup, P – Predator and PR – Predator + Roundup.

Table 6. ANOVAs to determine if predator cues and Roundup (including their interaction) influenced size (a; centroid), or shape (b and c) of tadpoles (MI_{geo}, for geometric morphometric measurements, or MI_{lin}, for linear measurements, respectively) with mesocosm nested within the interaction of predator cues and Roundup (i.e. mesocosm nested within treatment).

a) Centroid (size)	SS	d.f.	MS	F	p
Predator	19.9	1	19.91	0.97	0.326
Roundup	4.3	1	4.32	0.21	0.647
Predator*Roundup	521.7	1	521.7	25.38	<0.001
Mesocosm (Predator*Roundup)	423.0	12	35.25	1.72	0.06
Error	11386.2	554	20.55		
b) MI_{geo} (shape)	SS	d.f.	MS	F	p
Predator	11.77	1	11.766	12.21	<0.001
Roundup	5.17	1	5.172	5.37	0.021
Predator*Roundup	5.68	1	5.684	5.90	0.016
Mesocosm (Predator*Roundup)	12.37	12	1.031	1.07	0.383
Error	533.95	554	0.964		
c) MI_{lin} (shape)	SS	d.f.	MS	F	p
Predator	29.35	1	29.348	30.95	<0.001
Roundup	3.32	1	3.317	3.50	0.062
Predator*Roundup	2.46	1	2.463	2.60	0.108
Mesocosm (Predator*Roundup)	8.67	12	0.723	0.76	0.690
Error	525.32	554	0.948		

We used Sigma-restricted parameterization and Type III (Effective hypothesis) sum of squares.

herbicide and predator cues induced an increase that was not larger than predator cues alone. Therefore, since the combination of the herbicide and predators cues was not additive, in the presence of predator cues, exposure to the herbicide caused no change in burst speed.

The presence of either predator cues or the herbicide narrowed the performance breadth of the TPC while increasing maximal performance. As performance breadth is negatively correlated with maximal performance, we would expect a generalist-specialist trade-off. Tadpoles from a treatment which induced a more

Table 7. ANCOVA analysis using burst speed as dependent variable, shape variables MI_{geo} (a) or MI_{lin} (b) and tadpole size (centroid) as continuous predictors, alongside temperature, predator cues and Roundup as categorical predictors.

a)	SS	d.f.	MS	F	p
Predator	0.068	1	0.068	15.01	<0.001
Roundup	0.000	1	0.000	0.01	0.909
Temperature	0.197	5	0.039	8.66	<0.001
Size (Centroid)	0.167	1	0.167	36.70	<0.001
Shape (MI_{geo})	0.129	1	0.129	28.27	<0.001
Predator*Roundup	0.011	1	0.011	2.34	0.127
Error	2.544	559	0.005		
b)	SS	d.f.	MS	F	p
Predator	0.101	1	0.101	21.24	<0.001
Roundup	0.003	1	0.003	0.59	0.443
Temperature	0.507	5	0.101	21.30	<0.001
Size (Centroid)	0.410	1	0.410	86.09	<0.001
Shape (MI_{lin})	0.011	1	0.011	2.37	0.124
Predator*Roundup	0.012	1	0.012	2.45	0.118
Error	2.661	559	0.005		

Univariate tests of significance for burst speed. In both models, we used Sigma-restricted parameterization and Type III (Effective hypothesis) sum of squares.

specialist curve (as demonstrated by predator cues + herbicide) would perform better at the optimum temperature but gradually decrease in performance, as moving away from the optimum temperature, until reaching a point were tadpoles from a treatment which induced a more generalist curve (as demonstrated by control) would outperform them (see [31,51]; Table 4). However, we do not see a decline in performance at the extremes of the thermal performance curve, at the tested temperatures, as a result of this trade-off. This observation is confirmed by thermal tolerance data where none of the tadpoles raised in any of the treatments with predator cues or the herbicide had lower CT_{max} than those from the control treatment. Instead, it appears the expected decline in sub-optimal performance resulting from a generalist-specialist trade-off is compensated by the increase in overall performance, so that tadpoles raised in the control treatment always perform, on average, worse than herbicide- or predator-induced tadpoles, at least at the tested temperatures. Therefore, when comparing thermal performance curves, the resulting increase in overall performance was asymmetric, being greater around the optimum temperature and lower at the extreme temperatures.

Surprisingly, predator cues and the herbicide also produced changes in the optimum temperature, but in opposite directions. Of course, the small decrease in optimum temperature caused by the herbicide ($0.4°C$) may have little or no biological relevance. In contrast, the increase in optimum temperature promoted by predator cues (approximately $1.5°C$) may be important, especially when new assessments suggested that environmental impacts will require smaller degrees of global warming than previously thought [72]. Since predator cues increase optimum temperature, the difference between optimum temperature and the environmental temperature should also increase (i.e. thermal safety margins (TSM); see [70]), which would be beneficial to the tadpoles in the current scenario of increasing global temperatures.

Previous studies have demonstrated that changes in the shape or position of thermal performance curves can occur due to acclimation (e.g., [73,74,75]) or that thermal performance curves of different locomotor strategies for the same organism can have different shapes (e.g., [46,76]). In the present study, we demonstrate that the presence of sublethal concentrations of an herbicide and cues from predators can also produce changes in the thermal performance curves and therefore affect how tadpoles respond to environmental temperature changes.

Although it has been documented that predators can affect the behavioral thermoregulation of their prey (e.g., [25]), to our knowledge this is the first study to demonstrate a predator altering the thermal physiology of their prey by increasing CT_{max}, increasing the optimum temperature, and producing changes in the thermal performance curves. It has also been demonstrated that Roundup's lethality increases with competition stress [35] and that predator cues can improve tadpole survival when tadpoles are exposed to the herbicide under stratified water conditions [17]. Therefore, one could make the argument that acclimation to predator cues might be beneficial under warmer temperatures. However, we should also keep in mind that predation simultaneously has a negative effect on tadpole populations and can select for particular phenotypes (see [9]). To display a predator-induced phenotype, tadpoles need to detect chemical cues that are released when other tadpoles (particularly conspecifics) are consumed. So, the possible positive effects of predator cues on the thermal physiology, in a global warming scenario, would only be beneficial for those phenotypes that survive predation.

Predator cues in our study induced morphology changes (relative smaller bodies, deeper tails and deeper tail muscle) that were similar to those observed in previous studies (e.g., [77]). These morphological changes likely explain why tadpoles exposed to predator cues swam faster than control tadpoles. Exposure to the herbicide (see figure 3) induced relative smaller bodies, and the observed changes partially resembled the predator-induced phenotype (see also [17]). The induction of relatively deeper tadpole tails by the herbicide was less evident in the current work than in the study of Relyea [17]. However, this may be due to a number of differences in the experimental protocol including the duration of exposure and a substantially different experimental venue.

Predator cues and the herbicide caused interactive effects on tadpole size. Tadpoles exposed to predator cues + herbicide were smaller than those exposed only to the herbicide or only to predator cues. Tadpoles raised in the control treatment also tended to be smaller than those exposed only to the herbicide or only to predator cues. This may explain why tadpoles from the herbicide treatment also swam faster than tadpoles from the control treatment. As a result, all three treatments had better overall swimming performance than in control, with increase in burst speed related to the magnitude of morphology change (more induction, higher performance) and size. Furthermore, predator-induced morphology changes can be reversed if cues are removed [41]. As a result, some of the changes in the thermal performance curve may also be reversible. If so, in the absence of cues, the predator- and herbicide-induced TPC shapes would revert back to the original curve (i.e. the control curve).

The mechanism underlying the ability of the herbicide to induce morphological changes in tadpoles is still unknown. It has been suggested that the herbicide may be interfering with the stress hormones that induce anti-predator defenses [78] or that herbicides and predator cues activate shared endocrinological pathways [17]. We have demonstrated that predator cues and the herbicide can affect the thermal physiology of tadpoles, although not all changes occur in the same direction. However, the mechanisms behind these thermal physiology changes are also unknown, with possible scenarios arising from our results: a) herbicide interferes only with the stress hormones that induce anti-predator defenses; b) they do not share the same physiological pathways, or at least not all of them; c) they both activate shared endocrinological pathways but predator cues also indirectly activate temperature-stress response mechanisms; or d) stress response mechanisms are more general than previous thought and predator-induced stress produces similar physiological responses as temperature-induced stress.

Conclusions

Apart from inducing morphology changes, predator cues promoted an increase in CT_{max} and optimum temperature of *Hyla versicolor* tadpoles. As such, in the presence of predators, we can expect tadpoles to have greater warming tolerance and broader thermal safety margins. These changes might indirectly help tadpoles cope with increasing environmental temperatures. The herbicide Roundup is not only toxic to amphibians (and lethal over certain concentrations), but it also produces changes in morphology [17]. With this work, we now know that it also interferes, to some extent, with the thermal physiology of tadpoles (in particular in the thermal performance curves), although the effect on warming tolerance and thermal safety margins appears to be marginal. However, Roundup is just one of hundreds of chemicals currently used in anthropogenic activities (e.g., agriculture) and tadpoles can face predation by a wide variety of predator species. Because combinations of pesticides, which are a common situation in natural environments, can have greater impacts than each pesticide alone [79], future studies should test whether combinations of pesticides and predators could have different effects on the thermal physiology of organisms.

In the current scenario of climate change, it is important that we understand the physiological mechanisms underlying tolerance to abiotic stress [80,81] and the sensitivity of organisms to changes in the environment [80,82]. However, it also is important that we understand the indirect effects of physiological responses (in particular thermal physiology) on species interactions, such as predation, competition and disease transmission [2]. Therefore, understanding the plasticity of thermal performance curves and thermal limits (CT_{max} and CT_{min}) and how these parameters are altered by environmental stressors may be critical to understanding how physiological variation can influence a species' response to climate change [83].

Supporting Information

Methods S1 Appendices 1–4. Appendix 1, Detailed information on laboratory conditions for rearing tadpoles during acclimation for the experiments. Appendix 2, Description of method and apparatus used for measuring CT_{max}. Appendix 3, TMV method equation for calculating thermal performance

curve's parameters. Appendix 4, Description of the side-view landmarks and semi-landmarks, and linear measurements in a hypothetical tadpole.

Acknowledgments

Our thanks to Aaron Stoler, Jessica Hua, Will Brogan, R.J. Bendis, Keri Simonette, Zach Zbinden, Lindsey Freed, Patrick Noyes, Beverly French, Chris Hensley, Elena Valdés, Sara Tripodi and Casilda Maldonado for their assistance with the experiment. We also thank the anonymous reviewers for comments on previous versions of this manuscript.

Author Contributions

Conceived and designed the experiments: MK JH RAR. Performed the experiments: MK JH. Analyzed the data: MK CC. Contributed reagents/materials/analysis tools: MK MT RAR. Wrote the paper: MK JH HD MT CC RAR.

References

1. Kearney M, Porter W (2009) Mechanistic niche modelling: combining physiological and spatial data to predict species' ranges. Ecol Lett 12: 334–350.
2. Helmuth B, Broitman BR, Yamane L, Gilman SE, Mach K, et al. (2010) Organismal climatology: analyzing environmental variability at scales relevant to physiological stress. J Exp Biol 213: 995–1003.
3. Chown SL, Terblanche JS (2007) Physiological diversity in insects: ecological and evolutionary contexts. Adv In Insect Phys 33: 50–152.
4. Pörtner HO, Farrell AP (2008) Physiology and climate change. Science 322: 690–692.
5. Hoffmann AA (2010) Physiological climatic limits in Drosophila: patterns and implications. J Exp Biol 213: 870–880.
6. Chown SL, Hoffmann AA, Kristensen TN, Angilletta Jr MJ, Stenseth NC, et al. (2010) Adapting to climate change: a perspective from evolutionary physiology. Clim Res 43: 3–15.
7. Ghalambor CK, McKay JK, Carroll SP, Reznick DN (2007) Adaptive versus non-adaptive phenotypic plasticity and the potential for contemporary adaptation in new environments. Funct Ecol 21: 394–407.
8. Arnqvist G, Johansson F (1998) Ontogenic reaction norms of predator-induced defensive morphology in dragonfly larvae. Ecology 79: 1847–1858.
9. Relyea RA (2002) The many faces of predation: how induction, selection, and thinning combine to alter prey phenotypes. Ecology 83: 1953–1964.
10. Domenici P, Turesson H, Brodersen J, Brönmark C (2008) Predator-induced morphology enhances escape locomotion in crucian carp. Proc R Soc B Biol Sci 275: 195–201.
11. Brönmark C, Lakowitz T, Hollander J (2011) Predator-induced morphological plasticity across local populations of a freshwater snail. PLoS One 6: e21773.
12. Poupardin R, Reynaud S, Strode C, Ranson H, Vontas J, et al. (2008) Cross-induction of detoxification genes by environmental xenobiotics and insecticides in the mosquito Aedes aegypti: impact on larval tolerance to chemical insecticides. Insect Biochem Mol Biol 38: 540–551.
13. Hua J, Morehouse N, Relyea RA (2013) Pesticide resistance in amphibians: Induced resistance in susceptible populations, constitutive tolerance in tolerant populations. Evol Appl 6: 1028–1040.
14. Hanazato T (1991) Pesticides as chemical agents inducing helmet formation in Daphnia ambigua. Freshw Biol 26: 419–424.
15. Barry MJ (1998) Endosulfan-enhanced crest induction in Daphnia longicephala: evidence for cholinergic innervation of kairomone receptors. J Plankton Res 20: 1219–1231.
16. Oda S, Kato Y, Watanabe H, Tatarazako N, Iguchi T (2011) Morphological changes in Daphnia galeata induced by a crustacean terpenoid hormone and its analog. Environ Toxicol Chem 30: 232–238.
17. Relyea RA (2012) New effects of Roundup on amphibians: predators reduce herbicide mortality; herbicides induce antipredator morphology. Ecol Appl 22: 634–647.
18. Hanazato T (1999) Anthropogenic chemicals (insecticides) disturb natural organic chemical communication in the plankton community. Environ Pollut 105: 137–142.
19. Barry MJ (1999) The effects of a pesticide on inducible phenotypic plasticity in Daphnia. Environ Pollut 104: 217–224.
20. Barry MJ (2000) Effects of endosulfan on Chaoborus-induced life-history shifts and morphological defenses in Daphnia pulex. J Plankton Res 22: 1705–1718.
21. Sakamoto M, Chang K-H, Hanazato T (2006) Inhibition of development of anti-predator morphology in the small cladoceran Bosmina by an insecticide: impact of an anthropogenic chemical on prey-predator interactions. Freshw Biol 51: 1974–1983.
22. Hutchison VH (1961) Critical thermal maxima in salamanders. Physiol Zool 34: 92–125.
23. Lutterschmidt WI, Hutchison VH (1997) The critical thermal maximum: history and critique. Can J Zool 75: 1561–1574.
24. Heath AG, Cech JJ, Brink L, Moberg P, Zinkl JG (1997) Physiological responses of fathead minnow larvae to rice pesticides. Ecotoxicol Environ Saf 37: 280–288.
25. Angilletta Jr MJ (2009) Thermal Adaptation: A Theoretical and Empirical Synthesis. Oxford: Oxford University Press, USA. 320 p.
26. Van Buskirk J (2002) A comparative test of the adaptive plasticity hypothesis: relationships between habitat and phenotype in anuran larvae. Am Nat 160: 87–102.
27. Miner BG, Sultan SE, Morgan SG, Padilla DK, Relyea RA (2005) Ecological consequences of phenotypic plasticity. Trends Ecol Evol 20: 685–692.
28. Relyea RA (2007) Getting out alive: how predators affect the decision to metamorphose. Oecologia 152: 389–400.
29. Hayes TB, Falso P, Gallipeau S, Stice M (2010) The cause of global amphibian declines: a developmental endocrinologist's perspective. J Exp Biol 213: 921–933.
30. Wagner N, Reichenbecher W, Teichmann H, Tappeser B, Lötters S (2013) Questions concerning the potential impact of glyphosate-based herbicides on amphibians. Environ Toxicol Chem 32: 1688–1700.
31. Huey RB, Kingsolver JG (1989) Evolution of thermal sensitivity of ectotherm performance. Trends Ecol Evol 4: 131–135.
32. Van Buskirk J, McCollum SA (2000) Influence of tail shape on tadpole swimming performance. J Exp Biol 203: 2149–2158.
33. Hopkins WA, Winne CT, DuRant SE (2005) Differential swimming performance of two natricine snakes exposed to a cholinesterase-inhibiting pesticide. Environ Pollut 133: 531–540.
34. Relyea RA (2005) The impact of insecticides and herbicides on the biodiversity and productivity of aquatic communities. Ecol Appl 15: 618–627.
35. Jones DK, Hammond JI, Relyea RA (2011) Competitive stress can make the herbicide Roundup more deadly to larval amphibians. Environ Toxicol Chem 30: 446–454.
36. Petranka JW, Kats LB, Sih A (1987) Predator-prey interactions among fish and larval amphibians: use of chemical cues to detect predatory fish. Anim Behav 35: 420–425.
37. Kats LB, Petranka JW, Sih A (1988) Antipredator defenses and the persistence of amphibian larvae with fishes. Ecology 69: 1865–1870.
38. Relyea RA, Auld JR (2005) Predator- and competitor-induced plasticity: how changes in foraging morphology affect phenotypic trade-offs. Ecology 86: 1723–1729.

39. Jones DK, Hammond JI, Relyea RA (2010) Roundup and amphibians: the importance of concentration, application time, and stratification. Environ Toxicol Chem 29: 2016–2025.

40. Brattstrom BH (1968) Thermal acclimation in anuran amphibians as a function of latitude and altitude. Comp Biochem Physiol 24: 93–111.

41. Relyea RA (2003) Predators come and predators go: the reversibility of predator-induced traits. Ecology 84: 1840–1848.

42. Gosner KL (1960) A simplified table for staging anuran embryos and larvae with notes on identification. Herpetologica 16: 183–190.

43. Sherman E (1980) Ontogenetic change in thermal tolerance of the toad Bufo woodhousii fowleri. Comp Biochem Physiol Part A 65A: 227–230.

44. Terblanche JS, Hoffmann AA, Mitchell KA, Rako L, le Roux PC, et al. (2011) Ecologically relevant measures of tolerance to potentially lethal temperatures. J Exp Biol 214: 3713–3725.

45. Gvoždík L, Van Damme R (2006) Triturus newts defy the running-swimming dilemma. Evolution 60: 2110–2121.

46. Gvoždík L, Van Damme R (2008) The evolution of thermal performance curves in semi-aquatic newts: Thermal specialists on land and thermal generalists in water? J Therm Biol 33: 395–403.

47. Learning in Motion, Inc (2004) Measurement in Motion software, v3.0.

48. Arendt JD (2009) Influence of sprint speed and body size on predator avoidance in New Mexican spadefoot toads (Spea multiplicata). Oecologia 159: 455–461.

49. Arendt J (2010) Morphological correlates of sprint swimming speed in five species of spadefoot toad tadpoles: Comparison of morphometric methods. J Morphol 271: 1044–1052.

50. Van Damme R, Van Dooren TJ (1999) Absolute versus per unit body length speed of prey as an estimator of vulnerability to predation. Anim Behav 57: 347–352.

51. Izem R, Kingsolver JG (2005) Variation in continuous reaction norms: quantifying directions of biological interest. Am Nat 166: 277–289.

52. David JR, Gibert P, Gravot E, Petavy G, Morin J-P, et al. (1997) Phenotypic plasticity and developmental temperature in Drosophila: analysis and significance of reaction norms of morphometrical traits. J Therm Biol 22: 441–451.

53. Huey RB, Stevenson RD (1979) Integrating thermal physiology and ecology of ectotherms: a discussion of approaches. Am Zool 366: 357–366.

54. The MathWorks, Inc.. (2009) MATLAB v7.8 and Statistics Toolbox Release 2009a. Natick, Massachusetts, United States.

55. Dayton GH, Saenz D, Baum KA, Langerhans RB, DeWitt TJ (2005) Body shape, burst speed and escape behavior of larval anurans. Oikos 111: 582–591.

56. Rohlf FJ (2010) tpsDig2, digitize landmarks and outlines, version 2.16. Department of Ecology and Evolution, State University of New York at Stony Brook.

57. Rohlf FJ (2010) tpsRelw, relative warps analysis, version 1.49. Department of Ecology and Evolution, State University of New York at Stony Brook.

58. Bookstein FL (1991) Morphometric tools for landmark data. Cambridge/New York/Port Chester/Melbourne/Sydney: Cambridge University Press.

59. Klingenberg CP (2011) MorphoJ: an integrated software package for geometric morphometrics. Mol Ecol Resour 11: 353–357.

60. Patra RW, Chapman JC, Lim RP, Gehrke PC (2007) The effects of three organic chemicals on the upper thermal tolerances of four freshwater fishes. Environ Toxicol Chem 26: 1454–1459.

61. Carrier R, Beitinger TL (1988) Reduction in thermal tolerance of Notropis lutrensis and Pimephales promelas exposed to cadmium. Water Res 22: 511–515.

62. Lydy MJ, Wissing TE (1988) Effect of sublethal concentrations of copper on the critical thermal maxima (CTMax) of the fantail (Etheostoma flabellare) and johnny (E. nigrum) darters. Aquat Toxicol 12: 311–321.

63. Relyea RA, Jones DK (2009) The toxicity of Roundup Original Max to 13 species of larval amphibians. Environ Toxicol Chem 28: 2004–2008.

64. Navas CA, Gomes FR, Carvalho JE (2008) Thermal relationships and exercise physiology in anuran amphibians: integration and evolutionary implications. Comp Biochem Physiol Part A 151: 344–362.

65. Terblanche JS, Deere JA, Clusella-Trullas S, Janion C, Chown SL (2007) Critical thermal limits depend on methodological context. Proc R Soc London B 274: 2935–2942.

66. Chown SL, Jumbam KR, Sørensen JG, Terblanche JS (2009) Phenotypic variance, plasticity and heritability estimates of critical thermal limits depend on methodological context. Funct Ecol 23: 133–140.

67. Mitchell KA, Hoffmann AA (2010) Thermal ramping rate influences evolutionary potential and species differences for upper thermal limits in Drosophila. Funct Ecol 24: 694–700.

68. Rezende EL, Tejedo M, Santos M (2011) Estimating the adaptive potential of critical thermal limits: methodological problems and evolutionary implications. Funct Ecol 25: 111–121.

69. Mahoney JJ, Hutchison VH (1969) Photoperiod acclimation and 24-hour variations in the critical thermal maxima of a tropical and a temperate frog. Oecologia 2: 143–161.

70. Deutsch CA, Tewksbury JJ, Huey RB, Sheldon KS, Ghalambor CK, et al. (2008) Impacts of climate warming on terrestrial ectotherms across latitude. Proc Natl Acad Sci 105: 6668–6672.

71. Duarte H, Tejedo M, Katzenberger M, Marangoni F, Baldo D, et al. (2012) Can amphibians take the heat? Vulnerability to climate warming in subtropical and temperate larval amphibian communities. Glob Chang Biol 18: 412–421.

72. Smith JB, Schneider SH, Oppenheimer M, Yohe GW, Hare W, et al. (2009) Assessing dangerous climate change through an update of the Intergovernmental Panel on Climate Change (IPCC) "reasons for concern." Proc Natl Acad Sci 106: 4133.

73. Kingsolver JG, Huey RB (1998) Evolutionary analyses of morphological and physiological plasticity in thermally variable environments. Am Zool 545: 1–16.

74. Condon CH, Chenoweth SF, Wilson RS (2010) Zebrafish take their cue from temperature but not photoperiod for the seasonal plasticity of thermal performance. J Exp Biol 213: 3705–3709.

75. Lachenicht MW, Clusella-Trullas S, Boardman L, Le Roux C, Terblanche JS (2010) Effects of acclimation temperature on thermal tolerance, locomotion performance and respiratory metabolism in Acheta domesticus L. (Orthoptera: Gryllidae). J Insect Physiol 56: 822–830.

76. Gvoždík L, Puky M, Šugerková M (2007) Acclimation is beneficial at extreme test temperatures in the Danube crested newt, Triturus dobrogicus (Caudata, Salamandridae). Biol J Linn Soc 90: 627–636.

77. Relyea RA (2001) Morphological and behavioral plasticity of larval anurans in response to different predators. Ecology 82: 523–540.

78. Glennemeier KA, Denver RJ (2002) Small changes in whole-body corticosterone content affect larval Rana pipiens fitness components. Gen Comp Endocrinol 127: 16–25.

79. Relyea RA (2009) A cocktail of contaminants: how mixtures of pesticides at low concentrations affect aquatic communities. Oecologia 159: 363–376.

80. Gilman SE, Wethey DS, Helmuth B (2006) Variation in the sensitivity of organismal body temperature to climate change over local and geographic scales. Proc Natl Acad Sci 103: 9560–9565.

81. Tewksbury JJ, Huey RB, Deutsch CA (2008) Putting the Heat on Tropical Animals. Science (80-) 320: 1296–1297.

82. Porter WP, Gates DM (1969) Thermodynamic equilibria of animals with environment. Ecol Monogr 39: 227–244.

83. Buckley LB (2008) Linking traits to energetics and population dynamics to predict lizard ranges in changing environments. Am Nat 171: E1–E19.

Insecticide Resistance Status of United States Populations of *Aedes albopictus* and Mechanisms Involved

Sébastien Marcombe[1¤a], Ary Farajollahi[1,2], Sean P. Healy[3¤b], Gary G. Clark[4], Dina M. Fonseca[1]*

1 Center for Vector Biology, Rutgers University, New Brunswick, New Jersey, United States of America, **2** Mercer County Mosquito Control, West Trenton, New Jersey, United States of America, **3** Monmouth County Mosquito Extermination Commission, Eatontown, New Jersey, United States of America, **4** Mosquito and Fly Research Unit, Agriculture Research Service, United States Department of Agriculture, Gainesville, Florida, United States of America

Abstract

Aedes albopictus (Skuse) is an invasive mosquito that has become an important vector of chikungunya and dengue viruses. Immature *Ae. albopictus* thrive in backyard household containers that require treatment with larvicides and when adult populations reach pest levels or disease transmission is ongoing, adulticiding is often required. To assess the feasibility of control of USA populations, we tested the susceptibility of *Ae. albopictus* to chemicals representing the main insecticide classes with different modes of action: organochlorines, organophosphates, carbamates, pyrethroids, insect growth regulators (IGR), naturalytes, and biolarvicides. We characterized a susceptible reference strain of *Ae. albopictus*, ATM95, and tested the susceptibility of eight USA populations to five adulticides and six larvicides. We found that USA populations are broadly susceptible to currently available larvicides and adulticides. Unexpectedly, however, we found significant resistance to dichlorodiphenyltrichloroethane (DDT) in two Florida populations and in a New Jersey population. We also found resistance to malathion, an organophosphate, in Florida and New Jersey and reduced susceptibility to the IGRs pyriproxyfen and methoprene. All populations tested were fully susceptible to pyrethroids. Biochemical assays revealed a significant up-regulation of GSTs in DDT-resistant populations in both larval and adult stages. Also, β-esterases were up-regulated in the populations with suspected resistance to malathion. Of note, we identified a previously unknown amino acid polymorphism (Phe → Leu) in domain III of the VGSC, in a location known to be associated with pyrethroid resistance in another container-inhabiting mosquito, *Aedes aegypti* L. The observed DDT resistance in populations from Florida may indicate multiple introductions of this species into the USA, possibly from tropical populations. In addition, the mechanisms underlying DDT resistance often result in pyrethroid resistance, which would undermine a remaining tool for the control of *Ae. albopictus*. Continued monitoring of the insecticide resistance status of this species is imperative.

Editor: Zach N. Adelman, Virginia Tech, United States of America

Funding: This work was funded by a cooperative Agreement between the United States Department of Agriculture (USDA) and Rutgers University (USDA-ARS-58-6615-8-105) entitled "Area-wide Pest Management Program for the Asian Tiger Mosquito in New Jersey." The funders had no role in study design, data collection and analysis, decision to publish, or preparation of the manuscript.

Competing Interests: The authors have declared that no competing interests exist.

* Email: dinafons@rci.rutgers.edu

¤a Current address: Pasteur Institute, Vientiane, Laos
¤b Current address: Department of Entomology, Louisiana State University Agricultural Center, Baton Rouge, Louisiana, United States of America

Introduction

Aedes (Stegomyia) albopictus (Skuse), the Asian tiger mosquito, is an aggressive human- and day-biting species native to Asia that has recently expanded to at least 28 countries outside its native range, and now occurs in all inhabitable continents [1]. Detailed theoretical analyses indicate that the spread of *Ae. albopictus* may well continue into many more regions of the world [1–3]. Although this species is often considered mostly an urban nuisance, it was the principal dengue vector in Hawaii and other areas were *Aedes aegypti* L. populations have been controlled [4] and in the summer of 2013, an autochthonous case of dengue in Suffolk County, New York has been attributed to thriving populations of *Ae. albopictus* [5]. Furthermore, since recent mutations in the chikungunya virus (CHIKV) increased the vector competence of *Ae. albopictus* for the viral agent [6,7], chikungunya

has become epidemic in Africa and the Indian Ocean Basin [8]. Although chikungunya fever has not spread broadly in the temperate zone, an epidemic in northern Italy in 2007 sickened over 200 people [9] and small numbers of locally transmitted CHIKV cases were identified in southern France in 2010 [10], both of which were driven by local populations of *Ae. albopictus*. The European expansion of CHIKV would not have been possible without the prior invasion of that continent by *Ae. albopictus* [11].

Aedes albopictus is a container-inhabiting mosquito strongly associated with human habitats (especially outside its native range) and capable of ovipositing diapause-destined eggs that survive even in cold northern latitudes in parts of its native (*e.g.,* northern Japan, China) and introduced (*e.g.,* Europe and northeastern USA) ranges [12]. The first line of control against *Ae. albopictus* is often source reduction [13], but when containers cannot be removed or emptied, larvicides are used [13]. If adults become a serious nuisance, or disease outbreaks are ongoing or imminent, insecticides targeting the adults are applied [14].

Unfortunately, the development and spread of insecticide resistance represents a serious threat as it can lead to a reduction of the efficacy of larvicide or adulticide-based control programs, as demonstrated in the control of the main dengue vector *Ae. aegypti* [15,16]. In contrast to *Ae. aegypti*, there have been only a few reports of insecticide resistance in *Ae. albopictus* worldwide [16,17]. Several studies implemented in the 1960s and summarized by Mouchet et al. [18] showed that several populations of *Ae. albopictus* from Southeast Asia and India were resistant to some of the insecticides used at the time for vector control (*i.e.*, DDT, dieldrin and fenthion). A recent review by Ranson et al. [16] updated by Vontas et al. [19] summarized the levels of insecticide resistance in *Ae. albopictus* worldwide. It is apparent that resistance to the main families of insecticides currently or historically used for vector control across the world (*i.e.*, DDT, organophosphates and pyrethroids) has been found in *Ae. albopictus* [20–25]. In the USA, to our knowledge, only four studies have reported on insecticide resistance in *Ae. albopictus*: one population in Florida was resistant to the organophosphate malathion [26], populations in Texas and Illinois were also resistant to malathion [25,27], and resistance to a pyrethroid (deltamethrin) was found in a population from Alabama [28].

Insecticide resistance can be associated with mutations in the sequence of the target protein that induce insensitivity to the insecticide (target-site resistance), and/or to the up-regulation of detoxification enzymes (metabolic-based resistance). The main target site resistance mechanisms known in mosquitoes involve 1) amino acid substitutions in the voltage gated sodium channel that cause a resistance phenotype to pyrethroid (DDT) insecticides known as knockdown resistance (Kdr, [29] and 2) mutations in the acetylcholine esterase sequence that lead to insensitivity of this enzyme to organophosphates [30]. Metabolic-based resistance involves the bio-transformation of the insecticide molecule by enzymes and is now considered a key resistance mechanism in insects [31,32]. Three large enzyme families, the cytochrome P450 monooxygenases (P450s), glutathione S-transferases (GSTs), and carboxy/cholinesterases (CCEs) have been implicated in the metabolism of insecticides [32–34]. So far, compared to other mosquito species of importance such as *Anopheles* spp., *Culex* spp., and *Ae. aegypti*, very little is known about the molecular or biochemical basis of resistance in *Ae. albopictus* and, in particular, to our knowledge, no studies have specifically examined the underlying mechanisms of resistance in USA *Ae. albopictus*.

The objective of the present study was to determine the insecticide resistance status of *Ae. albopictus* across the full latitudinal range of the species in the USA. Specifically, we examined populations from New Jersey, Pennsylvania, and Florida (Table 1). We chose eleven chemicals that represent the main classes of insecticides historically or currently used for mosquito control (Table 2), including some that have only recently been adopted. We compared the levels of resistance of field-collected specimens to a susceptible strain of *Ae. albopictus* that we characterized for this purpose (reference strain ATM95). In addition, we used biochemical and molecular assays to identify putative resistance mechanisms in *Ae. albopictus* such as target-site mutations and up-regulation of detoxifying enzymes.

Materials and Methods

Ethics statement

No specific permits were required for collection of field specimens, which were performed in urban and suburban backyards in the US states of New Jersey, Pennsylvania, and Florida with homeowners assent by professional county mosquito control personnel. These studies did not involve endangered or protected species. In the laboratory, mosquito colonies were blood fed on quail, *Colinus virginianus*, under the guidelines of the Rutgers University Animal Use Protocol# 86–129 that was approved by the Rutgers IACUC.

Mosquito strains and collection

We characterized a reference laboratory strain (ATM95) and tested eight field populations of *Ae. albopictus* (Table 1). *Aedes albopictus* was first detected in New Jersey (NJ) on August 1, 1995 in a standard NJ light trap collection in Keyport [35]. Surveillance at a marina 300 m from the trap site yielded *Ae. albopictus* larvae from one discarded bucket and 2 tires and a colony started from this population, now named ATM95, has been continuously reared in the laboratory at the Center for Vector Biology at Rutgers University in New Brunswick, NJ without exposure to insecticides. Preliminary bioassays on the ATM95 strain showed that this strain could be considered susceptible in comparison to previous results from the literature. The field caught *Ae. albopictus* samples were collected as larvae, pupae, or eggs (ovitraps) in one site in Bergen county, NJ (NJBer, N 40°47′33″, W 74°1′32″), two replicate sites (less than 5 km apart) in Mercer county, NJ (NJMer1, NJMer2, N 40°13′1″ W 74°44′35″), two sites in Monmouth county, NJ (NJMon1 and NJMon2, N 40°26′36″ W 74°13′5″), one site in York county, Pennsylvania (PA, N 39°57′46″ W 76°43′41″) and two sites in St. Johns county, Florida (FL1 and FL2, N 29°53′39″ W 81°18′48″) during the 2011 active mosquito season (Figure 1). All stages were reared to adults in the laboratory on a diet of powdered cat food. After emergence of female *Ae. albopictus* they were provided restrained quails (*Colinus virginianus*) as sources of blood for egg development following the Rutgers University Animal Use protocol# 86–129. Larvae and adults obtained from the F_1 progeny were used for bioassays and biochemical and molecular studies.

Bioassays

We chose to test the susceptibility of *Ae. albopictus* to a range of insecticides representative of those historically and currently used for mosquito control in the USA from all main families of insecticides with different modes of action (Table 2).

Larval bioassays. Larval bioassays were carried out using the water-dispersible granule formulation (VectoBac WDG, Valent BioSciences, Libertyville, IL, USA) of *Bacillus thuringiensis* var. *israelensis* (*Bti*) (37.4% ai, 3000 ITU/mg). The remaining insecticides were tested by diluting the active ingredients (ai) purchased from Sigma-Aldrich (Seelze, Germany) in ethanol to required levels according to WHO guidelines [36]. We tested temephos (97.3% active ingredient [ai]), propoxur (99.8%), spinosad (97.6%), methoprene (95.6%), and pyriproxyfen (99.1%). All bioassays were performed using late third and early fourth-instars of *Ae. albopictus*.

To determine the activity range of the larvicides in *Ae. albopictus*, larvae of the susceptible laboratory strain, ATM95, were exposed to 3 replicates of a wide range of test concentrations. For each bioassay, 25 larvae of each population were transferred to plastic cups containing 99 mL of distilled water with 1 mL of the insecticide at the desired concentration. The appropriate volume

Table 1. Detailed description with geographic and socio-economic information of the sources of mosquito populations.

State	County	Municipality	Mosquito population name abbreviations	Coordinates	Altitude	Human density inhabitants/ Km^2
New Jersey	Bergen	Elmwood Park	NJBer	40°54'N74°70'W	14 m	2,829
	Mercer	Trenton	NJMer1	40°13'N74°45'W	15 m	4,286
		Ewing	NJMer2	40°15'N74°47'W	38 m	906
	Monmouth	Middletown	NJMon1	40°24'N74°04'W	30 m	626
		Belmar	NJMon2	40°10'N74°01'W	4 m	2,140
Pennsylvania	York	York	PA	39°57'N76°43'W	121 m	3,061
Florida	St John's	St Augustine south	FL1	29°50'N81°18'W	7 m	1,118
		St Augustine Beach	FL2	29°53'N81°18'W	0 m	936

Population name abbreviations are used throughout the text.

of dilution from the stock solution was added to the water in the cups to obtain the desired target dosage, starting with the lowest concentration. Four cups per concentration (100 larvae) and 4 to 8 concentrations in the activity range of the insecticide (between 10% and 95% mortality) were used to determine LC_{50} and LC_{90} values (LC: lethal concentration). Control treatments were made with 99 mL of distilled water and 1 mL of ethanol. Larval mortality was recorded after 24 h exposure except for pyriproxyfen and methoprene for which mortality was recorded every 24 h until emergence due to the delayed action of these insect growth regulators. In this case, larvae were provided with food at a concentration of 100 mg/L every day. For each bioassay, temperature was maintained at 27°C in an incubator with a 16L:8D photoperiod.

Adult bioassays. Adult bioassays also followed WHO protocols [37], with 3 to 5 day old females of each F_1 progeny used for tarsal contact tests with insecticide-treated filter paper and compared with the susceptible ATM95 strain. We started with technical grade (Pestanal Sigma-Aldrich, Seelze, Germany) deltamethrin (99.7% ai, type II pyrethroid), prallethrin (96.2%, type I), phenothrin (94.4%, type I), malathion (97.2%), and DDT (99.7%). Insecticide was applied to filter paper by dripping evenly onto the paper 2 mL of technical grade chemical dissolved in acetone and silicone oil to the appropriate concentration [37]. Concentrations were expressed in w/w percentage of the active

ingredient in silicone oil. Filter papers were dried for 24 h before the test. The resistance status of *Ae. albopictus* populations from each locality was determined by using WHO discriminating dosages (DD; double concentration of LC_{99}) of deltamethrin (0.05%), malathion (0.8%), and DDT (4%). Preliminary bioassays conducted on the ATM95 strain displayed that the discriminating dosages for prallethrin and phenothrin were 1% and 1.5%, respectively. Those two pyrethroids are used in combination in the newly available Duet dual-action adulticide formulation (Clarke Mosquito Control, Roselle, Illinois, USA) for adult mosquito control. For each strain, five batches of 20 non-blood fed females (2–5 days old; n = 100) were exposed to the insecticides in WHO test kits for 60 min to estimate the knock down effect (KDT_{50} and KDT_{90}) of the insecticides. The number of knocked down mosquitoes in the tubes was counted every 2 minutes. The adults were then transferred into holding tubes, were provided with sugar solution (10%), and kept at 27°C with a relative humidity of 80%. Mortality was recorded 24 h later. Mosquitoes exposed for 1 h to paper impregnated with the carrier (silicone oil) mixed with acetone were used as controls. Tests were replicated twice when the number of available mosquitoes was suitable. Following WHO criteria a population is considered resistant if the mortality after 24 h is under 90%, resistance is suspected with mortality between 90 and 98% and a population is susceptible with mortality over 98%.

Table 2. Name, class, and mode of action of all insecticides tested in this study.

Status	Insecticide	Family	Mode of action
Larvicide	*Bti*	Biolarvicide	Cell membrane destruction
	Spinosad	Naturalyte	Nicotinic acetylcholine receptor
	Temephos	Organophosphate	Acetylcholinesterase inhibitor
	Propoxur	Carbamate	
	Methoprene	Insect Growth Regulator	Juvenile hormone mimics
	Pyriproxyfen		
Adulticide	Malathion	Organophosphate	Acetylcholinesterase inhibitor
	DDT	Organochlorine	Sodium channel modulator
	Deltamethrin	Pyrethroid	
	Prallethrin		
	Phenothrin		

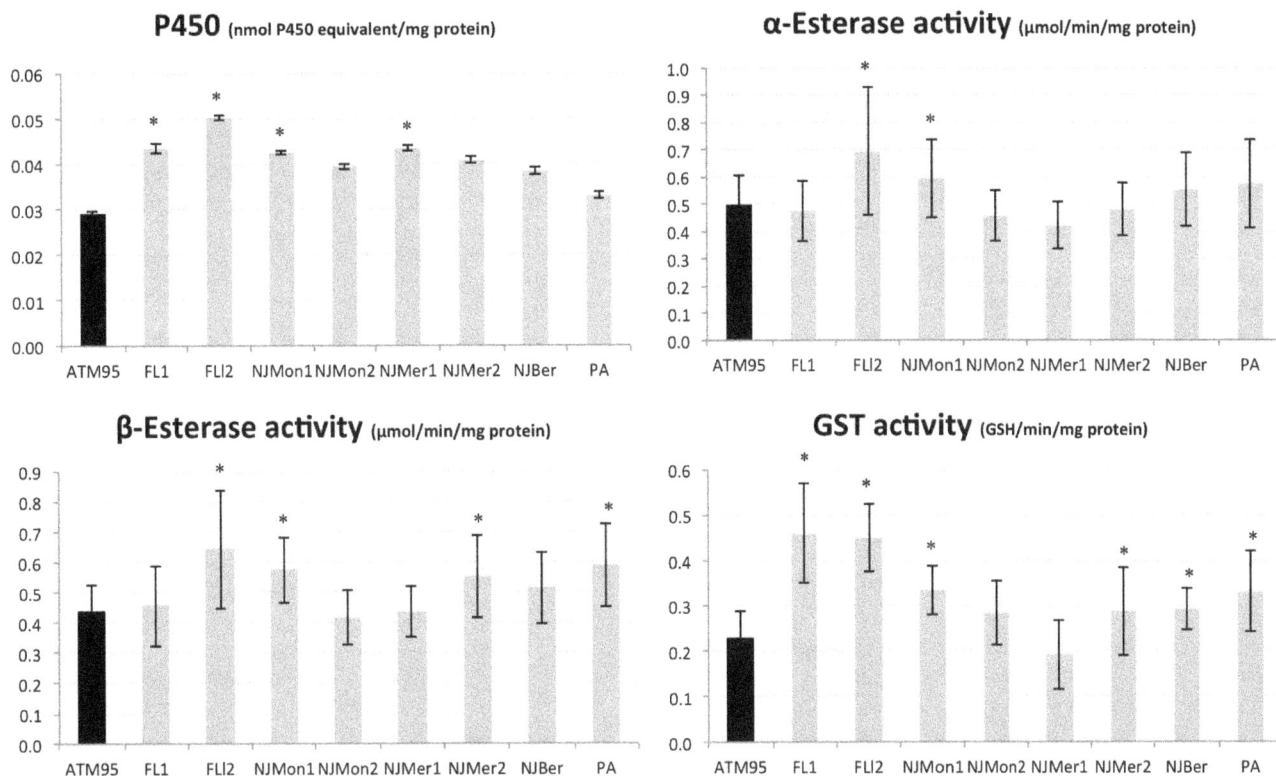

Figure 1. Global amount or activity of detoxification enzymes in *Aedes albopictus* larvae from field populations and the laboratory strain (ATM95): cytochrome P450 monooxygenases (P450s), Esterase (α and β-CCEs), and Glutathione-S transferases (GSTs). Sample sizes are 47 specimens/population (15 for P450, n = 3). Confidence intervals are one standard deviation of the mean. An asterisk (*) denotes significantly up-regulated values compared to the susceptible reference strain ATM95, Tukey-Kramer test.

Larval and adult knock down times (KDT) were analyzed with the log-probit method of Finney [38] using the Sakuma Probit software [39]. Data from all replicates were pooled for analysis. Lethal concentrations (LC_{50} and LC_{95} for larvae) and knock-down time (KDT_{50} and KDT_{95} for adults) were calculated together with their 95% confidence intervals. Adult mortality after 24 h exposure was also recorded for each population. Compared to the susceptible ATM95 strain field populations were considered as having some resistance to a given insecticide when their $LC_{50/95}$ or $KDT_{50/95}$ ratios (resistance ratio: $RR_{50/95}$) had confidence limits that excluded the value 1. We considered resistance to be moderate to strong when $RR_{50/95}$ values rose above 2.

Biochemical assays

The levels of P450 monooxygenases (P450s), and the activities of carboxy/cholinesterases (CCEs) and glutathione S-transferases (GSTs) were assayed from single 3 days-old F_1 females (n = 47) following microplate methods described by Hemingway [32] and Brogdon [40] on an Epoch spectrophotometer (BioTek, Vermont, USA). Total protein quantification of mosquito homogenates was performed using Bradford reagent with bovine serum albumin as the standard protein [41] to normalize enzyme activity levels by protein content. For P450 assays, the OD values were measured at 620 nm after 30 min incubation of individual mosquito homogenate with 200 µL of 2 mM 3, 3', 5, 5'-tetramethylbenzidine dihydrochloride (TMBZ) and 25 µL of 3% hydrogen peroxide and the quantity was determined from cytochrome-c standard curve. Nonspecific α- and β-CCEs activities were assayed by 10 min incubation of mosquito homogenate in each well with 100 µL of

3 mM napthyl acetate (either α- or β-) at room temperature and the OD values were measured at 540 nm. The activity was determined from α- or β-naphtol standard curves. Glutathione-S-transferases activity was measured in the reaction containing 2 mM reduced glutathione and 1 mM 1-chloro-2,4-dinitrobenzene (CDNB). The reaction rates were measured at 340 nm after 20 min, and the activity was expressed in nmoles GSH conjugated/min/mg protein.

Statistical comparisons of detoxification enzyme levels between ATM95 and the field populations were assessed with Tukey-Kramer tests in JMP8.0.1 (SAS Institute, Cary, North Carolina, USA) using a P value threshold of 0.05. Tukey-Kramer HSD (honestly significant difference) test is a highly conservative test that accounts for multiple comparisons [42].

Kdr genotyping

We extracted DNA from 14 adult *Ae. albopictus* collected in Florida (FL1 and FL2) using DNAeasy tissue kits (Qiagen, Valencia, California, USA). We chose 6 survivors and 6 dead specimens following DDT exposure and amplified portions of domains II, III, and IV of the voltage-gated sodium channel (VGSC), a known target of DDT and pyrethroid insecticides, using primers from Kasai et al. [43]. Specifically we amplified and sequenced domain II with aegSCF20 and aegSCR21, domain II with aegSCF7 and aegSCR8, and domain IV with albSCF6 and albSCR8. Our PCR was composed of 1× PCR buffer, 2.5 mM of $MgCl_2$ (2.0 mM for Domain III), 200 µM of each dNTP, 0.2 mg/ml of BSA, 0.2 µM of each primer, and 1 unit of *TaqGold* (Applied Biosystems, Foster City, California, USA). The PCR cycle started with a 10 min denaturation (and *TaqGold* activation) at 96°C followed by 40 cycles of 30 s at 96°C, 30 s at 55°C (Domain II and IV) or 53°C (Domain III) and 45 s at 72°C, and a final extension of 10 min at 72°C. The PCR products were cleaned with ExoSAP-IT (USB, Cleveland, Ohio, USA) and cycle sequenced for analyses on an ABI 3100 automated sequencer (Applied Biosystems). Sequences were cleaned and checked with Sequencher 5.0 (Gene Codes, Ann Harbor, Michigan, USA).

Enzymatic phenotyping of Ache1

The phenotypes of the acetylcholine esterase AChE1, encoded by the ace-1 gene, were examined in each population (n = 24) using the previously described TDP test [44] adapted for *Ae. albopictus* with both dichlorvos and propoxur concentrations of 1.10^{-2} M. The TDP test identifies all possible phenotypes containing the G119S, F290V and wild-type (susceptible) alleles.

Results

Larval and adult bioassays

Larval bioassays resulted in low resistant ratios (RRs) indicating that none of the eight USA populations of *Ae. albopictus* were resistant to the larvicides tested (Table 3). However, one of the populations from Florida, FL2, showed significant resistance to both methoprene and pyriproxyfen (IGRs) with RRs of 3.72 and 2.36 fold, respectively. Further, all the populations had values of RRs for propoxur that excluded 1, ranging from 1.47 (NJMon1) to 2.8 fold (FL1 and FL2); the latter indicating significant resistance to this carbamate in Florida populations. The insecticidal activities of the larvicides used against the ATM95 strain (Table 3) can be ranked as follows: pyriproxyfen > methoprene > temephos > *Bti* > spinosad > propoxur with LC$_{50}$ of 9.4E-6, 1.4E-4, 5.4E-3, 0.07, 0.1 and 1.02 mg/L, respectively.

The knockdown times (KDT) for *Ae. albopictus* exposed to DDT indicated that most KDT$_{50}$ values from field populations were higher (non overlapping 95% CIs) than those of the reference strain, ATM95, except for NJMer1 and NJBer that showed lower KDT$_{50}$ (Table 4). The two populations from Florida, FL1 and FL2, showed the highest RRs (1.61 and 1.88 respectively) for DDT. For deltamethrin the RRs ranged from 1.13 (NJMer2) to 1.74 (NJMon2) indicating that all the populations were susceptible. Likewise, for phenothrin the KDTs were lower than those of the susceptible strain and for prallethrin the RR$_{50}$ did not exceed 1.18 (FL1). Of note, the two populations from Florida (FL1 and FL2) showed RRs with values of 2.16 and 2.34, respectively, for malathion. The RR$_{50}$ for malathion for the remaining populations were low but significantly higher than 1 and ranged from 1.15 to 1.67.

Adult mortality after a 24 h exposure to the pyrethroid insecticides (deltamethrin, prallethrin, and phenothrin) at discriminating doses indicated that, like the ATM95 strain, all the field populations tested can be considered susceptible (99–100% mortality; Table 4). However, the two populations from Florida (FL1 and FL2) showed resistance to DDT (75 and 54% mortality, respectively) and a population from New Jersey (NJMon2) also showed resistance to this organochlorine (87% mortality). In addition, resistance to malathion was found in the two populations from Florida (FL1 and FL2) with 86 and 80% mortality, respectively. Finally, the populations from New Jersey (NJMon2, NJMer1, and NJBer) showed suspected resistance to malathion with 95, 96 and 93% mortality, respectively (Table 4).

Table 3. Resistance status of larvae Aedes albopictus.

Population	*Bti* LC$_{50}$ (95% CI)	RR$_{50}$	Temephos LC$_{50}$ (95% CI)	RR$_{50}$	Propoxur LC$_{50}$ (95% CI)	RR$_{50}$	Spinosad LC$_{50}$ (95% CI)	RR$_{50}$	Methoprene LC$_{50}$ (95% CI)	RR$_{50}$	Pyriproxyfen LC$_{50}$ (95% CI)	RR$_{50}$
ATM95	0.07 (0.066–0.071)	1	5.4E-03 (5.1E-03–5.7E-03)	1	1.02 (0.93–1.09)	1	0.10 (0.036–0.106)	1	1.4E-04 (9.9E-05–1.7E-04)	1	9.4E-06 (3.6E-06–2.5E-05)	1
FL1	0.07 (0.01–0.07)	0.99	5.3E-03 (4.9E-03–5.3E-03)	0.93	2.87 (2.67–3.15)	2.82	0.15 (0.145–0.162)	1.51	-	-	1.5E-05 (1.2E-05–1.9E-05)	1.57
FL2	0.06 (0.043–0.085)	0.84	5E-03 (5E-03–5.6E-03)	0.99	2.83 (2.59–3.29)	2.77	0.16 (0.151–0.165)	1.56	5.1E-04 (3.1E-04–1E-03)	3.72	2.2E-05 (1.7E-05–2.9E-05	2.36
NJMon1	0.12 (0.108–0.145)	1.78	4.7E-03 (4.5E-03–4.9E-03)	0.87	1.50 (1.32–1.79)	1.47	0.14 (0.138–0.147)	1.42	1.6E-04 (1.2E-04–2.1E-04)	1.15	4.7E-06 (2.6E-06–6.5E-06)	0.50
NJMon2	0.11 (0.10–0.13)	1.68	6.1E-03 (5.7E-03–6.6E-03)	1.14	1.62 (1.44–1.9)	1.59	0.10 (0.091–.112)	1.01	7.4E-05 (3.1E-05–1.2E-4)	0.54	3.6E-06 (2.2E-06–8.3E-06)	0.38
NJMer1	0.08 (0.06–0.1)	1.16	6.1E-03 (5.6E-03–7E-03)	1.13	1.73 (1.62–1.89)	1.69	0.14 (0.109–0.29)	1.38	1.7E-04 (1E-4–2.6E-04)	1.22	5.7E-06 (3.3E-06–8.2E-06)	0.60
NJMer2	0.08 (0.073–0.089)	1.19	6.3E-03 (5.8E-03–6.8E-03)	1.17	2.09 (1.68–3.26)	2.05	0.08 (0.068–0.089)	0.79	9.9E-05 (8.8E-6–3.E-04)	0.71	1.3E-05 (9.5E-06–1.7E-05)	1.37
NJBer	0.05 (0.047–0.057)	0.76	6.9E-03 (6.4E-03–7.4E-03)	1.27	2.13 (1.83–2.82)	2.09	0.16 (0.142–0.189)	1.56	4.5E-05 (1.4E-05–8.4E-05)	0.33	1.7E-05 (1.2E-05–2.4E-05)	1.81
PA	0.08 (0.069–0.085)	1.13	7.6E-03 (6.8E-03–8.8E-03)	1.41	1.94 (1.66–2.41)	1.90	0.18 (0.144–0.309)	1.73	1.1E-03 (5.7E-05–2.3E-04)	0.78	1.0E-05 (7.8E-06–1.4E-05)	1.11

ATM95: susceptible reference strain; LC$_{50}$: Lethal Concentration that kills 50% of the population (mg/L); RR$_{50}$: Resistant Ratio = LC$_{50}$ susceptible strain (ATM95)/LC$_{50}$ field population; CI: Confidence Interval. Significant RRs are shown in bold (P<0.05).

Table 4. Knock down times (min), Resistant ratio, and mortality rates (after 24 h) of *Aedes albopictus* after exposure to insecticides at the diagnostic doses (WHO tube test).

Population	DDT (4%)			Deltamethrin (0.05%)			Phenothrin (1.5)			Prallethrin (1%)			Malathion (0.8%)		
	KDT$_{50}$ (95% CI)	RR$_{50}$	Mortality (%)	KDT$_{50}$ (95% CI)	RR$_{50}$	Mortality (%)	KDT$_{50}$ (95% CI)	RR$_{50}$	Mortality (%)	KDT$_{50}$ (95% CI)	RR$_{50}$	Mortality (%)	KDT$_{50}$ (95% CI)	RR$_{50}$	Mortality (%)
ATM95	33 (30–35)	1	100	6 (6.3–6.6)	1	100	8.4 (8.2–8.7)	1	100	1.37 (1.29–1.44)	1	100	23 (22.2–23.8)	1	100
FL1	53 (51–56)	**1.61**	72	9.8 (9.6–10.1)	**1.57**	100	7.9 (7.6–8.3)	0.94	100	1.61 (1.46–1.82)	**1.18**	100	49.6 (47.4–52.3)	**2.16**	86
FL2	62 (59–68)	**1.88**	54	10.7 (10.4–10.9)	**1.70**	99	7.7 (7.4–8)	**0.91**	100	1.35 (1.28–1.43)	0.99	99	53.5 (50.5–57.6)	**2.34**	80
NJMon1	42 (41–44)	**1.28**	100	8.4 (7.9–8.9)	**1.34**	100	6.6 (6.4–6.8)	**0.79**	100	0.81 (0.73–0.88)	**0.59**	99	27.9 (26.9–28.8)	**1.22**	100
NJMon2	42 (40–45)	**1.28**	87	11.2 (10.9–11.5)	**1.78**	100	7.6 (7.3–7.9)	**0.9**	100	1.37 (1.25–1.5)	1.04	100	36.3 (35.1–37.5)	**1.59**	95
NJMer1	28 (24–30)	0.83	95	10.2 (9.4–11)	**1.62**	100	7.1 (6.9–7.4)	**0.85**	100	1.48 (1.29–1.78)	1.08	100	34.3 (33.5–35.2)	**1.5**	96
NJMer2	47 (45–49)	**1.4**	100	7.1 (6.8–7.4)	1.13	100	6.6 (5.9–7.2)	**0.78**	100	1.12 (1–1.21)	**0.83**	100	26.4 (25.5–27.3)	1.15	99
NJBer	27 (26–27.3)	0.8	99	10.3 (9.9–10.8)	**1.65**	99	8.3 (7.9–8.8)	0.98	100	1.13 (0.99–1.25)	**0.83**	100	38.3 (36–40.4)	**1.67**	93

ATM95: susceptible reference strain. KDT$_{50}$: Knock down time where 50% of the mosquitoes are knocked down (min); RR$_{50}$: Resistant Ratio = KDT$_{50}$ ATM95/KDT$_{50}$ field population. Significant RRs are shown in bold.

Detoxification enzyme levels

Comparison of constitutive detoxification enzyme activities between ATM95 and the field strains revealed significant differences in both larval and adult stages (Figure 1 and 2). The P450s levels were significantly higher in larvae from Florida (both FL1 and FL2), NJMon1, and NJMer1 populations. The FL2 and NJMon1 had significantly higher α- and β-ESTs activities and GSTs activities were significantly higher in most populations, particularly in FL1 and FL2, but not in NJMon2 and NJMer1 (Figure 1). In adults, only NJMer2 showed significantly up-regulated P450s, and only NJMer2 had significantly higher α-ESTs activities. The two populations from Florida and NJMer2 had significantly higher β-ESTs activities. Finally, except for NJMer1 and NJBer, all populations had significantly higher GSTs activities (Figure 2) than the susceptible strain.

Kdr genotyping

We obtained clean sequences of exonic regions in domains II (480 bp), III (exon 1 and 2, 347 bp), and IV (280 bp) of the voltage-gated sodium channel. Of note, in approximately half of the specimens in domains II and III we were not able to span the introns due to the presence of insertions or deletions and therefore we could not obtain both forward and reverse exonic sequences. We compensated by sequencing twice in each direction. Although a few silent mutations at codon positions 2 and 3 were seen, no amino-acid changing mutations were detected in the exons of domains II and IV of the mosquitoes tested. However, in one individual, a mutation was found in domain III at position 1534 (base pair positions are numbered according to the amino acid sequence of the most abundant splice variant of the house fly sodium channel, GenBank accession nos. AAB47605 and AAB47604) where a substitution occurred (TTC to CTC), changing the wild type Phenylalanine into a Leucine. The mutation in residue 1,534 that has been associated with pyrethroid resistance in *Ae. aegypti* is F1,534C, resulting in a Cysteine [62].

Enzymatic phenotyping of Ache1

All mosquito test populations from New Jersey, Pennsylvania, and Florida showed similar percentages of AChE inhibition with dichlorvos and propoxur compared to the susceptible ATM95 strain (data not shown), indicating they are all of the susceptible type.

Discussion

The purpose of this study was to evaluate the insecticide resistance status of *Ae. albopictus* populations in several states along the eastern coast of the USA. Insecticides representing the major classes of insecticide (OC, OP, CA, PYR), bio-insecticides (*Bti* and spinosad), and IGRs were used in this study against larvae and adult mosquitoes following WHO protocols. We investigated the possible insecticide resistance mechanisms involved (detoxification enzyme and target site mutations) with biochemical and molecular assays.

For both bioassays and biochemical assays, the eight populations tested were compared to the ATM95 strain, which we first characterized for insecticide susceptibility. The ATM95 strain had similar or higher susceptibilities to the insecticides tested than other *Ae. albopictus* populations used as a reference in previous studies. For example, Ali et al. [26] showed higher LC$_{50}$ for an *Ae. albopictus* strain from Florida maintained for 2 yrs in colony for temephos, *Bti*, methoprene, and pyriproxyfen of 0.01, 0.181, 0.0022, and 0.00011 mg/L respectively, than the ATM95 strain with LC$_{50}$ for the same insecticides of 0.00054, 0.07, 0.00014, and

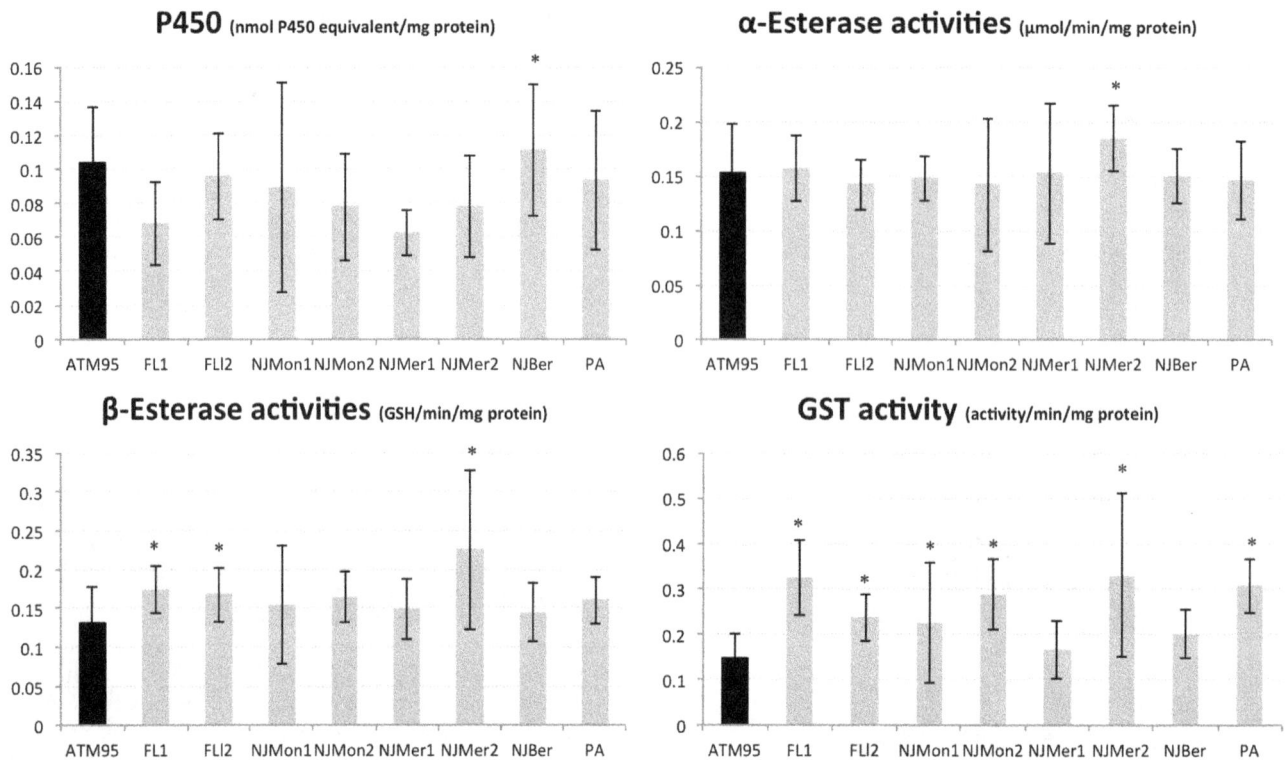

Figure 2. Global amount or activity of detoxification enzymes in adult *Aedes albopictus* **from field populations and the laboratory strain (ATM95): cytochrome P450 monooxygenases (P450s), Esterase (α and β-CCEs) and Glutathione-S transferases (GSTs).** Sample sizes are 47 specimens/population. Confidence intervals are one standard deviation of the mean. An asterisk (*) denotes significantly up-regulated values compared to the susceptible reference strain ATM95, Tukey-Kramer test.

$9.4 \ 10^{-6}$ mg/L. The susceptible reference strain Ikaken used for the study by Liu et al. [28] presented higher LC_{50} for *Bti*, propoxur, and spinosad (0.1, 3.3, and 0.3 mg/L, respectively) than the LC_{50} of ATM95 (0.07, 1.2, and 0.1 mg/L respectively). Furthermore, the larvae of the ATM95 strain showed higher susceptibility to deltamethrin, permethrin, and malathion than the Ikaken strain or the susceptible strain used by Selvi et al. [45]. In light of these results, we consider the ATM95 as a valid susceptible reference strain for the present study and propose it should be adopted as a reference in future studies of insecticide resistance in temperate *Ae. albopictus*. Reference strains such as the Rockefeller or Bora-Bora used for *Ae. aegypti* studies are essential for the quantification of insecticide resistance across studies [46].

The larval bioassays showed that none of the eight populations examined were strongly resistant to the larvicides tested. Likely because of their specific modes of action, resistance to *Bti*, spinosad, or pyriproxyfen has not been described in mosquitoes, except for a single case of putative resistance to *Bti* in a *Culex pipiens* L. population from New York [47], making these insecticides promising tools for the control of *Ae. albopictus* in the USA. However, we note that spinosad resistance has been reported in several insect pests previously, indicating that it is possible that resistance may occur over time in *Ae. albopictus* if intensive use occurs [48]. Our results showed that temephos was still effective against all the populations tested, although several studies have suggested that temephos resistance selection can develop in *Ae. albopictus* after laboratory selection or prolonged field exposure [49,50]. Indeed, resistant populations have been detected in South-East Asia, South America, and in Europe, where this larvicide is used against *Aedes* species [16,19]. The use of temephos

for control of *Ae. albopictus* larvae in the USA should therefore be carefully evaluated since adult populations from Florida and New Jersey showed resistance or suspected resistance to malathion (OP). Also, the low but significant resistance to propoxur (CA) exhibited by the Florida and New Jersey populations ($RR_{50} > 2$) should be taken into consideration since cross-resistance is known to occur between OPs and CAs.

Methoprene has been used for vector control in Florida for more than 3 decades [51] and even when *Ae. albopictus* is not been the primary control target in this area, populations may have been exposed to this insecticide and developed tolerance over time. One Florida population showed suspected resistance to both methoprene and pyriproxyfen and the adults showed resistance to the adulticide malathion. Previous authors have reported similar findings in mosquitoes exhibiting high resistance to OPs. Specifically, Marcombe et al. [52] and Andrighetti et al. [53] showed that *Ae. aegypti* populations with high resistance to the organophosphate temephos were less susceptible to pyriproxyfen, indicating a possible cross resistance in mosquitoes between these two insecticides families.

The adult bioassays revealed resistance to malathion in Florida and suspected resistance in New Jersey. Resistance to this insecticide, which is used in space spraying treatments was already a concern for the public health authorities in the 1980's [54] when malathion resistance in *Ae. albopictus* was described in Texas only a few years after *Ae. albopictus* became established. Furthermore, other studies report resistance to malathion in populations from Louisiana, Illinois, Alabama, and additional locations in Texas [25,27,28]. Worldwide *Ae. albopictus* resistance to malathion has been extensively reported in Asia, the presumed origin of the USA populations of this species, since the 1960's [55], and it is possible that the introduced populations were already resistant. However, since malathion and other OPs are still being used for mosquito control in the USA, it is also possible that resistance developed locally and is being maintained in this region.

All the populations were susceptible to the three pyrethroids tested at the diagnostic doses. Prallethrin and phenothrin are the components of the Duet formulation that showed promising efficacy in ultra-low volume adulticide applications against *Ae. albopictus* [14]. All the populations were also susceptible to deltamethrin, showing that this insecticide can still be an effective tool for *Ae. albopictus* control. However deltamethrin or pyrethroid resistance has already been detected in China, Japan, and South-East Asia [16,19,22,56] and also more recently in Florida and Alabama, USA [28].

Although we were initially surprised to detect DDT resistance in Florida populations of *Ae. albopictus*, DDT resistance is widespread in *Ae. albopictus* populations worldwide especially in Asia. Since the 1960's very high levels of resistance have been reported from India to the Philippines and from China to Malaysia [18,22]. So as for malathion resistance, it is also likely that the selection for resistance may have occurred in Asia, prior to USA introductions. However, since the use of DDT was terminated in the USA in 1972, before the introduction and establishment of *Ae. albopictus*, the observed levels of resistance in Florida may be explained by a regular exposure of the populations to pyrethroids or other xenobiotics that have the same mode of action as DDT. Alternatively, it is possible that DDT resistance in these populations does not impact fitness and therefore is simply being maintained neutrally or finally, that there have been more recent introductions of DDT resistant *Ae. albopictus* from Asia (Fonseca et al. unpublished data). This last scenario is supported by the study of Kamgang and colleagues [23] that reported DDT resistance in recently introduced populations in Cameroon. The high levels of resistance against DDT found in Florida and the suspected resistance in the populations from New Jersey also underscore the threat of pyrethroid resistance in USA *Ae. albopictus*. Cross resistance mechanisms between DDT and pyrethroids can negatively impact control strategies.

Regarding the various mechanisms of insecticide resistance, we found significant differences in detoxification enzyme activities in several USA resistant *Ae. albopictus* populations suggesting the involvement of metabolic based resistance mechanism. The malathion resistant populations from Florida and New Jersey showed significantly over-expressed β-ESTs and GSTs, which include two detoxification enzyme families known to play a role in organophosphate resistance in mosquitoes [32]. However, because several studies have showed that carboxylesterases do not play a role in resistance to organophosphate in *Ae. albopictus* [45,57], it remains unclear whether one or both of the enzyme families are involved in the resistance at the adult stage. Complementary studies with the use of specific enzyme inhibitors should be implemented to discriminate their roles in malathion resistance in the USA *Ae. albopictus*.

Larvae from Florida populations showed the highest RR_{50} against propoxur but were not resistant to temephos, confirming the absence of insensitive AChE responsible for the cross-resistance between OP and carbamates in mosquitoes. Of note, insensitive AChE was recently detected in *Ae. albopictus* populations in Malaysia [20], underscoring the importance of regular monitoring of this mechanism in the USA. All the populations tested showed a reduced susceptibility against propoxur and all had a significantly increased amount of P450s. It is therefore

possible that P450s may be involved in carbamate resistance in *Ae. albopictus* as in other mosquito species [58].

One population from Florida showed significant resistance against the two IGRs, methoprene and pyriproxyfen. The same population also presented over-expressed P450s, ESTs, and GSTs. The P450s are primarily involved in pyrethroid (DDT) resistance and may also be involved in IGR resistance in insects [59]. Indeed, recently the product of the *Ae. aegypti* CYP6Z8 detoxification gene, belonging to the P450s family, was shown to metabolize pyriproxyfen [60]. There are many reports demonstrating elevated P450 activity in insecticide resistant mosquitoes, frequently in conjunction with altered activities of other enzymes [32]. The global overexpression of the four detoxification enzyme families in *Ae. albopictus* from Florida may therefore be leading to a reduced susceptibility to IGRs.

In all populations that presented DDT resistance, GSTs were significantly overexpressed in the adults. This is not surprising since GST-overexpression is the major metabolic mechanism inducing DDT resistance [32,61] and the involvement of the DDT-dehydrochlorinase, now classified in the GST family, has been demonstrated in DDT resistant *Ae. albopictus* populations in China. The GSTs probably play an important role in DDT resistance in *Ae. albopictus* in the USA and this should be confirmed by the use of synergists in future studies. The other possible mechanism involved in DDT but also in pyrethroid resistance is a target site modification such as the *kdr* mutation [29]. Although none of the populations showed resistance to pyrethroids we identified a previously unknown amino acid polymorphism (F1534L) in domain III of the VGSC, in a location known to be associated with pyrethroid resistance in *Ae. aegypti* [62], in one of the Florida specimens. Kasai et al. [43] found at the same location a mutation leading to a cytosine in *Ae. albopictus* collected from Singapore (F1534C) but besides the fact that the area where the colony originated was treated with permethrin in the 1980s, there was no information about the current resistance status of this population against pyrethroids. This is the first time such a mutation is detected in *Ae. albopictus* and given the increasing use of pyrethroids for vector control in the USA [63,64] it is important to pursue studies on the global distribution of this allele and its involvement in pyrethroid resistance.

In conclusion, our studies have generated a fully characterized susceptible reference population for temperate *Ae. albopictus*, ATM95, which is available upon request from dinafons@rutgers.edu. We have also uncovered a complex landscape of populations of *Ae. albopictus* in the USA that are broadly susceptible to larvicides and adulticides. Unexpectedly, we found significant resistance to DDT in two Florida populations and in a New Jersey population. We also found resistance to malathion, an organophosphate, in Florida and suspected resistance in New Jersey plus suspected resistance to several insect growth regulators. Several detoxification enzyme families seemed to be involved in resistance as well, but further studies with the use of synergists should be performed to confirm these findings. All populations tested were fully susceptible to pyrethroids, however, we identified a previously unknown amino acid polymorphism (Phe $\rightarrow$ Leu) in domain III of the VGSC, in a location known to be associated with pyrethroid resistance in *Ae. aegypti*. We developed a rapid diagnostic PCR to detect this mutation (Marcombe and Fonseca unpublished data) but further studies should be conducted to confirm its implication in DDT/pyrethroid resistance and to assess the frequency of this mutation in *Ae. albopictus*.

This study showed standard larvicides and pyrethroids used for mosquito control are still effective against USA populations of *Ae. albopictus*, but it also demonstrates the importance of research on insecticide resistance and the constant need to develop new tools, new insecticides, and innovative strategies to prevent the development of insecticide resistance in these critical vectors of human diseases. Other strategies such as control using genetically modified male mosquitoes [65], or the use of *Wolbachia* to block disease transmission [66] are very promising because they do not use insecticides but the cost-effectiveness of these strategies and their long term success should be evaluated when compared with conventional control methods.

Acknowledgments

We appreciate the assistance of Linda McCuiston, responsible for the mosquito colonies at the Center for Vector Biology, Rutgers University, and vector control personnel from Mercer and Monmouth counties, particularly Isik Unlu and Taryn Crepeau. We also thank Warren Staudinger, Rui-de Xue, Andrew Kyle, and Mike Hutchinson for providing field-collected specimens from Bergen County, New Jersey, St. Johns County, Florida, and York County, Pennsylvania, respectively. This work was funded by Cooperative Agreement USDAARS-58-6615-8-105 between USDA-ARS and Rutgers University (PI: GGC; PI at Rutgers: DMF).

Author Contributions

Conceived and designed the experiments: SM GGC DMF. Performed the experiments: SM DMF. Analyzed the data: SM DMF. Contributed reagents/materials/analysis tools: AF SPH. Wrote the paper: SM DMF AF GGC SPH.

References

1. Benedict MQ, Levine RS, Hawley WA, Lounibos LP (2007) Spread of the tiger: global risk of invasion by the mosquito *Aedes albopictus*. Vector Borne Zoonotic Dis 7: 76–85.
2. Medley KA (2010) Niche shifts during the global invasion of the Asian tiger mosquito, *Aedes albopictus* Skuse (Culicidae), revealed by reciprocal distribution models. Global Ecology and Biogeography 19: 122–133.
3. Rochlin I, Ninivaggi DV, Hutchinson ML, Farajollahi A (2013) Climate change and range expansion of the Asian tiger mosquito (*Aedes albopictus*) in Northeastern USA: implications for public health practitioners. PLoS One 8: e60874.
4. Rezza G (2012) *Aedes albopictus* and the reemergence of Dengue. BMC Public Health 12: 72.
5. Health Commissioner Reports Dengue Virus Case (November 20, 2013). In: Government SC, editor. Suffolk County Press releases. Available: http://www.suffolkcountyny.gov/SuffolkCountyPressReleases/tabid/1418/itemid/1939/amid/2954/health-commissioner-reports-dengue-virus-case.aspx.
6. Ng LF, Ojcius DM (2009) Chikungunya fever - Re-emergence of an old disease. Microbes Infect. 11(14–15):1163–1164.
7. Tsetsarkin KA, Chen R, Sherman MB, Weaver SC (2011) Chikungunya virus: evolution and genetic determinants of emergence. Current Opinion in Virology 1: 310–317.
8. Enserink M (2007) Infectious diseases. Chikungunya: no longer a third world disease. Science 318: 1860–1861.
9. Moro ML, Gagliotti C, Silvi G, Angelini R, Sambri V, et al. (2010) Chikungunya virus in North-Eastern Italy: a seroprevalence survey. Am J Trop Med Hyg 82: 508–511.
10. Grandadam M, Caro V, Plumet S, Thiberge JM, Souares Y, et al. (2011) Chikungunya virus, southeastern France. Emerging Infectious Diseases 17: 910–913.
11. Lo Presti A, Ciccozzi M, Cella E, Lai A, Simonetti FR, et al. (2012) Origin, evolution, and phylogeography of recent epidemic CHIKV strains. Infection, genetics and evolution: journal of molecular epidemiology and evolutionary genetics in infectious diseases.
12. Mogi M, Armbruster P, Fonseca DM (2012) Analyses of the Northern Distributional Limit of *Aedes albopictus* (Diptera: Culicidae) With a Simple Thermal Index. Journal of Medical Entomology 49: 1233–1243.
13. Fonseca DM, Unlu I, Crepeau T, Farajollahi A, Healy S, et al. (2013) Area-wide management of *Aedes albopictus*: II. Gauging the efficacy of traditional integrated pest control measures against urban container mosquitoes. Pest Manag Sci 69(12): 1351–1361.
14. Farajollahi A, Healy SP, Unlu I, Gaugler R, Fonseca DM (2012) Effectiveness of ultra-low volume nighttime applications of an adulticide against diurnal *Aedes albopictus*, a critical vector of dengue and chikungunya viruses. PLoS One 7: e49181.
15. Marcombe S, Carron A, Darriet F, Etienne M, Agnew P, et al. (2009) Reduced efficacy of pyrethroid space sprays for dengue control in an area of Martinique with pyrethroid resistance. The American Journal of Tropical Medicine and Hygiene 80: 745–751.
16. Ranson H, Burhani J, Lumjuan N, Black WC IV (2010) Insecticide resistance in dengue vectors. TropIKAnet 1: ISSN 2078–8606.
17. McAllister JC, Godsey MS, Scott ML (2012) Pyrethroid resistance in *Aedes aegypti* and *Aedes albopictus* from Port-au-Prince, Haiti. Journal of Vector Ecology 37: 325–332.
18. Mouchet J (1972) [Survey of potential vectors of yellow fever in Tanzania]. Bulletin of the World Health Organization 46: 675–684.
19. Vontas J, Kioulos E, Pavlidi N, Morou E, Torre Ad, et al. (2012) Insecticide resistance in the major dengue vectors *Aedes albopictus* and *Aedes aegypti*. Pesticide Biochemistry and Physiology 104: 126–131.
20. Chen L, Zhao T, Pan C, Ross J, Ginevan M, et al. (2013) Absorption and excretion of organophosphorous insecticide biomarkers of malathion in the rat: implications for overestimation bias and exposure misclassification from environmental biomonitoring. Regul Toxicol Pharmacol 65: 287–293.
21. Chuaycharoensuk T, Juntarajumnong W, Boonyuan W, Bangs MJ, Akratanakul P, et al. (2011) Frequency of pyrethroid resistance in *Aedes aegypti* and *Aedes albopictus* (Diptera: Culicidae) in Thailand. Journal of Vector Ecology 36: 204–212.
22. Cui F, Raymond M, Qiao CL (2006) Insecticide resistance in vector mosquitoes in China. Pest Management Science 62: 1013–1022.
23. Kamgang B, Marcombe S, Chandre F, Nchoutpouen E, Nwane P, et al. (2011) Insecticide susceptibility of *Aedes aegypti* and *Aedes albopictus* in Central Africa. Parasites & Vectors 4: 79.
24. Ponlawat A, Scott JG, Harrington LC (2005) Insecticide susceptibility of *Aedes aegypti* and *Aedes albopictus* across Thailand. Journal of Medical Entomology 42: 821–825.
25. Wesson DM (1990) Susceptibility to organophosphate insecticides in larval *Aedes albopictus*. Journal of the American Mosquito Control Association 6: 258–264.
26. Ali A, Nayar JK, Xue RD (1995) Comparative toxicity of selected larvicides and insect growth regulators to a Florida laboratory population of *Aedes albopictus*. Journal of the American Mosquito Control Association 11: 72–76.
27. Khoo BK, Sutherland DJ, Sprenger D, Dickerson D, Nguyen H (1988) Susceptibility status of *Aedes albopictus* to three topically applied adulticides. Journal of the American Mosquito Control Association 4: 310–313.
28. Liu H, Cupp EW, Guo A, Liu N (2004) Insecticide resistance in Alabama and Florida mosquito strains of *Aedes albopictus*. Journal of Medical Entomology 41: 946–952.
29. Brengues C, Hawkes NJ, Chandre F, McCarroll L, Duchon S, et al. (2003) Pyrethroid and DDT cross-resistance in *Aedes aegypti* is correlated with novel mutations in the voltage-gated sodium channel gene. Medical and Veterinary Entomology 17: 87–94.
30. Raymond M, Berticat C, Weill M, Pasteur N, Chevillon C (2001) Insecticide resistance in the mosquito *Culex pipiens*: what have we learned about adaptation? Genetica 112–113: 287–296.
31. Hemingway J, Field L, Vontas J (2002) An overview of insecticide resistance. Science 298: 96–97.
32. Hemingway J, Hawkes NJ, McCarroll L, Ranson H (2004) The molecular basis of insecticide resistance in mosquitoes. Insect Biochemistry and Molecular Biology 34: 653–665.
33. Hemingway J, Karunaratne SH (1998) Mosquito carboxylesterases: a review of the molecular biology and biochemistry of a major insecticide resistance mechanism. Medical and Veterinary Entomology 12: 1–12.
34. Ranson H, Hemingway J (2005) Mosquito glutathione transferases. Methods in Enzymology 401: 226–241.
35. Crans WJ, Chomsky MS, Guthrie D, Acquaviva A (1996) First record of *Aedes albopictus* from New Jersey. Journal of the American Mosquito Control Association 12: 307–309.
36. WHO (2005) Guidelines for laboratory and field testing of mosquito larvicides. In: WHO/CDS/WHOPES/GCDPP/13, editor. Geneva, Switzerland: World Health Organization.
37. WHO (2006) Guidelines for testing mosquito adulticides for indoor residual spraying and treatment of mosquito nets. In: WHO/CDS/NTD/WHOPES/GCDPP/3, editor. Geneva, Switzerland: World Health Organization.
38. Finney DJ (1971) Probit Analysis. Cambridge, UK: Cambridge University Press.
39. Sakuma M (1998) Probit analysis of preference data. Applied Entomology 33: 339–347.
40. Brogdon WG, McAllister JC, Vulule J (1997) Heme peroxidase activity measured in single mosquitoes identifies individuals expressing an elevated oxidase for insecticide resistance. Journal of the American Mosquito Control Association 13: 233–237.
41. Bradford MM (1976) Rapid and sensitive method for the quantitation of microgram quantities of protein utilizing the principle of protein-dye binding. Anal Biochem 72: 248–254.
42. Hayter AJ (1984) A proof of the conjecture that the Tukey-Kramer multiple comparisons procedure is conservative. The Annals of Statistics 12: 1–401.
43. Kasai S, Ng LC, Lam-Phua SG, Tang CS, Itokawa K, et al. (2011) First detection of a putative knockdown resistance gene in major mosquito vector, *Aedes albopictus*. Japanese Journal of Infectious Diseases 64: 217–221.

44. Alout H, Labbe P, Berthomieu A, Pasteur N, Weill M (2009) Multiple duplications of the rare ace-1 mutation F290V in *Culex pipiens* natural populations. Insect Biochem Mol Biol 39: 884–891.

45. Selvi S, Edah MA, Nazni WA, Lee HL, Tyagi BK, et al. (2010) Insecticide susceptibility and resistance development in malathion selected *Aedes albopictus* (Skuse). Tropical biomedicine 27: 534–550.

46. Kuno G (2010) Early history of laboratory breeding of *Aedes aegypti* (Diptera: Culicidae) focusing on the origins and use of selected strains. J Med Entomol 47: 957–971.

47. Paul A, Harrington LC, Zhang L, Scott JG (2005) Insecticide resistance in *Culex pipiens* from New York. J Am Mosq Control Assoc 21: 305–309.

48. Sparks TC, Dripps JE, Watson GB, Paroonagian D (2012) Resistance and cross-resistance to the spinosyns – A review and analysis. Pesticide Biochemistry and Physiology 2012: 1–10.

49. Hamdan H, Sofian-Azirun M, Nazni WA, Lee HL (2005) Insecticide resistance development in *Culex quinquefasciatus* (Say), *Aedes aegypti* (L.) and *Aedes albopictus* (Skuse) larvae against malathion, permethrin and temephos. Tropical Bomedicine 22: 45–52.

50. Romi R, Toma L, Severini F, Di Luca M (2003) Susceptibility of Italian populations of *Aedes albopictus* to temephos and to other insecticides. Journal of the American Mosquito Control Association 19: 419–423.

51. Nayar JK, Ali A, Zaim M (2002) Effectiveness and residual activity comparison of granular formulations of insect growth regulators pyriproxyfen and s-methoprene against Florida mosquitoes in laboratory and outdoor conditions. J Am Mosq Control Assoc 18: 196–201.

52. Marcombe S, Darriet F, Agnew P, Etienne M, Yp-Tcha MM, et al. (2011) Field efficacy of new larvicide products for control of multi-resistant *Aedes aegypti* populations in Martinique (French West Indies). Am J Trop Med Hyg 84: 118–126.

53. Andrighetti MTM, Cerone F, Rigueti M, Galvani KC, Macoris MdLdG (2008) Effect of pyriproxyfen in *Aedes aegypti* populations with different levels of susceptibility to the organophosphate temephos. Dengue Bulletin: 186–198.

54. Robert LL, Olson JK (1989) Suceptibility of female *Aedes albopictus* from Texas to commonly used insecticides. Journal of the American Mosquito Control Association 5: 251–253.

55. Hawley WA, Reiter P, Copeland RS, Pumpuni CB, Craig GB Jr (1987) *Aedes albopictus* in North America: probable introduction in used tires from northern Asia. Science 236: 1114–1116.

56. Kawada H, Maekawa Y, Abe M, Ohashi K, Ohba SY, et al. (2010) Spatial distribution and pyrethroid susceptibility of mosquito larvae collected from catch basins in parks in Nagasaki city, Nagasaki, Japan. Japanese Journal of Infectious Diseases 63: 19–24.

57. Chen CD, Nazni WA, Lee HL, Sofian-Azirun M (2005) Weekly variation on susceptibility status of *Aedes* mosquitoes against temephos in Selangor, Malaysia. Tropical Biomedicine 22: 195–206.

58. Coleman M, Hemingway J (2007) Insecticide resistance monitoring and evaluation in disease transmitting mosquitoes. Journal of Pesticide Science 32: 69–76.

59. Brogdon WG, McAllister JC (1998) Insecticide resistance and vector control. Emerg Infect Dis 4: 605–613.

60. Chandor-Proust A, Bibby J, Regent-Kloeckner M, Roux J, Guittard-Crilat E, et al. (2013) The central role of mosquito cytochrome P450 CYP6Zs in insecticide detoxification revealed by functional expression and structural modelling. Biochem J 455: 75–85.

61. Neng W, Yan X, Fuming H, Dazong C (1992) Susceptibility of *Aedes albopictus* from China to insecticides, and mechanism of DDT resistance. Journal of the American Mosquito Control Association 8: 394–397.

62. Harris AF, Rajatileka S, Ranson H (2010) Pyrethroid resistance in *Aedes aegypti* from Grand Cayman. Am J Trop Med Hyg 83: 277–284.

63. Davis RS, Peterson RKD, Macedo PA (2007) An ecological risk assessment for insecticides used in adult mosquito management. Integrated Environmental Assessment and Management 3: 373–382.

64. Peterson RKD, Macedo PA, Davis RS (2006) A Human-Health Risk Assessment for West Nile Virus and Insecticides Used in Mosquito Management. Environ Health Perspect 114: 366–372.

65. Harris AF (2011) Field performance of engineered male mosquitoes. Nature Biotechnology 29: 1034–1037.

66. Hoffman AA (2011) Successful establishment of *Wolbachia* in *Aedes* populations to suppress dengue transmission. Nature 476: 454–457.

6

Development of Composite Indices to Measure the Adoption of Pro-Environmental Behaviours across Canadian Provinces

Magalie Canuel[1]*, Belkacem Abdous[2,3], Diane Bélanger[2,4], Pierre Gosselin[1,2,4]

1 Institut national de santé publique du Québec (INSPQ), Québec City, Canada, 2 Centre de recherche du Centre hospitalier universitaire de Québec, Québec City, Canada, 3 Département de médecine sociale et préventive de l'Université Laval, Québec City, Canada, 4 Institut national de la recherche scientifique, Centre Eau Terre Environnement, Québec City, Canada

Abstract

Objective: The adoption of pro-environmental behaviours reduces anthropogenic environmental impacts and subsequent human health effects. This study developed composite indices measuring adoption of pro-environmental behaviours at the household level in Canada.

Methods: The 2007 Households and the Environment Survey conducted by Statistics Canada collected data on Canadian environmental behaviours at households' level. A subset of 55 retained questions from this survey was analyzed by Multiple Correspondence Analysis (MCA) to develop the index. Weights attributed by MCA were used to compute scores for each Canadian province as well as for socio-demographic strata. Scores were classified into four categories reflecting different levels of adoption of pro-environmental behaviours.

Results: Two indices were finally created: one based on 23 questions related to behaviours done inside the dwelling and a second based on 16 questions measuring behaviours done outside of the dwelling. British Columbia, Quebec, Prince-Edward-Island and Nova-Scotia appeared in one of the two top categories of adoption of pro-environmental behaviours for both indices. Alberta, Saskatchewan, Manitoba and Newfoundland-and-Labrador were classified in one of the two last categories of pro-environmental behaviours adoption for both indices. Households with a higher income, educational attainment, or greater number of persons adopted more indoor pro-environmental behaviours, while on the outdoor index, they adopted fewer such behaviours. Households with low-income fared better on the adoption of outdoors pro-environmental behaviours.

Conclusion: MCA was successfully applied in creating Indoor and Outdoor composite Indices of pro-environmental behaviours. The Indices cover a good range of environmental themes and the analysis could be applied to similar surveys worldwide (as baseline weights) enabling temporal trend comparison for recurring themes. Much more than voluntary measures, the study shows that existing regulations, dwelling type, households composition and income as well as climate are the major factors determining pro-environmental behaviours.

Editor: Judi Hewitt, University of Waikato (National Institute of Water and Atmospheric Research), New Zealand

Funding: This study was funded by the Green Fund for Action 21 of the 2006-2012 Climate Change Action Plan of the Quebec government. The funders had no role in study design, data collection and analysis, decision to publish, or preparation of the manuscript.

Competing Interests: The authors have declared that no competing interests exist.

* Email: magalie.canuel@inspq.qc.ca

Introduction

A significant source of pollution to our natural environment comes from domestic activities and behaviours. For example household-generated waste in Canada accounts for around a third of total waste and household energy use and municipal water consumption for 17% and 57%, respectively [1-3]. Also, 46% of greenhouse gas emissions (GHG), which contribute to climate change, come from direct and indirect household emissions [4]. The impacts of such household pollution can be important.

Municipal waste can impact the environment in various ways including soil and water contamination from leachate in landfills disposal and the production of greenhouse gas emissions (GHG) and air pollution, either from landfills or the incineration process. When solid waste are recycled or composted instead of being landfilled or incinerated, the demand for energy and new-resources can be reduced significantly [3].

The production of energy can impact the environment in various ways, depending on the technology. In Canada, energy production and consumption accounts for around 80% of all GHG emission [5]. A household can reduce its emission of GHG by reducing electric power use. For instance high energy efficiency electronic devices or cleaner energy sources will generate less pollution and GHG.

Water shortages are happening worldwide and one way to limit their occurrence is through water conservation behaviours. In most homes, more than 60% of water use comes from toilet flushing, showers and baths, making water-saving devices like low-flow shower head an efficient way of reducing water consumption. In summer, water use can increase by 50% for yard activities such as watering the lawn. There are behaviours that households can implement to decrease their water consumption in summer time like using sprinklers with a timer or adopting the use of a rain barrel [6].

It thus becomes clear that addressing sustainability concerns has to take into account not only industry or agriculture, but also household behaviours, their impacts on ecosystems and ultimately on human health. Monitoring trends of household behaviours can inform policy and research agendas on the development of incentives or other mechanisms such as information campaigns to reduce domestic pollution and facilitate adaptative measures to minimize related health risks. The adoption of several pro-environmental behaviours, i.e. actions that contribute to the preservation of the environment, should be encouraged to significantly reduce the anthropic impact on the environment.

In Canada, the Households and the Environment Survey (HES) was designed to measure household behaviours with respect to the environment. The HES is a periodic survey conducted by Statistics Canada, the federal government statistical agency, and adminis-tered across Canadian provinces. The survey covers 12 broad themes including energy use and heating, water use, transportation decisions, motor vehicle use, recycling and composting (Figure 1) [7]. While this survey provides various estimations of up to 83 Canadian practices (Figure 1) as well as some information on their socio-demographic characteristics, survey reports are limited to analyses of simple cross-tabulation frequencies for some of the 83 separate behaviours [7-13].

It is difficult to follow up on such a wide array of relevant behaviours and their trends over time, unless they are summarized in some way. A composite index is a tool which can be useful to that purpose as it incorporates several aspects of an issue and allow for monitoring across several themes simultaneously, thus facili-tating the measurement of trends [14]. While other environmental indices exist, such as the environmental sustainability index [15], to our knowledge no index currently exists to reflect trends of pro-environmental behaviours at the household level in Canada.

This study thus sets out to develop a composite index that summarizes pro-environmental behaviours at the household level across Canadian provinces based on the HES (2007) given the periodicity and geographical coverage of the survey. Pro-environmental behaviours are defined as actions that contribute to the preservation of the environment and can have a positive impact on the health of the population. This study will serve as baseline of the trend of the composite index over time, given the periodicity and geographical coverage of the survey.

Materials and Methods

Ethics statement

This research did not require the approval of an ethics review board as we used an existing and anonymized database made available to universities by Statistics Canada. Statistics Canada obtained consent previous to survey administration. No new data was collected for this study.

Survey

The Households and the Environment Survey (HES) is conducted by Statistics Canada. It was designed to address the needs of the Canadian Environmental Sustainability Indicators project. The project reports on air quality, water quality and greenhouse gas emissions in Canada using indicators to identify areas of importance to Canadians and monitor progress [16].

The survey aimed Canadian households with at least one person aged 18 year or older. The HES covers all 10 of the provinces and excludes the 3 northern territories, Indian reserves and members of the Canadian Armed Forces. The survey was first conducted in 1991 and since 2005 has been carried out biennially. In the present study, the 2007 HES database was used in its Public Use Microdata Files format (PUMF) [16]. As a sub-sample of the dwellings that were part of the Canadian community health survey (CCHS), the sampling allocation for the HES followed that of the CCHS closely. The CCHS used a multistage stratified cluster design in which the dwelling is the final sampling unit. Three sampling frames were used to select the sample of households: 50% of the sample came from an area frame, 49% came from a list frame of telephone numbers and 1% came from a Random Digit Dialing sampling frame [16].

From the 40 584 households selected in the 2007 CCHS, a sub-sample of 29 957 households were selected for the HES. Of those, 21 690 households responded to the survey resulting in an overall response rate of 72%. The survey is representative of 12 932 350 households, corresponding to 97% of all Canadian households [16]. The questionnaire was administered to the 21 690 households by telephone interview spread over a 6-month period, from October 2007 to February 2008.

Questionnaire

The person with the best knowledge of environmental household practices was asked to respond on behalf of the household. The main questionnaire covered 12 themes and included 121 questions (figure 1) [16]. Among the questions, 83 measured behaviours and 7 measured socio-demographic charac-teristics. The other 31 questions covered knowledge, reasons for not adopting the behaviour, or served to specify some character-istics (e.g. of a good) or to filter for the next question.

Database

The PUMF was used for the analysis and unlike the master file, applies privacy measures to protect personal information [16]. In the PUMF, data were mostly coded as categorical variables. Three different labels (don't know, not stated, and refusal) were used to classify households who did not participate despite eligibility or to protect the anonymity of the household. A 'valid skip' label was used when the provision of a response was not appropriate. For example, a household who answered 'no' to the question for car ownership was allocated a 'valid skip' label for subsequent questions on the characteristics of the car.

Sampling weights

Sampling weights were applied to ensure that any derived composite index is representative of the study population. They were used when proportions and averages were estimated and to weight the relative frequencies of the Burt matrix in the MCA (see Statistical analysis below).

Variables selection

This study focuses on everyday pro-environmental behaviours, defined as actions that contribute to the preservation of the

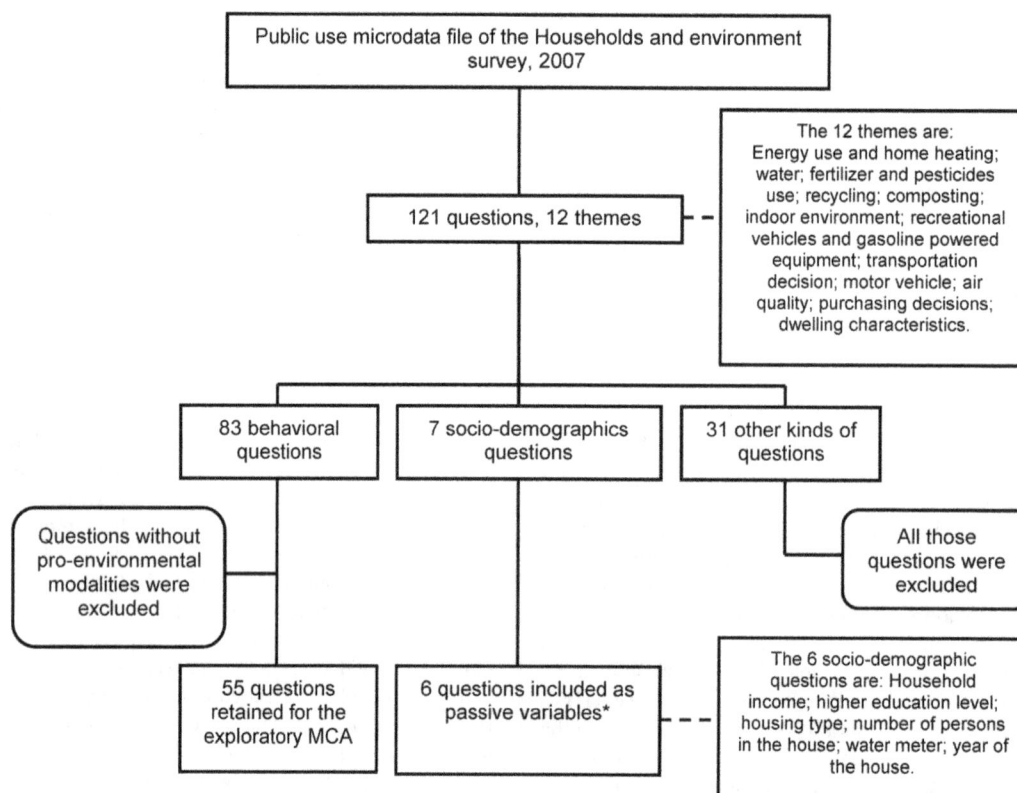

Figure 1. Number and type of questions selected to develop the composite index. Legend: *The composite index was also created for the 7th socio-demographic variable, the census metropolitan area (n = 33), but is not presented in this article.

environment and can have a positive impact on the health of the population. For example, air pollutants can be reduced when households adopt behaviours that decrease their energy consumption such as the use of energy-efficient appliances or when they use more sustainable transport options such as public or active transport.

Based on the above definition, a panel of four environmental health experts applied progressive development consensus after iterations, based on a nominal group technique [17] to evaluate HES variables for exclusion. These were either: variables not measuring a behaviour or questions with no clearly pro-environmental response option. Socio-demographic variables were kept as passive variables with zero mass and no influence on the analysis. They support and complement the interpretation of the map representation of the active variables [18].

Statistical analysis

Given that the data was mostly categorical, the indices in this paper were developed by multiple correspondence analysis (MCA) [18]. Several authors have used Multiple Correspondence Analysis (MCA) as a weighting method for the construction of a composite index [19–23]. MCA is a data reduction procedure for categorical variables (nominal or ordinal) as much as Principal Components Analysis is for quantitative variables [18]. It enables the exploration of associations within a set of variables by transforming the whole data set into dummy variables to form an indicator matrix or upon construction of a matrix from all two-way cross-tabulations among the variables (Burt matrix). This transformed data is treated as a cloud in a space equipped with the classical Chi-square distance. This distance is used in the assessment of

homogeneity and variance (inertia) of rows or columns of the indicator or Burt matrix. The most crucial step of MCA is its use of singular value decomposition and weighted least squares techniques to find low-dimensional best fitting subspaces with minimal inertia and information loss [18].

MCA was conducted using the 'ca' package of the R statistical software [24]. First, the HES database was converted to a Burt matrix taking into consideration the sampling weights. A Burt matrix is a square symmetric categories-by-categories matrix formed from all two-way contingency tables of pairs of variables [18].

Then, an exploratory MCA was performed to project data onto maps where potential outliers were identified and excluded from subsequent analyses. MCA was then applied again to determine the most relevant factorial axes that would serve to build the composite index. There are no universal rules for the determination of the number of dimensions to retain in MCA. However, since the first factorial axis captures the most important part of the total inertia, it plays a central role in the computation of a composite index.

As recommended by Asselin [23], we sought questions having the property of First Axis Ordering Consistency (FAOC). To this end, we projected all the questions on the first axis and tried to identify those having an ordinal structure consistent with respect to this axis, i.e. all questions with pro-environmental responses improving from left to right (or conversely).

The computation of the index score was performed as follows: first, the score of any household was obtained by taking the average of its category-weights generated by the MCA. Then for each province we took the average over all household scores as the value of its composite index. The sampling weight was used in this final step. Coordinates were missing for excluded responses.

The 10 average provincial scores were grouped into categories reflecting different levels of adoption of pro-environmental behaviours. First we applied a cluster analysis and then we used a dendrogram plot using SAS version 9.2 (SAS Institute, Cary, NC) to determine such groups. The categories limits generated for the provincial index were used as reference categories for indices on other socio-demographic variables.

Finally, others indices based on various socio-demographic variables were constructed (Figure 1). Household scores were calculated by taking the average of its category-weights generated by the MCA. Then the index score of the socio-demographic category (e.g. household with annual income less than $40,000) is set as the average of the corresponding household scores.

Results

Multiple Correspondence Analysis

Of the 121 questions in the survey, 55 were kept by the Expert Panel for use in the MCA. These represented 285 response possibilities. On the MCA map projection there was a clear opposition between the missing data (don't know, refusal, not stated) located far from the map center and the other responses which gathered close to the center (Figure S1). Excluding the missing data rebalanced the model (179 remaining responses) (Figure 2). However, since pro-environmental behaviours were spread over both sides of the first axis, we failed to find any meaning to this first dimension.

We then screened the projected responses to identify questions following an ordinal structure, (i.e all pro-environmental responses of a question have negative coordinates on the first factorial axis (or conversely)). Twenty-three such questions with pro-environmental responses deteriorating from left to right on the first axis (group A), and 16 questions with opposite ordinal structure (group B) were identified. The remaining 16 (of 55) questions were excluded from the analysis because their responses were not sufficiently discriminating (i.e. the pro- and anti-environmental responses were on the same side of the axis or they were grouped close together on the map). As well the majority of these questions (10/16) had at least two responses with a contribution of zero to the first axis (Table S1).

These exploration steps led us to consider two separate composite indices. Group A included 96 responses but after excluding missing data, 52 responses were used in the MCA. The majority of excluded responses had frequencies lower than 2.0% and two responses had frequencies of 4.6% and 4.7%. After exclusion of missing data, some responses still looked like extreme values on the map (Figure 3). They were kept in the analysis as they are 2 of the 3 responses for all questions concerning recycling. Excluding these responses would have resulted in the exclusion of all recycling questions. Responses used in this analysis had a frequency of 7.5% or higher, except for two responses with frequencies of 2.5% and 3.5% (responses on recycling).

For group A, the first dimension explained 32.6% of the inertia while the second explained 16.1% (Table 1). Given that the first factorial axis plays a central role in the construction of this composite index, only the first dimension was selected to construct the index. This group respects the FAOC as pro-environmental responses are located on the left of the first axis as opposed to others responses deteriorating to the right (Figure 3). Also, we noted that the retained questions were associated with five themes of the survey: energy use and home heating, water, recycling, composting and, purchasing decisions. All 23 questions assessed

behaviours practiced inside the dwelling and thus the first axis measures these behaviours. Twelve of 15 responses contributing the most to the first factorial axis concerned recycling (Table S2).

The second group of 16 questions (group B) consisted of 86 responses, 41 of which were missing values. The 45 remaining responses used for the MCA had frequencies of 7.0% or higher while excluded responses had frequencies lower than 3.5%. For group B, the first dimension explained 62.1% of the inertia while the second explained only 12.4% (Table 2). Again, the first dimension was selected for the construction of the index and pro-environmental responses were located on the right of the first axis with other responses deteriorating from right to left (Figure 4). The 16 questions cover five themes of the survey: water, fertilizer and pesticide use, recreational vehicles and gasoline powered equipment, transport decisions and air quality, all behaviours being practiced outdoors. Of note, 9 of the 15 responses contributing the most to the first factorial axis concern households with no lawn or garden (i.e. the application of fertilizers or pesticides, yard waste and watering of the lawn or the garden) (Table S3).

Because two distinct behavioural categories resulted from the MCA, two composite indices were created instead of one. The first index (group A) is named the 'Indoor Index' and the second one (group B) the 'Outdoor Index'. Questions included for each index are presented in supporting information, Table S4 and Table S5.

Composite indices by province

The map representations of the final coordinates generated by the MCA are shown in Figure 3 and Figure 4. Coordinates and other results of the MCA are available in supporting information, Table S2 and Table S3. The coordinates of the first dimension were used to construct each of the two composite indices. Coordinates are missing for responses that have been excluded. Only 0.9% and 0.4% of coordinates are missing for the indoor and outdoor indices, respectively.

For the Indoor Index, the households belonging to a province with negative coordinates tend to adopt more pro-environmental behaviours than those of a province with positive coordinates. In contrast, for the Outdoor Index provinces with positive coordinates adopt more outdoor pro-environmental behaviours than those with negative coordinates.

The cluster analysis and dendrogram plot resulted in the classification of each province into one of four categories reflecting different levels of adoption of pro-environmental behaviours: 1) adopting the most; 2) adopting slightly fewer; 3) adopting much fewer and; 4) adopting the fewest. The provincial coordinates and the categories generated from the cluster analysis are shown in Table 3 and Table 4. Maps of the Canadian provinces with their categories of pro-environmental behaviours are shown in Figure 5 and Figure 6.

None of the 10 provinces were classified in both indices as adopting the most pro-environmental behaviours. For the Indoor Index, Ontario (ON), Prince Edward Island (PEI) and Nova Scotia (NS) rated in the top category, British Columbia (BC) and Québec (QC) in the next, the three Prairie provinces and New Brunswick (NB) in the third and Newfoundland and Labrador (NL) in "adopting the fewest" category (Figure 5). For the Outdoor Index, QC scored in the top category with BC, NS, NB and PEI following in second, and Manitoba (MN), ON and NL in third, followed by Alberta (AB) and Saskatchewan (SK) in the bottom category (Figure 6).

Four provinces (BC, QC, NS and PEI) were classified in the top two categories for both indices while four provinces (AB, SK, MN and NL) were classified for both indices, in the two lower categories.

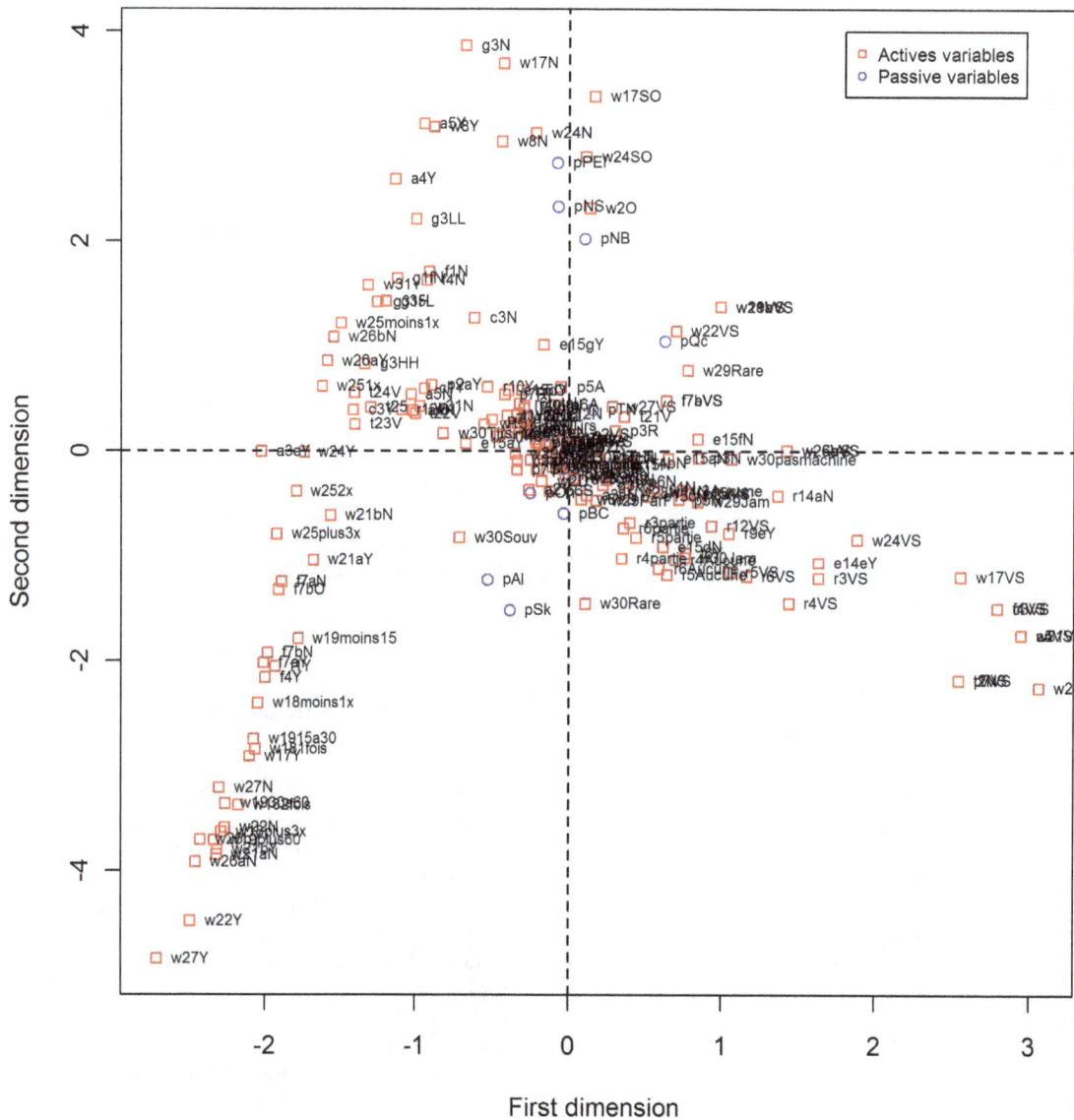

Figure 2. Map representation of the MCA results on the 55 questions without extreme responses.

Composite indices by socio-demographic variables

The coordinates and the classification for the six comparison variables are shown in Table 5. For household income, educational attainment and number of persons in the household, there were oppositions in the classification of the responses. Households with a higher income, or higher educational attainment, or greater number of persons adopted more indoor pro-environmental behaviours, while those with a lower household income, educational attainment, or number of people, adopted more outdoor such behaviours. As well, households with water meters tended to adopt more indoor pro-environmental behaviours than those without, but for outdoors behaviours, the opposite applied – not having a water meter was associated with better adoption of pro-environmental behaviours. And finally, the dwelling's year of construction did not influence the adoption of pro-environmental behaviours as there was no trend on either index (Table 5).

Discussion

This study sought to develop a composite index which measures the overall adoption of pro-environmental behaviours among Canadian households. MCA, our main analytical technique, was used to aggregate survey data and to provide weights to the responses in the construction of the index. Our approach is similar to other studies in different fields [19–23]. This was followed by a cluster analysis to classify the provinces, as well as an exploration of relationships with socio-demographic factors.

The MCA generated two indices based on 39 of the 55 behavioural questions, an Indoor Index and an Outdoor Index, each reflecting environmental behaviours for 5 of the 12 survey themes. Retaining both indices allowed for better representation of the survey; together they cover 9 themes out of 12 (water use is in both) whereas one single index would have covered only 5, excluding important environmental themes such as fertilizer and pesticide use. As well, because the provincial classifications were different for each index and varied as well in the classification by

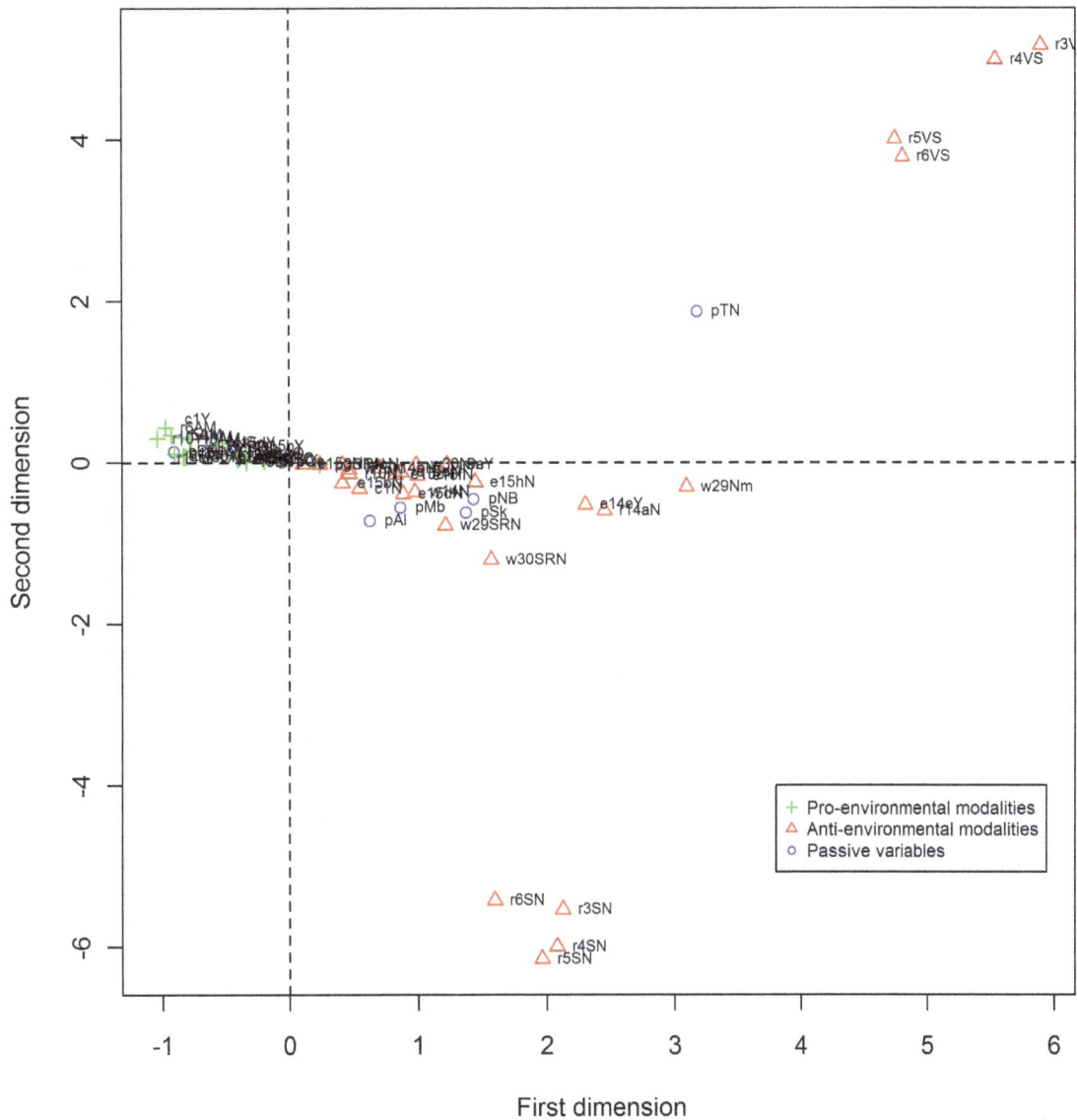

Figure 3. Map representation of the MCA results on the 23 questions of the group A (Indoor).

socio-demographics factors for each index (e.g., household income) it was deemed justifiable to keep both indices.

Most (19/23) questions included in the Indoor Index were asked to all households with the exception of questions on recycling where only those households with access to a program were asked to respond. For the Outdoor Index, most questions (11/16) concerned watering of the lawn or the garden, and the use of fertilizers or pesticides. These (11) questions were answered only by households having a yard. However, even if households living in an apartment did not have to answer these questions, they were still recorded in the Index as households adopting pro-environmental behaviours (i.e., most valid skips were classified as pro-environmental responses and some as anti-environmental ones).

The Indoor Index

One likely explanation for PEI and NS being classified in the top category for the Indoor Index is that nearly 100% of their households recycle and the proportion that compost is substantially above the Canadian average, as reported by Statistics

Canada. In these two provinces, households are obligated by law to recycle and compost [25]. Moreover, questions regarding recycling contributed the most to the Indoor Index.

Recycling and composting are also common in ON but its good ranking is also related to the proportion of households that adopt water conservation behaviours (i.e. use water-efficient shower heads and toilets, run dishwasher and washing machine only when full) [7]. Provinces have been slowly adopting a provincial plumbing code requiring that new buildings use water-saving fixtures, with the exception of NL [26;27]. ON however was the first to adopt such a code in 1996 [26;28], and saw an increase in new residential construction from 1996 to 2002 [29], likely contributing to the higher proportion of households practicing water conservation behaviours [28]. This is an example where building codes may be effective in beneficially influencing the passive uptake of pro-environmental practices.

QC's good classification in the Indoor Index is in part due to its proportion of households adopting recycling behaviours being higher than the Canadian average. There were four questions on

Table 1. Explained inertia by each dimension for group A: Indoor Index, 2007.

Dimension	Inertia	Inertia (%)	cumulative Inertia (%)	scree plot
1	0,0270	32,6	32,6	*************************
2	0,0133	16,1	48,8	************
3	0,0078	9,4	58,2	*******
4	0,0032	3,8	62,0	***
5	0,0030	3,6	65,6	***
6	0,0026	3,2	68,8	**
7	0,0024	2,9	71,7	**
8	0,0021	2,5	74,2	**
9	0,0019	2,3	76,5	**
10	0,0018	2,1	78,7	**
11	0,0017	2,0	80,7	**
12	0,0016	1,9	82,6	*
13	0,0015	1,8	84,4	*
14	0,0014	1,7	86,1	*
15	0,0014	1,7	87,7	*
16	0,0013	1,6	89,3	*
17	0,0012	1,5	90,8	*
18	0,0012	1,4	92,2	*
19	0,0011	1,3	93,5	*
20	0,0011	1,3	94,8	*
21	0,0010	1,2	96,1	*
52	0	0,0	100,0	

recycling which contributed significantly to the first dimension, thus contributing to QC's classification. Despite QC having the lowest proportion of households that compost [9] or participate in alternative recycling activities such as donations of furniture and clothing, QC's classification was only slightly affected as these behaviours had only moderate or low contributions to the Index.

In AB, MN, SK and NB, the proportion of households that adopted indoor pro-environmental behaviours is below the Canadian average (data not shown), explaining their lower classification in the Indoor Index. NL had only a few variables above the Canadian average and had most often the lowest proportion of all provinces. For example, the proportion is below the average for all four questions on water conservation and for all questions on recycling. In this province, there is no provincial plumbing code requiring the use of water-saving fixtures in new buildings [26;27]. Also, the proportion of households with access to a recycling program is only 71% [25].

The Outdoor Index

Results for the Outdoor Index show a pattern with respect to Coastal proximity, with coastal provinces, with the exception of NL, rating in the two higher categories, and the continental provinces in the two lower categories with the two lowest rated provinces situated in the Prairies. The climate of the Prairies grasslands is characterized by hot summers combined with low precipitation and periodic drought. The climatic region of the Maritimes however is the one with the greatest annual precipitation [30-32], a pattern which is likely reflected in the frequency of watering lawn and or garden. Although watering of the lawn or garden is around the Canadian average in NL, its inhabitants own more recreational vehicles, use more gas and burn more yard waste on the property (data not shown) which may explain its lower classification than the other coastal provinces.

Also, there was an important difference in the proportion of households that used fertilizers and pesticides and QC had, by far, the lowest proportion. QC was the first province to adopt a provincial law in 2006 prohibiting the sale of pesticides for cosmetic purposes [7;33]. The Prairies on the other hand had the highest proportions of households that used pesticides or fertilizers in 2007 [7;10]. Subsequently, other jurisdictions have adopted similar laws begging the question of whether their classifications in the Outdoor Index will change over time.

It should also be noted that QC and BC have the highest proportion of households living in an apartment [10]. Given that most households living in an apartment do not have a backyard, they do not water neither lawn nor garden, nor do they use pesticides outdoors. Hence, they passively adopt pro-environmental behaviours and are considered as such by the MCA. In fact, these responses, recorded as 'valid skip', had the highest contribution to the Index, likely contributing to the higher classification for BC and QC on the Outdoor Index. Such passive behaviours or external factors were not excluded from the Index as they significantly contribute to the preservation of environmental resources.

Indices for socio-demographic variables

For most socio-demographic variables, there were oppositions in the classification of the modalities, which means that it is not the

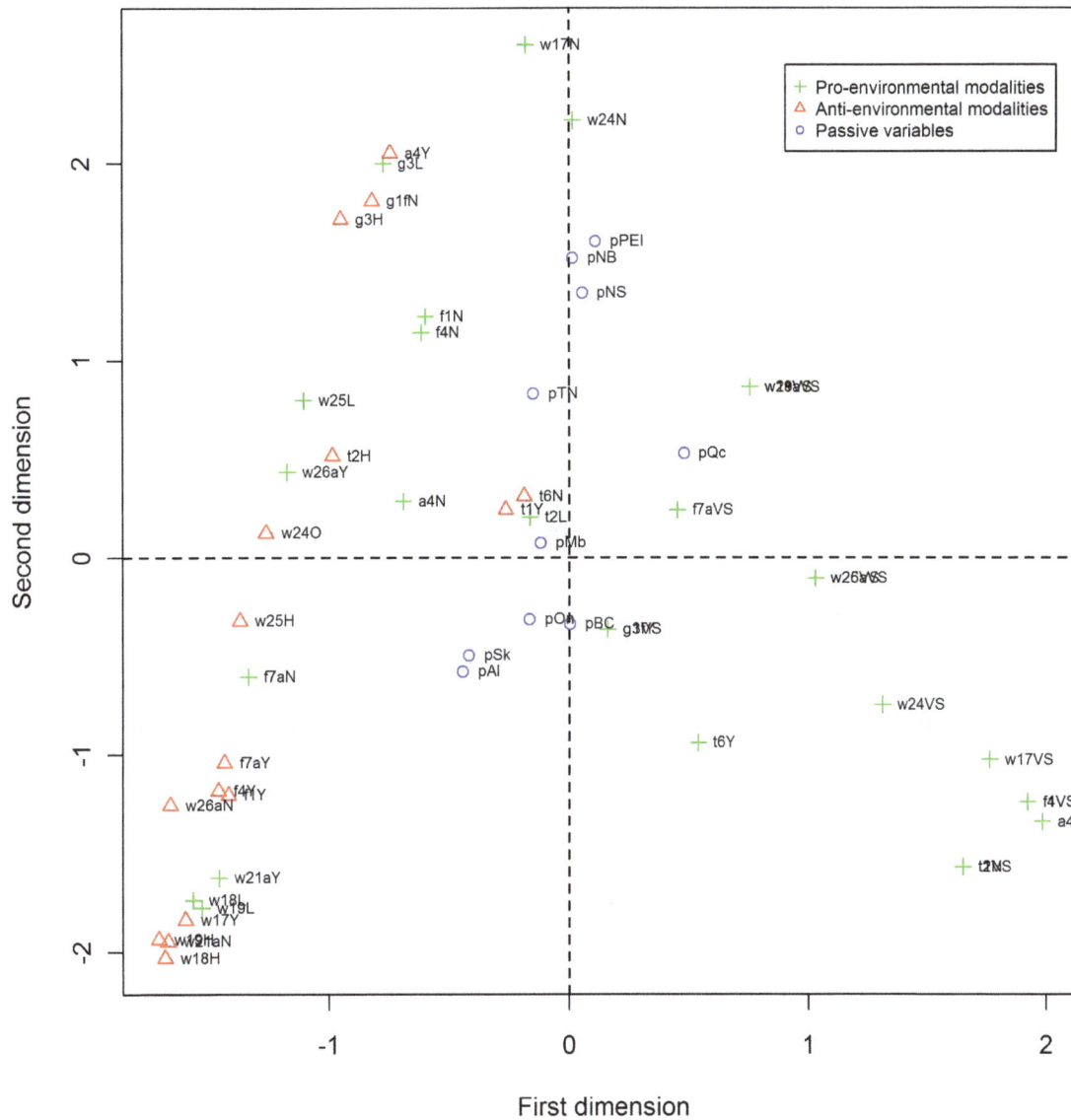

Figure 4. Map representation of the MCA results on the 16 questions of the group B (Outdoor).

Table 2. Explained inertia by each dimension for group B: Outdoor Index, 2007.

Dimension	Inertia	Inertia (%)	Cumulative inertia (%)	scree plot
1	0,2173	62,1	62,1	*************************
2	0,0434	12,4	74,5	*****
3	0,0162	4,6	79,1	**
4	0,0129	3,7	82,8	*
5	0,0111	3,2	86,0	*
6	0,0094	2,7	88,7	*
7	0,0066	1,9	90,6	*
8	0,0061	1,7	92,3	*
9	0,0041	1,2	93,5	
45	0	0,0	100,0	

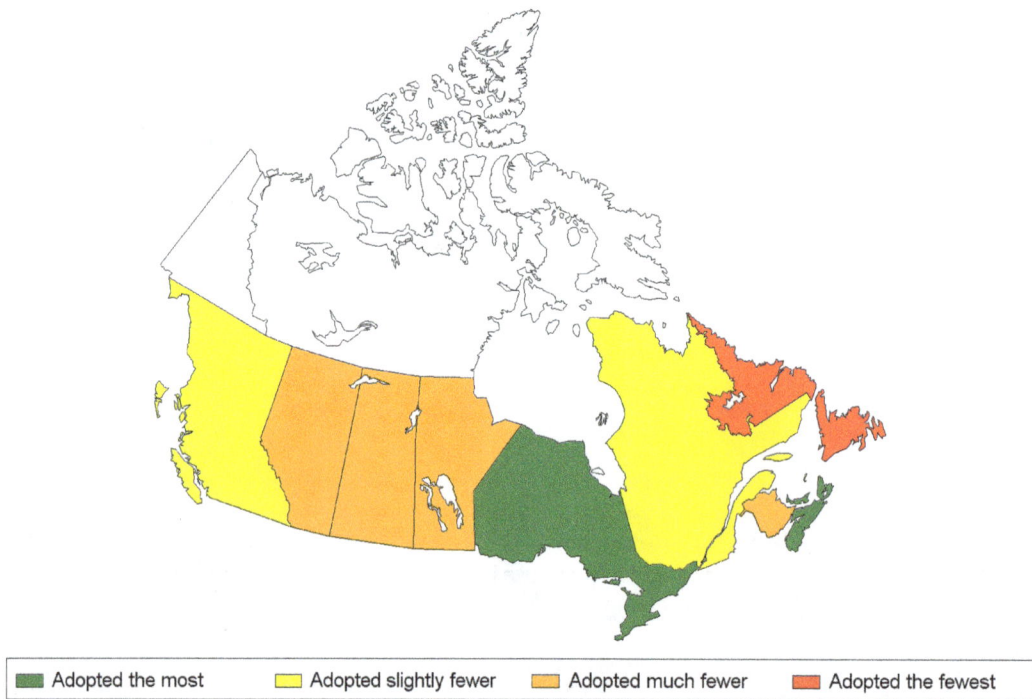

Figure 5. Provinces' classification according to the four categories of pro-environmental behaviours, Indoor Index, 2007. Legend: from left to right – British-Columbia, Alberta, Saskatchewan, Manitoba, Ontario, Quebec, New-Brunswick, Nova-Scotia. Prince-Edward-Island is North of the two latter provinces and Newfoundland-and-Labrador is located North-East of Quebec.

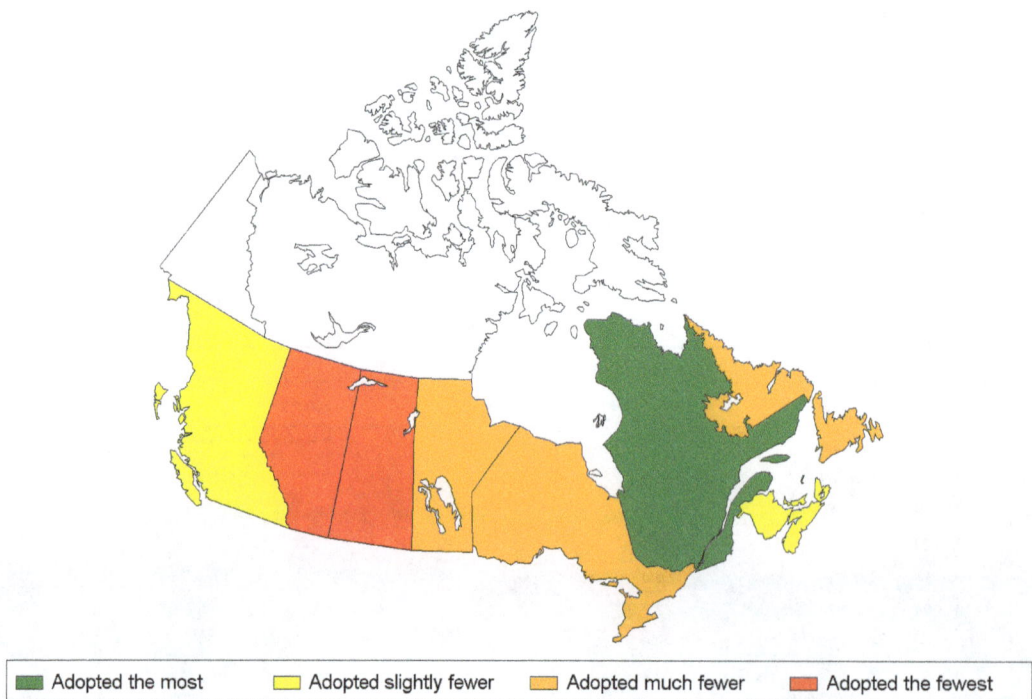

Figure 6. Provinces' classification according to the four categories of pro-environmental behaviours, Outdoor Index, 2007. Legend: from left to right – British-Columbia, Alberta, Saskatchewan, Manitoba, Ontario, Quebec, New-Brunswick, Nova-Scotia. Prince-Edward-Island is North of the two latter provinces and Newfoundland-and-Labrador is located North-East of Quebec.

Table 3. Provinces' coordinates on the Indoor Index, 2007.

Provinces	Coordinates	Categories[a]
Prince-Edward-Island	−0,0262	++
Nova-Scotia	−0,0179	++
Ontario	−0,0130	++
British-Columbia	−0,0055	+
Quebec	−0,0029	+
Alberta	0,0159	−
Manitoba	0,0225	−
Saskatchewan	0,0363	−
New-Brunswick	0,0381	−
Newfoundland-and-Labrador	0,0853	−

[a]Categories are: adopted the most pro-environmental behaviours (++), adopted slightly fewer (+), adopted much fewer (−) and adopted the fewest (−).

same households that adopt pro-environmental behaviours on both indices. Higher income households may be more able to maintain and repair their housing and also invest in environmentally friendly products such as water and energy efficient appliances or fixtures, which can be more expensive than their regular counterparts [34;35]. Access to such products may contribute to the better classification on the Indoor Index for higher income households. On the other hand, lower income households may be less willing to pay water taxes linked to consumption levels, or to buy chemical products for their lawn or garden. Furthermore, those lower income households live more frequently in apartments where they do not have a yard, and they also own fewer recreational vehicles (data not shown). All these factors likely weigh in on the higher classification attributed to lower versus higher income households on the Outdoor Index.

In Canada, income is usually positively associated to educational attainment [36]. Also, the number of persons in a household will influence the household income. In the HES database, there was a significant correlation between households' income and education level as well as one with the households' income and the number of persons in the households (data not shown). This may explain why the indices by educational level and by number of

Table 4. Provinces' coordinates on the Outdoor Index, 2007.

Provinces	Coordinates	Categories[a]
Quebec	0,1038	++
Prince-Edward-Island	0,0261	+
Nova-Scotia	0,0123	+
New-Brunswick	0,0052	+
British-Columbia	0,0012	+
Manitoba	−0,0243	−
Newfoundland-and-Labrador	−0,0273	−
Ontario	−0,0350	−
Saskatchewan	−0,0887	−
Alberta	−0,0962	−

[a]Categories are: adopted the most pro-environmental behaviours (++), adopted slightly fewer (+), adopted much fewer (−) and adopted the fewest (−).

persons in a household are similar to the one by household income. Any one of these three socio-demographic variables could potentially be used as a surrogate for the other two for future data collection for following Index trends over time.

Studies have shown that water meters with appropriate pricing are an incentive to reduce water consumption [2]. The US EPA estimated a 20% reduction in water consumption with universal metering [37] and a Canadian study also estimated a similar reduction according to structured water pricing [38]. While our results showed that households with water meters tended to score higher on the Indoor Index, households without a water meter scored higher on the Outdoor Index, which is in contrast to the other studies. We estimated that only 9% of households living in an apartment have water meters as opposed to 58% for all other types of dwellings in Canada (data not shown). As stated earlier, a household living in an apartment passively adopts more outdoor pro-environmental behaviours for lack of a lawn or garden to maintain with only a few having a water meter, possibly explaining the discrepancy between our results and those of other studies.

Factors that can lead to pro-environmental behaviours

There is a wide variety of measures or instruments than can be introduced by governments to influence households behaviours, from economic instruments to direct regulation, labeling, information campaigns and provision of environment-friendly public goods such as public transportation or bicycle paths [39].

This study has identified factors which seem to influence the uptake of beneficial environmental behaviours at the household level. Investment in infrastructure is one of them. The physical or material possibility to act pro-environmentally must indeed be available [40], such as what might be needed for Newfoundlanders to improve their recycling profile.

Regulation is frequently used to efficiently influence the environmental impacts of household decision-making [39] and in our study it also seems to be an important incentive for the adoption of pro-environmental behaviours. This was seen both in the case of building codes requiring the installation of water efficient shower heads and toilets, and in the case of the ban on pesticides for lawn care. In Ontario, the ban of cosmetic pesticides decreased significantly the concentration of some pesticides, mainly herbicides, in the majority of streams under surveillance near urban areas with limited agriculture activities [41].

To encourage a reduction in water consumption, both price and non-price policies should be used. Volumetric water charges are associated with both water-saving behaviours and adoption of water-efficient devices [39]. However, in a study in several OECD countries, Canada had the highest proportion of households that did not know how they were charged for residential water consumption, thus reducing the price effect on water-saving behaviours [39]. In our study, presence of water meters was an incentive to water-saving behaviours only for indoor behaviours. Climate was also another factor that could be influential. Hence, public information on the environmental impact of water consumption and on measures households can adopt to save water should be combined to economic measures according to the OECD [39] and this study.

Other than governmental measures, household characteristics may play a role in the adoption of environment-related behaviours such as income, household composition and dwelling characteristics [39]. According to the OECD survey, low income households and tenants households make fewer financial investments in water efficiency, as can be expected. Grants targeted at those households to correct the economic imbalance are thus recommended by the agency. Moreover, our study showed

Table 5. Coordinates and categories of pro-environmental behaviours for other socio-demographic variables, Indoor and Outdoor Indices, 2007.

	Indoor Index		Outdoor Index	
	Coordinates	Categories[a]	Coordinates	Categories[a]
Household income				
Less than $40,000	0,0253	−	0,1533	++
$40,000 to less than $80,000	−0,0080	+	−0,0133	−
$80,000 and over	−0,0273	++	−0,1567	−
Highest education level				
Secondary diploma or less	0,0228	−	0,0927	++
Postsecondary certificate or diploma	−0,0030	+	−0,0134	−
University	−0,0159	++	−0,0451	−
Dwelling type				
Apartment	0,0370	−	0,4335	++
Others	−0,0144	++	−0,1501	−
Number of persons in the dwelling				
One	0,0260	−	0,2058	++
Two	−0,0073	+	−0,0255	−
Three	−0,0104	++	−0,0723	−
Four or more	−0,0163	++	−0,1383	−
Water meter				
Yes	−0,0201	++	−0,1682	−
No	0,0123	−	0,1497	++
Year the dwelling was built				
Before 1946	−0,0089	+	0,0034	+
Between 1946 and 1960	−0,0040	+	−0,0130	−
Between 1961 and 1977	−0,0009	+	0,0025	+
Between 1978 and 1983	−0,0054	+	−0,0218	−
Between 1984 and 1995	−0,0084	+	−0,0364	−
Between 1996 and 2000	−0,0088	+	−0,0434	−
Between 2001 and 2005	−0,0099	++	−0,0948	−
2006 or latter	0,0042	+	0,0384	+

[a]Categories are: adopted the most pro-environmental behaviours (++), adopted slightly fewer (+), adopted much fewer (−) and adopted the fewest (−).

households from both income groups (high or low) or dwelling type (owned or rented) have to improve their act in different domains and that programs should target them accordingly. In short, Canadians remain very dependent for many such actions on where they live and what the climate brings to their yards, or not.

Limits of the study

We used data from a survey that has been created to address the needs of Statistics Canada and the federal government. Thus, we were limited to its content. The questionnaire does not cover all behaviours that can impact the environment and public health. Also, the indices developed here measure the behaviours available in the survey and retained after the analysis by an expert group for their potential positive impacts on health, and not all existing pro-environmental behaviours. The classification could have been different if other behaviours had been included.

Three themes of the survey were not covered by the indices, namely dwelling characteristics, motor vehicle and indoor environment. However, we believe they would not have much impact in the indices. First, there were no behaviours measured in

the dwelling characteristics theme and some of the characteristics were included as passive variables in the indices (e.g. year the building was built). The same happened for the motor vehicle theme (focused on the characteristic of the car), yet we used another theme to include the number of vehicles owned by the households in the outdoor index. For the indoor environment theme, only 2 of the 5 questions measured behaviours and they both concerned the type of chemical products used to clean windows and the dwelling. Although every small action is important for the environment, those questions were excluded as some other practices, such as agriculture, use similar products in much larger quantities [42].

A good standing in the classification does not mean that there is no place for improvement. Indeed, a high proportion of households that adopt pro-environmental behaviours on one question can compensate for a lower proportion on another question of the same index. Also, the provinces were compared to each other and not classified in relation to a gold standard.

Furthermore, it was the MCA that attributed the weight for each modality. Thus, a modality with a higher weight has more impact in the index. For instance, all four recycling questions had the highest contribution to the indoor index. Further studies should investigate if the inclusion of only one of those recycling question or a composite index of those four questions would be more appropriate. The same reasoning should also be applied to questions related to the watering of the lawn or of the garden. Households without a garden or a lawn are rewarded for every question on that subject which at the end can impact greatly the province classification. For example, they were not only rewarded for not watering their lawn, but they were also rewarded for the question concerning the watering duration and the number of time they water. Because the MCA attributed the weights, those household without a garden or lawn had a higher 'reward' that households with a garden or lawn that did not water them.

One general limit of MCA is that the first dimension usually explains a low proportion of the total inertia in the data set and the other dimensions explain less than the first [18]. In this study, the first dimension explained 33% and 62% of the total inertia for the indoor and outdoor index respectively. By using only the first dimension, these indices might not properly reflect all of the behaviours, especially for the indoor one. However, using more than one dimension to build the index would not considerably increase the total inertia explained but would in return increase its complexity. Composite indices are indeed built to simplify the analysis.

Because of the study design, based on households, it was also not possible to evaluate the impact of personal attributes, like age and gender, on the adoption of environmental behaviours. The association between environmental behaviours and age is not clear. Studies have observed all possible trends, from older people adopting more pro-environmental behaviours to the opposite trends or no trend at all [43–45]. Also, women would be more likely to take pro-environmental actions than men, although some studies have found the opposite depending on behaviour and region [43–46]. In our study we found that socio-demographic characteristics like household income and a higher level of education did not have the same influence on outdoor behaviours compared to indoor behaviours. Hence, some differences could also be expected between indoor and outdoor behaviours for age and gender.

Because the survey was not meant to measure attitudes or values, we cannot associate the classification of the province to any difference in values or perception. However, others studies have showed cultural differences across Canadian provinces [47–51]. In Canada, French speaking people are at majority in the province of Quebec but a minority in the rest of Canada as opposed to English speaking Canadian that are a majority in the rest of Canada [52;53]. Several studies have observed differences of values and attitudes in terms of personality, political perspective, priorities and social issues between English-Canadians and French-Canadians [47–51]. Differences in those values could also explain some differences in the adoption of pro-environmental behaviours but further studies are required to confirm it.

Attitudes and values can also be different in immigrants compared to the native born. The former usually have a smaller ecological footprint [54–57]. For example, several studies, mostly from United States, have observed that immigrants have lifestyles that are less demanding on the environment: they consume less, possess fewer luxury items like SUVs, they carpool or use public transportation more often and live in smaller houses [54–57]. In 2006, around 55% of all Canadian immigrants were in Ontario, followed by 18% in British-Columbia and 14% in Quebec [58]. British-Columbia and Quebec had a good classification on both indices. However, we could not estimate the impact of immigra-

tion on these classifications, as immigration rules and influx have changed significantly over the last decades [58].

Despite those limits, the indices still give a good idea of the global adoption of pro-environmental behaviours with potential positive impacts on health in Canada and remain easy to explain and understand. The main sectors in which households can have an impact are covered by the indices, like air and soil quality as well as water conservation. The weighting methods used (i.e. MCA) are also more appropriate to assign weights as opposed to an equal weights or expert opinion approach that are often criticized for being arbitrary or simplistic [22]. Others similar indices could be created as the survey is performed every two years. The results obtained with the 2007 indices could serve as the baseline for surveillance purposes, as the survey has been more comprehensive since that date.

Conclusion

MCA was successfully applied in creating Indoor and Outdoor composite Indices of environmental health relevance based on a readily available periodic Statistics Canada dataset. The Indices cover a good range of environmental themes at the household level and the analysis, particularly the indices weights obtained in the MCA, could be applied to similar surveys worldwide (as baseline weights) enabling temporal trend comparisons for recurring themes. Results uncovered provincial patterns of pro-environmental behaviours adoption with certain provinces scoring consistently higher and others consistently lower, as well as the associations between socio-demographic factors and the indices. Much more than voluntary measures, this study shows that existing regulations, dwelling type, household composition and income as well as climate are the major factors determining pro-environmental behaviours.

Supporting Information

Figure S1   Map representation of the MCA results on the 55 questions with extreme responses.

Table S1   Results of the MCA on the 55 questions without extreme responses (exploratory analysis). **Legend:** N/A: Results are not available for supplementary variables. Qlt: Quality (i.e. the sum of the squared correlations for the first two dimensions in this case). Inr: Inertias. K: Principal coordinates for the first dimension. Cor: Squared correlation with the first dimension. Ctr: Contributions of the modality to the explained inertia of the first dimension. All cells are multiplied by 1000. Results are the same on rows and on column when a Burt table is used.

Table S2   Results of the MCA for the Indoor Index, 2007. **Legend:** N/A: Results are not available for supplementary variables. Qlt: Quality (i.e. the sum of the squared correlations for the first two dimensions in this case). Inr: Inertias. K: Principal coordinates for the first dimension. Cor: Squared correlation with the first dimension. Ctr: Contributions of the modality to the explained inertia of the first dimension. All cells are multiplied by 1000. Results are the same on rows and on column when a Burt table is used.

Table S3   Results of the MCA for the Outdoor Index, 2007. **Legend:** N/A: Results are not available for supplementary variables. Qlt: Quality (i.e. the sum of the squared correlations for the first two dimensions in this case). Inr: Inertias. K: Principal coordinates for the first dimension. Cor: Squared correlation with the first dimension. Ctr: Contributions of the modality to the explained inertia of the first dimension. All cells are multiplied by 1000. Results are the same on rows and on column when a Burt table is used.

Table S4 Questions and responses selected for the Indoor Index, 2007.

Table S5 Questions and responses selected for the Outdoor Index, 2007.

References

1. Natural Resources Canada (2011) Energy Efficiency trends in Canada, 1990 to 2009Ottawa (On)54 p.
2. Environment Canada (2011) Ottawa (On)Municipal Water Use Report24 p.
3. Mustapha I, Tait M, Trant D (2012) Human Activity and the Environment. Waste management in Canada. Statistics CanadaOttawa (On)Report No: 16-201-x, 46 p.
4. Milito AC, Gagnon G (2008) Greenhouse gas emissions-a focus on Canadian households. EnviroStats 2(4):3–6.
5. Statistics Canada (2008) Human Activity and the Environment: Annual Statistics 2007 and 2008Ottawa (On)159 p.
6. Environment Canada (2013) Wise Water Use. Available: http://www.ec.gc.ca/eau-water/default.asp?lang = En&n = F25C70EC-1. Accessed 6 may 2014.
7. Statistics Canada (2009). Households and the environment, 2007.Ottawa (On)report no: 11-526-x,102 p.
8. Hardie D, Alasia A (2009) Domestic Water Use: The relevance of Rurality in Quantity Used and Perceived Quality. Rural and Small Town Canada Analysis Bulletin 7(5):1–31.
9. Mustapha I (2013) Composting by households in Canada. EnviroStats 7(11):1–6.
10. Lynch MF, Hofmann N (2007) Canadian lawns and gardens: Where are they the greenest? EnviroStats 1(2): 9–14.
11. Birrell C (2008) Energy-efficient holiday lights. EnviroStats 2(4):19–20.
12. Nelligan T (2008) Household's use of water and wastewater services. EnviroStats 2(4):17–8.
13. Babooram A (2008) Canadian participation in an environmentally active lifestyle. EnviroStats 2(4):7–12.
14. Nardo M, Saisana M, Saltelli A, Tarantola S, Hoffman A, et al. (2005) Handbook on Constructing Composite Indicators: Methodology and User GuideParis (Fr)Organisation for Economic Co-operation and Development Publishing162 p.
15. Esty DC, Levy M, Srebotnajk T, de Sherbinin A (2005) 2005 Environmental Sustainability Index: Benchmarking National Environmental Stewardship. New Haven: Yale Center for Environmental Law & Policy, 403 p.
16. Statistics Canada (2010) Microdata User Guide – Households and the environment survey, 2007Ottawa (On)53 p.
17. Stewart DW, Shamdasani PN, Rook DW (2007) Focus Groups: Theory and Practise. 2nd ed.Thousand OaksSAGE Publications200 p.
18. Greenacre M (2007) Correspondence analysis in practice. 2nd EditionNew YorkChapman & Hall/CRC284 p.
19. Dossa LH, Buerkert A, Schlecht E (2011) Cross-Location Analysis of the Impact of Household Socioeconomic Status on Participation in Urban and Peri-Urban Agriculture in West Africa. Hum Ecol Interdiscip J 39(5): 569–581.
20. Charreire H, Casey R, Salze P, Kesse-Guyot E, Simon C, et al. (2010) Leisure-time physical activity and sedentary behaviour clusters and their associations with overweight in middle-aged French adults. Int J Obes (Lond) 34(8):1293–1301.
21. Cortinovis I, Vella V, Ndiku J (1993) Construction of a socio-economic index to facilitate analysis of health data in developing countries. Soc Sci Med 36(8): 1087–1097.
22. Howe LD, Hargreaves JR, Huttly SR (2008) Issues in the construction of wealth for the measurement of socio-economic position in low-income countriesEmerg Themes Epidemiol 5(3): 14 p.
23. Asselin LM (2002) Composite indicator of Multidimensional Poverty - TheoryQuébec (Qc)Institut de Mathématique Gauss33 p.
24. Greenacre M, Nenadic O (2010) Package 'ca' - Simple, Multiple and Joint Correspondence AnalysisR project, 20 p.
25. Munro A (2010) Recycling by Canadian Households, 2007.Statistics Canada, Ottawa (On)34 p.
26. Oaks(2012) Province and Territory Water Efficiency and Conservation Policy Information. Available: http://www.allianceforwaterefficiency.org/2012-Province-Information.aspx. Accessed 13 November 2013.
27. Kinkead J, Boardley A, Kinkead M (2006) An analysis of Canadian and other water conservation practices and initiativesMississauga (On)Canadian Council of Ministers of the Environment274 p.
28. Gibbons WD (2008) Who uses water-saving fixture in the home? EnviroStats 2(3): 8–12.
29. Statistics Canada (2011) CANSIM Table 027-0017: Canada Mortgage and Housing Corporation, mortgage loan approvals, new residential construction and existing residential properties, monthly. Available: http://www5.statcan.gc.ca/cansim/a26?lang = eng&retrLang = eng&id = 0270017&paSer = &pattern = &stByVal = 1&p1 = 1&p2 = 37&tabMode = dataTable&csid = . Accessed 13 November 2013.
30. Bonsal B, Koshida G, O'Brien EG, Wheaton E (2013). Droughts. Available: http://www.ec.gc.ca/inre-nwri/default.asp?lang = En&n = 0CD66675-1&offset = 8&toc = hide . Accessed 13 july 2012.
31. Environment Canada (2010) Water and climate change. Available: http://www.ec.gc.ca/eau-water/default.asp?lang = En&n = 3E75BC40-1. Accessed 13 November 2013.
32. Mekis É, Vincent LA (2011) An overview of the second generation adjusted daily precipitation dataset for trend analysis in Canada. Atmosphere-Ocean 49(2): 163–77.
33. Ministère du Développement durable, de l'Environnement, de la Faune et des Parcs (2011) The pesticides Management Code - Highlights. Available: http://www.mddep.gouv.qc.ca/pesticides/permis-en/code-gestion-en/index.htm. Accessed 13 July 2012.
34. Canada mortgage and Housing Corporation (2013) Reducing energy cost. Available: https://www.cmhc-schl.gc.ca/en/inpr/afhoce/afhoce/afhostcast/afhoid/opma/reenco/index.cfm. Accessed 22 November 2013.
35. BChydro (2013) Buy, build, or rent an efficient home. Available: http://www.bchydro.com/powersmart/residential/guides_tips/green-your-home/whole_home_efficiency/energy_efficient_home.html. Accessed 22 November 2013.
36. Human Resources and Skills Development Canada (2007) What difference does learning make to financial security. Indicators of Well-Being – Special ReportGovernment of Canada14 p.
37. U.S. Environmental Protection Agency (1998) Washington (DC)Water Conservation Plan Guidelines208 p.
38. Reynaud A, Renzetti S, Villeneuve M (2005) Residential water demand with endogenous pricing: The Canadian CaseWater Resour Res 41(w11409)11 p.
39. OECD (2013) Greening Household Behaviour: Overview from the 2011 Survey, OECD Studies on Environmental Policy and Household Behaviours.OECD Publishing306 p.
40. Kollmuss A, Agyeman J (2002) Mind the Gap: Why Do People Act Environmentally and What Are the Barriers to Pro-Environmental Behaviour? Environmental Education Research Aug;8(3):239.
41. Todd A, Struger J (2014) Changes in acid herbicide concentrations in urban streams after a cosmetic pesticides ban. Challenges, 5:138–151.
42. Environment Canada (2013) Ammonia Emissions. Available: https://www.ec.gc.ca/indicateurs-indicators/default.asp?lang = en&n = FE578F55-1. Accessed 30 January 2014.
43. Mainieri T, Barnett EG, Valdero TR, Unipan JB, Oskamp S (1997) Green buying: The influence of environmental concern on consumer behavior. The Journal of Social Psychology Apr;137(2):189–204.
44. Melgar N, Mussio I, Rossi M (2013) Environmental Concern and Behavior: Do Personal Attributes Matter?Facultad de Ciencias Sociales, Universidad de la Republica; 21 p.
45. Xiao C, Hong D (2010) Gender differences in environmental behaviours in China. Population and Environment Sep;32(1):88–104.
46. Lopez A, Torres CC, Boyd B, Silvy NJ, Lopez RR (2007) Texas Latino College Student Attitudes Toward Natural Resources and the Environment. Journal of Wildlife Management Jun;71(4):1275–80.
47. Baer DE, Curtis JE (1984) French Canadian-English Canadian Differences in Values: National Survey Findings. Canadian Journal of Sociology/Cahiers canadiens de sociologie 9(4):405–27.
48. Baillargeon JP (1994) The Cultural Practices of Anglophones in Quebec. Recherches Sociographiques May;35(2):255–71.
49. Gibson KL, McKelvie SJ, Man AF (2008) Personality and Culture: A Comparison of Francophones and Anglophones in Québec. The Journal of Social Psychology Apr;148(2):133–65.

Acknowledgments

The authors thank Mr. Yves Lafortune of Statistics Canada for relevant comments on a preliminary version of this study and Ms Sandra Owens for her contribution to the redaction of this article. Also, thanks to Mr. Gaston Quirion of Laval University Library for facilitating access to the Statistics Canada survey database.

Author Contributions

Conceived and designed the experiments: MC BA DB PG. Analyzed the data: MC BA. Wrote the paper: MC BA DB PG.

50. Wu Z, Baer DE (1996) Attitudes toward family and gender roles: A comparison of English and French Canadian women. Journal of Comparative Family Studies 27(3):437–52.

51. Young N, Dugas E (2012) Comparing climate change coverage in Canadian English and French-language print media: environmental values, media cultures, and the narration of global warming. Canadian journal of sociology 37(1):25–54.

52. Corbeil JP (2012) Ottawa (On)French and the francophonie in Canada. Census in brief no. 1, Statistics Canada12 p.

53. Corbeil JP (2012) Linguistic Characteristics of Canadians. Language, 2011 Census of Population, Statistics CanadaOttawa (On)22 p.

54. Atiles JH, Bohon SA (2003) Camas Calientes: Housing Adjustments and Barriers to Social and Economic Adaptation among Georgia's Rural Latinos. Southern Rural Sociology 19(1):97–122.

55. Blumenberg E, Smart M (2010) Getting by with a little help from my friends and family: immigrants and carpooling. Transportation May;37(3):429–46.

56. Bohon SA, Stamps K, Atiles JH (2008) Transportation and Migrant Adjustment in Georgia. Population Research and Policy Review Jun;27(3):273–91.

57. Price CE, Feldmeyer B (2012) The Environmental Impact of Immigration: An Analysis of the Effects of Immigrant Concentration on Air Pollution Levels. Population Research and Policy Review Feb;31(1):119–40.

58. Statistics Canada (2011) Immigration in Canada: A portrait of the Foreign-born Population, 2006 Census: Data tables, figures and maps. Available: http://www12.statcan.ca/census-recensement/2006/as-sa/97-557/tables-tableaux-notes-eng.cfm. Accessed 30 January 2014.

7

Environmental Risk Factors and Amyotrophic Lateral Sclerosis (ALS): A Case-Control Study of ALS in Michigan

Yu Yu[1], Feng-Chiao Su[1], Brian C. Callaghan[2], Stephen A. Goutman[2], Stuart A. Batterman[1]*, Eva L. Feldman[2]*

1 Environmental Health Sciences, University of Michigan, Ann Arbor, Michigan, United States of America, 2 Department of Neurology, University of Michigan, Ann Arbor, Michigan, United States of America

Abstract

An interim report of a case-control study was conducted to explore the role of environmental factors in the development of amyotrophic lateral sclerosis (ALS). Sixty-six cases and 66 age- and gender-matched controls were recruited. Detailed information regarding residence history, occupational history, smoking, physical activity, and other factors was obtained using questionnaires. The association of ALS with potential risk factors, including smoking, physical activity and chemical exposure, was investigated using conditional logistic regression models. As compared to controls, a greater number of our randomly selected ALS patients reported exposure to fertilizers to treat private yards and gardens and occupational exposure to pesticides in the last 30 years than our randomly selected control cases. Smoking, occupational exposures to metals, dust/fibers/fumes/gas and radiation, and physical activity were not associated with ALS when comparing the randomly selected ALS patients to the control subjects. To further explore and confirm results, exposures over several time frames, including 0–10 and 10–30 years earlier, were considered, and analyses were stratified by age and gender. Pesticide and fertilizer exposure were both significantly associated with ALS in the randomly selected ALS patients. While study results need to be interpreted cautiously given the small sample size and the lack of direct exposure measures, these results suggest that environmental and particularly residential exposure factors warrant close attention in studies examining risk factors of ALS.

Editor: David R. Borchelt, University of Florida, United States of America

Funding: Funding support was provided by the Agency for Toxic Substances and Disease Registry (contract # 200-2013-56856), the National Institutes of Health (National Institute of Environmental Health Sciences (NIEHS) P30ES017885 and National Institute for Occupational Safety and Health (NIOSH) 5T42OH008455-08), the University of Michigan MCubed funding program, and the A. Alfred Taubman Medical Research Institute. The funders had no role in study design, data collection and analysis, decision to publish, or preparation of the manuscript.

Competing Interests: The authors have declared that no competing interests exist.

* Email: efeldman@umich.edu (ELF); stuartb@umich.edu (SAB)

Introduction

Amyotrophic lateral sclerosis (ALS) is a progressive neurodegenerative disorder involving primarily upper and lower motor neurons in the cerebral cortex, brainstem, and spinal cord [1]. ALS patients often experience difficulty talking, swallowing, breathing and walking, and often develop respiratory insufficiency which is the main cause of death. The prevalence of ALS is 2–9 per 100,000 persons overall and increases with age [2–6], and the incidence rate is 1–3 per 100,000 person-years [3,4,7]. Although approximately 5–10% of ALS cases can be attributed to genetic factors, the underlying cause of ALS remains largely unknown [8].

A number of epidemiologic studies have suggested that ALS patients have been exposed to environmental toxins [9–13]. Environmental risk factors investigated have included, among others, exposures to agriculture chemicals, heavy metals, solvents, electrical magnetic fields (EMF), and exercise [9,12,14–16]. Results have varied widely in the 40+ epidemiologic studies that have investigated such factors, and an understanding of causal links to environmental agents remains elusive.

The objective of this paper is to evaluate potential environmental risk factors for ALS using a case-control study conducted in the State of Michigan. Additional objectives include describing and evaluating the validity and efficiency of the survey instruments, and discussing key data and methodological issues. Elements of this study form portions of an ongoing study designed to account for interactions among covariates and to utilize biomarkers and other techniques to extend the exposure assessment.

Materials and Methods

Recruitment and Case Ascertainment

ALS subjects were recruited through the University of Michigan ALS Clinic with the following inclusion criteria: (1) age greater than 18 years, (2) possible, probable, probable lab-supported, or definite ALS by the revised El Escorial criteria [17], (3) able to provide informed consent, and (4) able to understand and communicate in English. Age- and gender-matched controls were recruited through postings and a University of Michigan website (umclinicaltrials.org) that described the study and eligibility criteria, which included: (1) no diagnosis of ALS or other neurodegenerative condition, (2) no family history of ALS in a first or second degree blood relative, (3) ability to understand and communicate in verbal and written English, and (4) no underlying health condition or medication use that made participation risky, as determined by the study team. Age was matched by decade and gender (e.g., males ages 50–59, females ages 50–59). The study was approved by the Institutional Review Board (Protocol # HUM00028826) at the University of Michigan. All participants provided written informed consent.

Participant Survey

Participants completed a detailed self-administered written questionnaire that encompassed occupational and residential exposures, residence location, exercise and sports, body weight, tobacco use, military experience, and family history. Questionnaires were mailed to subjects and telephone follow-up was conducted for clarifications, if needed. For patients who had difficulty communicating, the next of kin completed the instrument. Typically, about 90 minutes was needed to complete the fairly lengthy (200 questions, 28 page) questionnaire, which was divided into several sections and used a structured query approach, allowing participants to skip irrelevant sections. Portions of this questionnaire were adapted from those used in the National Health and Nutrition Examination Survey Questionnaire [18].

After requesting standard demographic information, residential history was collected using 54 questions for the current dwelling and 18 questions for each of the three previous dwellings. Questions addressed dates of building construction and occupancy, building type (including detached single family, duplex, multi-family/apartment, mobile home or trailer, other, or unknown) and features (including building materials, floor coverings, presence of basement, presence of outdoor storage, and presence of garage), weatherization, storage of chemicals (e.g., pesticides, solvents, gasoline, and paint among others), and drinking water source (e.g., well or city water). A detailed smoking history was collected, including whether subjects ever smoked and, if applicable, the year smoking started, stopped, the type of tobacco used, and the frequency of smoking. Participants were asked about hobbies, particularly those that might include chemical exposure, such as wood working, metal working, home remodeling, lawn care, automobile repair, small engine repair, and painting. Participants were asked about their physical activity, including the frequency of ten categories of activities (jogging/running, bicycling, swimming, aerobic dancing, recreational dancing, calisthenics, gardening or yard work, weightlifting, playing soccer/football/baseball/field hockey/golf, playing ice hockey/tennis/boxing/wresting) using a five year recall period. In addition, we asked about other activities not mentioned above. Summary or composite measures of physical activity were developed and classified as low, medium, and high based on the number of activities by an individual (defined as 0–3, 4–6, and 7+ activities, respectively).

A detailed occupational history was also requested, and the questionnaire included 24 questions for each of four previous jobs; specifically, their current or most recent job, the previous job, and the two other jobs that were held for the longest period of time. Requested information included job title, industry type, and dates of employment. In addition, subjects were asked to identify occupational exposures, using lists of potential workplace hazards and exposures (e.g., specific chemicals, particles, radiation), the use and presence of personal protective equipment, and hygiene habits (e.g., hand washing).

Data analysis

Following data entry and consolidation, initial analyses included data cleaning and verification. Dubious or missing entries were checked by either reviewing the original questionnaire, telephone follow up, or as a last resort and if practical, imputed manually based on related questions. Next, using the residence and occupational histories, residence and employment durations were calculated and compared as a test of internal validity. If the residence or job information was inconsistent, such as residing in a different state compared to the workplace, then respondents were queried by telephone and the information revised as indicated. If this attempt failed, then the questioned information was set as missing. Job titles were coded using the Dictionary of Occupational Titles, and workplace types were coded using the North American Industry Classification System. Several new composite variables for specific risk factors were also created; for example, a binary (yes/no) variable called "occupational exposure to heavy metals" combined any positive response to separate questions concerning workplace exposure to arsenic, beryllium, cadmium, chromate, lead, mercury, nickel, and welding fumes. Such variables were created for each job, and also across all jobs. Similarly, composite variables were created for risk factors pertaining to residences and hobbies.

The four exposure time frames that were considered were: (1) no exposure (in the past 30 years), (2) exposure in the last 10 years only (not earlier), (3) exposure 10 to 30 years ago and not in the past 10 years, and (4) continuous exposure over the past 30 years. These time frames were referenced to survey completion in the control population and ALS symptom onset for the case population.

Questions not answered by subjects were interpreted as missing. To obtain a full dataset needed in the conditional logistic regression (CLR) models (described below), missing values were imputed with five replacements. The consistency of imputed data was confirmed with manual checks (for example, the imputed number of years of smoking should be smaller than the subject's age). Subsequent sensitivity analyses confirmed that results obtained using original and imputed data sets were similar.

The survey data were used to generate approximately 100 variables as potential risk factors for each of the four exposure time windows, and a subset of variables for further analysis was selected for use in the CLR models. For continuous variables, univariate statistics (range, mean, medium, quartiles) were calculated and differences between cases and controls were tested using Student t-tests (normally distributed variables) and Kruskal-Wallis tests (non-normal distributions). For categorical variables, cross-tabulations and chi-square tests were used to detect differences between cases and controls. Fisher's exact test was used if the expected count was less than five. Variables showing variability and significant differences were retained for further analyses, as were several variables identified in prior research, such as smoking, physical activity, exposures to metals, and exposures to radiation. Related exposure variables that were moderately to highly correlated ($r > 0.5$) were consolidated as a new variable (e.g., occupational exposures to radiation, x-rays, and electromagnetic fields were grouped into a radiation variable). Stepwise regression was then used to generate final models. A sensitivity analysis for variable entry (with p-values from 0.1 to 0.3) and removal (p-values from 0.15 to 0.35 for variables) showed that in most cases the same variables were selected. Then, odds ratios (OR) and p-values (significance level: p-value <0.1) for potential risk factors were estimated using CLR models for case/control pairs matched on age and gender. Each model included covariates to control for demographics, smoking (cigarette packs per day), physical activity status (low, medium, and high physical intensity groups), and educational attainment. Four models with different exposure conditions (model 1: exposure in the last 30 years, model 2: exposure in the last 10 years, model 3: exposure in the period from 30 years ago to 10 years ago, model 4: continuous exposure in the last 30 years) were constructed to test potential risk factors such that

$$\text{logit}(p) = \alpha_0 + \alpha_1 x_{s1} + \ldots + \alpha_m x_{sm} + \beta_1 x_1 + \ldots + \beta_k x_k \quad (1)$$

where $\alpha =$ effect of the stratum (the matching variables age and gender), m = number of matched sets (strata), $\beta =$ log-odds ratio of interested risk factors, x = risk factor, and k = number of variables. Several interaction effects were tested (e.g., two-way interactions between education level and exposure risk factors). Because interaction effects were not statistically significant, they were not retained in the final models. Several sensitivity analyses were used to investigate the robustness of the model results to the exposure window.

Data was entered and managed using RedCap [19]. Some analyses used Excel (Microsoft, Redmond, WA, USA). Multiple imputation IVEware 2.0 (Survey Research Center, Institute for Social Research, University of Michigan, Ann Arbor, MI, USA) was used to replace missing data. Statistical analyses were performed using SAS 9.3 (SAS Institute Inc., Cary, NC, USA). The following data represent an interim analysis of an ongoing larger study.

Results

Participant Characteristics

The study population was 31% female and averaged (± SD) 61.6±9.0 years of age. Cases and controls had the same age and gender distribution (Table 1), reflecting successful matching. Controls had a higher level of education attainment than cases (p<0.001), with 67% of cases and 95% of controls receiving education beyond high school. Controls were also less likely to be married (p = 0.008, Table 2). When stratified by gender, differences in education were maintained for both males and females, but differences in marital status were found only among males (Table S1). Cases and controls had similar smoking patterns: 55–56% were current or past smokers, 15–18% smoked less than 1 pack per day, and 38–39% smoked greater than1 pack per day (Table 2). Cases had a slightly, but not statistically, longer duration of smoking than controls (mean (SD) years of smoking: 15.7±18.4 for cases and 12.0±14.5 for controls, p = 0.20). Similarly, cases had higher cigarette pack-years (mean = 20.5±33.6) than controls (mean = 13.9±20.4), but again, the difference was not statistically significant (p = 0.18). The average cigarette packs per day for cases (0.64±0.88) and controls (0.61±0.69) was similar (p = 0.85). When stratified by gender, smoking status, cigarette packs per day, and the number of years of smoking did not differ between cases and controls; however, female cases had more cigarette pack-years than female controls, due to both slightly higher numbers of heavy smokers (>1 pack/day) and a longer duration of smoking (Table S1). No significant differences were seen for other measures of smoking, including 25th and 75th percentile values of ever-smokers and cigarette packs per day, years of smoking, and cigarette pack-years.

Involvement in individual sports and other physical activities (Table 2) did not differ between cases and controls, with the exception of recreational dancing (p = 0.029). Similarly, no differences were seen for the number of activities engaged in by participants. Among males, cases were more likely than controls to

engage in gardening (p = 0.023) and also in a high intensity of activities when other sports were included (p = 0.054). Among females, the only difference seen was that cases were less likely to dance recreationally (p = 0.038; Table S1). Swimming showed marginally significant differences, but in different directions, when stratified by gender.

ALS Risk Factors

Results of the final CLR models for the entire sample (matched by age and gender) are given in Table 3. Multiple regression model results, stratified by age (adjusted by gender) and gender (adjusted by age), are in Table S2 and Table S3, respectively. Each table includes models for four exposure windows, and each model includes education, smoking, activity, and environmental variables.

Higher educational attainment was associated with a lower odds of ALS. ALS was not associated with smoking or physical activity. This applied to both categorical (e.g., smoking status) and continuous (e.g., cigarette packs per day) indicators, and to analyses stratified by age or gender. One residential factor associated with ALS was exposure to fertilizer by treating private yards/gardens (for all participants: ORs (95% CI) = 2.97 (0.81–10.9) and 2.97 (1.01–8.76) for exposure windows 1 (n = 40/29, cases/controls) and 3 (n = 31/15, cases/controls), respectively; Table 3). While this relationship (treating private yards/gardens) was marginally significant in cases, significance was maintained for three of the exposure time conditions, in analyses for younger individuals (<60 years of age), and in males. The second residential factor associated with ALS was living near industry/sewage treatment plants/farms for individuals below 60 years of age (n = 18/12, cases/controls; OR (95% CI) = 4.56 (0.75–27.7); Table S2), and for females (n = 23/16, cases/controls; OR (95% CI) = 5.18 (0.98–27.3); Table S3) for exposure window 3. The sample size did not permit further stratification or examination of interactions, such as analysis of both younger and female subjects.

Occupational exposure to pesticides was associated with ALS (n = 38/19, cases/controls; OR (95% CI) = 6.95 (1.23–39.1)) for exposures occurring over the last 30 years but not in other time frames (Table 3). Pesticide exposure was also associated with ALS in age- and gender-stratified analyses, although significance decreased (p = 0.037–0.090; Table S2 and Table S3). Occupational exposure to metals was uncommon among study participants. The most frequent metal exposures were welding fumes (n = 11/8, cases/controls), lead (n = 9/6), nickel (n = 3/5), mercury (n = 4/2), and chromates (n = 1/3). Given these small numbers, we considered exposure to any heavy metal. No association was observed with this composite measure in the present study. Likewise, no association between exposure to dust/fibers/fumes or gas was found in the final models for all participants. An association between dust/fibers/fumes or gas exposure and ALS was found among males for continuous exposures over the past 30 years (OR (95% CI) = 15.6 (1.38–177); Table S3). Women had an increased odds of ALS with exposure to occupational radiation occurring 10 to 30 years earlier (OR (95% CI) = 67.7 (2.50–999); Table S3). This was not found in men or for other exposure windows.

Discussion

ALS Risk Factors

Education. Our finding that educational attainment was associated with a lower odds of ALS is consistent with case-control studies in Boston [14] and Pennsylvania [20], and may represent an interaction with exposure since subjects with more education

Table 1. Demographic characteristics of cases and controls.

Demographics		Cases (n = 66)		Controls(n = 66)		p-value
Variable	Group	Frequency	%	Frequency	%	
Age of consent	40–49	8	12.1	8	12.1	1.000
	50–59	17	25.8	17	25.8	
	60–69	27	40.9	27	40.9	
	70–79	14	21.2	14	21.2	
	80–89	0	0.0	0	0.0	
Gender	Female	31	47.0	31	47.0	1.000
	Male	35	53.0	35	53.0	
Education	≤ High school	22	33.3	3	4.6	<0.001
	> High school	44	66.7	63	95.5	
Marital status	Married*	45	68.2	30	45.5	0.044
	Widowed	7	10.6	5	7.6	
	Divorced	8	12.1	16	24.2	
	Separated	1	1.5	1	1.5	
	Never married	3	4.6	12	18.2	
	Living with partner	2	3.0	2	3.0	

*, significant difference between married and non-married (p = 0.008).

are more likely to work and live in environments with lower exposure. Interaction terms between these variables (education and exposure risk factors) did not attain statistical significance in model 1 (exposure in the last 30 years), possibly due to the limited sample size for this analysis.

Smoking. Although cigarette smoking and tobacco smoke exposure may increase the odds of ALS via inflammation, oxidative stress, and neurotoxicity induced by heavy metals or other chemicals present in cigarette smoke [21–24], we did not find an association between ALS and smoking. A large prospective study in the United States (414,493 male and 572,736 female participants; 617 ALS deaths among men and 539 among women) reported an increased risk of ALS with formaldehyde exposure, a component of cigarette smoke (women: RR = 1.67, 95% CI = 1.24–2.24, p = 0.002; men: RR = 0.69, 95% CI = 0.49–0.99, p = 0.04) [25]. A population-based case-control study in western Washington State (161 cases, 321 controls) showed an increased odds of ALS for ever-smokers (alcohol-adjusted OR = 2.0, 95% CI = 1.3–3.2), current smokers (OR = 1.5, 95% CI = 0.9–2.4), and for smoking duration and cigarette pack-years [26]. Conversely, negative results were obtained in a prospective study in the United Kingdom, although ever-smoking females had an increased risk of ALS (RR = 1.53, 95% CI = 1.04–2.23) [27]. Another prospective study in the United States (459,360 male and 638,849 female participants; 330 ALS deaths among men and 291 among women) associated cigarette smoking with increased ALS mortality in women, but not men (RR = 1.67, 95% CI = 1.24–2.24) [28]. Such inconsistencies likely arise from limitations related to sample design (most were case-control studies, not prospective cohort studies), small sample sizes, the use of hospital controls with a high frequency of smoking, interactions with other variables such as educational attainment, the use of spouses and/or friends as controls that may overmatch with respect to tobacco use, and other unaccounted for factors [29]. Nevertheless, in an updated literature review of 7 articles meeting the author's inclusion criteria, it was concluded that "smoking may be considered an established risk factor of sporadic ALS" [30].

Physical Activity. The reported disproportionate increase in ALS incidence among professional athletes, which has received considerable attention, has prompted many investigations into the potential role of physical activity in the development of ALS, which may result through increased risk of exposure to toxins, increased transport of the toxins, and increased susceptibility of target cells to injury from the toxins [31]. The present study did not find an association between ALS physical activity. Similarly, studies that have separated work and leisure-related physical activity have also been inconsistent [14,32]. Accurate and quantitative assessment of past physical activity is challenging, as the number or type of activities reported in a survey may not reflect energy expenditures. Longstreth et al used a more sophisticated metabolic measure, metabolics equivalents for each activity weighted by the annual number of hours spent at each activity, but still found no association with ALS [32]. The present analysis did not include physical activity information prior to the five year recall period or direct measures of work-related physical activity. Several case reports have associated physical trauma and ALS development [33]. An older review identified heavy labor as a risk factor [12], and a case-control study among New England construction workers (109 cases, 253 controls) found elevated odds (OR = 2.9, 1.2–7.2) [15]. However, negative findings were reported in western Washington State (174 cases, 348 controls) [32] and in a prospective case-control study (95 cases, 106 controls) examining physical activity and trauma [34]. Like the environmental factors discussed earlier, evidence pertaining to the role of physical activity in ALS causation remains inconclusive.

Residential Factors and Chemicals. ALS has been associated with exposure to a number of chemicals, with most of the supporting evidence implicating agricultural chemicals such as pesticides, fertilizers, herbicides, and insecticides. Our study showed an association between an exposure to fertilizer and ALS. Information regarding residential exposure to fertilizers was obtained for only the current dwelling, which likely explains the stronger association for the last 10 years compared to the other time frames. It is also reasonable that younger males are more

Table 2. Smoking and physical activity status by case type.

Variable	Group	Cases (n = 66)		Controls (n = 66)		p-value
		Number	%	Number	%	
Smoker	Never smoker	30	45.5	29	43.9	0.861
	Ever smoker	36	54.6	37	56.1	
Smoking status	Never smoker	30	45.5	29	43.9	0.929
	Former smoker	27	40.9	29	43.9	
	Current smoker	9	13.6	8	12.1	
Cigarette packs/day	Never	30	45.5	29	43.9	0.897
	<1 pack	10	15.2	12	18.2	
	≥1 pack	26	39.4	25	37.9	
Number of years of smoking	Never	30	45.5	29	43.9	0.395
	<20 years	11	16.7	17	25.8	
	≥20 years	25	37.9	20	30.3	
Cigarette pack-years	Never	30	45.5	29	43.9	0.416
	<20 pack-years	13	19.7	19	28.8	
	≥20 pack-years	23	34.8	18	27.3	
Physical activities	Jogging, running	18	27.3	15	22.7	0.547
	Bicycling	33	50.0	33	50.0	1.000
	Swimming	17	25.8	17	25.8	1.000
	Aerobic dancing	10	15.2	9	13.6	1.000
	Recreational dancing	17	25.8	8	12.1	0.029
	Calisthenics	24	36.4	27	40.9	0.592
	Gardening, yard work	54	81.8	46	69.7	0.104
	Weightlifting	17	25.8	18	27.3	0.844
	Soccer, football, baseball, field hockey, golf	24	36.4	17	25.8	0.188
	Ice hockey, tennis, boxing, wresting	9	13.6	5	7.6	0.258
	Other*	15	22.7	16	24.2	1.000
Physical activity intensity (excluding others)	Never	5	7.6	5	7.6	0.245
	Low (0–3 activities)	36	54.5	36	54.6	
	Medium (4–6 activities)	19	28.8	24	36.4	
	Highs (7+ activities)	6	9.1	1	1.5	
Physical activity intensity (including others)	Never	5	7.6	4	6.1	0.090
	Low (0–3 activities)	32	48.5	32	48.5	
	Medium (4–6 activities)	18	27.3	27	40.9	
	Highs (7+ activities)	11	16.7	3	4.6	

*, other activities include ski, fishing, hunting, bowling, yoga, etc.

likely to perform yard and gardening work. Similar findings were reported in Australia for 179 case-control pairs [35]. Regular gardening (non-occupational exposure) was significantly associated with ALS (OR = 6.64, 95% CI = 1.61–27.4) for all subjects. After stratified by gender, the significant association only was shown in males (OR = 4.90, 95% CI = 1.11–21.7).

An association between a residence near industry and sewage treatment plants or farms for individuals less than 60 years of age was also demonstrated. Living near such facilities may involve exposure to a variety of air, water and soil pollutants. Possibly women in the cohort were less likely to be occupationally exposed to such pollutants and instead living near such facilities might provide exposures otherwise not encountered. Also, women may spend more time at home and thus experience greater exposure from nearby emission sources.

Pesticides initially aroused interest due to the increased risk of ALS observed among United States veterans exposed to pesticides [36,37]. In addition, pesticides have been implicated in other neurodegenerative diseases like Parkinson's and Alzheimer's disease [38,39]. Our study shows an association between pesticide exposure and ALS. Pesticide exposure has been associated with ALS in a multiple studies [11,40]. The same association was seen in a Boston study using self-reported exposures [14], and an Australian study also using self-reported exposures of both industrial and residential use of herbicides and pesticides [35]. Recently, a study with 66 pairs of age-, race-, and gender-matched cases and controls found a very similar result to the present study.

Table 3. Results of multiple conditional regression models at four exposure windows for all participants.

Risk Factors	1. Exposure in the last 30 years OR (95% CI)	2. Exposure in the last 10 years OR (95% CI)	3. Exposure in the period from 30 years ago to 10 years ago OR (95% CI)	4. Continuous Exposure in the last 30 years OR (95% CI)
Education ≥ high school	0.05 (0.01–0.36)**	0.07 (0.01–0.44)**	0.10 (0.02–0.55)**	0.08 (0.01–0.46)**
Cigarette pack per day#	0.74 (0.34–1.64)	0.69 (0.32–1.48)	0.82 (0.42–1.59)	0.63 (0.32–1.27)
Low activity intensity	1.46 (0.22–9.61)	1.40 (0.24–8.25)	1.25 (0.22–7.22)	1.13 (0.21–6.05)
Medium activity intensity	0.46 (0.06–3.61)	0.49 (0.07–3.52)	0.46 (0.07–3.33)	0.39 (0.06–2.71)
High activity intensity	5.98 (0.38–93.3)	6.26 (0.44–89.6)	5.03 (0.38–67.4)	5.22 (0.38–71.9)
Using fertilizer to treat gardens	2.97 (0.81–10.9)*	2.44 (0.73–8.17)	2.97 (1.01–8.76)**	2.43 (0.72–8.23)
Living near industry/sewage treatment plant/farm	1.16 (0.41–3.30)	1.15 (0.40–3.28)	1.87 (0.69–5.11)	2.25 (0.69–7.27)
Occupational exposure to metal	4.76 (0.39–58.8)	2.04 (0.27–15.3)	0.69 (0.10–4.84)	0.78 (0.12–5.08)
Occupational exposure to pesticide	6.95 (1.23–39.1)**	2.64 (0.47–14.8)	2.66 (0.47–14.9)	0.88 (0.12–6.69)
Occupational exposure to dust/fibers/ fumes or gas	0.47 (0.06–3.77)	1.54 (0.29–8.12)	1.94 (0.37–10.2)	4.44 (0.69–28.4)
Occupational exposure to radiation	1.25 (0.35–4.47)	1.73 (0.37–8.14)	1.73 (0.46–6.50)	1.96 (0.37–10.3)

*, p<0.1;
**, p<0.05; OR, odds ratio.
#Cigarette packs per day is a continuous variable.

A significant association between occupational exposure to pesticides and ALS (OR = 6.50, 95% CI = 1.78–23.77) was shown after adjusting education, smoking, and other occupational exposures, including metals, solvents, and electromagnetic fields [20]. Pesticide exposure itself is associated with farming, living on or near farms, and the use of well water, and higher incidence rates of ALS have been found in United States rural farm areas west of the Mississippi River [41] and among farm workers and shepherds in Greece and Italy [42,43]. A systematic review study concluded that the pesticide exposure and ALS association was significant [13].

Metals. The role of heavy metals, especially lead, in neurodegenerative diseases has received considerable attention. We did not see an association in this study. Lead may play several roles in the onset and progression of ALS [44], and like pesticides, lead exposure has been widespread. Lead exposure has been linked to ALS in many case-control studies. In New England (109 cases and 256 controls), elevated blood and bone lead levels were associated with increased odds of ALS (OR = 1.9, 95% CI = 1.4–2.6) for each μg dl^{-1} increase in lead, with OR = 3.6 (95% CI = 0.6–20.6) for each unit increase in log-transformed patella Pb and OR = 2.3 (95% CI = 0.4–14.5) for each unit increase in log-transformed tibia lead [9]. In United States veterans (184 cases and 194 controls), elevated blood Pb levels were associated with ALS (OR = 1.9, 95% CI = 1.3–2.7) [16]. As a third example, in Boston (95 cases and 106 controls), self-reported lead exposure was associated with ALS (p = 0.02) [14]. Overall, however, results have not been consistent and many studies show negative outcomes [45–47]. In addition, other metals, especially mercury and cadmium, have been studied, but again results have been inconsistent [48–50]. Malek et al [20] showed that a composite measure of heavy metal exposure (lead and mercury) was significantly associated with increased risk of ALS (OR = 3.65). While lead exposure has been associated with ALS [14,16], associations and causal mechanisms have not been shown for mercury, cadmium or other metals. In a small Japanese study (21 cases, 36 controls), mercury and selenium levels in plasma and

blood cells were significantly lower for ALS patients in the late stage than controls due to their disability, including consuming a liquid diet [51]. A very small study (9 cases) in Italy showed significantly higher cadmium blood levels in cases than in controls (after excluding advanced cases with the worst functional impairment) [50]. Grouping exposure to different metals together may result in exposure misclassification and diminished ability to detect associations, a limitation of the present as well as earlier studies [13].

Dust/Fibers/Fumes. Several studies have indirectly implicated occupational exposure to particulate matter with ALS [11,40,52]. We did not see an association in our study. Airborne dust, fumes, and fibers found in some occupational settings may represent an important exposure to airborne particulate matter. Exposure to particulate matter in ambient air, much of which arises from combustion sources, is widespread but concentrations are typically far below those in occupational settings. Particulate matter exposures have been examined with respect to neurological outcomes in a number of studies [53–55], and associated with ALS in several occupational settings [11,40,52]. The occupational settings investigated (veterinarians, hairdressers, graders and sorters (non-agricultural)), however, were likely to have elevated co-exposures of solvents, metals, and possibly other agents. No study has directly evaluated associations between environmental exposure of particulate matter and ALS.

Radiation/Electromagnetic Fields. Radiation has been considered as a potential risk factor for ALS since a myeloradiculopathy presentation can be caused by electrical injuries with a long latency period [56]. We saw an association with ALS in women who had an exposure to occupational radiation 10 to 30 years earlier; however, the very large OR, the small number of exposure cases (n = 10/2 for cases/controls), and gender effect specificity suggest that results may not be meaningful. Three studies previously reported associations between radiation or electromagnetic field exposure [57–59]; thus, further investigation of such exposures is warranted. Radiation has been investigated in many case-control studies, again with inconsistent results. Electrical-related occupations (OR = 1.3, 95% CI = 1.1–1.6) [59], and exposure to electromagnetic fields (OR = 2.3, 95% CI = 1.29–4.09) [57] were associated with ALS. In a cohort mortality study at five large United States electric utility companies (139,905 men), mortality from ALS was associated with the duration of work in jobs with electromagnetic field exposure (RR = 2.0, 95% CI = 1.0–9.8) [60]. A prospective case-control study in Denmark of utility workers (n = 30,631, 81% male), however, did not find significant linkages between electromagnetic field exposure and neurological diseases [61].

Study Limitations and Recommendations

Case-control studies assessing environmental risk factors in ALS are pursued due to low cost, efficiency, disease latency, and tendency to affect older individuals [16,57]. While a few prospective cohort studies have been conducted [25,34,62], such designs are challenging given the often long period between exposure and ALS diagnosis [63]. Limitations of any case-control study, including our own, include small sample size, selection bias, environment exposure misclassification, recall bias, and confounding.

Typical sizes for ALS case-control studies are about 100 to 400 participants [13]. Larger studies benefit from greater power and enhanced ability to evaluate small effects and interactions. Thus, another limitation of this study is the relatively small sample size; however, our current goal is to recruit 600 participants which should address this concern.

While cases and controls were matched on both age and gender, several differences are worth mention. Recruiting patients via University of Michigan resources may attract more educated control compared to case volunteers. As educational attainment may be related to working and living in an environment with fewer exposures and risk factors, our data may have selection bias [64,65].

Exposure misclassification is also a concern in epidemiology studies. Data was collected using self-reported questionnaires that queried an exhaustive range of chemicals and potential exposures, designed to aid in subject recall. Nonetheless, individuals may be unaware of exposures and potential risks. In addition, two important challenges in estimating exposure are the long latency and recall period and the large number of toxins of interest. This suggests that a blended approach that combines questionnaires, exposure modeling (e.g., using residence information to evaluate past air pollution exposure), biological measurements (e.g., bone or blood lead measurements), and possibly environmental monitoring data (e.g., water quality measurements) could provide greater specificity of exposures and accuracy.

Recall bias is also a limitation when using questionnaires, especially when seeking long term or historical information regarding exposure as these and other factors of possible significance may have occurred many years before ALS diagnosis and study enrollment, and thus participants may have difficulty remembering potential data. We used self-reported exposure, several time periods, and considered exposures up to 30 years earlier. Future studies are needed to confirm exposures using biomarkers.

Risk factors such as educational attainment and smoking status may act as confounders or effect modifiers that alter results of statistical models. As previously noted, controls were more likely to have a higher educational attainment, which may affect exposure. Including education variables in the regression models should help adjust for potential differences, but does not eliminate the potential for confounding; however, we obtained comparable results when analyses were stratified by educational attainment, suggesting this was not an issue (Table S1).

Conclusions

This interim analysis of a larger ongoing case-control study shows an association of fertilizers and pesticides with an increased odds of developing ALS in a randomly selected ALS subjects compared to controls. These results were largely consistent over multiple time frames, as well as in analyses stratified by age and gender. Smoking, occupational exposures to metals, dust/fibers/fumes/gas and radiation, and physical activity were not associated with ALS. While consistent with earlier literature, these associations should be interpreted cautiously given the relatively modest sample size and other study limitations. The study is innovative in its use of different exposure periods and the wide range of exposures and covariates considered. Future studies could build on our methods by increasing sample size, using face-to-face interviews and trained interviewers (possibly with pictorial methods to increase awareness of exposures), obtaining exposure information using exposure models and biomonitoring, quantifying physical activity, and including information regarding income and alcohol consumption.

Acknowledgments

The authors would like to thank Jayna Duell, RN for assistance with subject recruitment and Dr. Stacey A. Sakowski for critical review of the manuscript.

Author Contributions

Conceived and designed the experiments: YY SAB BCC ELF. Performed the experiments: YY FCS. Analyzed the data: YY FCS SAB BCC SAG ELF. Contributed reagents/materials/analysis tools: SAB BCC ELF. Contributed to the writing of the manuscript: YY FCS SAB BCC SAG ELF.

References

1. Brooks BR (1994) El Escorial World Federation of Neurology criteria for the diagnosis of amyotrophic lateral sclerosis. Subcommittee on Motor Neuron Diseases/Amyotrophic Lateral Sclerosis of the World Federation of Neurology Research Group on Neuromuscular Diseases and the El Escorial "Clinical limits of amyotrophic lateral sclerosis" workshop contributors. J Neurol Sci 124 Suppl: 96–107.

2. Bobowick AR, Brody JA (1973) Epidemiology of motor-neuron diseases. N Engl J Med 288: 1047–1055.

3. Annegers JF, Appel S, Lee JR, Perkins P (1991) Incidence and prevalence of amyotrophic lateral sclerosis in Harris County, Texas, 1985–1988. Arch Neurol 48: 589–593.

4. Traynor BJ, Codd MB, Corr B, Forde C, Frost E, et al. (1999) Incidence and prevalence of ALS in Ireland, 1995–1997: a population-based study. Neurology 52: 504–509.

5. Ahmed A, Wicklund MP (2011) Amyotrophic lateral sclerosis: what role does environment play? Neurologic Clinics 29: 689–711.

6. Chancellor A, Warlow C (1992) Adult onset motor neuron disease: worldwide mortality, incidence and distribution since 1950. Journal of Neurology, Neurosurgery & Psychiatry 55: 1106–1115.

7. McGuire V, Longstreth WT Jr, Koepsell TD, van Belle G (1996) Incidence of amyotrophic lateral sclerosis in three counties in western Washington state. Neurology 47: 571–573.

8. NINDS (2012) Amyotrophic Lateral Sclerosis (ALS) Fact Sheet. National Institute of Neurological Disorders and Stroke, National Institutes of Health.

9. Kamel F, Umbach DM, Munsat TL, Shefner JM, Hu H, et al. (2002) Lead exposure and amyotrophic lateral sclerosis. Epidemiology 13: 311–319.

10. Armon C, Kurland LT, Daube JR, O'Brien PC (1991) Epidemiologic correlates of sporadic amyotrophic lateral sclerosis. Neurology 41: 1077–1084.

11. McGuire V, Longstreth WT, Nelson LM, Koepsell TD, Checkoway H, et al. (1997) Occupational exposures and amyotrophic lateral sclerosis. A population-based case-control study. American Journal of Epidemiology 145: 1076–1088.

12. Nelson L (1995) Epidemiology of ALS. Clinical Neuroscience 3: 327–331.

13. Sutedja N, Veldink JH, Fischer K, Kromhout H, Heederik D, et al. (2009) Exposure to chemicals and metals and risk of amyotrophic lateral sclerosis: A systematic review. Amyotrophic Lateral Sclerosis 10: 302–309.

14. Qureshi MM, Hayden D, Urbinelli L, Ferrante K, Newhall K, et al. (2006) Analysis of factors that modify susceptibility and rate of progression in amyotrophic lateral sclerosis (ALS). Amyotrophic Lateral Sclerosis 7: 173–182.

15. Fang F, Quinlan P, Ye W, Barber MK, Umbach DM, et al. (2009) Workplace exposures and the risk of amyotrophic lateral sclerosis. Environ Health Perspect 117: 1387–1392.

16. Fang F, Kwee LC, Allen KD, Umbach DM, Ye W, et al. (2010) Association between blood lead and the risk of amyotrophic lateral sclerosis. Am J Epidemiol 171: 1126–1133.

17. Brooks BR, Miller RG, Swash M, Munsat TL (2000) El Escorial revisited: Revised criteria for the diagnosis of amyotrophic lateral sclerosis. Amyotrophic Lateral Sclerosis 1: 293–299.

18. CDC NCHS (2013) National Health and Nutrition Examination Survey Questionnaire. Hyattsville, MD: US Department of Health and Human Services, Centers for Disease Control and Prevention.

19. Harris PA, Taylor R, Thielke R, Payne J, Gonzalez N, et al. (2009) Research electronic data capture (REDCap) - A metadata-driven methodology and workflow process for providing translational research informatics support. Journal of Biomedical Informatics 42: 377–381.

20. Malek AM, Barchowsky A, Bowser R, Heiman-Patterson T, Lacomis D, et al. (2013) Environmental and Occupational Risk Factors for Amyotrophic Lateral Sclerosis: A Case-Control Study. Neurodegenerative Diseases.

21. Chiba M, Masironi R (1992) Toxic and trace elements in tobacco and tobacco smoke. Bull World Health Organ 70: 269–275.

22. Morrow JD, Frei B, Longmire AW, Gaziano JM, Lynch SM, et al. (1995) Increase in circulating products of lipid peroxidation (F2-isoprostanes) in smokers. Smoking as a cause of oxidative damage. N Engl J Med 332: 1198–1203.

23. Yanbaeva DG, Dentener MA, Creutzberg EC, Wesseling G, Wouters EF (2007) Systemic effects of smoking. Chest 131: 1557–1566.

24. Weisskopf MG, Ascherio A (2009) Cigarettes and amyotrophic lateral sclerosis: only smoke or also fire? Ann Neurol 65: 361–362.

25. Weisskopf MG, Morozova N, O'Reilly EJ, McCullough ML, Calle EE, et al. (2009) Prospective study of chemical exposures and amyotrophic lateral sclerosis. Journal of Neurology, Neurosurgery & Psychiatry 80: 558–561.

26. Nelson LM, McGuire V, Longstreth WT Jr, Matkin C (2000) Population-based case-control study of amyotrophic lateral sclerosis in western Washington State. I. Cigarette smoking and alcohol consumption. Am J Epidemiol 151: 156–163.

27. Alonso A, Logroscino G, Jick SS, Hernan MA (2010) Association of smoking with amyotrophic lateral sclerosis risk and survival in men and women: a prospective study. BMC Neurol 10: 6.

28. Weisskopf MG, McCullough ML, Calle EE, Thun MJ, Cudkowicz M, et al. (2004) Prospective study of cigarette smoking and amyotrophic lateral sclerosis. Am J Epidemiol 160: 26–33.

29. Savettieri G, Salemi G, Arcara A, Cassata M, Castiglione MG, et al. (1991) A Case-Control Study of Amyotrophic Lateral Sclerosis. Neuroepidemiology 10: 242–245.

30. Armon C (2009) Smoking may be considered an established risk factor for sporadic ALS. Neurology 73: 1693–1698.

31. Longstreth W, Nelson L, Koepsell T, Van Belle G (1991) Hypotheses to explain the association between vigorous physical activity and amyotrophic lateral sclerosis. Medical hypotheses 34: 144–148.

32. Longstreth WT, McGuire V, Koepsell TD, Wang Y, van Belle G (1998) Risk of amyotrophic lateral sclerosis and history of physical activity: a population-based case-control study. Arch Neurol 55: 201–206.

33. Riggs JE (2001) The latency between traumatic axonal injury and the onset of amyotrophic lateral sclerosis in young adult men. Military medicine 166: 731–732.

34. Muddasir Qureshi M, Hayden D, Urbinelli L, Ferrante K, Newhall K, et al. (2006) Analysis of factors that modify susceptibility and rate of progression in amyotrophic lateral sclerosis (ALS). Amyotrophic Lateral Sclerosis 7: 173–182.

35. Morahan JM, Pamphlett R (2006) Amyotrophic Lateral Sclerosis and Exposure to Environmental Toxins: An Australian Case-Control Study. Neuroepidemiology 27: 130–135.

36. Horner RD, Grambow SC, Coffman CJ, Lindquist JH, Oddone EZ, et al. (2008) Amyotrophic lateral sclerosis among 1991 Gulf War veterans: evidence for a time-limited outbreak. Neuroepidemiology 31: 28–32.

37. Kasarskis EJ, Lindquist JH, Coffman CJ, Grambow SC, Feussner JR, et al. (2009) Clinical aspects of ALS in Gulf War veterans. Amyotroph Lateral Scler 10: 35–41.

38. Elbaz A, Dufouil C, Alperovitch A (2007) Interaction between genes and environment in neurodegenerative diseases. C R Biol 330: 318–328.

39. Stozicka Z, Zilka N, Novak M (2007) Risk and protective factors for sporadic Alzheimer's disease. Acta Virol 51: 205–222.

40. Park RM, Schulte PA, Bowman JD, Walker JT, Bondy SC, et al. (2005) Potential occupational risks for neurodegenerative diseases. American Journal of Industrial Medicine 48: 63–77.

41. Bharucha NE, Schoenberg BS, Raven RH, Pickle LW, Byar DP, et al. (1983) Geographic distribution of motor neuron disease and correlation with possible etiologic factors. Neurology 33: 911–915.

42. Kalfakis N, Vassilopoulos D, Voumvourakis C, Ndjeveleka M, Papageorgiou C (1991) Amyotrophic lateral sclerosis in southern Greece: an epidemiologic study. Neuroepidemiology 10: 170–173.

43. Giagheddu M, Puggioni G, Masala C, Biancu F, Pirari G, et al. (1983) Epidemiologic study of amyotrophic lateral sclerosis in Sardinia, Italy. Acta Neurol Scand 68: 394–404.

44. Callaghan B, Feldman D, Gruis K, Feldman E (2011) The association of exposure to lead, mercury, and selenium and the development of amyotrophic lateral sclerosis and the epigenetic implications. Neurodegenerative Diseases 8: 1–8.

45. Kapaki E, Segditsa J, Zournas C, Xenos D, Papageorgiou C (1989) Determination of cerebrospinal fluid and serum lead levels in patients with amyotrophic lateral sclerosis and other neurological diseases. Experientia 45: 1108–1110.

46. Stober T, Stelte W, Kunze K (1983) Lead concentrations in blood, plasma, erythrocytes, and cerebrospinal fluid in amyotrophic lateral sclerosis. J Neurol Sci 61: 21–26.

47. Bergomi M, Vinceti M, Nacci G, Pietrini V, Bratter P, et al. (2002) Environmental exposure to trace elements and risk of amyotrophic lateral sclerosis: a population-based case-control study. Environ Res 89: 116–123.

48. Fang F, Quinlan P, Ye W, Barber M, Umbach D, et al. (2009) Workplace Exposures and the Risk of Amyotrophic Lateral Sclerosis. Environmental Health Perspectives 117: 1387–1392.

49. Moriwaka F, Satoh H, Ejima A, Watanabe C, Tashiro K, et al. (1993) Mercury and selenium contents in amyotrophic lateral sclerosis in Hokkaido, the northernmost island of Japan. Journal of the Neurological Sciences 118: 38–42.

50. Vinceti M, Guidetti D, Bergomi M, Caselgrandi E, Vivoli R, et al. (1997) Lead, cadmium, and selenium in the blood of patients with sporadic amyotrophic lateral sclerosis. Italian Journal of Neurological Sciences 18: 87–92.

51. Moriwaka F, Satoh H, Ejima A, Watanabe C, Tashiro K, et al. (1993) Mercury and selenium contents in amyotrophic lateral sclerosis in Hokkaido, the northernmost island of Japan. J Neurol Sci 118: 38–42.

52. Strickland D, Smith SA, Dolliff G, Goldman L, Roelofs RI (1996) Amyotrophic lateral sclerosis and occupational history: A pilot case-control study. Archives of Neurology 53: 730–733.

53. Kreyling WG, Semmler M, Erbe F, Mayer P, Takenaka S, et al. (2002) Translocation of ultrafine insoluble iridium particles from lung epithelium to extrapulmonary organs is size dependent but very low. J Toxicol Environ Health A 65: 1513–1530.

54. Oberdorster G, Sharp Z, Atudorei V, Elder A, Gelein R, et al. (2002) Extrapulmonary translocation of ultrafine carbon particles following whole-body inhalation exposure of rats. J Toxicol Environ Health A 65: 1531–1543.

55. Lockman PR, Koziara JM, Mumper RJ, Allen DD (2004) Nanoparticle surface charges alter blood-brain barrier integrity and permeability. J Drug Target 12: 635–641.

56. Farrell DF, Starr A (1968) Delayed neurological sequelae of electrical injuries. Neurology 18: 601–606.

57. Noonan CW, Reif JS, Yost M, Touchstone J (2002) Occupational exposure to magnetic fields in case-referent studies of neurodegenerative diseases. Scand J Work Environ Health 28: 42–48.

58. Davanipour Z, Sobel E, Bowman JD, Qian Z, Will AD (1997) Amyotrophic lateral sclerosis and occupational exposure to electromagnetic fields. Bioelectromagnetics 18: 28–35.

59. Savitz DA, Loomis DP, Tse CK (1998) Electrical occupations and neurodegenerative disease: analysis of U.S. mortality data. Arch Environ Health 53: 71–74.

60. Savitz DA, Checkoway H, Loomis DP (1998) Magnetic field exposure and neurodegenerative disease mortality among electric utility workers. Epidemiology 9: 398–404.

61. Johansen C (2000) Exposure to electromagnetic fields and risk of central nervous system disease in utility workers. Epidemiology 11: 539–543.

62. Gallo V, Bueno-De-Mesquita HB, Vermeulen R, Andersen PM, Kyrozis A, et al. (2009) Smoking and risk for amyotrophic lateral sclerosis: Analysis of the EPIC cohort. Annals of Neurology 65: 378–385.

63. Brown RC, Lockwood AH, Sonawane BR (2005) Neurodegenerative diseases: an overview of environmental risk factors. Environ Health Perspect 113: 1250–1256.

64. Sutedja N, Veldink JH, Fischer K, Kromhout H, Wokke JHJ, et al. (2007) Lifetime occupation, education, smoking, and risk of ALS. Neurology 69: 1508–1514.

65. Weisskopf MG, McCullough ML, Morozova N, Calle EE, Thun MJ, et al. (2005) Prospective Study of Occupation and Amyotrophic Lateral Sclerosis Mortality. American Journal of Epidemiology 162: 1146–1152.

Molecular Mechanisms of Reduced Nerve Toxicity by Titanium Dioxide Nanoparticles in the Phoxim-Exposed Brain of *Bombyx mori*

Yi Xie[1,2♙], Binbin Wang[1,2♙], Fanchi Li[1,2♙], Lie Ma[1,2♙], Min Ni[1,2], Weide Shen[1,2], Fashui Hong[1]*, Bing Li[1,2]*

1 School of Basic Medicine and Biological Sciences, Soochow University, Suzhou, Jiangsu, P.R. China, **2** National Engineering Laboratory for Modern Silk, Soochow University, Suzhou, Jiangsu, P.R. China

Abstract

Bombyx mori (*B. mori*), silkworm, is one of the most important economic insects in the world, while phoxim, an organophosphorus (OP) pesticide, impact its economic benefits seriously. Phoxim exposure can damage the brain, fatbody, midgut and haemolymph of *B. mori*. However the metabolism of proteins and carbohydrates in phoxim-exposed *B. mori* can be improved by Titanium dioxide nanoparticles (TiO$_2$ NPs). In this study, we explored whether TiO$_2$ NPs treatment can reduce the phoxim-induced brain damage of the 5th larval instar of *B. mori*. We observed that TiO$_2$ NPs pretreatments significantly reduced the mortality of phoxim-exposed larva and relieved severe brain damage and oxidative stress under phoxim exposure in the brain. The treatments also relieved the phoxim-induced increases in the contents of acetylcholine (Ach), glutamate (Glu) and nitric oxide (NO) and the phoxim-induced decreases in the contents of norepinephrine (NE), Dopamine (DA), and 5-hydroxytryptamine (5-HT), and reduced the inhibition of acetylcholinesterase (AChE), Na$^+$/K$^+$-ATPase, Ca^{2+}-ATPase, and Ca^{2+}/Mg^{2+}-ATPase activities and the activation of total nitric oxide synthase (TNOS) in the brain. Furthermore, digital gene expression profile (DGE) analysis and real time quantitative PCR (qRT-PCR) assay revealed that TiO$_2$ NPs pretreatment inhibited the up-regulated expression of *ace1*, *cytochrome c*, *caspase-9*, *caspase-3*, *Bm109* and down-regulated expression of *Bmlap* caused by phoxim; these genes are involved in nerve conduction, oxidative stress and apoptosis. TiO$_2$ NPs pretreatment also inhibited the down-regulated expression of *H$^+$ transporting ATP synthase* and *vacuolar ATP synthase* under phoxim exposure, which are involved in ion transport and energy metabolism. These results indicate that TiO$_2$ NPs pretreatment reduced the phoxim-induced nerve toxicity in the brain of *B. mori*.

Editor: Vipul Bansal, RMIT University, Australia

Funding: This work was supported by the National High Technology Research and Development Program of China (863 Program) (Grant No. 2013AA102507), the transformation project of agriculture scientific and technological achievements (2013GB2C100180), the projects sponsored by the national cocoons silk development funds in 2014, the Priority Academic Program Development of Jiangsu Higher Education Institutions, the Doctoral Fund of Ministry of Education of China (20113201110008), the China Agriculture Research System (CARS-22-ZJ0305), and the Science & Technology support Program of Suzhou (ZXS2012005, SNG201352). The funders had no role in study design, data collection and analysis, decision to publish, or preparation of the manuscript.

Competing Interests: The authors have declared that no competing interests exist.

* Email: hongfsh_cn@sina.com (FH); lib@suda.edu.cn (BL)

♙ These authors contributed equally to this work.

Introduction

Silkworm, *Bombyx mori* (*B. mori*, *Bombycidae: Lepidoptera*), is one of the most important economic insects in Asia, Africa, Europe and Latin America. *B. mori* has been domesticated for about 5,700 years in China, and it produces more than 80% of raw silk around the world [1]. However, *B. mori* is highly sensitive to adverse environmental conditions, especially pesticides. Every year, pesticide contamination causes as much as 30% of the reduction in raw silk production in China [2]. Due to its short growth cycle and pesticide sensitivity, *B. mori* has been a widely used model insect for pesticide toxicology studies. Phoxim is an efficient broad-spectrum organophosphorus (OP) pesticide, but its indiscriminate use has generated serious environmental problems.

Phoxim may trigger oxidative stress, which is mainly reflected in altered Malondialdehyde (MDA) content and Glutathione S-transferase (GST) activity in the fat body and midgut of *B. mori* [3]. Our previous study demonstrated that phoxim destroyed the

carbohydrate and lipid metabolism in the haemolymph of *B. mori* [4]. Exposure of *B. mori* to phoxim also affected the activities of acetylcholinesterase (AChE) and detoxification enzymes, which play important roles in organophosphorus pesticide resistance and metabolism [5]. The catalytic substrate of AChE, Acetylcholine (ACh), is a chemical transmitter of cholinergic neurons that are exclusively in the central nervous system (CNS) of insects [6]. However, clear understanding of phoxim's effects on the the brain of *B. mori* is still lacking. We hypothesized that nerve toxicity of phoxim in *B. mori* is associated with brain damages and gene expression profile alterations.

Titanium dioxide nanoparticles (TiO$_2$ NPs) are widely used as whitening agents in paper, cosmetics, and food industries because of their whitening effects. TiO$_2$ NPs may also be used for photocatalytic degradation of pesticide in water, soil, and air [7–9]. The growth of plants can be promoted by TiO$_2$ NPs that improve their antioxidative capacity [10,11]. Recently, it was reported that TiO$_2$ NPs increased the cold-tolerance of *Chickpea*

[12]. Our previous studies have shown that TiO_2 NPs improve protein and carbohydrate metabolism to meet required energy demands and increase antioxidant capacity of midgut in *B. mori* exposed to phoxim [4,13]. It was also found TiO_2 NPs pretreatment decreased phoxim-induced toxicity to silkworms by greatly reducing the phoxim residue [14]. Therefore, we speculated that TiO_2 NPs treatments may relieve phoxim-induced damage by modulating gene expression and enzymatic activities in the brain of *B. mori*.

Digital Gene Expression Profile (DGE) with massive parallel sequencing has been shown to have high sensitivity and reproducibility for transcriptome profiling [15]. DGE is based on new generation high-throughput sequencing technologies and high-performance computing analyses. Nowadays DGE has been widely used in biological, medical and pharmaceutical research [16–18].

In this study, we investigated the nerve toxicity of phoxim and the effects of TiO_2 NPs in the brain of *B. mori*. To further explore the mechanisms of toxicity, we adopted DGE assay and real time quantitative PCR (qRT-PCR) to detect the alterations of genes participated in regulating neurotransmitter contents, oxidative stress and apoptosis. These findings may promote future mechanistic studies on the effects of TiO_2 NPs on the toxicity of insecticides in *B. mori*.

Results

Body weight and Survival rate

We observed that the fifth-instar larvae appeared as gastric juice spit, head nystagmus, body distortion, body shrink, paralysis and other symptoms after 48 h of phoxim exposure. However, the larvae in the control group, the TiO_2 NPs group, and the TiO_2 NPs + phoxim group did not show such symptoms. As shown in Figure 1, phoxim exposure significantly decreased the body weight

($P<0.05$) and survival rate ($P<0.001$) of the larvae, while TiO_2 NPs promoted their body weights and survival rate.

Histopathological evaluation

The brains of the larvae of both the control group (Fig. 2a) and the TiO_2 NPs-treated group (Fig. 2b) had normal morphology. In the phoxim-exposed group, we observed widespread gaps among plasma membrane, breakage of nerve fibers, protein aggregation, adipose degeneration, and cell debris (Fig. 2c). However, the TiO_2 NPs + phoxim-treated group did not show such pathological changes (Fig. 2d). It demonstrated that phoxim exposure caused brain damages, while TiO_2 NPs treatments were able to reduce such damages.

Brain ultrastructure evaluation

As shown in Figure 3, the ultrastructure of cells in the control group and the TiO_2 NPs group was normal with well distributed chromatin and integral mitochondria crista (Fig. 3a, 3b), compared with karyopyknosis, chromatin marginalization, and mitochondria swelling in the phoxim exposure group (Fig. 3c) at 48 h after phoxim exposure. However, only chromatin marginalization was observed in the TiO_2 NPs + phoxim group (Fig. 3d), indicating that TiO_2 NPs reduced the damage in *B. mori* brain cells caused by phoxim exposure.

Neurotransmitter contents and enzyme activities in the brain

The contents of neurotransmitters, including ACh, Glutamate (Glu), and nitric oxide (NO), in the brains of fifth-instar larvae in the phoxim-exposed group were higher than those of the control, while the contents of norepinephrine (NE), dopamine (DA), and 5-hydroxytryptamine (5-HT) were otherwise decreased significantly by phoxim exposure (Fig. 4a). Pretreatments with TiO_2 NPs reversed the changes in the contents of NE, DA, 5-HT, ACh, Glu, and NO (Fig. 4a). We also observed that phoxim exposure

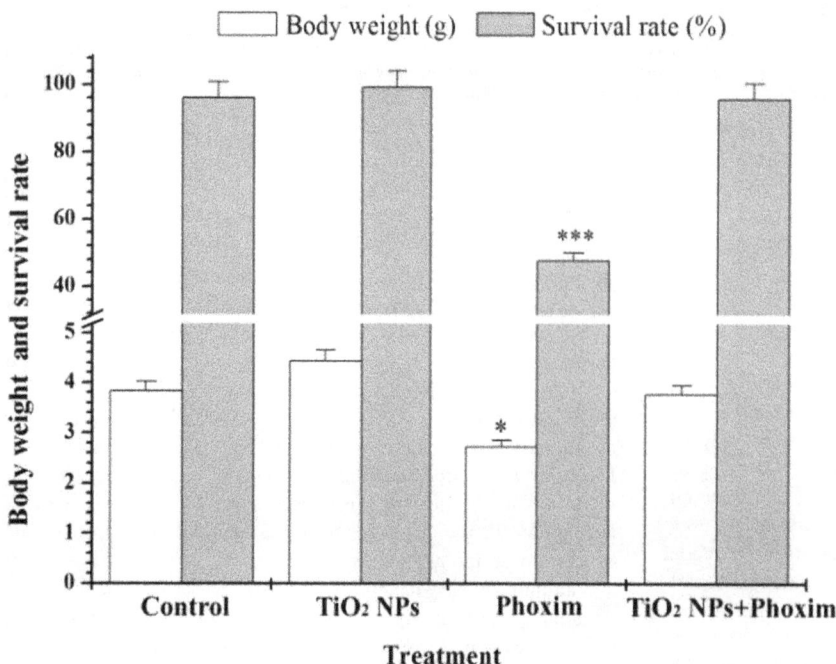

Figure 1. Effects of TiO₂ NPs on body weight, survival of phoxim-exposed fifth-instar larvae. *$P<0.05$, and ***$P<0.001$. Values represent means $\pm$ SEM ($n=5$).

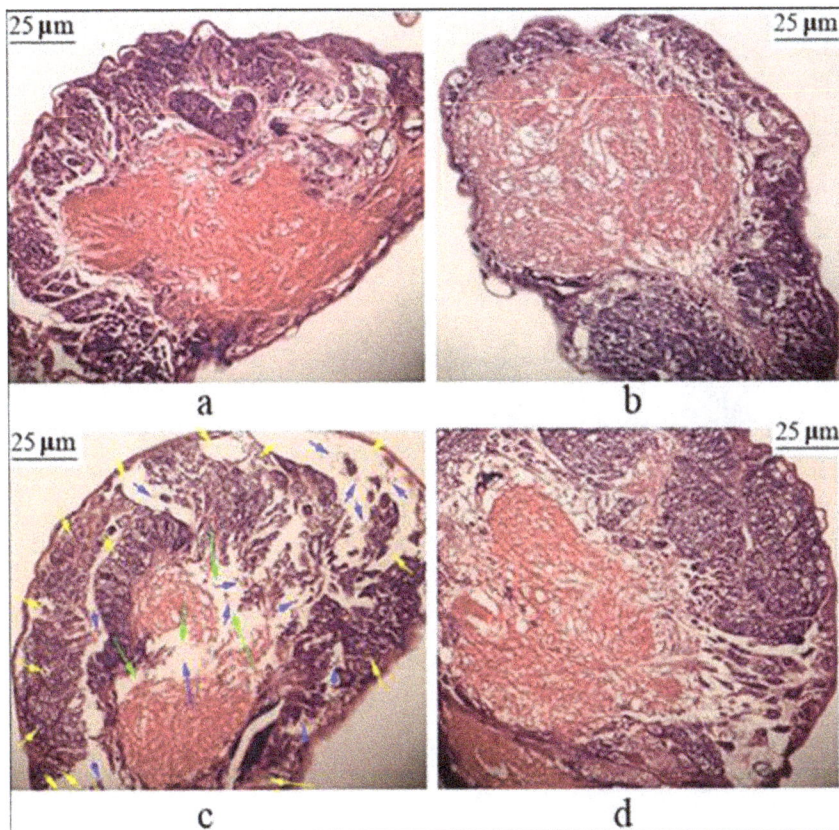

Figure 2. Histopathology of the brain tissue in fifth-instar larvae after phoxim exposure 48 h. (a) Control; (b) TiO$_2$ NPs; (c) Phoxim; (d) TiO$_2$ NPs + Phoxim. Green arrows indicate breakage of nerve fibers, yellow arrows show adipose degeneration, blue arrows indicate cell debris.

significantly inhibited the activities of AChE, Na$^+$/K$^+$-ATPase, Ca^{2+}-ATPase, and Ca^{2+}/Mg^{2+}-ATPase and promoted the activity of total nitric oxide synthase (TNOS) in the brain, while TiO$_2$ NPs significantly promoted the activities of AChE, Na$^+$/K$^+$-ATPase, Ca^{2+}-ATPase, Ca^{2+}/Mg^{2+}-ATPase, and AChE and inhibited the activity of TNOS (Fig. 4b). These results demonstrated that phoxim exposure altered the releases of neurotransmitters and the activities of several important enzymes in the nerve conduction in *B. mori* larvae brain, while TiO$_2$ NPs were able to reverse such changes.

Oxidative stress

As shown in Figure 5a, phoxim exposure significantly promoted the production of ROS species, such as O$_2^-$ and H$_2$O$_2$, in larval brains ($P<0.001$) at 48 h, while TiO$_2$ NPs attenuated such enhancement in ROS production ($P<0.05$). The ROS production was further demonstrated by the measurements of the levels of lipid peroxidation (MDA), protein peroxidation (protein carbonyl, PC), and DNA damage (8-hydroxy deoxyguanosine, 8-OHdg) in the larval brain (Fig. 5b). Significantly increased MDA, protein carbonyl, and 8-OHdG were observed in the phoxim-exposed midguts, but the increases became much lower with the combined treatments (Fig. 5b). It suggested that TiO$_2$ NPs treatments decreased ROS accumulation, which may lead to attenuated peroxidation of lipids, proteins, and DNAs in the larval brains under phoxim-induced toxicity.

Digital Gene Expression Profile

To investigate the molecular mechanisms of reduced nerve toxicity by TiO$_2$ NPs under phoxim stress in the brain of *B. mori*, we adopted the DGE method to detect the differences in gene expression in the brain among the control-, TiO$_2$ NPs-, phoxim-, and TiO$_2$ NPs + phoxim-treated larvae at 48 h. Compared with those of the control group, 288, 295, and 472 genes were expressed significantly differently in the TiO$_2$ NPs group (Fig. S1), the phoxim group (Fig. S2), and the TiO$_2$ NPs + phoxim group (Fig. S3), respectively, with 117, 64, and 48 genes being up-regulated, respectively, and 171, 231, and 424 genes being down-regulated, respectively. The genes with differential expression were classified by Gene Ontology (GO) classification analysis into 12 groups, which were oxidative stress, stress response, metabolic process, cell component, transport, transcription-related, translation-related, growth and development, nerve conduction, immune response, cell cycle, and apoptosis (Tab. S1, S2, S3).

Gene Expression Detection by qRT-PCR

Combine with DGE assay, histopathological and ultrastructure evaluation, we hypothesized TiO$_2$ NPs pretreatment might decrease the expression changes of genes essential in maintain normal physiological activity. To validate the hypothesis, we performed qRT-PCR for several genes that are involved in neurotoxicity, ion transport, oxidative stress and apoptosis. In the present study, *actin3* was used as the internal reference gene. As shown in Table 1, the expression level of *ace1* was significantly increased by 19.45-fold after 48 h of phoxim exposure, while the expression levels of *H$^+$ transporting ATP synthase, vacuolar ATP*

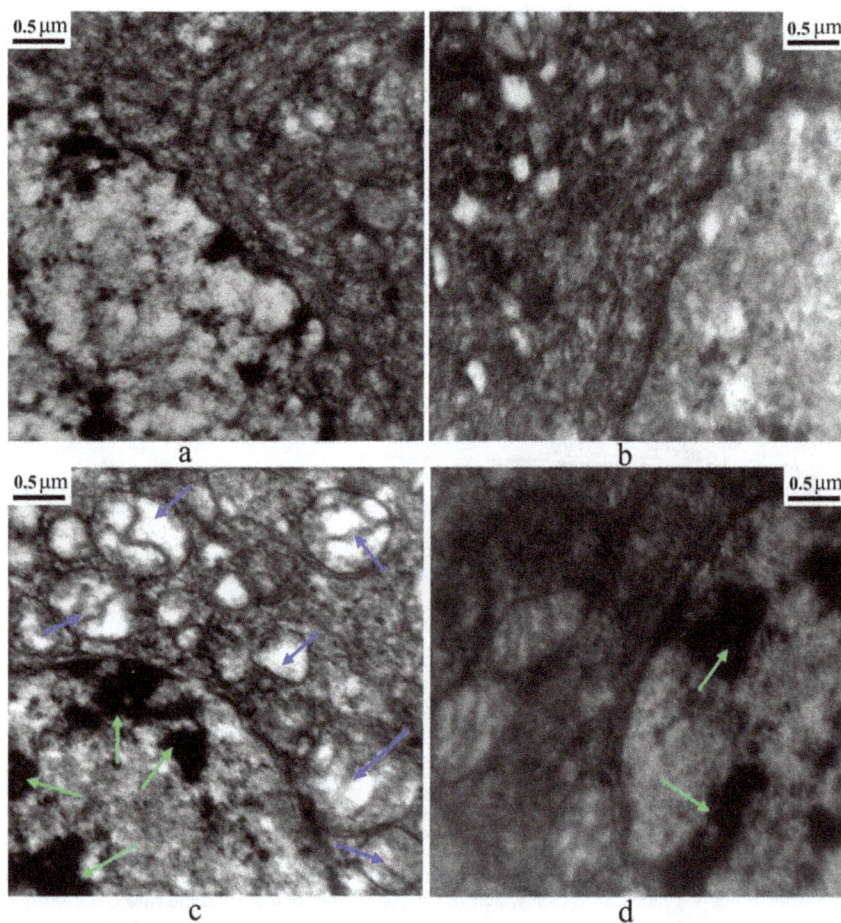

Figure 3. Ultrastructure of the brain tissue in fifth-instar larvae after phoxim exposure 48 h. (a) Control; (b) TiO₂ NPs; (c) Phoxim; (d) TiO₂ NPs + Phoxim. Green arrows indicate karyopyknosis and chromatin marginalization, blue arrows show mitochondria swelling and became deformed, crest broken.

synthase, *SOD* and *TPx* were significantly reduced by 63%, 55%, 74% and 35% respectively. However, the expression changes of *ace1*, *H⁺ transporting ATP synthase* and *vacuolar ATP synthase* were 6.69-fold, 0.78-fold, 0.93-fold, 0.81-fold and 0.93-fold respectively for TiO₂ NPs + phoxim-treated brains. Moreover, the expression of *cytochrome-c* was up-regulated by 1.21-fold in the phoxim-exposed brain, but by 1.02-fold in the TiO₂ NPs + phoxim treated group. All the qRT-PCR data were consistent with those of DGE assay (Tab. 1). In order to explore whether phoxim stress induced apoptosis through the mitochondria/cytochrome-c pathway, the expression of four additional genes regulating mitochondria apoptosis pathway were determined by qRT-PCR. As shown in Table 1, compared with the control group, the expression levels of three pro-apoptotic genes, *Bm109*, *caspase-9*, and *caspase-3* were changed by 3.0-fold, 2.51-fold, and 3.07-fold, respectively in the phoxim-exposed group, and by 2.42-fold, 2.2-fold, and 2.45-fold, respectively in the TiO₂ NPs + phoxim-treated group. On the other hand, the mRNA levels of *BmIAP*, an apoptosis inhibitor gene, were down-regulated by 0.785-fold under phoxim stress, but by 0.91-fold in the TiO₂ NPs + phoxim-treated group, respectively. These results indicated that TiO₂ NPs treatment decreased expression alterations of these genes involved in neurotoxicity, ion transport, oxidative stress and apoptosis in the brain under phoxim stress.

Discussion

The insect brain, a part of CNS, is essential in regulating nerve conduction, growth and development. It has been reported that the brain of *B. mori* is the target organ of nerve agent phoxim. In the present study, the body weight and survival rate of *B. mori* were significantly reduced by phoxim (Fig. 1), and severe brain damage was observed (Fig. 2), while pretreatment with TiO₂ NPs protected the brain (Fig. 2). In addition, TiO₂ NPs decreased the severe apoptosis of brain cells after phoxim exposure (Fig. 3) and protected larvae from anomalous nerve conduction (Fig. 4) and excessive ROS production (Fig. 5). Furthermore, we adopted DGE assay and qRT-PCR method to explore the molecular mechanisms of reduced nerve toxicity by TiO₂ NPs in the phoxim-exposed brain of *B. mori*, the main results were divided into three parts and discussed below.

Nerve conduction

It has been reported that vacuolar-type ATPases (V-ATPases) produce proton-motives that are indispensable for ion transports and the energization of membrane transport in insect systems [19]. In the present study, the expression of *H⁺ transporting ATP synthase* and *vacuolar ATP synthase* was down-regulated under phoxim stress in the brain of *B. mori* larvae, which was reversed by TiO₂ NPs. Furthermore, the activity of Na⁺/K⁺-ATPase that maintains the

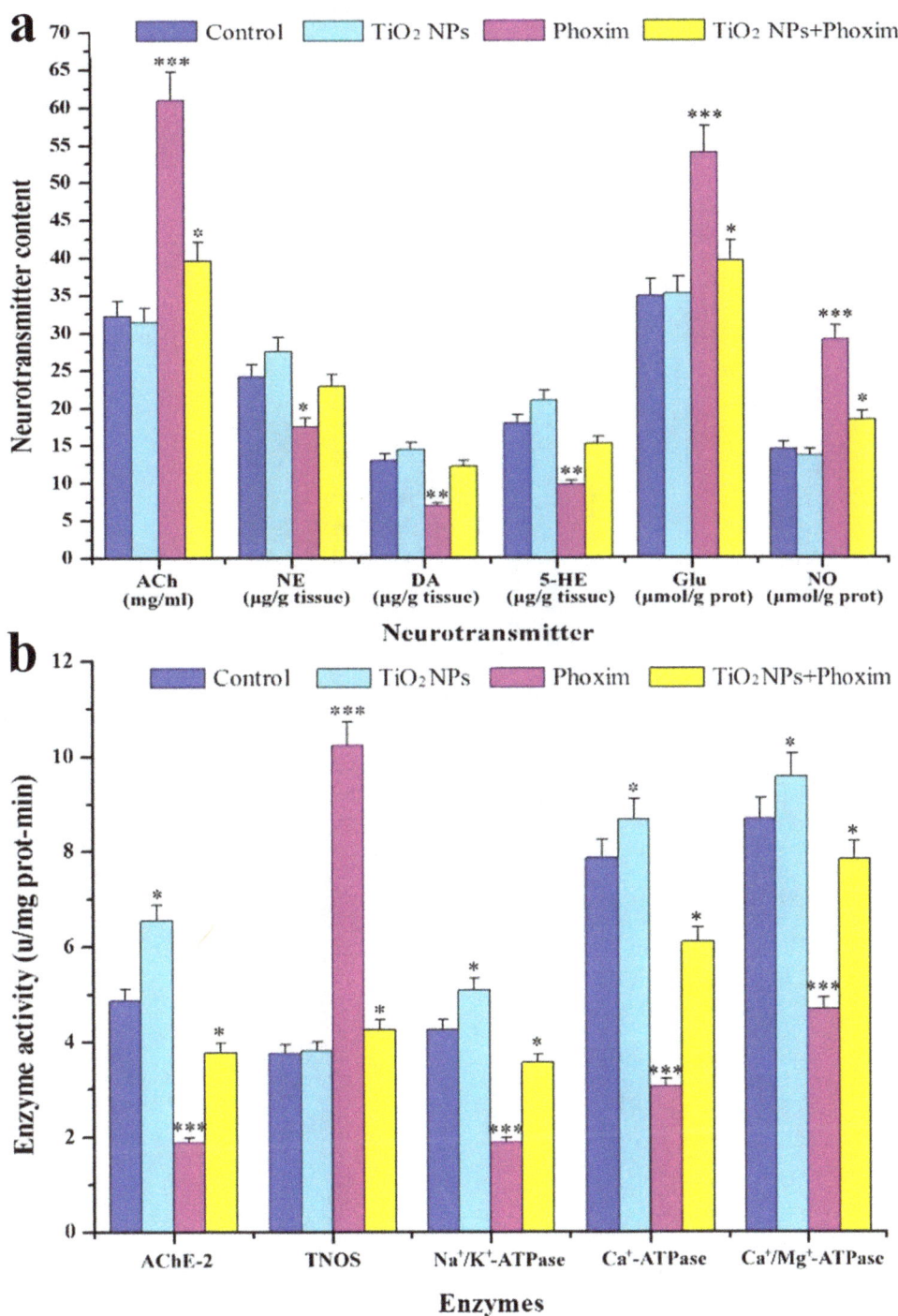

Figure 4. Effects of TiO$_2$ NPs on nerve conduction in the brain of phoxim-exposed fifth-instar larvae. *$p<0.05$, **$p<0.01$, and ***$p<$ 0.001. Values represent means $\pm$ SEM ($N=5$). (a) Neurotransmitter contents, (b) Enzyme activity.

balance of K$^+$ and Na$^+$ concentrations in the organisms was inhibited in the brain under phoxim stress, which resulted in physiological damages and cellular homeostasis disturbance [20]; these changes could also be mitigated by TiO$_2$ NPs pretreatments. Besides, Ca^{2+} concentration that is essential for ion transport and nerve conduction is regulated by Ca^{2+}-ATPase and Ca^{2+}/Mg^{2+}-ATPase in eukaryotic cells, and defects in these enzymes seriously compromise the normal functions of cells [21]. Similar to the

finding of inhibited activity of Ca^{2+}/Mg^{2+}-ATPase by pyrethroids [22], we observed inhibited activities of Ca^{2+}-ATPase and Ca^{2+}/ Mg^{2+}-ATPase by phoxim in the brain of *B. mori* in our study. However, the activities of the two enzymes were only slightly inhibited in the TiO$_2$ NPs + phoxim group. These changes in gene expression and enzymatic activity in the brain are expected to lead to abnormal concentrations of neurotransmitters that are important in nerve conduction. Monoamine neurotransmitters, such as

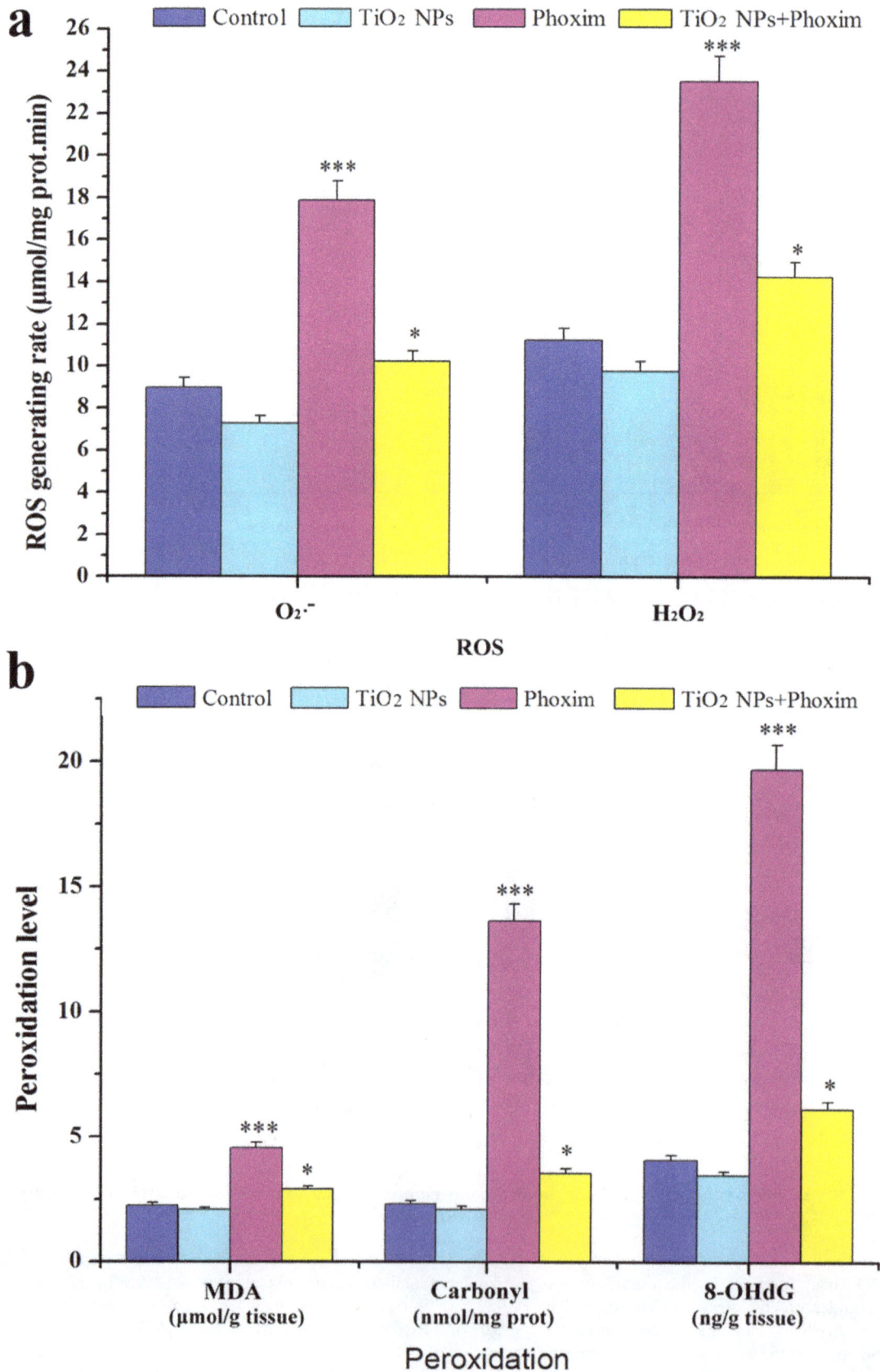

Figure 5. Effects of TiO$_2$ NPs on oxidative stress in brain of phoxim-exposed fifth-instar larvae. *$p<0.05$, and ***$p<0.001$. Values represent means $\pm$ SEM ($N=5$). (a) ROS production, (b) Levels of lipid, protein, and DNA peroxidation.

Table 1. Comparison between fold-difference with qRT-PCR results and DGE assay in each group.

Gene	TiO$_2$ NPs/Control		Phoxim/Control		TiO$_2$ NPs + Phoxim/Control	
	qRT-PCR (Fold)	DGE (log$_2$ value)	qRT-PCR (Fold)	DGE (log$_2$ value)	qRT-PCR (Fold)	DGE (log$_2$ value)
ace1	0.823	0.068	19.453***	0.956	6.689***	0.583
H+ tATPase	2.143	1.324	0.373***	−0.303	0.779**	−0.148
vATPase	1.065	0.547	0.453**	−1.236	0.933	−1.061
SOD	1.317	0.218	0.263**	−0.627	0.808	−0.472
TPx	1.204	0.484	0.649*	−0.680	0.930	−0.406
Bm109	0.849	No difference	2.999**	No difference	2.416**	No difference
Bmlap	1.026	No difference	0.785*	No difference	0.905	No difference
caspase-9	0.979	No difference	2.513**	No difference	2.204**	No difference
caspase-3	0.970	No difference	3.065***	No difference	2.451**	No difference
cytochrome c	0.953	−0.085	1.214*	0.294	1.021	0.099

*p<0.05, **p<0.01, and ***p<0.001.
Values represent means ± SEM (n = 5).

5-HT, DA, and NE, are closely related to learning, memory, and normal behaviors [14,23]. In this study, the contents of 5-HT, DA, and NE were decreased significantly by phoxim exposure, while those in the TiO$_2$ NPs + phoxim-treated larvae were similar to the control. The contents of several amino acid neurotransmitters, such as ACh, Glu, and NO, were increased significantly by phoxim exposure.

It was reported that inhibition of the amino acid neurotransmitter AChE is the main mechanism of OP pesticides [2]. AChE catalyzes the hydrolysis of the excitatory neurotransmitter ACh into choline and acetic acid, which terminates nerve impulses on postsynaptic membrane [24]. Therefore, inhibited AChE activity results in increased ACh contents and continuing nerve impulses. In the current study, a significant inhibition of AChE activity was observed in the brain of phoxim-exposed larvae (Fig. 4b). However, the expression of ace1 was actually up-regulated, likely a compensation for the inhibited AChE activity. This is consistent with the finding in a previous study [1]. TiO$_2$ NPs treatments mitigated the inhibition of AChE activity (Fig. 4b) and down-regulated the ace1 expression.

Glu, another excitatory amino acid neurotransmitter, binds to NMDA receptors to promote the influx of extracellular Ca^{2+} [25,26] and enhance the activity of calcium-dependent proteases, such as TNOS that is involved in the release of NO. We observed that phoxim exposure significantly increased Glu contents, TNOS activity, and NO contents, while TiO$_2$ NPs pretreatments reversed such increases. The free radical NO modulates neuronal functions by increasing the release of neurotransmitters [27], and NO can be oxidized to peroxinitrite (ONOO-) that may cause neuronal damages and induce apoptosis [26]. Therefore, the reversed changes in Glu contents, TNOS activity, and NO contents may further explain the protective effects of TiO$_2$ NPs against phoxim-induced damages in the brain of B. mori.

Oxidative stress

Previous study had shown mitochondria may be the primary target of OP-initiated cytotoxicity [28]. They play crucial roles in oxidative stress [29], as the levels of ROS species, such as O$_2^-$ and H$_2$O$_2$, are related to the respiratory chain, substrate dehydrogenases in the matrix, monoamine oxidase, and cytochrome P450 [30]. When mitochondria are damaged, the ROS levels are

usually increased significantly, causing oxidative damages to lipids, proteins, and DNA. These oxidative damages generate peroxidation products, such as MDA, PC, and 8-OHdG [31–33], which may induce apoptosis and necrosis. However, many stress response proteins are associated with the removal of ROS in insects, such as SOD that is the primary antioxidant enzyme catalyzing the dismutation of superoxide radicals to hydrogen peroxide and oxygen [34] and TPx that removes hydrogen peroxide and alkyl hydroperoxides [35]. In the present study, the expression of both SOD and TPx was significantly inhibited after 48 h of phoxim exposure, along with significantly increased contents of O$_2^{\bullet-}$ and H$_2$O$_2$ (Fig. 5a), significantly improved levels of MDA, PC, and 8-OHdG (Fig. 5b), and swelling mitochondria and broken mitochondria crista (Fig. 3c). However, in the TiO$_2$ NPs group and the TiO$_2$ NPs + phoxim group, the morphology of cells and mitochondria were normal, indicating that TiO$_2$ NPs protected the brain of B. mori from phoxim stress by regulating the expression of genes important in oxidative stress and mitochondria respiratory chain.

Apoptosis

Many pesticides have been shown to cause apoptosis and necrosis [36]. The accumulation of ROS and peroxidation may also promote the release of mitochondrial cytochrome c [37]. Once released to the cytoplasm, cytochrome c binds to the apoptotic protease activating factor-1 (Apaf-1) and pro-caspase-9 to form a tripolymer protein complex. The tripolymers form apoptosomes that activate caspase-9, an initiator caspase in the mitochondria/cytochrome-c pathway, and caspase-3 [38]. In the present study, over-expression of cytochrome-c, caspase-9 and caspase-3 was observed in phoxim-exposed brains, while TiO$_2$ NPs treatments mitigated the over-expression. Moreover, apoptosis is regulated by many apoptotic associated proteins, such as Bcl-2 and the inhibitor of apoptosis proteins (IAP) [39,40]. In B. mori, Bm109 is homologous to the anti-apoptotic Bcl-2 family proteins within the four conserved BH regions. Bm109 has been reported to up-regulate apoptosis by participating in the translocation of Bax to mitochondria and the release of cytochrome c [41,42]. In this study, TiO$_2$ NPs mitigated the over-expression of Bm109 and the down-expression of BmIAP, a specific inhibitor of caspase-9 [43], under phoxim stress in the brain of B. mori. We also observed cell debris, swelled mitochondria, and broken mitochondria crista by histological and ultrastructure photomicrographs (Figs. 2c, 3c), indicating that phoxim induces apoptosis through the mitochondria/cytochrome-c pathway, and that TiO_2 NPs treatments can mitigate mitochondrial damages and block apoptosis in the brain of *B. mori* under phoxim stress.

Conclusion

The results from this study indicate that TiO_2 NPs can reduce the phoxim-induced changes in the expression of genes and the activity of enzymes that regulate nerve conduction, oxidative stress and apoptosis, and relieve phoxim-induced physiological disorders and brain damages in *B. mori*. Our study may promote the application of TiO_2 NPs in reducing pesticide toxicity in *B. mori* in the future, although further investigations are needed to reveal the specific mechanisms of the effects of TiO_2 NPs on phoxim exposure.

Materials and Methods

Insects and Chemicals

The larvae of *B. mori* (*Bombyx mori* L Qiufeng × baiyu), which were maintained in our laboratory, were reared on mulberry leaves under 12-h light/12-h dark cycles for this study.

Nanoparticulate anatase TiO_2 was prepared through controlled hydrolysis of titanium tetrabutoxide. Detailed synthesis and characterization of TiO_2 NPs have been described previously [44].

Phoxim was purchased from Sigma Co. at 98.1% purity.

Preparation of phoxim and TiO₂ NPs solutions

TiO_2 NPs powder was dispersed onto the surface of 0.5% Hydroxypropylmethylcellulose (HPMC) (w/v), suspended, sonicated for 30 min, and mechanically vibrated for 5 min. Phoxim was dissolved in acetone to prepare the stock solution, which was diluted with water into different concentrations for analysis. 0.5% HPMC was used as the suspending agent.

Resistance measurement

In the pre-experiment, we tried 1, 2, 5, 10, and 15 mg/L TiO_2 NPs suspensions in fifth-instar larvae and determined that the optimum concentration for larvae growth was 5 mg/L for further experiments. The Lethal Concentration 50 (LC_{50}) of phomix in *B. mori* was 7.86 µg/mL, and 4 µg/mL was used as the concentration in further experiments. 100 g of fresh mulberry, *Morus albus (L.)*, leaves were dipped in 5 mg/L TiO_2 NPs suspension for 1 min, followed by dipping in the solution of 4 µg/mL phoxim for 1 min.

After being air-dried, TiO_2 NPs-treated leaves were used to feed *B. mori* instar larvae three times a day until the second day of fifth-instar larvae. Fresh leaves treated with 0.5% HPMC served as controls. Later, phoxim-treated leaves were used to feed *B. mori* larvae three times a day from the third day, before the silkworms (1,000 in each group) were fed with either TiO_2 NPs-treated leaves or the control fresh leaves under long-day photoperiods (16 h light: 8 h dark) at 25°C and about 75% relative humidity. Each experiment was repeated three times with 200 larvae. The mortality of larvae was counted 48 h later.

To measure the resistance of larvae, the body weight and survival rate of larvae was counted 48 h later by the method of our previous study [4] with minor modifications.

Brain tissue collection

Forty eight hours after phoxim treatments, 100 fifth-instar larvae were selected randomly from each group. Larval brains were collected and frozen at −80°C for subsequent antioxidant assay.

Histopathological evaluation of brain

All histopathological examinations were performed using standard laboratory procedures. Brains of five larvae of each group were embedded in paraffin, sliced (5 µm thickness), placed onto glass slides, and stained with hematoxylin–eosin (HE). Stained sections were evaluated by a histopathologist who was unaware of the treatments using an optical microscope (Nikon U-III Multi-point Sensor System, Japan).

Observation of brain ultrastructure

Brains of five larvae of each group were fixed in freshly prepared 0.1 M sodium cacodylate buffer with 2.5% glutaraldehyde and 2% formaldehyde, before being treated at 4°C with 1% osmium tetroxide in 50 mM sodium cacodylate (pH 7.2–7.4) for 2 h. Staining was performed overnight with 0.5% aqueous uranyl acetate. After serial dehydration with ethanol (75, 85, 95, and 100%), the specimens were embedded in Epon 812 and sliced. Ultrathin sections were treated with uranyl acetate and lead citrate and observed with a HITACHI H600 TEM (HITACHI Co., Japan). The apoptosis in brain was determined by observing the changes in nuclear morphology (e.g., chromatin condensation and fragmentation).

Oxidative stress assay of brain

ROS (O_2^- and H_2O_2) production, MDA levels, protein carbonyl (PC), and 8-OHdG in brain tissues were assayed using commercial enzyme-linked immunosorbent assay (ELISA) kits (Nanjing Jiancheng Bioengineering Institute, Jiangsu, China) following the manufacturer's instructions.

Assay of enzymatic activities

For enzymatic activity determinations, brain tissues were homogenized in 10 volumes of 0.15 M NaCl. A quantity of homogenate was used to the activies of different enzymes. The activities of AChE, Ca^{2+}-ATPase, Ca^{2+}/Mg^{2+}-ATPase, Na^+/K^+-ATPase, and TNOS in the brain were spectrophotometrically measured with commercial kits (Nanjing Jiancheng Bioengineering Institute, China) targeting the oxidation of oxyhaemoglobin to methaemoglobin by nitric oxide.

Determination of neurochemicals

The homogenate of brains was centrifuged at 12,000 g for 20 min at 4°C. The concentrations of DA, 5-HT, NE, and ACh were spectrophotometrically measured with commercially kits (Nanjing Jiancheng Bioengineering Institute, China).

Glu concentrations were measured using commercial kits (Nanjing Jianchen Biological Institute, China), and the standard curves were produced by using standard Glu stock solutions. Glu levels in the samples were detected using a spectrophotometer at 340 nm and expressed as µmol/g prot. The concentration of NO in the brain was measured using a commercial kit (Nanjing Jiancheng Bioengineering Institute, China). The OD value was determined by a spectrophotometer (U-3010, Hitachi, Japan). Results of NO were read with OD value at 550 nm. The results were calculated using the following formula: NO (µmol/L) = (Asample−Ablank)/(Astandard− Ablank)×20(µmol/L).

Total RNA isolation

Total RNA was extracted from brain samples using the Trizol reagent (Takara, Japan) and treated with DNase to remove potential genomic DNA contamination. The quality of RNA was assessed by formaldehyde agarose gel electrophoresis and was quantitated spectrophotometrically.

DGE library preparation, sequencing, tag mapping and evaluation of DGE libraries

For RNA library construction and deep sequencing, equal quantities of brain RNA samples (n = 3) were pooled for the control group and the treated group, respectively. Approximately 6 μg of RNA representing each group were submitted to Solexa (now Illumina Inc.) for sequencing. The detailed methodology were performed in our previous study [45].

qRT-PCR analysis

The specific primers for the 11 genes are listed in Table S4. The internal reference gene was *actin3*. qRT-PCR was performed using the 7500 Real-time PCR System (ABI) with SYBR Premix Ex Taq^{TM} (Takara, Japan) according to the manufacturer's instructions. The qRT-PCR analysis was carried out following the method described in previous studies [1,5].

Statistical Analysis

Statistical analyses were performed using the SPSS 19 software. Data are expressed as means ± standard error of the mean (SEM). One-way analysis of variance (ANOVA) was carried out to compare the differences of means of the multigroup data. Dunnett's test was performed when each dataset was compared with the solvent-control data. Statistical significance for all tests was judged at a probability level of 0.05 ($P<0.05$).

Supporting Information

Figure S1 Functional categorization of 295 genes which significantly altered by phoxim exposure. Genes were functionally classified based on the ontology-driven clustering approach of PANTHER.

Figure S2 Functional categorization of 288 genes which altered by TiO$_2$ NPs pretreatment. Genes were functionally classified based on the ontology-driven clustering approach of PANTHER.

Figure S3 Functional categorization of 472 genes which altered by TiO$_2$ NPs + phoxim exposure. Genes were functionally classified based on the ontology-driven clustering approach of PANTHER.

Table S1 Genes related to oxidative stress, stress response, metabolic process, cell component, transport, transcription, translation, growth and development, signal transduction, immune response, cell cycle and apoptosis altered significantly by phoxim exposure.

Table S2 Genes related to oxidative stress, stress response, metabolic process, cell component, transport, transcription, translation, growth and development, signal transduction, immune response, cell cycle and apoptosis altered significantly by TiO$_2$ NPs exposure.

Table S3 Genes related to oxidative stress, stress response, metabolic process, cell component, transport, transcription, translation, growth and development, signal transduction, immune response, cell cycle and apoptosis altered significantly by TiO$_2$ NPs + phoxim exposure.

Table S4 Primer pairs for qRT-PCR in the gene expression analysis.

Author Contributions

Conceived and designed the experiments: YX BBW FCL LM MN WDS FSH BL. Performed the experiments: YX BBW FCL LM. Analyzed the data: YX BBW FCL LM. Wrote the paper: YX BBW FCL LM.

References

1. Peng GD, Wang JM, Ma L, Wang YH, Cao YQ, et al. (2011) Transcriptional characteristics of acetylcholinesterase genes in domestic silkworms (*Bombyx mori*) exposed to phoxim. Pesticide Biochemistry and Physiology 101: 154–158.
2. Li B, Wang D, Zhao HQ, Shen WD (2010) Comparative analysis of two acetylcholinesterase genes of *Bombyx mandarina* and *Bombyx mori*. Afr J Biotechnol 9: 8477–8485.
3. Yu QY, Fang SM, Zuo WD, Dai FY, Zhang Z, et al. (2011) Effect of organophosphate phoxim exposure on certain oxidative stress biomarkers in the silkworm. J Econ Entomol 104: 101–106.
4. Li B, Hu RP, Cheng Z, Cheng J, Xie Y, et al. (2012) Titanium dioxide nanoparticles relieve biochemical dysfunctions of fifth-instar larvae of silkworms following exposure to phoxim insecticide. Chemosphere 89: 609–614.
5. Wang YH, Gu ZY, Wang JM, Sun SS, Wang BB, et al. (2013) Changes in the activity and the expression of detoxification enzymes in silkworms (*Bombyx mori*) after phoxim feeding. Pesticide Biochemistry and Physiology 105: 13–17.
6. Surendra Nath B, Surendra Kumar RP (1999) Toxic impact of organophosphorus insecticides on acetylcholinesterase activity in the silkworm, *Bombyx mori* L. Ecotoxicology and environmental safety 42: 157–162.
7. Higarashi MM, Jardim WE (2002) Remediation of pesticide contaminated soil using TiO$_2$ mediated by solar light. Catal Today 76: 201–207.
8. Konstantinou IK, Albanis TA (2004) TiO$_2$-assisted photocatalytic degradation of azo dyes in aqueous solution: kinetic and mechanistic investigations-A review. Appl Catal B-Environ 49: 1–14.
9. Esterkin CR, Negro AC, Alfano OM, Cassano AE (2005) Air pollution remediation in a fixed bed photocatalytic reactor coated with TiO$_2$. Aiche J 51: 2298–2310.
10. Hong FH, Yang F, Liu C, Gao Q, Wan ZG, et al. (2005) Influences of nano-TiO$_2$ on the chloroplast aging of spinach under light. Biological trace element research 104: 249–260.
11. Zheng L, Hong FS, Lu SP, Liu C (2005) Effect of nano-TiO$_2$ on strength of naturally and growth aged seeds of spinach. Biological trace element research 104: 83–91.
12. Mohammadi R, Maali-Amiri R, Abbasi A (2013) Effect of TiO$_2$ Nanoparticles on *Chickpea* Response to Cold Stress. Biological trace element research 152: 403–410.
13. Su JJ, Li B, Cheng S, Zhu Z, Sang XZ, et al. (2013) Phoxim-induced damages of *Bombyx mori* larval midgut and titanium dioxide nanoparticles protective role under phoxim-induced toxicity. Environmental toxicology. DOI:10.1002/tox.21866.
14. Li B, Yu XH, Gui SX, Xie Y, Hong J, et al. (2013) Titanium Dioxide Nanoparticles Relieve Silk Gland Damage and Increase Cocooning of *Bombyx mori* under Phoxim-Induced Toxicity. J Agric Food Chem 61: 12238–12243.
15. Asmann YW, Klee EW, Thompson EA, Perez EA, Middha S, et al. (2009) 3′ tag digital gene expression profiling of human brain and universal reference RNA using Illumina Genome Analyzer. Bmc Genomics 10: 531.
16. Nishiyama T, Miyawaki K, Ohshima M, Thompson K, Nagashima A, et al. (2012) Digital gene expression profiling by 5′-end sequencing of cDNAs during reprogramming in the moss Physcomitrella patens. Plos One 7(5): 1.
17. Nordlund J, Kiialainen A, Karlberg O, Berglund EC, Goransson-Kultima H, et al. (2012) Digital gene expression profiling of primary acute lymphoblastic leukemia cells. Leukemia 26: 1218–1227.
18. Zhang Y, Luoh SM, Hon LS, Baertsch R, Wood WI, et al. (2007) GeneHub-GEPIS: digital expression profiling for normal and cancer tissues based on an integrated gene database. Nucleic Acids Res 35: W152–158.
19. Azuma M, Ohta Y (1998) Changes in H+-translocating vacuolar-type ATPase in the anterior silk gland cell of *Bombyx mori* during metamorphosis. The Journal of experimental biology 201: 479–486.
20. Kopecka-Pilarczyk J (2010) Effect of exposure to polycyclic aromatic hydrocarbons on Na$^+$-K$^+$-ATPase in juvenile gilthead seabream Sparus aurata. Journal of fish biology 76: 716–722.

21. Brini M, Carafoli E (2009) Calcium pumps in health and disease. Physiological reviews 89: 1341–1378.
22. Alrajhi DH (1990) Properties of $Ca^{2+}+Mg^{2+}$-Atpase from Rat-Brain and Its Inhibition by Pyrethroids. Pesticide Biochemistry and Physiology 37: 116–120.
23. Meneses A, Liy-Salmeron G (2012) Serotonin and emotion, learning and memory. Reviews in the neurosciences 23: 543–553.
24. Fournier D, Mutero A (1994) Modification of acetylcholinesterase as a mechanism of resistance to insecticides. Comparative Biochemistry and Physiology Part C: Pharmacology, Toxicology and Endocrinology 108: 19–31.
25. Fu W, Ruangkittisakul A, Mactavish D, Baker GB, Ballanyi K, et al. (2013) Activity and metabolism related Ca^{2+} and mitochondrial dynamics in co-cultured human fetal cortical neurons and astrocytes. Neuroscience 250: 520–535.
26. Hu RP, Gong XL, Duan YM, Li N, Che Y, et al. (2010) Neurotoxicological effects and the impairment of spatial recognition memory in mice caused by exposure to TiO_2 nanoparticles. Biomaterials 31: 8043–8050.
27. Prast H, Philippu A (2001) Nitric oxide as modulator of neuronal function. Progress in neurobiology 64: 51–68.
28. Carlson K, Ehrich M (1999) Organophosphorus compound-induced modification of SH-SY5Y human neuroblastoma mitochondrial transmembrane potential. Toxicol Appl Pharmacol 160: 33–42.
29. Sinha K, Das J, Pal PB, Sil PC (2013) Oxidative stress: the mitochondria-dependent and mitochondria-independent pathways of apoptosis. Arch Toxicol 87: 1157–1180.
30. Chernyak BV, Izyumov DS, Lyamzaev KG, Pashkovskaya AA, Pletjushkina OY, et al. (2006) Production of reactive oxygen species in mitochondria of HeLa cells under oxidative stress. Biochimica et biophysica acta 1757: 525–534.
31. Yang ZP, Dettbarn WD (1998) Lipid peroxidation and changes in cytochrome c oxidase and xanthine oxidase activity in organophosphorus anticholinesterase induced myopathy. Journal of physiology, Paris 92: 157–161.
32. Stadtman ER, Levine RL (2003) Free radical-mediated oxidation of free amino acids and amino acid residues in proteins. Amino acids 25: 207–218.
33. Tokiwa H, Sera N, Nakanishi Y, Sagai M (1999) 8-Hydroxyguanosine formed in human lung tissues and the association with diesel exhaust particles. Free radical biology & medicine 27: 1251–1258.
34. Park SY, Nair PM, Choi J (2012) Characterization and expression of superoxide dismutase genes in *Chironomus riparius* (*Diptera, Chironomidae*) larvae as a potential biomarker of ecotoxicity. Comparative biochemistry and physiology Toxicology & pharmacology: CBP 156: 187–194.
35. Lee KS, Kim SR, Park NS, Kim I, Kang PD, et al. (2005) Characterization of a silkworm thioredoxin peroxidase that is induced by external temperature stimulus and viral infection. Insect biochemistry and molecular biology 35: 73–84.
36. Astiz M, de Alaniz MJ, Marra CA (2009) Effect of pesticides on cell survival in liver and brain rat tissues. Ecotoxicology and environmental safety 72: 2025–2032.
37. Zamzami N, Larochette N, Kroemer G (2005) Mitochondrial permeability transition in apoptosis and necrosis. Cell death and differentiation 12 Suppl 2: 1478–1480.
38. Stennicke HR, Deveraux QL, Humke EW, Reed JC, Dixit VM, et al. (1999) Caspase-9 can be activated without proteolytic processing. The Journal of biological chemistry 274: 8359–8362.
39. Yano M, Terada K, Gotoh T, Mori M (2007) In vitro analysis of Bcl-2 proteins in mitochondria and endoplasmic reticulum: similarities in anti-apoptotic functions and differences in regulation. Experimental cell research 313: 3767–3778.
40. Deveraux QL, Reed JC (1999) IAP family proteins–suppressors of apoptosis. Genes & development 13: 239–252.
41. Tambunan J, Kan Chang P, Li H, Natori M (1998) Molecular cloning of a cDNA encoding a silkworm protein that contains the conserved BH regions of Bcl-2 family proteins. Gene 212: 287–293.
42. Wu WX, Wei W, Ablimit M, Ma Y, Fu T, et al. (2011) Responses of two insect cell lines to starvation: autophagy prevents them from undergoing apoptosis and necrosis, respectively. J Insect Physiol 57: 723–734.
43. Huang QH, Deveraux QL, Maeda S, Stennicke HR, Hammock BD, et al. (2001) Cloning and characterization of an inhibitor of apoptosis protein (IAP) from *Bombyx mori*. Biochimica et biophysica acta 1499: 191–198.
44. Hu RP, Zheng L, Zhang T, Gao GD, Cui YL, et al. (2011) Molecular mechanism of hippocampal apoptosis of mice following exposure to titanium dioxide nanoparticles. Journal of hazardous materials 191: 32–40.
45. Gu ZY, Zhou YJ, Xie Y, Li FC, Ma L, et al. (2013) The adverse effects of phoxim exposure in the midgut of silkworm, Bombyx mori. Chemosphere 96: 33–38.

9

Imidacloprid Alters Foraging and Decreases Bee Avoidance of Predators

Ken Tan[1,2]*, Weiwen Chen[2], Shihao Dong[2], Xiwen Liu[2], Yuchong Wang[2], James C. Nieh[3]

1 Key Laboratory of Tropical Forest Ecology, Xishuangbanna Tropical Botanical Garden, Chinese Academy of Science, Kunming, Yunnan Province, China, 2 Eastern Bee Research Institute, Yunnan Agricultural University, Heilongtan, Kunming, Yunnan Province, China, 3 Division of Biological Sciences, Section of Ecology, Behavior, and Evolution, University of California San Diego, La Jolla, California, United States of America

Abstract

Concern is growing over the effects of neonicotinoid pesticides, which can impair honey bee cognition. We provide the first demonstration that sublethal concentrations of imidacloprid can harm honey bee decision-making about danger by significantly increasing the probability of a bee visiting a dangerous food source. *Apis cerana* is a native bee that is an important pollinator of agricultural crops and native plants in Asia. When foraging on nectar containing 40 µg/L (34 ppb) imidacloprid, honey bees (*Apis cerana*) showed no aversion to a feeder with a hornet predator, and 1.8 fold more bees chose the dangerous feeder as compared to control bees. Control bees exhibited significant predator avoidance. We also give the first evidence that foraging by *A. cerana* workers can be inhibited by sublethal concentrations of the pesticide, imidacloprid, which is widely used in Asia. Compared to bees collecting uncontaminated nectar, 23% fewer foragers returned to collect the nectar with 40 µg/L imidacloprid. Bees that did return respectively collected 46% and 63% less nectar containing 20 µg/L and 40 µg/L imidacloprid. These results suggest that the effects of neonicotinoids on honey bee decision-making and other advanced cognitive functions should be explored. Moreover, research should extend beyond the classic model, the European honey bee (*A. mellifera*), to other important bee species.

Editor: Nicolas Desneux, French National Institute for Agricultural Research (INRA), France

Funding: This work was supported by the Key Laboratory of Tropical Forest Ecology, Xishuangbanna Tropical Botanical Garden, and the CAS 135 program (XTBGT01) of Chinese Academy of Science, China National Research Fund (31260585) to Ken Tan. The funders had no role in study design, data collection and analysis, decision to publish, or preparation of the manuscript.

Competing Interests: The authors have declared that no competing interests exist.

* Email: kentan@xtbg.ac.cn

Introduction

Bees play an important global role as pollinators of native plants and crops [1,2]. However, pollinator foraging behavior, and thus the ability of pollinators to pollinate, can be negatively influenced by pesticides at sublethal doses [3]. Impaired foraging may reduce colony fitness [4], contributing to bee population declines [5]. Research has focused on neonicotinoid pesticides, a class of compounds that became commercially available in the early 1990's [6] and which bind to insect cholinergic receptors, causing death at sufficient concentrations [7]. In fact, the European Union recently restricted the use of neonicotinoids because of concerns about their effects on bees [8]. However, neonicotinoids remain widely used around the world [9]. Continued research on the sublethal effects of these compounds is therefore important because assays that only test for lethality, the standard approach for determining safe doses, do not reveal how these compounds impair bee behaviors that are intimately involved in pollination and colony fitness [3,10,11].

Imidacloprid is a neonicotinoid pesticide: its metabolites are nicotinic acetylcholine receptor (nAChR) agonists in honey bee neurons [12] and therefore have widespread behavioral consequences. Field-realistic levels of imidacloprid in nectar reduce honey bee performance by 6–20%: decreasing hive entrance activity [13], learning ability [14], food uptake [15], and locomotion [16]. Imidacloprid also decreases foraging activity

[17] and the ability of bees to successfully return to the nest [18], perhaps because of navigational deficits [3].

Such cognitive impairments are particularly intriguing because honey bees have sophisticated decision-making skills [19], and the deficits elicited by neonicotinoids may therefore be complex and, perhaps, unexpected. For example, sublethal doses of imidacloprid reduce olfactory [13,20,21] and visual [22] learning. Imidacloprid also evidently alters bee internal thresholds, elevating the sucrose response threshold and reducing waggle dancing even for very rich nectar sources [16,23]. However, the role of imidacloprid in altering other types of decision-making remains poorly understood.

In addition to deciding whether a nectar source is sweet enough to collect or recruit for, bees also evaluate and respond to the risk of predation during foraging [24]. Bees face a wide variety of predator threats and can normally avoid dangerous food sources. For example, honey bees avoid flowers with living crab spiders [25] and live mantises [26]. In Asia, hornets within the genus *Vespa* are major honey bee predators [27,28], and the hornet *V. velutina* attacks *A. cerana* colonies [29–31]. Recently, we showed that this hornet species will capture *A. cerana* foraging on flowers [32]. *Apis cerana* foragers consequently show a strong aversion to a feeder with a *V. velutina* hornet and will reduce visitation to such a dangerous feeder by 78% but will not reduce visitation to a feeder with a harmless butterfly [32]. Given that neonicotinoids can alter *A. mellifera* foraging behavior [15], we hypothesized that neonicotinoids can also impair a bee's judgment about danger, altering its ability or its willingness to avoid predators.

The sublethal effects of neonicotinoids have been studied in relatively few bee species, and comparatively little is known about neonicotinoid effects on native bees in large areas of the world such as Asia. Studies have demonstrated detrimental sublethal effects of neonicotinoids in European honey bees (*A. mellifera*), European bumble bees (*Bombus terrestris*), and the South American stingless bee (*Melipona quadrifasciata anthidioides*) [3,4,33,34]. Recently, Arena et al. [35] conducted a meta-analysis and suggested that native Asian honey bees species (*A. cerana* and *A. florea*) may have a higher sensitivity to pesticides than *A. mellifera*. However, to the best of our knowledge, no published studies have systematically compared the effects of neonicotinoids on the foraging behaviors of multiple honey bee species. Such knowledge would be valuable. For example, the native Asian honey bee, *A. cerana*, plays a large role in agriculture [36,37], and is an important native pollinator of native Asian plants [37,38]. *Apis cerana* is widespread and is found throughout southern and eastern Asian, extending from India to China [39]. In China alone, more than two million managed colonies of *A. cerana* are used for honey production and crop pollination [37]. Previous studies have shown that pesticides such as carbamates [40,41], synthetic pyrethroids [42], and imidacloprid [43,44] are toxic to *Apis cerana*. Imidacloprid can bind to two different receptor sites within *A. cerana* nAChRs [45]. However, studies have not explored the sublethal effects of neonicotinoids on *A. cerana* even though imidacloprid is widely used in China [46]. We therefore tested the effects of field-realistic doses of imidacloprid on *A. cerana* foraging behavior and decision-making with respect to food source danger, the hornet *V. velutina*, a native Asian predator and an emerging threat to *A. mellifera* in Europe [47].

Materials and Methods

We conducted experiments from October to November 2013. This field season corresponds to the time of peak hornet activity, when *V. velutina* actively hunts honey bees at our field site Yunnan Agricultural University, Kunming, China (22°42′30 N, 100°56′01 E, 1890 m altitude). Our field season also corresponded to a period of floral dearth, which facilitated feeder training of bees. We used three colonies of *A. cerana*, each with four frames of bees and brood.

Imidacloprid concentrations

The scientific literature commonly uses four different units (μg/L, nmol/L, μg/Kg, and ppb) to express neonicotinoid concentrations. To facilitate comparisons, we provide conversions for all of these commonly used values in our figures and in our discussion, as appropriate. For conversions involving sucrose concentrations and density (at 20°C), we use tables published in Bubnik et al. [48]. In our Discussion, we convert concentration values reported by the original study to a common unit (μg/L) based upon the sucrose concentrations used in the original study) and give the value reported by the original study in parentheses.

We used 1.25 M sucrose solutions (37% sucrose w/w) with field-realistic doses of imidacloprid [6]: 10 μg/L (8.6 parts per billion), 20 μg/L (17.2 ppb), and 40 μg/L (34.0 ppb) of imidacloprid, corresponding to 39.1 nmol/L, 78.2 nmol/L, and 156.5 nmol/L (Fig. 1). We chose these concentrations to cover a range of field-realistic, sub-lethal doses. Imidacloprid residues have been found to occur at 1–50 ppb in the nectar and pollen of a variety of crop species [6]. In citrus trees treated with imidacloprid and grown within an enclosure, residues of 3–39 μg/L were detected in

nectar, with an average of 16.4 μg/L from floral nectar and 15.3 μg/L from bee crops (caged tunnel studies) [49]. In open field studies, total residues of imidacloprid were 5.0 μg/L in floral nectar and 3.5 μg/L in bee crops [49]. Field realistic doses of imidacloprid from a variety of crops and studies are 0.7–10 μg/L, corresponding to a 0.024–0.3 ng dose per nectar load [50].

The imidacloprid concentrations that we used (39.1–156.5 nmol/L) were sublethal. In *A. mellifera*, only imidacloprid concentrations ≥ 1000 nmol/L increased mortality: 10 and 100 nmol/L did not alter mortality [51]. The acute 48-hr oral LD_{50} for imidacloprid ranges widely and depends upon a variety of factors that include bee physiological condition and season [14,24,52]. Values range between 3.7 ng/bee [53] and >81 ng/bee [54]. The average nectar load at our highest treatment concentration (40 μg/L) contained a sublethal dose (0.52 ng of imidacloprid/bee). None of the foragers died from exposure to our imidacloprid treatments during the experiments.

Training bees

We used three colonies in all experiments. We trained bees by capturing departing foragers at the hive entrance in a vial and releasing them slowly at the training feeder placed 130 m from the focal colony. The feeder consisted of a 70 mL vial (8 cm high) with 18 holes (each 3 mm diameter) drilled around its lid. Once it was filled with sucrose solution, the vial was inverted on a blue plastic square and bees could slowly collect sugar solution through the drilled holes [32]. We filled this feeder with unscented 1.25 M sucrose solution (37% sucrose w/w). We marked all trained foragers with individually numbered honey bee queen tags attached to their thoraces with resin. We used a different set of bees for each of the following three experiments.

Foraging experiment

We trained bees to the unscented 1.25 M sucrose feeder (see above). After bees made 10 trips to the training feeder, it was replaced with an identical feeder containing one of four different treatment concentrations of imidacloprid (0, 10, 20, and 40 μg/L) in 1.25 M sucrose. We allowed the bees to sample the treatment feeder and then recorded which bees subsequently returned to this feeder. A bee that made a single return within 1 hr was counted as a returning bee. The proportion of bees that continued to forage at the different imidacloprid concentrations was then calculated. In total, we trained 360 bees, 120 from each colony, and 90 bees per treatment.

Nectar collection experiment

To determine the amount of the treatment solutions that bees collected, we trained bees as before. After 10 training trips, we changed the feeder to one containing one of the four treatment concentrations of imidacloprid. Each bee was allowed to visit the treatment feeder 10 times so that its collection volume would reach equilibrium. After the 10th visit, the bee was captured at the nest upon its return. We used CO_2 to anesthetize each bee, weighed it, gently squeezed its abdomen and absorbed the contents of its collected nectar onto tissue paper, and then immediately weighed it again to determine the mass of nectar it had collected. The volume collected per bee was then calculated. In total, we used 84 bees, 28 from each colony, and 21 bees per treatment.

Danger experiment

To test the effect of imidacloprid on bee decision making about danger, we trained bees as previously described. Once the marked bees made 10 consecutive visits to the training feeder, we

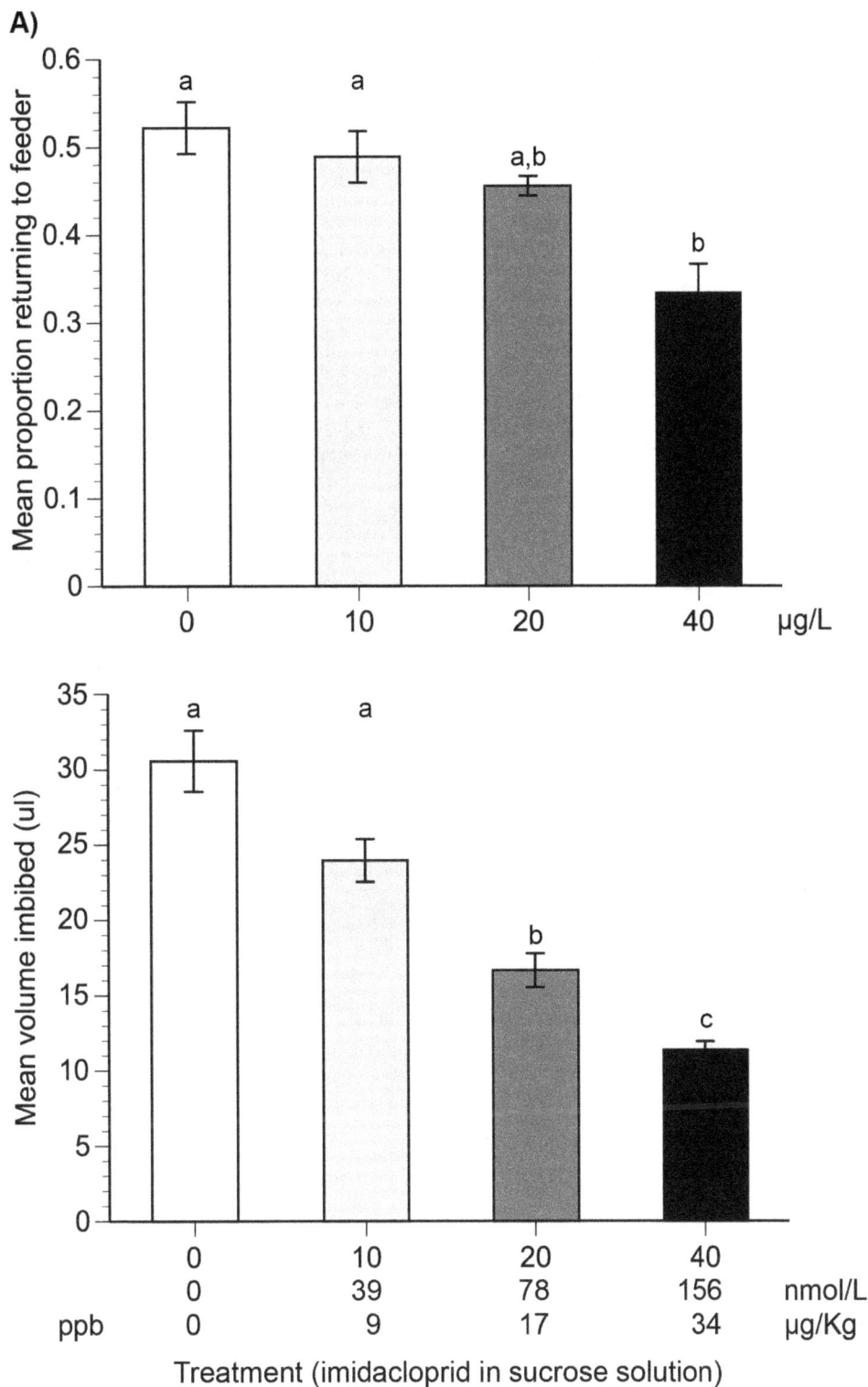

Figure 1. Effect of imidacloprid treatments on bee return rates and nectar collection (imbibing) volume. a) The mean proportion of trained bees that returned to the test feeders with different imidacloprid treatments. b) The mean volume of 1.25 M sucrose solution collected by bees trained to a safe feeder with no hornet predators. Standard error bars are shown. Different letters indicate treatments that are significantly different from each other (Tukey HSD tests, $P<0.05$). The imidacloprid concentrations are given in commonly used units, with ppb and µg/Kg shown in the same row because the numeric values are identical. Different shades of gray correspond to different imidacloprid concentrations. Sample sizes are given in the Methods.

surrounded this feeder with a cage (70×66×52 cm) with a single   exit and entrance. The cage allowed us to restrict bee access and to ensure independent choices by testing individual choices in the absence of other bees. Bees were exposed to the imidacloprid once the training feeder was placed inside the foraging cage. Each forager was randomly assigned to one imidacloprid treatment concentration and was exposed to this concentration for 10 collecting visits before we tested its choice of safe vs. dangerous feeders.

After the bees had learned to forage inside the cage for these 10 successive visits, we randomly selected seven bees from each colony for the experiment. We replaced the single feeder with an array of two identical unscented 1.25 M sucrose feeders spaced 40 cm apart inside the cage. These feeders both contained the same treatment solution. For example, if a bee was trained to a 40 μg/L imidacloprid solution, the safe and dangerous feeders also contained this solution. To create a "dangerous" feeder, we captured *V. velutina* hornets with insect nets and tethered a single hornet 10 cm above the feeder with a stiff wire wrapped between the thorax and abdomen. The "safe" feeder had a similar wire, but with no hornet. Tan et al [32] demonstrated that a significant majority of bees exposed to such a feeder array consistently chose the safe feeder.

We monitored five successive choices of our trained bees. Thus, each bee was exposed over approximately 1.5 hrs for 15 total visits to a given treatment. After each visit, we replaced the feeders to eliminate potential odor marks and randomly swapped dangerous and safe feeder positions to avoid potential site bias. In total, we used 80 bees (28 bees from colony 1, 28 bees from colony 2, and 24 bees from colony 3), 20 bees per treatment.

Statistics

We arcsin-square-root transformed the proportion of visiting bees, log-transformed the volume of sucrose collected, and used Analysis of Variance (ANOVA) to test for a significant effect of treatment, with Tukey Honestly Significant Difference (HSD) post-hoc tests to compare between treatments [55]. These data met parametric assumptions as determined by residual analyses. We used Chi-square tests to compare observed and expected bee choices. For each bee, we calculated the proportion of visits to the safe feeder and classified the bee as preferring the safe feeder if it chose the safe feeder $\geq$60% out of its five trips. To compare overall choices, we used a null hypothesis expectation of 50% of bees visiting each feeder. Finally, we used Chi-square tests to compare the choices of imidacloprid-treated bees against the choices of control bees. We used a Repeated-Measures Analysis of Variance (ANOVA) model with individual bee (random effect) nested within treatment to test the effect of two fixed effects, treatment and trial number (the visit number of each bee for the five consecutive visits tested), on bee choices (0 = dangerous feeder, 1 = safe feeder). We included colony identity as a random variable in the ANOVA models.

Ethics statement

This research was conducted in full compliance with the laws of the People's Republic of China. No specific permits were required for our field studies, which were conducted at Yunnan Agricultural University. Our studied involved colonies of *A. cerana* and hornets, *V. velutina*. Neither species is endangered or protected.

Results

Foraging experiment

There is a significant effect of pesticide concentration on the proportion of trained bees that returned to the feeder ($F_{3,6} = 7.67$, $P = 0.018$), with colony accounting for <1% of model variance. As compared to the control and the 10 μg/L feeders, significantly fewer bees (23% fewer) returned to the 40 μg/L feeder. There is no significant difference between control and 10 μg/L bees. There is no significant difference between 20 μg/L and 40 μg/L bees (Tukey HSD test, Fig. 1A).

Nectar collection experiment

Increasing pesticide concentration significantly decreases the amount of sucrose solution that bees collect ($F_{3,78} = 45.4$, $P<$ 0.0001, Fig. 1B). Colony accounts for 2% of model variance. A Tukey HSD test shows that bees significantly decreased the average volume collected by 46% and 63% for 20 μg/L and 40 μg/L imidacloprid solutions, respectively, as compared to the control solution. There is no significant difference between the volume collected from the control and 10 μg/L solutions (Tukey HSD test, Fig. 1B). Based upon the average nectar volume collected per trip, each bee collected (but did not necessarily absorb into its hemolymph) 0.27, 0.39, and 0.52 ng of imidacloprid per trip from solutions with imidacloprid concentrations of 10, 20, and 40 μg/L, respectively. However, these values likely do not represent the amount of pesticide that bees individually absorbed into their bodies per trip since the majority of this sugar solution was regurgitated to other bees and stored inside the nest.

Danger experiment

In the danger experiment, there is a significant effect of treatment ($F_{3,74} = 3.57$, $P = 0.03$) but no significant effect of trial ($F_{1,319} = 0.40$, $P = 0.53$) and no significant interaction between treatment and trial ($F_{3,316} = 1.63$, $P = 0.18$). Colony and individual bee account for respectively 0.5% and 7% of model variance. Individual foragers therefore exhibited preferences that did not significantly alter over multiple trials (Fig. 2).

Control bees that received no imidacloprid avoided the hornet: overall, 85% of control bees chose the feeder without the hornet ($\chi^2_1 = 9.8$, $P = 0.002$). Similarly, bees that received 10 μg/L ($\chi^2_1 = 7.2$, $P = 0.007$) or 20 μg/L ($\chi^2_1 = 9.8$, $P = 0.002$) also avoided the hornet: overall, 80% and 85% respectively chose the feeder without the hornet. However, bees that received the highest dose (40 μg/L) did not avoid the hornet: only 65% chose the feeder without the hornet, not significantly different from a random choice ($\chi^2_1 = 1.8$, $P = 0.18$, Fig. 3).

We then compared the distribution of bee choices to the observed distribution of control bee choices. Bees treated with the highest level of imidacloprid (40 μg/L) exhibited significantly different choices than control bees ($\chi^2_1 = 6.3$, $P = 0.012$). Bees treated with the lower imidacloprid levels of 10 μg/L ($\chi^2_1 = 0.4$, $P = 0.53$) and 20 μg/L ($\chi^2_1 = 0$, $P = 1.0$, observed and expected distributions were identical) did not make choices that were significantly different from control bees (Fig. 3).

Discussion

Neonicotinoid pesticides can impair honey bee cognitive abilities [52]. We provide the first demonstration that sublethal concentrations of a neonicotinoid, imidacloprid, can also impair honey bee decision-making about danger, significantly increasing the likelihood that a bee will visit a dangerous feeder with a hornet predator. At an imidacloprid nectar concentration of 40 μg/L (34 ppb), 1.8 fold more bees chose the dangerous feeder as compared to bees fed with no imidacloprid (Fig. 3). These preferences were consistent and did not significantly vary over multiple choices by the same bees (Fig. 2). In addition, we provide the first data demonstrating that foraging by *A. cerana*, a native bee and important pollinator of agricultural crops and native plants in

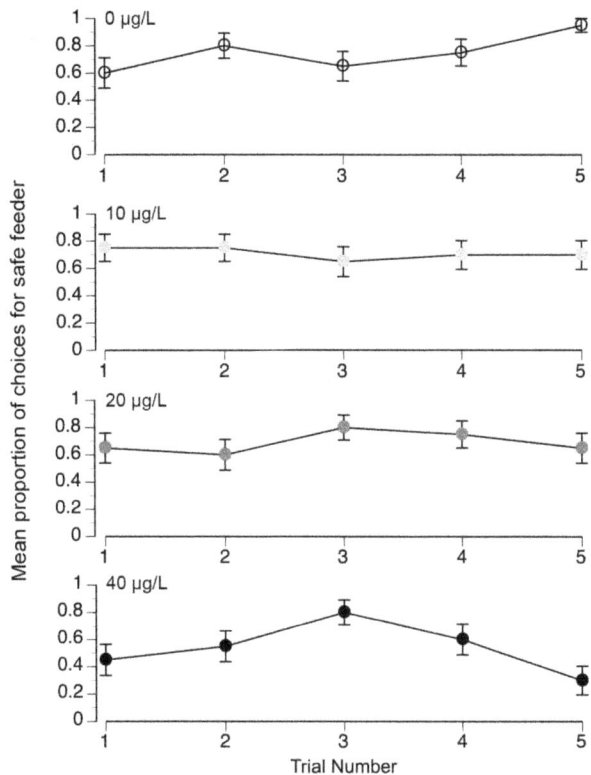

Figure 2. Mean proportion of choices for the safe feeder over five trials. The different treatments are identified above each plot (1 = all choices for safe feeder). Different shades of gray correspond to different imidacloprid concentrations. Standard error bars are shown.

Asia, can be influenced by sublethal doses of imidacloprid, which is also used widely in Asia. Imidacloprid reduced food collection: 23% fewer foragers returning to collect the 40 μg/L imidacloprid-laced nectar as compared with the control nectar. Compared to controls, bees that returned to the feeder respectively collected 46% and 63% less nectar containing 20 μg/L and 40 μg/L imidacloprid. These effects could arise if imidacloprid reduces bee motivation to forage, reduces their physical ability to imbibe (drink) nectar, or both.

Effects on visitation and nectar collection

The reduction in feeder visitation and collecting that we observed (Fig. 1) matches what other studies have found for the closely related European honey bee, *A. mellifera*. Imidacloprid at 7 μg/L (6 μg/kg) reduced *A. mellifera* foraging at a 1.38 M sucrose solution (40% w/w) [56]. In our experiment, *A. cerana* foragers were somewhat more resistant, and visitation for our 37% w/w sucrose solution did not significantly decline until we provided a concentration of 40 μg/L (Fig. 1A). Our result is a better match for the data of Yang et al [57], who showed that *A. mellifera* foragers significantly delayed their willingness to return to a 50% sucrose solution w/w containing imidacloprid at 50 μg/L (close to 40 μg/L). Similarly, a higher concentration of imidacloprid, 145 μg/L (115 ppb), reduced *A. mellifera* visits to a 2 M (55% w/w) sucrose feeder by 47% as compared to controls [17]. In our foraging experiment, we tested bees immediately after exposing them to imidacloprid. Thus, it is possible that their aversion was based upon their gustatory detection of the contaminant, not upon its neurological effects. However, our *A. cerana* foragers nonetheless exhibited the reduced visitation shown in *A. mellifera*.

In the collection experiment, we allowed each *A. cerana* forager to visit the treatment feeder 10 times in order to equilibrate its response to the treatment solution. These *A. cerana* foragers significantly decreased by 43% the volume of 20 μg/L imidacloprid solution (37% sucrose w/w) that they collected as compared to controls (Fig. 1B). This is similar to what is reported for *A. mellifera* foragers, which decreased the volume of 50% sucrose solution that they collected by 69% when it contained 29 μg/L (24 μg/Kg) of imidacloprid [13]. We note that the behavioral differences observed between these studies may arise from differences in bee species, sugar concentrations, and pesticides doses. For example, bees may tolerate a higher concentration of contaminants in sweeter nectar. Different environmental conditions may also play a role [3]. However, imidacloprid generally reduces the volume of sucrose solution collected by both species.

The overall exposure of each colony to imidacloprid was minimal. Based upon the number of bees trained and their visits to each treatment concentration in all three experiments over two months, each colony received only 3.6 ng of imidacloprid per day, on average. This dose was potentially shared among thousands of bees per colony. Control bees were evidently not affected by this potential exposure (Figs. 1 & 3).

Imidacloprid exposure in the danger experiment

By the time they were tested in the danger experiment, each bee had made 15 visits over approximately 1.5 hrs, sufficient time for imidacloprid absorption. For example, 20 min [17] or 60 min [23] is sufficient for orally administered imidacloprid to elicit strong effects: reduction in foraging activity and longer foraging flights [17] and reduced waggle dancing [23]. Because we allowed our bees to unload their collected sucrose solution to nestmates, it is unclear how much imidacloprid they absorbed before exchanging their collected nectar with nestmates. If bees absorbed a full nectar load over the 15 visits, each bee would have been exposed, on average, to 0.27, 0.39, and 0.52 ng of imidacloprid. These doses are based upon the average collection volume of the different imidacloprid concentrations (Fig. 1B) and therefore reflect decreased collection at higher imidacloprid concentrations. In addition, the tested bees completed all 15 visits to the feeder and were therefore exposed for the same number of trips. Thus, after compensating for reduced collection of higher imidacloprid concentrations, bees were exposed to 1.4 and 1.9 fold higher doses of imidacloprid at the 20 μg/L and 40 μg/L treatments as compared to the 10 μg/L treatment. We note that the effects of imidacloprid can be complex due to separate actions of imidacloprid and its toxic metabolites, 5-hydroxyimidacloprid and olefin [52]. In *A. mellifera*, imidacloprid has a metabolic half life of 4.5–5 hrs [58], and thus, over the average 1.5 hr exposure period of our experiment, imidacloprid was likely the main molecule responsible for altering bee behavior.

Effects on decision-making about danger

Apis cerana foragers that were not exposed to imidacloprid stayed clear of the dangerous feeder with the *V. velutina* hornet, in agreement with previous research [32]. However, the foragers that continued to visit the 40 μg/L feeder showed no significant avoidance of the dangerous feeder (Fig. 3). Did the imidacloprid-treated bees (1) suffer from a decision-making deficit, (2) were they unable to sense the predator, or (3) were they unable to control their flight and therefore randomly landed on the dangerous feeder? The latter two explanations seem unlikely. Although imidacloprid can affect visual learning in honey bees [22], there is no evidence that it degrades vision at sublethal doses. Moreover, the hornets are quite large (averaging 2 cm in length) and have

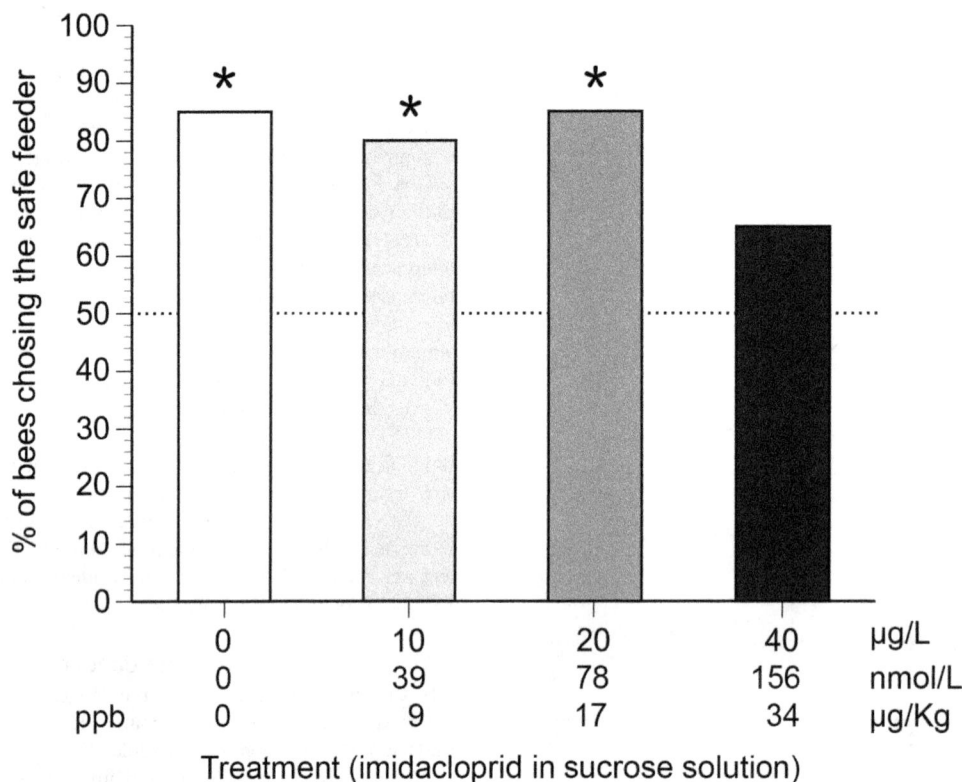

Figure 3. The effect of imidacloprid on the percentage of bees choosing a safe over a dangerous feeder. Stars above bars indicate treatments in which bees significantly avoided the dangerous feeder ($P<0.05$). Different shades of gray correspond to different imidacloprid concentrations. A dashed line shows the null hypothesis expectation: 50% of bees choose the safe feeder.

visually conspicuous aposematic coloration [29]. In addition, the foragers collecting 40 µg/L imidacloprid were still able to navigate between feeder and nest, capably orienting towards a feeder that was only 8 cm tall. This task requires good vision. The bees flew in a straight line and did not exhibit any signs of tremors as they collected the sucrose solution. At the nest, these foragers also found and unloaded their collected nectar without problems for five successive trips. Interestingly, the bee foragers that continued to visit even the highest concentrations of imidacloprid would, in other studies, have been considered bees that were relatively unaffected by the treatment because they continued to forage. However, in our experiment, we are able to demonstrate impairment: these bees did not distinguish the dangerous from the safe feeder. Whether this lack of discrimination arises from a decrease in "fear" of the hornet, an inability to make appropriate decisions, or from some other cognitive deficit remains unclear, but is a fascinating area for future research.

In general, we know relatively little about the effects of pesticides on bee information processing. Schricker et al. [59] hypothesized that bees sublethally poisoned with the organophosphate, parathion, had degraded integration of information in their central nervous system. Belzunces [52] suggested that pesticide exposure may lead to incorrect interpretation of external stimuli. Eiri and Nieh [23] proposed that imidacloprid can alter honey bee decision-making by reducing the number of recruitment dance circuits that foragers perform for a good food source. Bees increase the number of waggle dance circuits that they perform according to how valuable they consider a resource, and their perception of

resource value can be altered by a neuromodulator, octopamine [60]. Neonicotinoids may also alter bee perceptions about food. Eiri and Nieh [23] found that a single imidacloprid dose of 0.21 ng/bee (24 ppb) resulted in bees that continued to visit a rich 2.0 M (55% w/w) sucrose solution 24 hours later. These bees did not exhibit any impairments in flight, walking, or waggle dancing, a task requiring complex coordination [61]. Treated bees simply performed fewer dance circuits than controls, suggesting that they perceived the food, which was pure and free of any imidacloprid, as being less valuable for the colony [23].

Future research on sublethal doses of neonicotinoids may therefore reveal unsuspected effects on the complex decision-making shown by honey bees and contribute towards developing more sophisticated assays for determining safe application levels for neonicotinoids and future pesticides. Studies that also examine long term exposure or exposure at a sensitive developmental stage (such as in larvae [21]) would be beneficial. Fitness consequences of these impaired decisions should also be measured. For example, it is unclear if bees exhibiting risky behavior induced by imidacloprid suffer from increased predation. However, this possibility adds a new peril that deserves further study.

Author Contributions

Conceived and designed the experiments: KT JCN. Performed the experiments: KT WWC SHD XWL YCW. Analyzed the data: KT JCN. Contributed reagents/materials/analysis tools: JCN. Wrote the paper: KT JCN.

References

1. Lonsdorf E, Ricketts T, Kremen C, Winfree R, Greenleaf S, et al. (2011) Crop pollination services. In: Karieva P, Tallis H, Ricketts T, Daily GC, Polasky S, editors. Natural capital: theory and practice of mapping ecosystem services. Oxford: Oxford University Press. 168–187.

2. Potts SG, Biesmeijer JC, Kremen C, Neumann P, Schweiger O, et al. (2010) Global pollinator declines: trends, impacts and drivers. Trends Ecol Evol (Amst) 25: 345–353. doi:10.1016/j.tree.2010.01.007.

3. Desneux N, Decourtye A, Delpuech JM (2007) The sublethal effects of pesticides on beneficial arthropods. Annu Rev Entomol 52: 81–106. doi:10.1146/annurev.ento.52.110405.091440.

4. Whitehorn PR, O'Connor S, Wackers FL, Goulson D (2012) Neonicotinoid pesticide reduces bumble bee colony growth and queen production. Science 336: 351–352. doi:10.1126/science.1215025.

5. Cresswell JE, Desneux N, vanEngelsdorp D (2012) Dietary traces of neonicotinoid pesticides as a cause of population declines in honey bees: an evaluation by Hill's epidemiological criteria. Pest Manag Sci 68: 819–827. doi:10.1002/ps.3290.

6. Goulson D (2013) An overview of the environmental risks posed by neonicotinoid insecticides. J Appl Ecol 50: 977–987. doi:10.1111/1365-2664.12111.

7. Jeschke P, Nauen R (2008) Neonicotinoids–from zero to hero in insecticide chemistry. Pest Manag Sci 64: 1084–1098.

8. Gross M (2013) EU ban puts spotlight on complex effects of neonicotinoids. Curr Biol.

9. Abrol DP (2013) Safety of bees in relation to pest management. Asiatic Honeybee Apis cerana. Dordrecht: Springer Netherlands. 575–640. doi:10.1007/978-94-007-6928-1_14.

10. Johnson RM, Ellis MD, Mullin CA, Frazier M (2010) Pesticides and honey bee toxicity–USA. Apidologie 41: 312–331. doi:10.1051/apido/2010018.

11. Decourtye A, Henry M, Desneux N (2013) Environment: overhaul pesticide testing on bees. Nature 497: 188. doi:10.1038/497188a.

12. Schmuck R, Nauen R, Ebbinghaus-Kintscher U (2003) Effects of imidacloprid and common plant metabolites of imidacloprid in the honeybee: toxicological and biochemical considerations. Bull Insectol 56: 27–34.

13. Decourtye A, Devillers J, Cluzeau S, Charreton M, Pham-Delègue MH (2004) Effects of imidacloprid and deltamethrin on associative learning in honeybees under semi-field and laboratory conditions. Ecotox Environ Safe 57: 410–419. Available: http://www.sciencedirect.com/science/article/pii/S0147651303001477.

14. Decourtye A, Lacassie E, Pham-Delègue MH (2003) Learning performances of honeybees (Apis mellifera L) are differentially affected by imidacloprid according to the season. Pest Manag Sci 59: 269–278. doi:10.1002/ps.631.

15. Ramirez-Romero R, Chaufaux J, Pham-Delègue MH (2005) Effects of Cry1Ab protoxin, deltamethrin and imidacloprid on the foraging activity and the learning performances of the honeybee Apis mellifera, a comparative approach. Apidologie 36: 601–611. doi:10.1051/apido:2005039.

16. Lambin M, Armengaud C, Raymond S, Gauthier M (2001) Imidacloprid-induced facilitation of the proboscis extension reflex habituation in the honeybee. Arch Insect Biochem Physiol 48: 129–134. doi:10.1002/arch.1065.

17. Schneider CW, Tautz JR, Grünewald B, Fuchs S (2012) RFID tracking of sublethal effects of two neonicotinoid insecticides on the foraging behavior of Apis mellifera. PLOS ONE 7: e30023. doi:10.1371/journal.pone.0030023.g005.

18. Henry M, Beguin M, Requier F, Rollin O, Odoux JF, et al. (2012) A common pesticide decreases foraging success and survival in honey bees. Science 336: 348–350. doi:10.1126/science.1215039.

19. Pahl M, Tautz JR, Zhang SW (2010) Honeybee cognition. Animal Behaviour: Evolution and Mechanisms. Berlin: Springer. 87–120. doi:10.1007/978-3-642-02624-9_4.

20. Williamson SM, Baker DD, Wright GA (2012) Acute exposure to a sublethal dose of imidacloprid and coumaphos enhances olfactory learning and memory in the honeybee Apis mellifera. Invert Neurosci 13: 63–70. doi:10.1007/s10158-012-0144-7.

21. Yang EC, Chang HC, Wu WY, Chen YW (2012) Impaired olfactory associative behavior of honeybee workers due to contamination of imidacloprid in the larval stage. PLOS ONE 7: e49472.

22. Han P, Niu CY, Lei CL, Cui JJ, Desneux N (2010) Use of an innovative T-tube maze assay and the proboscis extension response assay to assess sublethal effects of GM products and pesticides on learning capacity of the honey bee Apis mellifera L. Ecotox 19: 1612–1619. doi:10.1007/s10646-010-0546-4.

23. Eiri DM, Nieh JC (2012) A nicotinic acetylcholine receptor agonist affects honey bee sucrose responsiveness and decreases waggle dancing. J Exp Biol 215: 2022–2029. doi:10.1242/jeb.068718.

24. Gonçalves-Souza T, Omena PM, Souza JC, Romero GQ (2008) Trait-mediated effects on flowers: artificial spiders deceive pollinators and decrease plant fitness. Ecology 89: 2407–2413.

25. Dukas R, Morse DH (2005) Crab spiders show mixed effects on flower-visiting bees and no effect on plant fitness components. Ecoscience 12: 244–247.

26. Bray A, Nieh JC (2014) Non-consumptive predator effects shape honey bee foraging and recruitment dancing. PLOS ONE 9: e87459. doi:10.1371/journal.pone.0087459.

27. Abrol DP (2006) Defensive behaviour of Apis cerana F. against predatory wasps. J Apic Sci 20: 39–46.

28. Burgett M, Akratanakul P (1982) Predation on the western honey bee, Apis mellifera L., by the hornet, Vespa tropica (L.). Psyche 89: 347–350. doi:10.1155/1982/37970.

29. Matsuura M, Yamane S (1990) Biology of the vespine wasps. Berlin, Germany: Springer Verlag. 1 p.

30. Tan K, Li H, Yang MX, Hepburn HR, Radloff SE (2010) Wasp hawking induces endothermic heat production in guard bees. J Insect Sci 10: 1–6. doi:10.1673/031.010.14102.

31. Tan K, Radloff SE, Li JJ, Hepburn HR, Yang MX, et al. (2007) Bee-hawking by the wasp, Vespa velutina, on the honeybees Apis cerana and A. mellifera. Naturwissenschaften 94: 469–472. Available: http://www.springerlink.com/index/10.1007/s00114-006-0210-2.

32. Tan K, Hu Z, Chen W, Wang Y, Wang Z, et al. (2013) Fearful foragers: honey bees tune colony and individual foraging to multi-predator presence and food quality. PLOS ONE 8: e75841.

33. Tomé HVV, Martins GF, Lima MAP, Campos LAO, Guedes RNC (2012) Imidacloprid-induced impairment of mushroom bodies and behavior of the native stingless bee Melipona quadrifasciata anthidioides. PLOS ONE 7: e38406. doi:10.1371/journal.pone.0038406.

34. Gill RJ, Ramos-Rodriguez O, Raine NE (2012) Combined pesticide exposure severely affects individual- and colony-level traits in bees. Nature: 1–5. doi:10.1038/nature11585.

35. Arena M, Sgolastra F (2014) A meta-analysis comparing the sensitivity of bees to pesticides. Ecotox 23: 324–334.

36. Verma LR, Partap U (1993) The Asian hive bee, Apis cerana, as a pollinator in vegetable seed production. International Centre for Integrated Mountain Development (ICIMOD).

37. Yang GH (2005) Harm of introducing the western honey bee, Apis mellifera L., to the Chinese honey bee Apis cerana F. and its ecological impact. Acta Entomol Sinica (Chinese language journal) 48: 401–406.

38. Corlett RT (2001) Pollination in a degraded tropical landscape: a Hong Kong case study. J Trop Ecol 17: 155–161. doi:10.1017/S0266467401001109.

39. Peng YS, Nasr ME, Locke SJ (1989) Geographical races of Apis cerana Fabricius in China and their distribution. Review of recent Chinese publications and a preliminary statistical analysis. Apidologie 20: 9–20.

40. Higo M (1980) Sensitivity of honeybees to some pesticides with special reference to the effects of age and season. Honeybee Sci 1: 177–180.

41. Suh YT, Shim JH (1988) A study on the enzyme activities of a honeybee (Apis cerana F.) associated with the degradation of some insecticides. Korean J Environ Agric 8: 47–54.

42. Mishra RC, Verma AK (1982) Relative toxicity of some insecticides to Apis cerana indica F. workers. Indian Bee J 44: 475–476.

43. Khan RB, Dethe MD (2005) Toxicity of new pesticides to honey bees. In: Kumar A, editor. Environment and Toxicology. New Delhi: APH Publishing Corporation. 59–62.

44. Yuchong W, Weiwen C, Shihao D, Xiwen L, Ken T (2014) Toxicity of imidacloprid to honeybee (Apis cerana). J Yunnan Agricul Univ: inpress.

45. Yu X, Wang M, Kang M, Liu L, Guo X, et al. (2011) Molecular cloning and characterization of two nicotinic acetylcholine receptor β subunit genes from Apis cerana cerana. Arch Insect Biochem Physiol 77: 163–178.

46. Sun ZJ, Fang XC, Du WZ (1995) Imidacloprid as a highly efficient pesticide. J Plant Protect 2: 44–45.

47. Monceau K, Bonnard O, Thiéry D (2014) Vespa velutina: a new invasive predator of honeybees in Europe. J Pest Sci 87: 1–16. doi:10.1007/s10340-013-0537-3.

48. Bubník Z, Kadlec P, Urban D, Bruhns M (1995) Sugar technologists manual: Chemical and physical data for sugar manufacturers and users. Verlag Dr Albert Bartens 120: 574–575.

49. Byrne FJ, Visscher PK, Leimkuehler B, Fischer D, Grafton-Cardwell EE, et al. (2013) Determination of exposure levels of honey bees foraging on flowers of mature citrus trees previously treated with imidacloprid. Pest Manag Sci. doi:10.1002/ps.3596.

50. Cresswell JE (2010) A meta-analysis of experiments testing the effects of a neonicotinoid insecticide (imidacloprid) on honey bees. Ecotox 20: 149–157. doi:10.1007/s10646-010-0566-0.

51. Williamson SM, Wright GA (2013) Exposure to multiple cholinergic pesticides impairs olfactory learning and memory in honeybees. J Exp Biol 216: 1799–1807. doi:10.1242/jeb.083931.

52. Belzunces LP, Tchamitchian S, Brunet JL (2012) Neural effects of insecticides in the honey bee. Apidologie 43: 348–370. doi:10.1007/s13592-012-0134-0.

53. Schmuck R, Schöning R, Stork A, Schramel O (2001) Risk posed to honeybees (Apis mellifera L, Hymenoptera) by an imidacloprid seed dressing of sunflowers. Pest Manag Sci 57: 225–238. doi:10.1002/ps.270.

54. Nauen R, Ebbinghaus-Kintscher U, Schmuck R (2001) Toxicity and nicotinic acetylcholine receptor interaction of imidacloprid and its metabolites in Apis mellifera (Hymenoptera: Apidae). Pest Manag Sci 57: 577–586. doi:10.1002/ps.331.

55. Zar JH (1984) Biostatistical analysis. Englewood Cliffs, New Jersey: Prentice-Hall.

56. Colin ME, Bonmatin JM, Moineau I, Gaimon C, Brun S, et al. (2004) A method to quantify and analyze the foraging activity of honey bees: relevance to the sublethal effects induced by systemic insecticides. Arch Environ Contam Toxicol 47. doi:10.1007/s00244-004-3052-y.

57. Yang EC, Chuang YC, Chen YL, Chang LH (2008) Abnormal foraging behavior induced by sublethal dosage of imidacloprid in the honey bee (Hymenoptera: Apidae). J Econ Entomol 101: 1743–1748.

58. Suchail S, Debrauwer L, Belzunces LP (2004) Metabolism of imidacloprid in *Apis mellifera*. Pest Manag Sci 60: 291–296.

59. Schricker B, Stephen WP (1970) The effect of sublethal doses of parathion on honeybee behaviour. I. Oral administration and the communication dance. Journal of Apicultural Research 9: 141–153.

60. Barron AB, Maleszka R, Vander Meer RK, Robinson GE (2007) Octopamine modulates honey bee dance behavior. Proc Natl Acad Sci USA 104: 1703–1707. doi:10.1073/pnas.0610506104.

61. Landgraf T, Rojas R, Nguyen H, Kriegel F, Stettin K (2011) Analysis of the waggle dance motion of honeybees for the design of a biomimetic honeybee robot. PLOS ONE 6: e21354. doi:10.1371/journal.pone.0021354.t001.

10

Prenatal and Postnatal Exposure to DDT by Breast Milk Analysis in Canary Islands

Oriol Vall[1,2,3], Mario Gomez-Culebras[4], Carme Puig[1,2], Ernesto Rodriguez-Carrasco[4], Arelis Gomez Baltazar[1,3], Lizzeth Canchucaja[1,3], Xavier Joya[1,2], Oscar Garcia-Algar[1,2,3]*

1 Unitat de Recerca Infància i Entorn (URIE), Institut Hospital del Mar d'Investigacions Mèdiques (IMIM), Barcelona, Spain, 2 Red de Salud Materno-Infantil y del Desarrollo (SAMID), Instituto Carlos III, Madrid, Spain, 3 Departament de Pediatria, Obstetricia, Ginecologia i Medicina Preventiva, Universitat Autònoma de Barcelona, Barcelona, Spain, 4 Departamento de Cirugía Pediátrica, Hospital de la Candelaria, Universidad de Tenerife, Santa Cruz de Tenerife, Spain

Abstract

Introduction: The use of *p,p'*-dichlorodiphenyltrichloroethane (DDT) has been banned since the late 1970s due to its toxicity. However, its long half-life makes it persistent in the environment and, consequently, almost everyone has DDT residues in the body. Human milk constitutes an ideal non-conventional matrix to investigate environmental chronic exposure to organochlorine compounds (OCs) residues. The study aimed to identify potential population risk factors of exposure to DDT due to the proximity to countries where it is still used.

Methods: Seventy-two consecutive lactating women were prospectively included in Tenerife, Canary Islands (Spain). A validated questionnaire was used to obtain socioeconomic, demographics data, and daily habits during pregnancy. DDT levels in breast milk were measured by gas chromatography with-electron capture detector (GC-ECD). Anthropometrics measurements in newborns were obtained.

Results: Thirty-four out of 72 (47.2%) of the analysed milk samples presented detectable levels of DDT (mean: 0.92 ng/g), ranging between 0.08 to 16.96 ng/g. The socio-demographic variables did not significantly differ between detectable DDT and non-detectable DDT groups. We found positive association between DDT levels and vegetables (OR (95%CI): 1.23 (1.01–1.50)) and poultry meat (OR (95%CI): 2.05 (1.16–3.60)) consumption, and also between the presence of DDT in breast milk and gestational age (OR (95%CI): 0.59 (0.40–0.90)).

Conclusions: DDT is present in breast milk of women at the time of delivery. Residual levels and the spread from countries still using DDT explain DDT detection from vegetables and from animal origin food. The presence of this compound in breast milk represents a pre- and postnatal exposure hazard for foetuses and infants due to chronic bioaccumulation and poor elimination, with possible deleterious effects on health. This data should be used to raise awareness of the risks of OCs exposure and to help establish health policies in order to avoid its use worldwide and thus, to prevent its propagation.

Editor: Cheryl S. Rosenfeld, University of Missouri, United States of America

Funding: These authors have no support or funding to report.

Competing Interests: The authors have declared that no competing interests exist.

* E-mail: 90458@parcdesalutmar.cat

Introduction

According to the Stockholm Convention, Organochlorine Compounds (OCs) are considered Persistent Organic Pollutants (POPs). Their production and usage is forbidden in many countries due to toxicity, persistence, mobility and bioaccumulation in the environment. Due to their high lipophilicity and persistency they can bioaccumulate through the food chain [1,2]. Since the discovery of its insecticidal properties in 1939, dichlorodiphenyltrichloroethane (DDT) was used extensively all over the world as a domestic and agricultural pesticide.

However, in Spain, as in most European countries, DDT was banned in the late 70 s [3,4]. Currently, it is still used in African countries, such as Morocco [5].

The Canary Islands are located in the Atlantic Ocean, only 100 km away from the North African coast (South-West of Morocco). This proximity facilitates OCs transport by air and

water, becoming sources of propagation. The economy of Canary Islands is based mainly on tourism and, to a much lesser extent, on farming and fishing. In the last decades farming in these Islands has turned into an intensive type of agriculture (plastic greenhouses) [6]. It is well known that intensive agriculture uses pesticides in large amounts [7]. According to calculations, in 2001 the Canary Islands consumed 12 times more pesticides by hectare than the rest of Spain [8]. Previous data confirms that the general population in the Canary Islands presents with dichlorodiphenyldichloroethylene (DDE), the main metabolite of DDT [6].

Except for the individuals exposed through their occupation, most of the exposure to these compounds occurs through diet, especially food with high content of fat [9] such as meat, fish, poultry and dairy products. In the organism, due to their lipophilic character, POPs accumulate in the adipose tissue. During the production of breast milk, the human body uses lipids from the adipose tissue, and subsequently the accumulated POPs from the adipose tissue can migrate to breast milk [10]. Therefore, while providing the necessary nutrients for the correct development of the infant, human milk is also a source of lipophilic environmental pollutants [10–13].

The purpose of this study was to determine the levels of DDT in breast milk from a group of lactating mothers from the Canary Islands, and ascertain possible associations between this exposure and socio-demographic coordinates, diet and mother's habits, and their potential impact on the newborn's health.

Materials and Methods

Subjects and study design

A cross sectional descriptive pilot study was conducted in Tenerife, the largest and most populated island of the seven islands that make up the archipelago of the Canary Islands (Spain). The economy of Tenerife is based on a few sectors: tourism (78%) industrial activities (9%) and, to a lesser extent, farming and fishing (2%). Agriculture is centred on the northern slopes. In the last decades, farming in these Islands has turned to an intensive kind of agriculture (plastic greenhouses).

The study was carried out in the Paediatric Department of the Hospital La Candelaria of Santa Cruz de Tenerife, Spain, the main hospital of the island. A total of 72 consecutive pregnant women agreed to participate in the study and signed an informed consent form. The inclusion criteria were to be pregnant, to accept enrolment in the study and to sign a mother–newborn dyad informed consent. The exclusion criteria were not to sign the informed consent and if a breast milk sample was not obtained or was inadequate for DDT analysis. The study was approved by the local Ethics Committee (Comité de Ética e Investigación Clínica (CEIC), Hospital la Candelaria, Santa Cruz de Tenerife) in accordance with the Helsinki Declaration. The enrolment period was between August 2006 and June 2007.

Determination and quantification of DDT in breast milk samples

On the day of delivery, breast milk was collected and stored at $-20°C$ until analysis. A simple, sensitive and efficient liquid-liquid extraction method prior to analytical detection was used in this study with minor modifications [14]. PCB#1 and PCB#209 were added as internal standards to 200 μL of breast milk sample previously homogenized by vortexing it. The breast milk sample was homogenized with anhydrous sodium sulphate and extracted with n-hexane. At this point 1 μL of the organic layer was injected into the gas chromatograph with flame ionisation detector (GC-FID) in order to determine the concentration of fatty acid methyl esters (FAMEs). Then, the sample was extracted with 1 ml H_2SO_4 (96%) three times, mixing and centrifuging each time. Analyses of OC were carried out in a gas chromatograph with electron capture detection (GC-ECD) equipped with a fused silica, 5% diphenyl–95% dimethylsilicone (SGE, Ringwood, Australia) open tubular column of 25 m, 0.53 mm I.D. and 0.25 μm film thickness. The initial temperature of the oven was 140°C, raised to 267°C at a rate of 2.5°C/min. Injector and detector temperature were 290°C and 310°C, respectively. Carrier gas (He) was set at a linear velocity of 25 cm/s. Make-up gas (N_2) was adjusted to a flow of 60 ml/min. Quantification of the OC compounds was made using the PCB#209 as internal standard. The concentration of DDT in breast milk is expressed in ng/g. The limit of detection (LOD) and limit of quantification (LOQ) were around 1 ng/g [15].

Study variables

A previously validated questionnaire [16] was administered the day after delivery by trained interviewers and was focused on socio-demographic, dietary and lifestyle information, with the following variables: age, socio-economic status, area of residence, country of origin, length of time living on the island, educational level, parity, working status, tobacco and alcohol consumption. . The dietary information regarding the period of pregnancy was collected using a food frequency questionnaire of 16 items. Participants were asked how many times per week, on average, had consumed each item during the last month of pregnancy. Also, they were asked about the origin of their drinking water. Obstetric history of the mother was collected and neonatal anthropometric characteristics using customized growth charts for birth weight and height and clinical examination at birth were also recorded.

Statistical analysis

Descriptive statistics of DDT skewed distribution in breast milk were performed using mean, median, and percentiles. The dependent variable was defined as dichotomous DDT level (detected vs non-detected). Preliminary association between socio-demographic, life style characteristics of pregnant women and neonatal outcomes with foetal exposure to DDT were done by Student's t test for continuous variables and Chi-square test for dichotomous variables. Multivariate logistic regression was adjusted for maternal age and gestational age and was used to find positive associations to DDT.. Potential determinants were analysed using dichotomous DDT levels, taking into account confounding variables. Smoking was included as a potential confounder. Statistical significance was set at $p<0.05$. Database management and statistical analysis were performed with SPSS v 20.0 (SPSS, Chicago, IL, USA).

Results

Seventy two breast milk samples (participation rate: 49.65%) were used to determine the presence of DDT. DDT was detected in 34 breast milk samples (47.22%). The mean concentration (SD) was 0.92 (2.40) ng/g, ranging between 0.08 to 16.96 ng/g. The percentile 25th–75th ranged between 0.00–1.19 ng/g. The distribution of parental socio-demographic characteristics in two categories (DDT detected and DDT non-detected) is shown in **Table 1**. More than 80% of mothers with detectable levels of DDT in breast milk were Spanish with an average age of 30.48 (5.31) years. The area of residence, length of time living on the island and maternal profession had no influence on breast milk values of DDT.

Anthropometrics characteristics and perinatal history data are shown in **Table 2**. The mean gestational age was 39.2 (1.59) weeks. Newborns from mothers with detectable levels of DDT presented a significantly lower gestational age in comparison to the newborns of mothers with non-detectable DDT levels (p = 0.002). The presence of DDT in breast milk showed a significant negative correlation with the newborn's gestation time. (Spearman's rho: −0.373; p = 0.001). Women with detectable levels of DDT had an 8.8% of premature birth, compared with 2.7% for the mothers with no detectable DDT levels. (p = 0.344). Weight, length and cranial perimeter at birth did not differ between the groups exposed or non exposed to DDT. There were no differences between the two groups with respect to other conditions such as chromosomic alterations, increased risk of perinatal infection, hypoglycemia and/or congenital hip dysplasia.

Table 1. Parental socio-demographics characteristics by prenatal and postnatal exposure to DDT detected in breast milk.

		Breast milk positive samples to DDT (n = 34)	Breast milk negative samples to DDT (n = 38)	p-value
Maternal age (years), mean (SD)		30.59 (5.83)	30.37 (4.85)	0.865
Parental country of origin (Spain/Other) (%)				
	Non-Spanish mothers	20.6	10.5	0.236
	Non-Spanish fathers	21.2	15.8	0.556
Parental Educational Level (%)				
Mother's	Non finalized elementary school studies	12.5	21.6	0.489
	Elementary school studies	40.6	43.2	
	University Studies	46.9	35.1	
Father's	Non finalized elementary school studies	21.9	33.3	0.568
	Elementary school studies	78.8	84.2	
	University Studies	21.2	15.8	
Parental employment (yes) (%)				
Unemployed mother		23.5	21.6	0.848
Unemployed father		0	2.6	1
Parental socioeconomic status (%)				
Mother's	Skilled	46.2	65.5	0.148
	Unskilled	53.8	34.5	
Father's	Skilled	58.6	51.4	0.556
	Unskilled	41.4	48.6	
Area of residence (%)				
	Rural (<10.000 inhab.)	17.6	15.8	0.975
	Semi-rural (10–100.000 inhab.)	26.5	26.3	
	Urban (>100.000) inhab.)	55.9	57.9	

Chi-square test; p<0.05.

Maternal lifestyle, chemical exposure and dietary habits during pregnancy are presented in **Table 3**. Tobacco consumption before pregnancy was documented in 43.3% of the women positive to DDT in comparison to 31% with negative DDT result (p = 0.329). A frequent intake of vegetables was associated with detectable DDT levels in breast milk (p = 0.014). The presence of DDT in breast milk correlated with the average consumption of vegetables (Spearman's rho: 0.301; p = 0.013). Moreover, a frequent intake of poultry meat was also associated with detectable DDT levels in breast milk (p = 0.025). Nevertheless, the presence of DDT in breast milk did not correlate with the average consumption of poultry meat (Spearman's rho: 0.231; p = 0.058). Detectable DDT levels in breast milk were not associated with other categories of food and fish consumption.

Multiple linear regression was performed but we didn't obtain any significance probably due the small size of the sample. **Table 4** shows the OR and 95%CI obtained in the adjusted logistic regression model. This model includes only the statistically significant variables from the univariate logistic regression model. The model was adjusted using confounding variables such as smoking, maternal age or newborn gender. The gestational age (OR (95%CI): 0.59 (0.40–0.90)), the intake of vegetables (OR (95%CI): 1.23 (1.01–1.50)) and poultry meat (OR (95%CI): 2.05 (1.16–3.60)) were significantly associated with DDT detection in breast milk.

Discussion

This study found detectable levels of DDT in 34 (47.2%) of 72 breast milk samples obtained from a population of lactating women from Tenerife (Canary Islands). Our data confirms that almost half of the pregnant women from our study performed in the Canary Islands, present unexpected DDT residues in breast milk. We found that the concentrations of DDT observed in the present study were within the range reported by other studies performed in similar areas [17–20]. However, DDT levels documented in this study were considerably lower than the levels found in human breast milk from China, South-Africa and other Mediterranean countries [21–27]. This variability observed in DDT concentration between different countries could be explained in part because some of them still produce and/or use DDT. This fact represents a source of propagation by air, water and exportation of products for human consumption. However, there is evidence that POPs concentrations have been decreasing significantly in the last four years due to bio-monitoring program interventions [28].

Taking into account that one of the main sources of DDT is the environment, we studied the possible differences in DDT levels associated with differences in urbanization of our population. In our study, DDT levels were not different in women living in urban or rural zones. This data differs from other studies in the same area. Zumbado M *et al.* found high levels of DDT in Tenerife and

Table 2. Obstetric and anthropometric characteristics of the newborns according to the results obtained.

	Breast milk positive samples to DDT (n = 34)	Breast milk negative samples to DDT (n = 38)	p-value
Previous pregnancies (%)			
0	47.1 (16)	55.3 (21)	0.244
1	20.6 (7)	28.9 (11)	
>2	32.4 (11)	15.8 (6)	
Previous premature infants (%)			
Yes	0	2.6 (1)	1
Previous abortions (%)			
Yes	35.3 (12)	18.9 (7)	0.119
Children characteristics at birth			
Gender; Female (%)	38.2 (13)	55.3 (21)	0.148
Gestational age (week), mean (SD)	38.4 (1.4)	39.4 (1.6)	0.008
Prematurity (%)	8.8 (3)	2.7 (1)	0.344
Weight at birth (g), mean (S.D.)	3335.29 (474.6)	3357.50 (466)	0.856
Length at birth (cm), mean (S.D.)	50.5 (2.3)	51.1 (3.2)	0.341
Cranial Perimeter (cm), mean (S.D.)	34.4 (1.4)	34.2 (14)	0.526
Clinical diagnosis at birth (yes), (%)			
Perinatal History	23.5 (8)	34.2 (13)	0.32
Chromosomic Alteration	0	0	NA
Loss of foetal well being	0	2.6 (1)	1
Risk of Perinatal Infection	8.8 (3)	23.7 (9)	0.091
Hypoglycemia	8.8 (3)	5.3 (02)	0.662
Developmental dysplasia of the hip	0	0	NA
Other outcomes	2.9 (1)	0	1

NA: Not applicable.
Chi-square test; $p < 0.05$.

Gran Canaria Island (with many more inhabitants, with the largest urban areas and the largest surface covered by plastic greenhouses) probably due to higher intake of dairy products and of lipids and saturated fatty acids [6,29].

It has been shown previously that DDT levels increased proportionally with age since OCs accumulate in adipose tissue over time [6,30,31]. We didn't find an association between maternal age and accumulation of DDT in breast milk. This fact can also be related to the so-called "cohort effect", because older subjects, born before restriction of the production and use of DDTs, would start with a higher body burden of POPs in comparison to younger people [31–34]. With respect to the perinatal effects of DDT exposure, it is well known that this pollutant is associated with negative reproductive outcomes in human studies [35,36]. We found a significant negative correlation between gestational age and DDT levels. This observation was in accordance with other authors that studied the exposure to POPs during prenatal life [37,38]. However, we didn't find any association between DDT exposure and birth outcomes such birth weight [39,40].

Food is considered a constant source of exposure; Schafer *et al.* claimed that every food group usually contained at least five OCs, with DDT and dieldrin being the most frequent ones. [41]. Although ingestion of legumes, vegetables and bread was related to lower serum and adipose tissue concentrations of DDT [31], our data showed a statistically significant correlation between DDT levels and vegetables consumption. DDT residues in soil due to its use in past decades and ground water pollution in the islands can explain the contamination. Canary Islands, like all vulcanic islands, have geological characteristics that facilitate soil and ground water contamination (low organic matter). This type of soil and water saturation by pesticides has also been shown in the Hawaii Islands [42]. This feature, plus the proximity of the islands to Africa and the high consumption of vegetables imported from countries where these compounds are still used, may explain the presence of some DDT residues in lactating women eating vegetables [32,43].

Poultry meat seems to be another intake source of DDT in our study. The importance of eggs, chicken and meat consumption as a source of OCP has been established worldwide [44–46]. Chicken meat is among the most popular food items in the diet of many communities and because of its high fat content it increased our concern related to human exposure to OCPs. Animals intended for human food may absorb pesticides from residues in their feed, water or during direct/indirect exposure in the course of pest control. The literature indicated that poultry feed could be one of the major sources of contamination through meat and eggs [47]. In order to minimize intake of OCP residues, people should adapt their pattern of food consumption (including a limitation of the consumption or including different types of meat).

Table 3. Maternal lifestyle, chemical exposure and dietary habits during pregnancy.

	Breast milk positive samples to DDT (n = 34)	Breast milk negative samples to DDT (n = 38)	p-value
Maternal tobacco smoke exposure (yes), (%)			
Smoking before pregnancy	43.3 (13)	31.0 (9)	0.329
Smoking during pregnancy	18.2 (6)	24.3(9)	0.532
Drug abuse (yes), (%)	3.0 (1)	5.4 (2)	0.240
Alcohol consumption (yes), (%)	0	0	NA
Medicine use (yes), (%)			
Antidepressants	10.0 (1)	5.7 (2)	0.284
Vitamin supplements	12.1 (4)	19.4 (7)	0.689
Antibiotics	21.2 (7)	16.7 (6)	0.233
Water source (%)			
Running Water	26.5 (9)	32.4 (12)	0.697
Private Well	2.9 (1)	0	
Mineral Water	70.6 (24)	67.6 (25)	
Diet (times per week), mean (SD)			
Full-Fat Milk	8.21 (9.2)	6.62 (8.40)	0.364
Fat-Free Milk	10.59 (11.21)	10.84 (10.59)	0.788
Fruit	8.65 (5.37)	8.89 (6.05)	0.807
Vegetables	7.50 (4.06)	6.30 (3.75)	0.014
Eggs	2.62 (1.65)	2.30 (1.45)	0.332
Butter and/or margarine	2.74 (2.68)	3.03 (2.64)	0.528
Legume	1.71 (1.19)	1.81 (1.15)	0.592
Nuts	1.24 (2.10)	0.62 (1.44)	0.282
Red meat	2.21 (2.56)	1.92 (2.25)	0.678
Processed meat	0.56 (0.89)	0.43 (0.83)	0.463
Poultry meat	2.65 (1.32)	2.0 (0.8)	0.025
Fish	1.62 (1.15)	1.57 (0.95)	0.905
Precooked food	0.35 (0.64)	0.57 (1.04)	0.491
Canned food	1.91 (1.84)	1.76 (1.64)	0.868
Commercial juice	4.59 (4.62)	6.3 (5.74)	0.217
Soft drinks	2.05 (3.69)	4.7 (4.0)	0.624

NA: Not applicable.
Chi-square test; p<0.05.

Table 4. Multivariate logistic regression model that associates gestational age, vegetables and poultry meat consumption with the presence of DDT in breast milk.

Variable	OR Adj	CI (95%)	p-value
Gestational age	0.598	0.397–0.900	0.014
Vegetables	1.230	1.007–1.502	0.042
Poultry meat	2.045	1.161–3.604	0.013

Adj: Adjusted for maternal age.
Calibration of the model (Hosmer & Lemeshow test) p = 0.161.
Discrimination power (AUC): 0.803 (0.693–0.912).

Limitations

The moderate participation rate (49.65%) and subjects selected from a single hospital, in spite of the fact that Hospital La Candelaria in Tenerife has the largest number of births on the island, precludes generalization of the results to the study area population as a whole. , Being only a preliminary study, . we can confirm a relationship between vegetables consumption and higher risk of high DDT levels, however increasing the sample size could provide results which using multivariate analysis could potentially determine personal characteristics associated with higher risk for elevated DDT levels. Additional data in dietary habits questionnaire are required to establish a relationship between diet patterns and positivity for DDT and concerning effects of foetal exposure to DDT. Another factor to be considered is that our dietary habits questionnaire shows the consumption only in the last month and in some cases, the food habits reported by the women may not have been representative for the long term food habits. . Several studies in Spain did not find significant associations between levels of these pesticides and dietary patterns [48]. DDT levels in food can indicate its recent use in agricultural areas. Nevertheless, due to the obvious benefits of eating vegetables and fruits for proper foetal development, caution should be exercised when making sweeping dietary recommendations for pregnant women.

Ongoing use of DDT for malaria prevention, its environmental persistence, documented adverse reproductive effects in animals, and inconsistent findings across human studies, justifies continued exploration of the DDT and DDE impact on human beings. The results of this study highlight the importance of carrying out similar studies in this field.

Conclusion

Despite DDT being banned in Spain since 1977, Canary Islands population is still exposed to this insecticide. It is important to emphasize that with one isolated exception, concentrations were under the acceptable criteria by WHO, suggesting that infants in the island are exposed to small quantities of DDT [49]. However, even after controlling for smoking, there was still a positive association of DDT and gestational age. There is evidence that POPs concentrations have decreased significantly, but, because of globalization, children eat a variety of vegetables from all over the world, including countries in which usage of DDT has not been restricted yet, due to its usefulness in eradicating diseases such as malaria. Virtually, all populations worldwide bear a body burden of POPs with large interindividual and inter-population differences [28,49]. Therefore, studies are essential to establish reference concentrations, to analyze predictors of exposure, to increase public awareness, to stimulate more energetic policies and population strategies and, hence, to diminish the burden of exposure, especially in infants. The accumulation of these compounds in the fatty tissue over the mother's life may be an important source of exposure for children, both during gestation and through breastfeeding. DDT levels measured in our study were not associated with perinatal outcomes such as weight or length at birth. However, further research regarding other OCs pesticides and especially the chemicals currently used and their doses is needed, with the goal to improve knowledge and implement health and policy strategies in order to regulate their use and to prevent its propagation.

Author Contributions

Conceived and designed the experiments: OGA OV MGC. Performed the experiments: OGA OV MGC ERC AGB XJ. Analyzed the data: OGA OV MGC CP ERC XJ. Contributed reagents/materials/analysis tools: CP ERC AGB XJ. Wrote the paper: OGA OV MGC CP ERC AGB LC XJ.

References

1. Ridal JJ, Mazumder A, Lean DR (2001) Effects of nutrient loading and planktivory on the accumulation of organochlorine pesticides in aquatic food chains. Environ Toxicol Chem 20: 1312–1319.
2. Ahlborg UG, Brouwer A, Fingerhut MA, Jacobson JL, Jacobson SW, et al. (1992) Impact of polychlorinated dibenzo-p-dioxins, dibenzofurans, and biphenyls on human and environmental health, with special emphasis on application of the toxic equivalency factor concept. Eur J Pharmacol 228: 179–199.
3. Ferrer A (2003) [Pesticide poisoning]. An Sist Sanit Navar 26 Suppl 1: 155–171.
4. Gomez-Catalan J, Lezaun M, To-Figueras J, Corbella J (1995) Organochlorine residues in the adipose tissue of the population of Navarra (Spain). Bull Environ Contam Toxicol 54: 534–540.
5. UNEP (2006) Report of the expert group on the assessment of the production and use of DDT and its alternatives for disease vector control to the Conference of the Parties of the Stockholm Convention at its third meeting.
6. Zumbado M, Goethals M, Alvarez-Leon EE, Luzardo OP, Cabrera F, et al. (2005) Inadvertent exposure to organochlorine pesticides DDT and derivatives in people from the Canary Islands (Spain). Sci Total Environ 339: 49–62.
7. Olea N, Olea-Serrano F, Lardelli-Claret P, Rivas A, Barba-Navarro A (1999) Inadvertent exposure to xenoestrogens in children. Toxicol Ind Health 15: 151–158.
8. Diaz-Diaz R, Loague K (2001) Assessing the potential for pesticide leaching for the pine forest areas of Tenerife. Environ Toxicol Chem 20: 1958–1967.
9. Covaci A, de Boer J, Ryan JJ, Voorspoels S, Schepens P (2002) Distribution of organobrominated and organochlorinated contaminants in Belgian human adipose tissue. Environ Res 88: 210–218.
10. LaKind JS, Amina Wilkins A, Berlin CM Jr (2004) Environmental chemicals in human milk: a review of levels, infant exposures and health, and guidance for future research. Toxicol Appl Pharmacol 198: 184–208.
11. Botella B, Crespo J, Rivas A, Cerrillo I, Olea-Serrano MF, et al. (2004) Exposure of women to organochlorine pesticides in Southern Spain. Environ Res 96: 34–40.
12. Minh NH, Someya M, Minh TB, Kunisue T, Iwata H, et al. (2004) Persistent organochlorine residues in human breast milk from Hanoi and Hochiminh City, Vietnam: contamination, accumulation kinetics and risk assessment for infants. Environ Pollut 129: 431–441.
13. Harcz P, De Temmerman L, De Voghel S, Waegeneers N, Wilmart O, et al. (2007) Contaminants in organically and conventionally produced winter wheat (Triticum aestivum) in Belgium. Food Addit Contam 24: 713–720.
14. Guitart R, Riu JL, Puigdemont A, Arboix M (1990) Organochlorine residues in adipose tissue of chamois from the Catalan Pyrenees, Spain. Bull Environ Contam Toxicol 44: 555–560.
15. Manosa S, Mateo R, Freixa C, Guitart R (2003) Persistent organochlorine contaminants in eggs of northern goshawk and Eurasian buzzard from northeastern Spain: temporal trends related to changes in the diet. Environ Pollut 122: 351–359.
16. Garcia Algar O, Pichini S, Basagana X, Puig C, Vall O, et al. (2004) Concentrations and determinants of NO2 in homes of Ashford, UK and Barcelona and Menorca, Spain. Indoor Air 14: 298–304.
17. Ribas-Fito N, Torrent M, Carrizo D, Munoz-Ortiz L, Julvez J, et al. (2006) In utero exposure to background concentrations of DDT and cognitive functioning among preschoolers. Am J Epidemiol 164: 955–962.
18. Gladen BC, Shkiryak-Nyzhnyk ZA, Chyslovska N, Zadorozhnaja TD, Little RE (2003) Persistent organochlorine compounds and birth weight. Ann Epidemiol 13: 151–157.
19. Burke ER, Holden AJ, Shaw IC, Suharyanto FX, Sihombing G (2003) Organochlorine pesticide residues in human milk from primiparous women in Indonesia. Bull Environ Contam Toxicol 71: 148–155.
20. Smith D (1999) Worldwide trends in DDT levels in human breast milk. Int J Epidemiol 28: 179–188.
21. Qu W, Suri RP, Bi X, Sheng G, Fu J (2010) Exposure of young mothers and newborns to organochlorine pesticides (OCPs) in Guangzhou, China. Sci Total Environ 408: 3133–3138.
22. Cok I, Mazmanci B, Mazmanci MA, Turgut C, Henkelmann B, et al. (2012) Analysis of human milk to assess exposure to PAHs, PCBs and organochlorine pesticides in the vicinity Mediterranean city Mersin, Turkey. Environ Int 40: 63–69.
23. Saleh M, Kamel A, Ragab A, El-Baroty G, El-Sebae AK (1996) Regional distribution of organochlorine insecticide residues in human milk from Egypt. J Environ Sci Health B 31: 241–255.
24. Alawi MA, Ammari N, al-Shuraiki Y (1992) Organochlorine pesticide contaminations in human milk samples from women living in Amman, Jordan. Arch Environ Contam Toxicol 23: 235–239.
25. Schinas V, Leotsinidis M, Alexopoulos A, Tsapanos V, Kondakis XG (2000) Organochlorine pesticide residues in human breast milk from southwest Greece: associations with weekly food consumption patterns of mothers. Arch Environ Health 55: 411–417.
26. Bouwman H, Kylin H, Sereda B, Bornman R (2012) High levels of DDT in breast milk: intake, risk, lactation duration, and involvement of gender. Environ Pollut 170: 63–70.
27. Ennaceur S, Gandoura N, Driss MR (2008) Distribution of polychlorinated biphenyls and organochlorine pesticides in human breast milk from various locations in Tunisia: levels of contamination, influencing factors, and infant risk assessment. Environ Res 108: 86–93.
28. Porta M, Puigdomenech E, Ballester F, Selva J, Ribas-Fito N, et al. (2008) [Studies conducted in Spain on concentrations in humans of persistent toxic compounds]. Gac Sanit 22: 248–266.
29. Serra Majem L, Ribas Barba L, Armas Navarro A, Alvarez Leon E, Sierra A (2000) [Energy and nutrient intake and risk of inadequate intakes in Canary Islands (1997–98)]. Arch Latinoam Nutr 50: 7–22.
30. Glynn AW, Granath F, Aune M, Atuma S, Darnerud PO, et al. (2003) Organochlorines in Swedish women: determinants of serum concentrations. Environ Health Perspect 111: 349–355.

31. Arrebola JP, Mutch E, Rivero M, Choque A, Silvestre S, et al. (2012) Contribution of sociodemographic characteristics, occupation, diet and lifestyle to DDT and DDE concentrations in serum and adipose tissue from a Bolivian cohort. Environ Int 38: 54–61.

32. Ahlborg UG, Lipworth L, Titus-Ernstoff L, Hsieh CC, Hanberg A, et al. (1995) Organochlorine compounds in relation to breast cancer, endometrial cancer, and endometriosis: an assessment of the biological and epidemiological evidence. Crit Rev Toxicol 25: 463–531.

33. Porta M, Lopez T, Gasull M, Rodriguez-Sanz M, Gari M, et al. (2012) Distribution of blood concentrations of persistent organic pollutants in a representative sample of the population of Barcelona in 2006, and comparison with levels in 2002. Sci Total Environ 423: 151–161.

34. Akkina J, Reif J, Keefe T, Bachand A (2004) Age at natural menopause and exposure to organochlorine pesticides in Hispanic women. J Toxicol Environ Health A 67: 1407–1422.

35. Kezios KL, Liu X, Cirillo PM, Cohn BA, Kalantzi OI, et al. (2013) Dichlorodiphenyltrichloroethane (DDT), DDT metabolites and pregnancy outcomes. Reprod Toxicol 35: 156–164.

36. Jusko TA, Koepsell TD, Baker RJ, Greenfield TA, Willman EJ, et al. (2006) Maternal DDT exposures in relation to fetal and 5-year growth. Epidemiology 17: 692–700.

37. Rylander L, Stromberg U, Hagmar L (1995) Decreased birthweight among infants born to women with a high dietary intake of fish contaminated with persistent organochlorine compounds. Scand J Work Environ Health 21: 368–375.

38. Rylander L, Hagmar L (1995) Mortality and cancer incidence among women with a high consumption of fatty fish contaminated with persistent organochlorine compounds. Scand J Work Environ Health 21: 419–426.

39. Farhang L, Weintraub JM, Petreas M, Eskenazi B, Bhatia R (2005) Association of DDT and DDE with birth weight and length of gestation in the Child Health and Development Studies, 1959–1967. Am J Epidemiol 162: 717–725.

40. Khanjani N, Sim MR (2006) Maternal contamination with dichlorodiphenyltri-chloroethane and reproductive outcomes in an Australian population. Environ Res 101: 373–379.

41. Schafer KS, Kegley SE (2002) Persistent toxic chemicals in the US food supply. J Epidemiol Community Health 56: 813–817.

42. Allen RH, Gottlieb M, Clute E, Pongsiri MJ, Sherman J, et al. (1997) Breast cancer and pesticides in Hawaii: the need for further study. Environ Health Perspect 105 Suppl 3: 679–683.

43. Cruz S, Lino C, Silveira MI (2003) Evaluation of organochlorine pesticide residues in human serum from an urban and two rural populations in Portugal. Sci Total Environ 317: 23–35.

44. Windal I, Hanot V, Marchi J, Huysmans G, Van Overmeire I, et al. (2009) PCB and organochlorine pesticides in home-produced eggs in Belgium. Sci Total Environ 407: 4430–4437.

45. Fontcuberta M, Arques JF, Villalbi JR, Martinez M, Centrich F, et al. (2008) Chlorinated organic pesticides in marketed food: Barcelona, 2001–06. Sci Total Environ 389: 52–57.

46. Corrigan PJ, Seneviratna P (1990) Occurrence of organochlorine residues in Australian meat. Aust Vet J 67: 56–58.

47. Tao S, Liu WX, Li XQ, Zhou DX, Li X, et al. (2009) Organochlorine pesticide residuals in chickens and eggs at a poultry farm in Beijing, China. Environ Pollut 157: 497–502.

48. Jakszyn P, Goni F, Etxeandia A, Vives A, Millan E, et al. (2009) Serum levels of organochlorine pesticides in healthy adults from five regions of Spain. Chemosphere 76: 1518–1524.

49. World Health Organization (2003) Health risks of persistent organic pollutants from long-range transboundary air pollution. Copenhagen, Denmark: Regional Office for Europe of the World Health Organization.

Persistence and Dissipation of Chlorpyrifos in Brassica Chinensis, Lettuce, Celery, Asparagus Lettuce, Eggplant, and Pepper in a Greenhouse

Meng-Xiao Lu[1,2], Wayne W. Jiang[3], Jia-Lei Wang[1], Qiu Jian[4], Yan Shen[2], Xian-Jin Liu[1,2], Xiang-Yang Yu[1,2]*

1 Pesticide Biology and Ecology Research Center, Nanjing, Jiangsu, China, **2** Key Laboratory of Food Safety Monitoring and Management of Ministry of Agriculture, Nanjing, Jiangsu, China, **3** Department of Entomology, Michigan State University, East Lansing, Michigan, United States of America, **4** Institute for the Control of Agrochemicals, Ministry of Agriculture, Beijing, China

Abstract

The residue behavior of chlorpyrifos, which is one of the extensively used insecticides all around the world, in six vegetable crops was assessed under greenhouse conditions. Each of the vegetables was subjected to a foliar treatment with chlorpyrifos. Two analytical methods were developed using gas chromatography equipped with a micro-ECD detector (LOQ = 0.05 mg kg^{-1}) and liquid chromatography with a tandem mass spectrometry (LOQ = 0.01 mg kg^{-1}). The initial foliar deposited concentration of chlorpyrifos (mg kg^{-1}) on the six vegetables followed the increasing order of brassica chinensis<lettuce<celery<asparagus lettuce<eggplant <pepper. The initial deposition of chlorpyrifos showed differences among the six selected vegetable plants, ranging from 16.5±0.9 mg kg^{-1} (brassica chinensis) to 74.0±5.9 mg kg^{-1} (pepper plant). At pre-harvest interval 21 days, the chlorpyrifos residues in edible parts of the crops were <0.01 (eggplant fruit), < 0.01 (pepper fruit), 0.56 (lettuce), 0.97 (brassica chinensis), 1.47 (asparagus lettuce), and 3.50 mg kg^{-1} (celery), respectively. The half-lives of chlorpyrifos were found to be 7.79 (soil), 2.64 (pepper plants), 3.90 (asparagus lettuce), 3.92 (lettuce), 5.81 (brassica chinensis), 3.00 (eggplant plant), and 5.45 days (celery), respectively. The dissipation of chlorpyrifos in soil and the six selected plants was different, indicating that the persistence of chlorpyrifos residues strongly depends upon leaf characteristics of the selected vegetables.

Editor: Youjun Zhang, Institute of Vegetables and Flowers, Chinese Academy of Agricultural Science, China

Funding: This work was financially supported by the Independent Innovation Fund of Agricultural Sciences in Jiangsu Province (cx (12) 3090) and the Natural Science Foundation of China (31071719). The funders had no role in study design, data collection and analysis, decision to publish, or preparation of the manuscript.

Competing Interests: The authors have declared that no competing interests exist.

* Email: yu98190@gmail.com

Introduction

Chlorpyrifos [O,O-diethyl O-(3,5,6-trichloro-2-pyridyl) phosphorothioate] is an organophosphorous insecticide, acaricide, and nematicide used to control a broad spectrum of foliage and soil-born insect pests on a variety of food and feed crops [1–2]. It is ranked as one of the most extensively used insecticides all around the world. In China, since the use of several highly-toxic organophosphorous insecticides was banned in 2006, chlorpyrifos has been recommended as one of the alternative insecticides and broadly used in agriculture. Extensive use of chlorpyrifos has led to a potential risk of residues in various crops. Chlorpyrifos is of great environmental concerns due to its widespread use in the past several decades and its potential toxic effects on human health. Thus, the degradation study of chlorpyrifos has become increasing important in recent years [3–5]. In a market monitoring study conducted between 2007 and 2010, chlorpyrifos was detected in approximately 22.8% of 2082 samples of 17 vegetable commodities collected from Zhejiang Province, China with a highest residue of 3.47 mg kg-1. The residue levels in 1.4% of vegetable

samples were found to be higher than the maximum residue limits (MRLs) of China [3].

Although most pesticides are effective to control pests in agricultural industry, inappropriate uses of pesticides may lead to public concerns on food safety and human health [5–10], environmental contamination [11–12], insect resistance and resurgence [12–14], etc. In China, the use of agrochemicals is critical to provide food supplies for its growing population with its limited arable land. Ideally, control the harmful organism efficiently, having no or minimum pesticide residues in the harvested crops, or at least lower than the statutory MRLs [3,15]. Therefore, the field dissipation studies of pesticide persistence in foods and pesticide residue behavior is of particular importance in order to find out which pesticide application strategies are efficient to control insect pests while leaving minimum residues [16,17]. There are many factors that influence the dissipation behavior of pesticides in plants, including the climate conditions (temperature, humidity, light intensity, etc) [18], the crop species [19–22] the nature of the chemicals, the formulations, and the application methods [23–24]. Due to the difference in the extension of the foliar, which resulted in the different initial pesticide deposition, and/or the difference in the metabolism system of different crops, dissipation of a pesticide on various crops may be markedly different [25–26]. A dissipation study of pesticides on leafy vegetables showed that among the tested vegetables, spinach and amaranth could incur higher pesticide deposition and that half-lives ($t_{1/2} = 1.37$–5.17 days) of chlorpyrifos were from on different leafy vegetables [22]. For chiral pesticide malathion, the calculated $t_{1/2}$ values of the enantiomers were relatively short, ranging from 0.83 to 1.43 days in five plants. The degradation of the two enantiomers in Chinese cabbage (brassica chinesis), rape, and sugar beet was highly selective, while non-enantioselectivity was found in paddy rice and wheat [20–21]. For better understanding the possible residue risk of a pesticide, dissipation studies for different crop species in the specific growing conditions are necessary to test if the established application strategies are suitable.

It is concluded that agrochemicals enter into plants via two major pathways, which are either via foliar treatment - foliar deposition followed by entering into the inner parts of the crops, or via soil treatment - root uptake from the soil [27–28]. For the foliar pesticide applications, the agrochemicals are deposited directly on foliar surfaces of the crops and the excess agrochemicals will precipitate in the soil. The resulting deposition fractions are determined predominantly by crop species, growth stage of crop, pesticide formulation, and spraying technology [29]. Generally, optimization of chlorpyrifos uses is foliar application with minimizing losses of the applied pesticides from plants to soil. Spray loss of chlorpyrifos may lead to the soil environment pollution and extend its residue duration in plants. The chlorpyrifos residues were found to persist in soils, as the half-lives ranged generally between 50 and 120 days [30–32]. It was reported in the literature that the chlorpyrifos residues were found in soils for over one year after the applications. Pesticide persistence in soils may depend on the formulation, rate of application, soil type, climate, and other conditions [33–37]. Plant rhizosphere plays an important role in the degradation of pesticides in soils [21,36].

In the present study, six vegetables were selected, including two fruit vegetables (pepper and eggplant) and 4 whole plant edibles leafy vegetables (brassica chinensis, lettuce, celery, and asparagus lettuce). They were treated with chlorpyrifos by foliar application under the controlled conditions in a greenhouse. The initial deposition of chlorpyrifos on the crop foliar was analyzed by GC

and on the fruits by LC/MS/MS and the dynamics of pesticide residues in the plants and rhizosphere soils were to be monitored. The work was to evaluate the persistence and dissipation behavior of chlorpyrifos in different vegetables and soil. The results would help to provide understanding of the residue characteristics of chlorpyrifos in vegetables, and help to guide proper and safe use of pesticides on vegetables to ensure food safety.

Materials and Methods

Instruments and Reagents

Gas chromatography was an Agilent GC 7890 (Agilent Technologies, Santa Clara, CA, USA) equipped with a micro-ECD detector and an analytical column HP-5ms J&W Ultra Inert capillary column (30 m length×0.25 mm I.D.×0.25 μm film thickness, Agilent Technologies, USA). LC/MS/MS (Agilent Technologies, USA) contained a 1200 SL HPLC system coupled to an Agilent G6410A triple quadrupole mass spectrometer. The column was an Agilent ZORBAX SB-C18 (2.1×150 mm, 5 μm) analytical column. Chlorpyrifos-EC 40% (Hubei Xian Long Chemical Co., LTD) was purchased from a local pesticide store. The analytical reference substance, chlorpyrifos (certified analytical standard, 99.7%) was purchased from National Standard Company (Tianjin, China). Two stock solutions (1000 mg L^{-1}) were prepared by dissolving the chlorpyrifos standard (100 mg) in 100 mL of acetone (for GC) and acetonitrile (for LC/MS/MS), respectively. Working solutions were prepared by diluting the stock solution or a working solution using the organic solvents (acetone for GC and acetonitrile for LC/MS/MS). Acetone, n-hexane, acetonitrile, sodium chloride, and anhydrous sodium sulfate were of analytical grades, purchased from Kermel Chemical Reagent Co., Ltd (Tianjin, China). The 0.22 μm SCAA-104 membranes and 500 mg florisil SPE cartridges were purchased from Anpel Scientific Instrument Co., Ltd (Shanghai, China).

The six selected vegetables were brassica chinensis (*Brassicachinensis L.*), lettuce (*Lactuca sativa* spp), pepper (*Capsicum annuum* spp), eggplant (*Solanum melongena* L), celery (*Apium graveolens*), and asparagus lettuce (*Asparagus Lettuce* spp). The pepper and eggplant seeds were grown in a nursery tray before being transplanted to the field at the 2–3 leaf stage seedlings, and the other four vegetables were directly sowed in the field. The row spacing and inter-plant spacing were set to be 30 cm for the vegetables, except for celery which was sowed in rows with row spacing of 30 cm.

Experiment design

Field experiments were conducted in a controlled environment in a greenhouse at the Experiment Station of Jiangsu Academy of Agricultural Science (JAAS), Nanjing, China from September to November 2013. JAAS permitted the study in its greenhouse and this study did not use protected area of land or sea, neither with relevant protected wildlife. Since the work was completed in the greenhouse, no specific permission was required. This study did not involve endangered or protected species. The soil was of sandy loam texture dried and its contents contained 30% of sand, 53% silt, 15% clay, and 2% organic matter. There were four 30 m^2 trial plots to be selected for each vegetable, i.e., three replicates and one control. Between the plots there was a buffer strip of 0.5 m in width. Chlorpyrifos-EC 40% was applied by foliar spraying at a rate of 0.97 kg a.i./ha on October 21, 2013. At the time of spraying, brassica chinensis, lettuce, and asparagus lettuce were at the stage when the leaves overspread; celery was 15–16 cm of height and the leaves scattered; pepper and eggplant were at the stages of 30–40 cm of height and the leaves of the adjacent plants touched each other and overlapped. In this study, the plants were

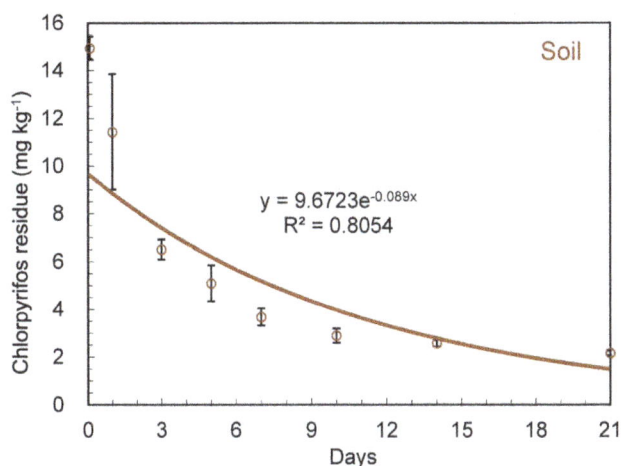

Figure 1. Dissipation dynamic of chlorpyrifos in soil.

$$y = 9.6723e^{-0.089x}$$
$$R^2 = 0.8054$$

Table 1. Dynamic equations, correlation coefficients and half-lives of chlorpyrifos in soil and six vegetable plants (foliar).

Matrix	Dynamic equation	Correlation coefficient (R^2)	Half-life (days)
Soil	$C_t = 9.672\,e^{-0.089t}$	0.8054	7.79
Brassica chinensis	$C_t = 11.74\,e^{-0.1192t}$	0.9618	5.81
Lettuce	$C_t = 17.30\,e^{-0.1769t}$	0.9359	3.92
Celery	$C_t = 40.62\,e^{-0.1271t}$	0.9648	5.45
Asparagus lettuce	$C_t = 31.97\,e^{-0.1775t}$	0.8850	3.90
Pepper	$C_t = 40.23\,e^{-0.2622t}$	0.9315	2.64
Eggplant	$C_t = 48.28\,e^{-0.2307t}$	0.9756	3.00

allowed to the leaves to overlap to cover most of the surface of soil and thus the loss of pesticides sprayed was minimum. The temperature inside the greenhouse was controlled between 16–26°C. Three representative samples of whole plant and rhizosphere soil were collected at 0 (2 h after application), 1, 3, 5, 7, 10, 14, and 21 days intervals after pesticides application. Pepper and eggplant fruit samples were collected on 0, 3, 7, 14, and 21 days.

Extraction and purification

All samples (plants and fruits) were homogenized using a Philips blender (Shanghai, China) and the ground samples were stored in a freezer (at −20°C) until analysis.

Plant samples. The extraction of chlorpyrifos residues from plants and fruits was carried out by the procedure as follows. Five (5.0) g of the homogenized sample was weighed into a 50 mL Teflon centrifuge tube. The extraction solvent (acetonitrile, 10 mL) was added. The samples were then mixed thoroughly for 1 min with a vortex mixer, followed by high-speed homogenizing for about 2 min. After addition of 2 g of sodium chloride, the samples were vortexed immediately for 1 min and centrifuged for 5 min at 5000 rpm. An aliquot of 1 mL of the supernatant was transferred into a 10 mL glass test tube, and then evaporated just

to dryness under a stream of nitrogen (40°C). The residue was dissolved in 1 mL of hexane and then subjected to Florisil SPE column clean-up. The SPE column was pre-conditioned by rinsing it with 5 mL of hexane. The extraction was added to the SPE column followed by eluting with 10 mL of a mixture containing acetone and n-hexane (9:1, v/v). The eluate was evaporated to dryness. The residues were redissolved with acetone to 1 mL. The final extract was filtered through a 0.2 μm SCAA-104 membrane followed by GC analysis.

Fruit samples. The method of sample extraction was a modification of a reference method [38]. Fruit sample (5.0 g) was weighted into a polypropylene centrifuge tube. An extraction solvent (10 mL of acetonitrile) was added. The sample was homogenized for 1 min using the homogenizer. The homogenizer probe was rinsed with a portion of 5 mL of the extraction solvent. The extracts were combined. The sample was centrifuged at 5000 rpm for 5 min. Transferred 2 mL of supernatant into a 15 mL centrifuge tube containing PSA (100 mg), ODS-C18 (100 mg) and florisil (100 mg). The sample was vortexed for 1 min and centrifuged at 5000 rpm for 2 min. After centrifugation, the supernatant was filtered using a 0.22 μm nylon filter into an autosampler vial for LC/MS/MS analysis.

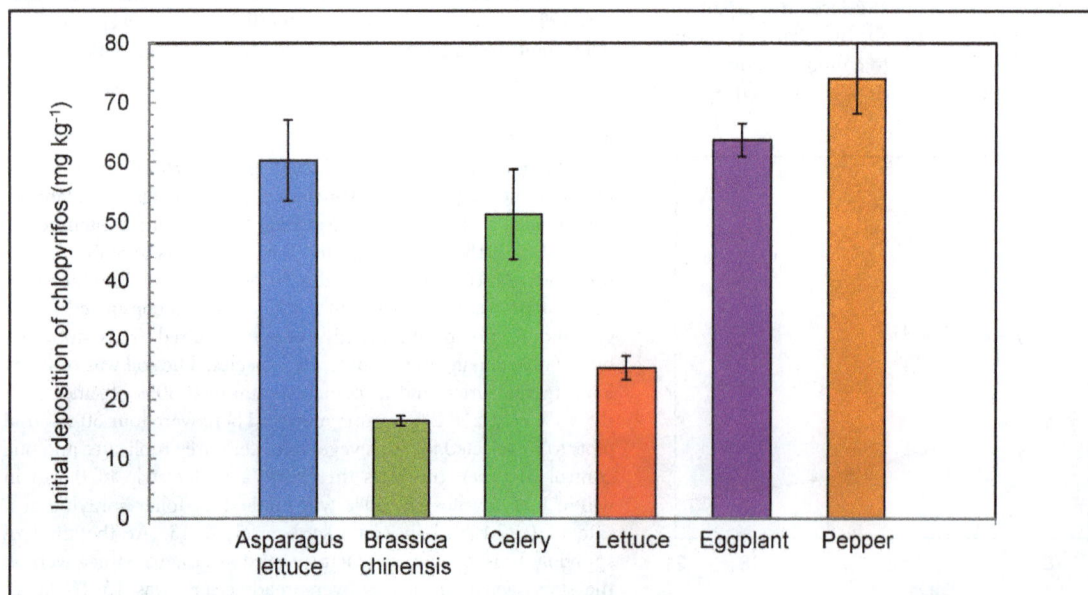

Figure 2. Initial depositions of chlorpyrifos on the six vegetable plants.

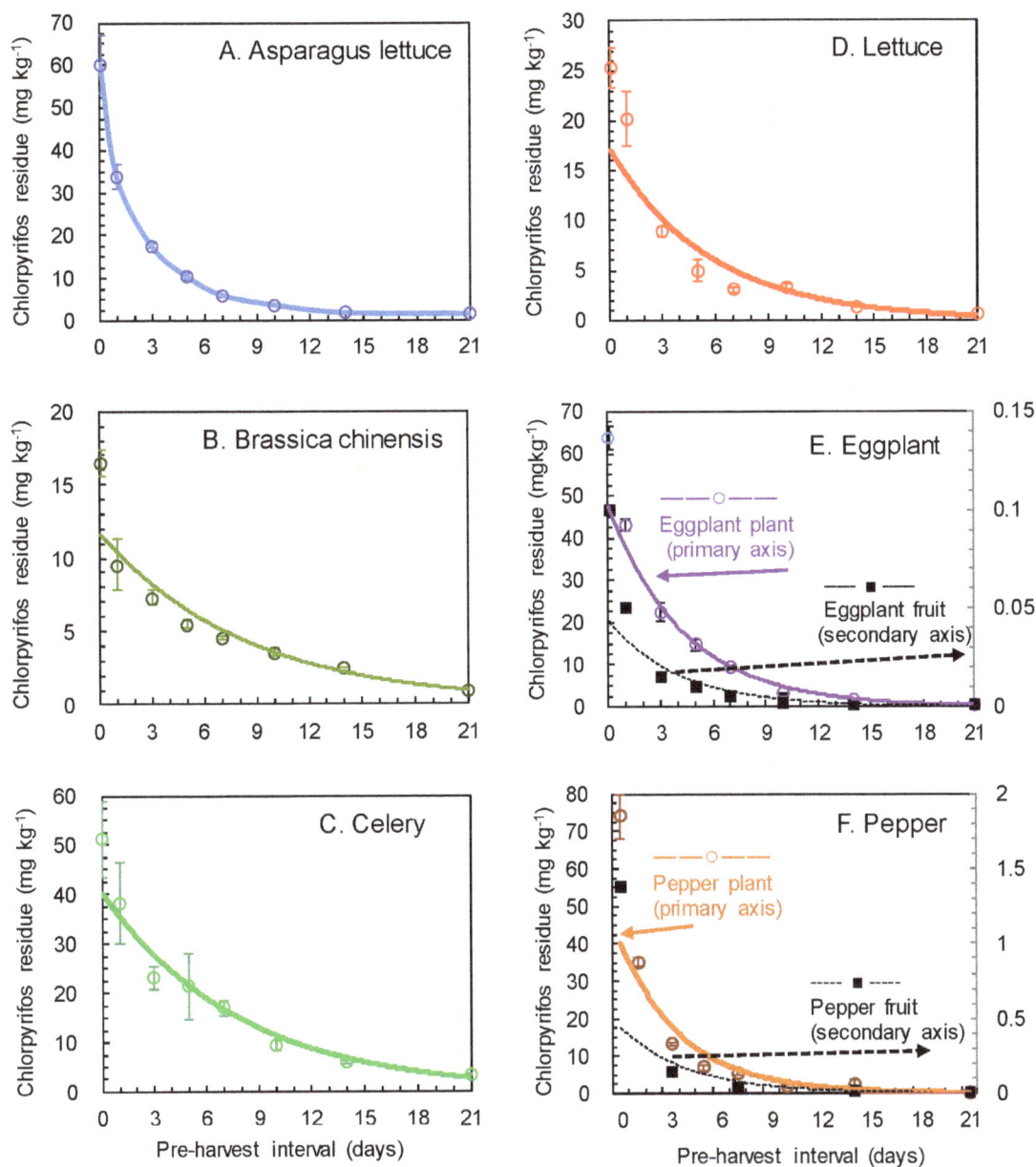

Figure 3. Dissipation dynamic of chlorpyrifos in vegetables and fruits. A. Asparagus lettuce, B. Brassica chinensis, C. Celery, D. Lettuce, E. Eggplant (eggplant plant - solid curve with marker ○ and primary axis; eggplant fruit – dashed curve with marker ■ and secondary axis), and F. Pepper (pepper plant - solid curve with marker ○ and primary axis; pepper fruit – dashed curve with marker ■ and secondary axis).

Soil. Five (5.0) g of soil was weighed into a 50 mL Teflon centrifuge tube and 5 g of sodium chloride was added. The contents were thoroughly mixed. Then, 20 mL of acetonitrile was added. The mixture was vortexed for 1 min, ultrasonically extracted at 40–45°C for 30 min, shaken on a rotary shaker for 2 h, and centrifuged at 5000 rpm for 5 min. Then 1 mL of the supernatant was transferred into a centrifuge tube and dried under a stream of nitrogen (40°C). The chlorpyrifos residues were redissolved in 1 mL of acetone followed by GC analysis.

GC and LC/MS/MS instrument analyses

GC. The conditions for the analysis were: detector temperature, 280°C; injector temperature 270°C; oven temperature program starting at 120°C, 3.67 min at 120–230°C (ramp 30°C min^{-1}), 5 min at 230°C, 2 min at 230–270°C (ramp 20°C min^{-1}), 2 min at 270°C; carrier gas, N_2 at 1 mL/min; injection volume 1.0 μL, in a splitless mode. A linear calibration curve was used and the calibration range was 0.01–5 mg kg^{-1}. Under these conditions chlorpyrifos retention times were approximately 7.52 min. The software was Agilent ChemStation Rev. B04.03 software for instrument control, data acquisition and processing.

LC/MS/MS. The instrument analysis method was an adaption of reference methods [38]. The HPLC conditions for the analysis were: mobile phase A: water containing 0.1% of formic acid (v/v); mobile phase B: acetonitrile containing 0.1% of formic acid (v/v); flow rate 0.4 mL min^{-1}; injection volume 10 μL; mobile phase gradients of binary pump: 10% B (0–1.50 min), 95% B (1.51–4.50 min), 95% B (4.51–6.00 min), and

Table 2. Comparison of chlorpyrifos residues in edible parts of brassica chinensis, lettuce, celery, asparagus lettuce, eggplant and pepper with Maximum Residue Limits (MRLs).*1.

Commodity	Crop group*2	Residue (mg kg⁻¹)		MRL (mg kg⁻¹)			
		PHI 7 days	PHI 21 days	CAC*3	China*4	USA*5	EU*6
Edible plants:							
Brassica chinensis	5B	4.55	0.97	1.0	0.1	1.0	0.5
Lettuce	4A	3.07	0.56	*7	0.1	*7	0.05
Celery	4B	16.9	3.50	*7	0.05	*7	0.05
Asparagus Lettuce	4B	5.96	1.47	*7	0.1	*7	*7
Edible fruits:							
Eggplant (fruit)	8–10B 8–10C	0.01	<0.01	*7	*7	*7	0.5
Pepper (fruit)	8–10B	0.03	<0.01	2.0 (sweet pepper)	*7	1.0	1.0

*1Please note that this work intended to study pesticide persistence instead of for MRL establishment. The comparison described in above the table is used to study the persistence and dissipation of chlorpyrifos in the selected crops. The high residues above were due to the high application rate and different formulations.

*2Code of Federal Regulations Title 40 Part 180.41 Crop group table (40 CFR 180.41). Group 4A/4B Leafy vegetables (except brassica vegetables), 5B Brassica leafy vegetable, 8–10B 8–10C Fruiting vegetable group.

*3CODEX Alimentarius: List of standards: http://www.codexalimentarius.org/standards/list-of-standards/en/?provide=standards&orderField=fullReference&sort=asc&num1=CAC/MRL.

*4Chinese National Standards (GB2763-2012): Maximum Residue Limits for Pesticides in Foods.

*5The United States Tolerances and Exemptions for Pesticide Chemical Residues in Food: http://www.ecfr.gov/cgi-bin/text-idx?SID=e33dfa87fab3f5ddecad25dafa7028ea&node=40:25.0.1.1.27.3.19.113&rgn=div8.

*6Pesticide EU-MRLs (Regulation (EC) No 396/2005, MRLs updated on 28/01/2014) http://ec.europa.eu/sanco_pesticides/public/index.cfm?event=substance.resultat&s=1.

*7MRLs not currently established or registrations canceled.

10% B (6.01–7.00 min). The MS/MS conditions were: gas temperature 350°C; gas flow 10 L/min; nebulizer pressure 45 psi; and capillary voltage 4000 V. In order to achieve the highest sensitivity, the fragment, voltage and the collision energy were optimized. MRM (chlorpyrifos) transitions were 350.1>198 (quantitation, collision energy, CE, 20 V) and 350.1>97 (identification, CE 30 V). The retention time was 3.89 min. The software was Agilent MassHunter software for instrument control, data acquisition and processing.

Data analysis

The degradation rate constant and half-life were calculated using a first-order rate equation:

$$C_t = C_0 e^{-kt}$$

where C_t and C_0 represent the concentrations of the chlorpyrifos residues at the day t and day 0 (2 h), respectively, and k is the degradation rate constant. The half-life ($t_{1/2}$) is defined as the time required for the pesticide residue level to fall to the half of the initial residue level of day 0 (i.e., C_0) and was calculated using the following equation:

$$t_{1/2} = (ln\,2)/k$$

Results and Discussion

Method validation

GC method. Quantification was accomplished by using the standard curve constructed by plotting analyte concentrations against peak areas. Good linearity was achieved with the chlorpyrifos concentration and the correlation coefficient was 0.9995. Recoveries of chlorpyrifos at different fortification levels, i.e., 0.05, 1, and 10 mg kg^{-1}, were determined in three replicates for validation of the method. The recoveries of chlorpyrifos in the soil were 72.5%–89.6% with the relative standard deviation (RSD) 2.1%–7.2% and $R^2 = 0.9846$. For the six vegetable plants, the recoveries were from 79.3%–97.0% with RSD 3.0%–15% and $R^2 = 0.9910$. The limit of quantification (LOQ) was defined as the lowest fortification concentration whose signal-to-noise (S/N) ratio was equal to or greater than 10 and thus the LOQ of the GC analysis was 0.05 mg kg^{-1}. The sample extracts in which the chlorpyrifos residues were greater than 10 mg kg^{-1} were diluted 10 times prior to the evaporation/cleanup steps. And the diluted samples were re-analyzed to ensure the residues were in the acceptable ranges.

LC/MS/MS method. Three fortification levels, i.e., 0.01, 0.1, and 2 mg kg^{-1}, were analyzed. For pepper fruits, the average recoveries of chlorpyrifos residue in the soil were 87.1%–106% with the relative standard deviation (RSD) from 1.1% to 7.4% and $R^2 = 0.9991$. For eggplant fruits, the average recoveries were from 86.1% to 97.0%, with RSD 2.9%–5.9% and R2 = 0.9978. The LOQ of the LC/MS/MS analysis was set to be 0.01 mg kg^{-1}.

Soil and whole plant samples of the six selected crops were analyzed using GC (LOQ = 0.05 mg kg^{-1}). The fruit samples of pepper and eggplant were analyzed by LC/MS/MS (LOQ = 0.01 mg kg^{-1}). These methods were capable of conducting the analyses in this study.

Degradation of chlorpyrifos

Soil. The rhizosphere soil samples were collected at different intervals after chlorpyrifos was applied. The chlorpyrifos residues in soil were analyzed by GC. The concentration of chlorpyrifos in the soil decreased over time (Figure 1). The average initial deposition of chlopyrifos was 14.9±0.5 mg kg^{-1} (i.e., 2 h, day 0) and the final residue was 2.2±0.1 mg kg^{-1} on day 21. The first-order kinetic equation of chlorpyrifos dissipation is $C_t = 4.84e^{-0.089t}$ (Table 1) with correlation coefficient $R^2 = 0.8054$ and the half-life $t_{1/2} = 7.79$ days. The half-lives of chlorpyrifos in soil were in the range of 3–7 days reported by Singh et al. [30]. Singh et al. observed that chlorpyrifos persisted in a low pH soil, i.e., less than 3% of the pesticide had degraded after 10 days and more than 50% of chlorpyrifos was dissipated at a higher pH soil (pH 8.5). Chai et al. reported that the half-lives in humid tropical soils from Malaysia were typically 7–120 days [32]. However, Chai, et. al. also reported that some half-lives were 257 days in the soils containing less soil microbial populations [32]. In the literature, it was reported long environmental dissipation half-lives of chlorpyrifos, i.e., up to 4 years, depending on application rate, ecosystem, and pertinent environments [33]. Since chlorpyrifos presented low water solubility and a higher log K_{ow}, it had a strong tendency to sorb to organic matter and soil. Stability and effectiveness had made chlorpyrifos one of the most popular pesticides worldwide but on the other side its persistence had raised environmental concerns [34].

Initial Deposition. After foliar application, the whole plants of the six selected crops were collected. The chlorpyrifos residues in plants were analyzed by GC. The initial depositions of chlorpyrifos in the six selected plants are compared in Figure 2. As can be seen in Figure 2, the initial depositions (2 h, day 0) on the six plants were in an increasing order: 16.5±0.87 mg kg^{-1} (brassica chinensis), 25.3±2.0 mg kg^{-1} (lettuce), 51.2±7.6 mg kg^{-1} (celery), 60.3±6.8 mg kg^{-1} (asparagus lettuce) < 63.7±2.8 mg kg^{-1} (eggplant) <74.0±5.9 mg kg^{-1} (pepper), respectively.

It is presumed that the initially deposited chlorpyrifos amount mainly depended upon the surface area of the foliar which the pesticide was sprayed on in spite of the leaf characteristics of the plants, such as leaf roughness, content of cuticular waxes, etc. which were assumed to contribute little to the initial depositions. Therefore, the concentration of the initial deposition is directly proportional to the foliar area and inversely proportional to the biomass of the whole plant. The foliages of pepper plant, eggplant plant, celery, and asparagus lettuce were overlapping to maximize the effective foliar surface area. The shorter plants (lettuce and brassica chinensis) had an area of uncovered soils between rows. As a result, lettuce and brassica chinensis had the lowest initial depositions. Since it was lightest in weight, the pepper plant had over all the largest initial deposition of chlorpyrifos.

Dissipation. The dynamic equations of chlorpyrifos degradation are given in Table 1. The curves of dissipation in the six plants and two fruits are described in Figure 3. As can be seen in Figure 3, the dynamics curves demonstrated that the chlorpyrifos residues dissipated significantly in the first a few days and persisted in the crops for extended period of time. For example, at pre-harvest interval (PHI) 21 days, the chlorpyrifos residues in the six plants decreased to 1.47±0.22 mg kg^{-1} ((A) asparagus lettuce), 0.97±0.03 mg kg^{-1} ((B) brassica chinensis), 3.50±0.27 mg kg^{-1} ((C) celery), 0.56±0.06 mg kg^{-1} ((D) lettuce), 0.53±0.06 mg kg^{-1} ((E) eggplant), and 0.15±0.01 mg kg^{-1} ((F) pepper), respectively. As can be seen in Table 1, the half-lives (from low to high) were found to be: 0.91 days (pepper plants) <3.92 days (lettuce) <3.92

days (asparagus lettuce) <5.82 days (brassica chinensis) <3.00 days (eggplant plants) <5.46 days (celery).

The possible mechanisms are thought to be due to the difference in the activities of pesticide degradation enzymes and/or pesticide degradation endophytes among these plants [25]. It is interesting to observe that the half-lives of the crop plants seemed to be a reverse order of the initial deposition. Pepper plants had the highest initial deposition but shortest half-live (0.92 day) while brassica chinensis had the lowest initial deposition but longest half-live (5.82 days). The difference in the calculated half-life values indicated that different vegetables had different degradation rates. One of the key factors could be photodegradation of chlorpyrifos [37]. The pepper plant had a greater effective foliar area which led to the highest pesticide position, and exposed to the ultraviolet lights. Also, after application, the leaf characteristics may affect how the pesticide would be retained on the surfaces of the leaves and then be penetrating into the plant tissues such as leaf surface roughness [39–40] and the content of water repellent cuticular waxes [41–44].

Edible parts of the vegetables. The edible parts are edible leaves and stems of brassica chinensis, lettuce, celery, and asparagus, and fruits of eggplant and pepper. The residue in pepper and eggplant fruits were analyzed by LC/MS/MS. The comparison of the residue data are given in Table 2. A number of existing Maximum Residue Limits (MRLs) and crop grouping are also included in Table 2. The higher residues of chlorpyrifos in the plants were due to high initial depositions. It was observed that the residues of foliages were significantly higher than those in the edible fruits (eggplant and pepper fruits), i.e., pepper had 0.03 mg kg^{-1} and <0.01 mg kg^{-1} of chlorpyrifos residues and eggplant had <0.10 mg kg^{-1} at PHI 7 days and PHI 21 days, respectively (Table 2). In Figure 3 (E and F), The half-lives were calculated to be 2.84 days (pepper fruits) and 3.15 days (eggplant fruits), respectively. It was observed that same residues exceeded the MRLs. However, it should be noted that the purpose of this work was to study the residues of chlorpyrifos change in the crops. Therefore, higher application rates were used in order to monitor such changes.

Because of chlorpyrifos' high hydrophobicity (high K$_{ow}$ value), the pesticide would readily enter into the inter parts from the surfaces resulting in high residue levels. A portion of the residues

may be transferred from leaves to the growing fruits. For eggplant and pepper fruits, the difference in chlorpyrifos residues mainly depended upon the composition of the surface waxes from pepper and eggplant [41,44]. Bauer et. al. [44] reported that the bell pepper contained 39% of fraction 1 (mainly C20–C35 of alkanes and aldehydes) and 61% of fraction 2 (15 various triterpenes) while the eggplant cultivars had 77% of fraction 1 and 23% of fraction 2. Fraction 2 consisted of 15 triterpenes, including α- and β-amyrin, lupeol, glutinol, 3β-friedelanol, friedelin, taraxerol, taraxasterol, δ-amyrin, germanicol, multi-florenol, ω-taraxasterol, isomultiflorenol, isobauerenol and bauerenol, as well as n-alkanoic acids 2-hydroxy-alkanoic acids [44]. These chemicals would enhance the dissipation of chlorpyrifos.

The comparisons of the chlorpyrifos residue data listed in Table 2 and presented in Figure 3 indicated that chlorpyrifos was relatively stable and persisted in the crops, especially leafy crops. The rate of degradation of pesticide residue is affected by environmental conditions, nature of the pesticide, application rate, formulation, and plant species, etc. [45]. The vegetables selected in this study are minor crops representing a range of various species and it was intended to promote the process. For the crop grouping, celery may be a good reprehensive species of leafy vegetable (group 4B, Table 2) with consideration of initial deposition and dynamic data.

Conclusions and Implications

The study investigated the residue behavior of chlorpyrifos in six vegetables in the greenhouse. The results of chlorpyrifos of initial depositions on different vegetables detected after pesticide application showed differences among the six selected crops. The half-lives of chlorpyrifos in the six vegetables were different indicating that different vegetables had different capacities for metabolizing chlorpyrifos.

Author Contributions

Conceived and designed the experiments: XYY MXL XJL. Performed the experiments: MXL. Analyzed the data: MXL WWJ QJ XYY. Contributed reagents/materials/analysis tools: MXL JLW YS. Contributed to the writing of the manuscript: WWJ MXL.

References

1. FAO (2000) Pesticide Residues in Food, 2000: Report of the Joint Meeting of the FAO Panel of Experts on Pesticide Residues in Food and the Environment and the WHO Core Assessment Group on Pesticide Residues, Geneva, Switzerland, 20–29 September 2000. 45–59 p.

2. Lemus R, Abdelghani A (2000) Chlorpyrifos: an unwelcome pesticide in our homes. Rev Env Health 15: 421–433.

3. Yuan Y, Chen C, Zheng C, Wang X, Yang G, et al. (2014) Residue of chlorpyrifos and cypermethrin in vegetables and probabilistic exposure assessment for consumers in Zhejiang Province, China. Food Control 36: 63–68.

4. Gao Y, Chen S, Hu M, Hu Q, Luo J, et al. (2012) Purification and Characterization of a Novel Chlorpyrifos Hydrolase from *Cladosporium cladosporioides* Hu-01. PLoS ONE 7(6): e38137.

5. Wentzell J, Cassar M, Kretzschmar D (2014) Organophosphate-Induced changes in the PKA regulatory function of Swiss cheese/NTE lead to behavioral deficits and neurodegeneration. PLoS ONE 9(2): e87526.

6. Rauh VA, Perera FP, Horton MK, Whyatt RM, Bansal R, et al. (2012) Brain anomalies in children exposed prenatally to a common organophosphate pesticide. PNAS 109: 7871–7876.

7. Janssens L, Stoks R (2013) Fitness Effects of Chlorpyrifos in the Damselfly *Enallagma cyathigerum* Strongly Depend upon Temperature and Food Level and Can Bridge Metamorphosis. PLoS ONE 8(6): e68107.

8. Canesi L, Negri A, Barmo C, Banni M, Gallo G, et al. (2011) The Organophosphate Chlorpyrifos Interferes with the Responses to 17β-Estradiol in the Digestive Gland of the Marine Mussel *Mytilus galloprovincialis*. PLoS ONE 6(5): e19803.

9. Sasikala C, Jiwal S, Rout P, Ramya M (2012) Biodegradation of chlorpyrifos by bacterial consortium isolated from agriculture soil. World J Microbio Biotech, 28(3), p1301.

10. Trunnelle KJ, Bennett DH, Tulve NS, Clifton MS, Davis MD, et al. (2014) Urinary pyrethroid and chlorpyrifos metabolite concentrations in northern California families and their relationship to indoor residential insecticide levels, Part of the study of use of products and exposure related behavior (SUPERB). Environ. Sci. Technol., 48 (3): 1931–1939.

11. Watts M (2012) Chlorpyrifos as a Possible Global Persistent Organic Pollutant. Pesticide Network North America, Oakland, CA, USA. Available: http://www.ipen.org/cop6/wp-content/uploads/2013/04/Chlorpyrifos_as_POP_final.pdf. Accessed on March 17, 2014.

12. Popp J, Peto K, Nagy J (2013) Pesticide productivity and food security. A review. Agron Sustain Dev 33: 243–255.

13. Zhang NN, Liu CF, Yang F, Dong SL, Han ZJ (2012) Resistance mechanisms to chlorpyrifos and F392W mutation frequencies in the acetylcholine esterase ace1 allele of field populations of the tobacco whitefly, Bemisia tabaci in China. J Insect Sci, Vol 12 Article 41.

14. Ouyang Y, Chueca P, Scott SJ, Montez GH, Grafton-Cardwell EE (2010) Chlorpyrifos Bioassay and Resistance Monitoring of San Joaquin Valley California Citricola Scale Populations. J Econ Ent 103(4): 1400–1404.

15. Mouron P, Heijne B, Naef A, Strassemeyer J, Hayer F, et al. (2012) Sustainability assessment of crop protection systems: SustainOS methodology and its application for apple orchards. Agri Syst 113: 1–15.

16. MacLachlan J, Hamilton D (2010) Estimation methods for maximum residue limits for pesticides. Reg Toxicol Pharmacol, 58: 208–218.

17. Malhat F, Kamel E, Saber A, Hassan E, Youssef A, et al. (2013) Residues and dissipation of kresoxim methyl in apple under field condition. Food Chem 140: 371–374.

18. Garau VL, Angioni A, Aguilera Del Real A, Russo MT, Cabras P (2002) Disappearance of azoxystrobin, cyprodinil, and fludioxonil on tomato in a greenhouse. J Agri Food Chem 50: 1929–1932.

19. Cabras P, Meloni M, Manca MR, Pirisi FM, Cabitza F, et al. (1988) Pesticide residues in lettuce. 1. Influence of the cultivar. J Agri Food Chem: 36: 92–95.

20. Wang M, Zhang Q, Cong L, Yin W, Wang M (2014) Enantioselective degradation of metalaxyl in cucumber, cabbage, spinach and pakchoi. Chemosphere 95: 241–256.

21. Sun H, Xu J, Yang S, Liu G, Dai S (2004) Plant uptake of aldicarb from contaminated soil and its enhanced degradation in the rhizosphere. Chemosphere. 54: 569–574.

22. Fan S, Zhang F, Deng K, Yu C, Liu S, et al. (2013) Spinach or amaranth contains highest residue of metalaxyl, fluazifop-p-butyl, chlorpyrifos, and lambda-cyhalothrin on six leaf vegetables upon open field application. J Agri Food Chem 61: 2039–2044.

23. Montemurro N, Grieco F, Lacertosa G, Visconti A (2002) Chlorpyrifos decline curves and residue levels from different commercial formulations applied to oranges. J Agri Food Chem 50: 5975–5980.

24. Cabras P, Meloni M, Gennari M, Cabitza F, ubeddu M (1989) Pesticide residues in lettuce. 2. Influence of formulations. J Agri Food Chem 37: 1405–1407.

25. Xia XJ, Zhang Y, Wu JX, Wang JT, Zhou YH, et al. (2009) Brassinosteroids promote metabolism of pesticides in cucumber. J Agri Food Chem 57: 8406–8413.

26. Itoiz ES, Fantke P, Juraske R, Kounina A, Vallejo AA (2012) Deposition and residues of zaoxystrobin and imidacloprid on greenhouse lettuce with impliacations of human consumption. Chemosphere 89: 1034–1041.

27. Collins C, Fryer M, Grosso A (2006) Plant uptake of non-ionic organic chemicals. Environ Sci Tech 40: 45–52.

28. Juraske R, Castells F, Vijay A, Muñoz P, Antón A (2009) Uptake and persistence of pesticides in plants: Measurements and model estimates for imidacloprid after foliar and soil application. J Hazard Mater 165: 683–689.

29. Hauschild M (2000) Estimating pesticide emissions for LCA of agricultural products. In: Weidema BP, Meeusen MJG. (Eds.), Agricultural Data for Life Cycle Assessments. Agricultural Economics Research Institute, The Hague, 64–79 p.

30. Singh BK, Walker A, Wright DJ (2006) Bioremedial potential of fenamiphos and chlorpyrifos degrading isolates: influence of different environmental conditions. Soil Biol Biochem 38: 682–93.

31. Chen S, Liu C, Peng C, Liu H, Hu M, et al. (2012) Biodegradation of Chlorpyrifos and Its Hydrolysis Product 3,5,6-Trichloro-2-Pyridinol by a New Fungal Strain Cladosporium cladosporioides Hu-01. PLoS ONE 7(10): e47205.

32. Chai LK, Wong MH, Hansen HCB (2013) Degradation of chlorpyrifos in humid tropical soils. J Environ Manage 125: 28–32.

33. Gebremariam SY, Beutel MW, Yonge DR, Flury M, Harsh JB (2012) Adsorption and Desorption of Chlorpyrifos to Soils and Sediments. Rev Environ Contam Tox 215: 123–175.

34. Kamrin MA (1997) Pesticide profiles toxicity, environmental impact, and fate. Lewis publishers: Boca Raton, FL 147–152 p.

35. US EPA (1999) Reregistration eligibility science chapter for chlorpyrifos fate and environmental risk assessment chapter. US EPA, office of prevention, pesticides and toxic substances, office of pesticide programs, environmental fate and effects division, US government printing office: Washington, DC.

36. Fang C, Radosevich M, Fuhrmann JJ (2001) Atrazine and phenanthrene degradation in grass rhizosphere soil. Soil Biol Biochem 33: 671–678.

37. Nieto LM, Hodaifa G, Vives SR, Casares JAG, Casanov MS (2009) Photodegradation of phytosanitary molecules present in virgin olive oil. J Photoch Photobio A 203: 1–6.

38. Liang Y, Wang W, Shen Y, Liu Y, Liu XJ (2012) Dynamics and residues of chlorpyrifos and dichlorvos in cucumber grown in greenhouse. Food Control 26: 231–234.

39. Gaskin RE, Steele KD, Foster WA (2005) Characterizing plant surfaces for spray adhesion and retention. New Zealand Plant Protection 58: 1790–183.

40. Hunche M, Bringe K, Schmitz-Eiberger M, Noga G (2006) Leaf surface characteristics of apple seedlings, bean seedlings and Kohlrabi plans and their impact on the retention and rainfastness of mancozeb. Pest Manage Sci 62: 839–847.

41. Bargel H, Koch K, Cerman Z, Neinhuis C (2006) Structure-function relationships of the plant cuticle and cuticular waxes – a smart material? Funct Plant Biol. 33: 893–910.

42. Wagner P, Furstner R, Barthlott W, Neinhuis C (2003) Quantitative assessment to the structural basis of water repellency in natural and technical surfaces. J Exp Bot 54: 1295–1303.

43. Malhat F, Badawy HMA, Barakat DA, Saber AN (2014) Residues, dissipation and safety evaluation of chromafenozide in strawberry under open field conditions. Food Chem 152: 18–22.

44. Bauer S, Schulte E, Their H-P (2005) Composition of the surface waxes from bell pepper and eggplant. Eur Food Res Technol 220: 5–10.

45. Fantke P, Juraske R (2013) Variability of Pesticide Dissipation Half-Lives in Plants. Environ Sci Tech 47: 3548–3562.

Trypsin-Catalyzed Deltamethrin Degradation

Chunrong Xiong[1,2¤a], **Fujin Fang**[1,2], **Lin Chen**[1,2], **Qinggui Yang**[1,2¤b], **Ji He**[1,2¤c], **Dan Zhou**[1,2], **Bo Shen**[1,2], **Lei Ma**[1,2], **Yan Sun**[1,2], **Donghui Zhang**[1,2]*, **Changliang Zhu**[1,2]

1 Department of Pathogen Biology, Nanjing Medical University, Nanjing, Jiangsu, China, **2** Jiangsu Province Key Laboratory of Modern Pathogen Biology, Nanjing, Jiangsu, China

Abstract

To explore if trypsin could catalyze the degradation of non-protein molecule deltamethrin, we compared *in vitro* hydrolytic reactions of deltamethrin in the presence and absence of trypsin with ultraviolet-visible (UV/Vis) spectrophotometry and gas chromatography-mass spectrometry (GC/MS). In addition, acute oral toxicity of the degradation products was determined in Wistar rats. The results show that the absorption peak of deltamethrin is around 264 nm, while the absorption peaks of deltamethrin degradation products are around 250 nm and 296 nm. In our GC setting, the retention time of undegraded deltamethrin was 37.968 min, while those of deltamethrin degradation products were 15.289 min and 18.730 min. The LD_{50} of deltamethrin in Wistar rats is 55 mg/kg, while that of deltamethrin degradation products is 3358 mg/kg in female rats and 1045 mg/kg in male rates (61-fold and 19-fold reductions in toxicity), suggesting that trypsin could directly degrade deltamethrin, which significantly reduces the toxicity of deltamethrin. These results expand people's understanding of the functions of proteases and point to potential applications of trypsin as an attractive agent to control residual pesticides in the environment and on agricultural products.

Editor: David L. McCormick, IIT Research Institute, United States of America

Funding: This work was supported by the Academic Natural Science Foundation of Jiangsu Province (No. 07KJD180137 to Donghui Zhang), the National Institutes of Health of the USA (No. R01AI075746 to Changliang Zhu) and the National Natural Science Foundation of China (Nos. 30972564 and 81171900 to Changliang Zhu, No. 30901244 to Yan Sun). The funders had no role in study design, data collection and analysis, decision to publish, or preparation of the manuscript.

Competing Interests: The authors have declared that no competing interests exist.

* E-mail: donghuizhang2004@gmail.com

¤a Current address: Jiangsu Institute of Parasitic Diseases, Wuxi, Jiangsu, China
¤b Current address: National Key Laboratory of Vector Biology, Jiangsu Entry-Exit Inspection and Quarantine Bureau, Nanjing, Jiangsu, China
¤c Current address: National Key Laboratory of Surveillance and Detection for Medical Vectors, Xiamen Entry–Exit Inspection and Quarantine Bureau, Xiamen, Fujian, China

Introduction

Serine proteases are involved in diverse physiological and cellular processes, including food digestion, protein maturation, blood coagulation, host immune responses, fibrinolysis, tissue remodeling, fertilization, and embryogenesis [1]. Trypsin, a widely studied serine protease, is present in the digestive systems of vertebrates and invertebrates [2]. As the most abundant protease, its physiological importance, especially in vertebrate digestive functions, has been well documented [3–5]. In addition, recent works are pointing to possible novel functions of trypsin. In our previous study, trypsin gene was cloned from deltamethrin-resistant *Culex pipiens pallens* mosquito using a combination of suppression subtractive hybridization (SSH) and microarray. We found that trypsin gene was significantly up-regulated in deltamethrin-resistant *Culex* mosquito [6]. Stable transfection of trypsin gene into mosquito cells conferred protection against deltamethrin treatment. On the other hand, RNA interference-mediated down-regulation of trypsin in mosquito cells decreased deltamethrin resistance. These results suggest that trypsin might contribute to deltamethrin resistance in *Culex* mosquito [7]. Interestingly, in subsequent toxicity test of trypsin-treated deltamethrin in mosquitoes, our data suggest that trypsin may directly degrade deltamethrin. However, this possibility needs to be further tested.

To explore the potential function of trypsin, we investigated trypsin-catalyzed *in vitro* hydrolytic reactions of deltamethrin with ultraviolet-visible spectrophotometry (UV/Vis) and gas chromatography-mass spectrometry (GC/MS). In addition, acute oral toxicity of the degradation products in Wistar rats was measured.

Materials and Methods

Chemicals

Deltamethrin. [(S)-alpha-cyano-3-phenoxybenzyl-(1R,*cis*)-2,2-dimethyl-3-(2,2-dibromovinyl)-cyclopropanecarboxylate] was purchased from Roussel Uclaf Company (Paris, France, with more than 98% purity). Trypsin extracted from bovine pancreas and Nα-Benzoyl-L-arginine 4-nitroanilide hydrochloride (L-BAPNA) were purchased from Sigma Chemical Company (St. Louis, MO, USA). Other chemicals used in this study were of analytical grade.

Animals

Wistar rats of both genders with body weight of 180–200 g were obtained from Shanghai Slac Laboratory Animal Co. [certificate no. SCXK (Hu) 2003–0003]. Animals were maintained in a barrier-sustained animal room with certified environmental condition [certificate no. SCXK (Su) 2002-0031]. Five rats of

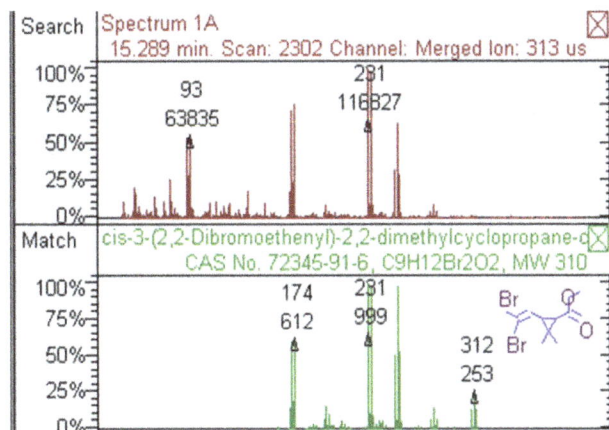

Figure 1. One of the main deltamethrin degradation products with the molecular formula of $C_9H_{12}Br_2O_2$.

the same gender were kept in each cage, with unrestricted food and water supply until dosing procedure. The animal facility was maintained at a temperature of $22 \pm 1°C$, with relative humidity of $50\% \pm 10\%$, and with a 12-h light/dark cycle. All the operation procedures of animals followed the NIH (US National Institutes of Health) guidelines and were approved by Nanjing Medical University Animal Care and Use Committee.

Degradation of deltamethrin

Trypsin and L-BAPNA were handled according to the manufacturer's instruction. An optimal pH was determined by plotting trypsin activity on L-BAPNA against a range of pH from 5 to 11. For the degradation of deltamethrin catalyzed by trypsin, reaction products were detected using the method described by Leng and Gries [8]. The optimal pH in our reaction system was 8.0, which was in accordance with results reported before [9,10]. Briefly, trypsin was dissolved in 50 mM Tris–HCl buffer (pH 8.0). One hundredth volume of 1 mg/mL deltamethrin stock solution in DMSO was added to trypsin solution and incubated at $37°C$ for 10 minutes. Remaining deltamethrin and its degradation products were extracted for the following measurements.

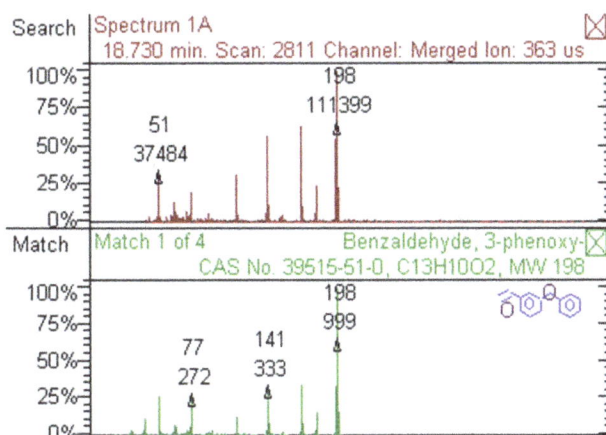

Figure 2. One of the main deltamethrin degradation products with the molecular formula of $C_{13}H_{10}O_2$.

Extraction

Degradation reaction mixtures were vortexed for 30 s upon the addition of 1 mL of acetone and 1 mL of cyclohexane for every 3 mL of reaction mixture, then centrifuged for 10 min at 2,000 rpm. The organic phase was transferred to a glass vial with screw top. The aqueous phase was then extracted twice with 1 mL of cyclohexane each time, and the organic phase was also transferred to a fresh glass vial. The organic phase was then combined and evaporated to dryness under a stream of nitrogen at room temperature. After completely drying, the residue was reconstituted in 200 μL of cyclohexane. A 1 μL sample was then analyzed by GC/MS [11].

Ultraviolet-visible (UV/Vis) spectrophotometry

Spectrograms were obtained using a Cary 5000 UV/Vis spectrophotometer (Varian, Inc., USA) according to the method described by Modi and LaCourse [12]. The time course of deltamethrin degradation and kinetics experiments were carried out in 3-mL UV-transparent cuvettes using Cary WinUV Bio Package software (Version 3.0). Dissolved trypsin (1 mg/mL) was added into 23.34 μM deltamethrin and incubated at $37°C$ using a thermostatic circulating water bath. The absorbance at 264 nm was detected every minute, and a trypsin internal standard was evaluated with the same procedure except in the absence of deltamethrin. For the kinetics experiment, trypsin solution (1 mg/mL) was dissolved in buffer as described above, and different amounts of trypsin were added to deltamethrin solution to make a total volume of 3 mL. The mixtures were incubated at $37°C$, and the increase in absorbance from 200 nm to 500 nm was measured over a time period of 10 min. Standard curve of deltamethrin assayed by ultraviolet spectroscopy was constructed at the same time.

Gas chromatography-mass spectrometry (GC/MS)

A Saturn 2200 GC/MS equipped with a CP-3800 GC and a split/splitless injector port (Varian, Inc., USA) was used in the experiment. The analytical column was a DB-5 [crosslinked (5% phenyl)-methylsiloxane, 30 m in length, 0.25 mm inner diameter, 0.25 μm film thickness] fitted with a retention gap (uncoated and deactivated, 1 m×0.25 mm i.d.) (Agilent Technologies, USA). The GC/MS was programmed to perform a 1.0 μL splitless injection. The injector port was set at $220°C$. The temperature in column oven was as follows: initial temperature was kept at $40°C$ for 2 min; then the temperature was increased to $260°C$ at $8°C$/min and maintained at $260°C$ for 20 min, and the total run time was 50 min. The flow of the carrier gas (helium) was maintained at 1.0 mL/min in constant flow mode. The MS was operated in full scan mode.

Acute oral toxicity study

The acute oral toxicity study was conducted using the up-and-down procedure according to OECD (Organisation for Economic Co-operation and Development) Test Guideline 425. A designated statistical software Acute Oral Toxicity (Guideline 425) Statistical Program (AOT425statPgm, version 1.0, developed by Westat) was used to guide the dosing procedure. The assumed Sigma value was 0.5. The amount of deltamethrin was measured before degradation reaction. After degradation and extraction, the amount remaining in aqueous phase and the degree of reaction completion were checked by vaporization of an aliquot, reconstitution in acetone and measurement by GC. We found negligible amount of deltamethrin and degradation products in aqueous phase, indicating their complete extraction into organic phase. The

Table 1. Percentage of deltamethrin degraded in trypsin-catalyzed reactions.

Trypsin concentration (mg/mL)	0	1.25	2.5	5	10	20
GC peak area	525867	270343	111366	48176	27196	12475
Amount of unconverted deltamethrin (µg)	20	10.28	4.23	1.83	1.03	0.47
Degree of reaction completion (%)	0	48.6	78.9	90.9	94.9	97.7

mixture of deltamethrin degradation products and unconverted deltamethrin from degradation reaction was reconstituted in corn oil and administered by gavage with a gastric feeding tube. The rats were fasted overnight before dosing. The dose progression followed the standard of OECD Guideline 425. After dosing, each rat was carefully observed for the first 5 min to make sure no fluid leaked out from the stomach. For short-term outcomes, rats were checked every 15 min for the first 4 h, every 30 min for the next 6 h, then every 6 h for the following 38 h. Animals surviving the first 48 h were observed for another 12 days with regular check every 6 h for long-term outcome. Any signs of intoxication including changes in skin and fur, eyes, mucosae were carefully observed. Behavioral and somatomotor manifestations of acute oral toxicity were also observed, including tremors, convulsions, short breath, salivation, diarrhea, lethargy, sleep and coma. Death was used as an endpoint, because the difference in lifespan provides important information about toxicity. All observations for each rat were systematically recorded. The LD_{50} were calculated by Maximum Likelihood Estimation (MLE).

Results

Deltamethrin degradation reactions catalyzed by trypsin

Both UV/Vis and GC/MS were employed to measure deltamethrin degradation products. In the UV/Vis test, the obtained spectra of deltamethrin before and after reaction were different, indicating at least some deltamethrin was converted to its degradation products. When the reaction mixture was subjected to GC/MS analysis, the retention times of crude products of deltamethrin degradation reaction (including undegraded deltamethrin) were 37.968 min, 15.289 min, and 18.730 min. Comparing the MS spectra with National Institute of Standards and Technology (NIST) Mass Spectral Database, it was found that the peak with 37.968 min retention time represented undegraded deltamethrin, and the other two represented degradation products of deltamethrin with molecular formula of $C_9H_{12}Br_2O_2$ and

$C_{13}H_{10}O_2$ (Fig. 1 and Fig. 2). The degrees of reaction completion were determined by GC as the ratios of converted deltamethrin to initial amount of deltamethrin, i.e. $1-$(GC peak area of deltamethrin from a reaction)/(GC peak area of deltamethrin from unreacted control). At a trypsin concentration of 10 mg/mL, 94.9% deltamethrin was converted; at a trypsin concentration of 20 mg/mL, 97.7% deltamethrin was converted in our experimental setting (Table 1).

Acute oral toxicity study

The short- and long-term outcomes of the main test of deltamethrin and its degradation mixture are shown in Table 2. According to GC/MS results, a control group dosed with the amount of deltamethrin equal to the undegraded deltamethrin in the degradation reactions was also included. The toxicity symptoms included hyperspasmia, diaphoresis, salivation, and myasthenia in the posterior limbs. The onset of the symptoms was observed 2–3 h after administration, and the surviving rats recovered in 2–3 days. The LD_{50} of deltamethrin was 55 mg/kg in both male and female Wistar rats, while the observed LD_{50} of deltamethrin reaction mixture in male and female rats were 550 and 838.9 mg/kg, respectively (Table 2). After a correction for residual 5% deltamethrin in the reaction mixture according to the equation (1), the LD_{50} of deltamethrin degradation products were 1045 mg/kg and 3358 mg/kg in male and female rats, respectively. The acute oral toxicity of the degradation products was significantly lower than that of deltamethrin.

$$\frac{1}{LD_{50mix}} = \frac{w_{deltamethrin}}{LD_{50deltamethrin}} + \frac{w_{products}}{LD_{50products}} \quad (1)$$

Where w_i is the mass fraction of substance i.

Table 2. The survival and death of Wistar rats after treatment of deltamethrin or its degradation products.

Group	Dosage (mg/kg) and outcome						LD_{50}
Male rats treated with deltamethrin	55 (×)	17.5 (√)	55 (×)	17.5 (√)	55 (√)	175 (×)	55
Female rats treated with deltamethrin	55 (√)	175 (×)	55 (×)	17.5 (√)	55 (×)	17.5 (√)	55
Male rats treated with degradation mixture	175 (√)	550 (√)	2000 (×)	550 (√)	2000 (×)	550 (×)	550
Female rats treated with degradation mixture	175 (×)	55 (√)	175 (√)	550 (√)	2000 (×)	550 (√)	838.9

√ survival, × death.

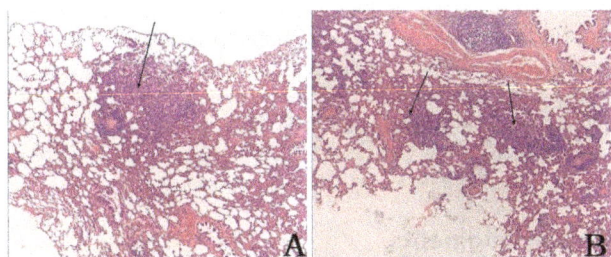

Figure 3. Acute toxicity-associated pathological changes in Wistar rats manifested as atelectasis. A) The group receiving deltamethrin degradation mixture; B) The control group receiving residual deltamethrin.

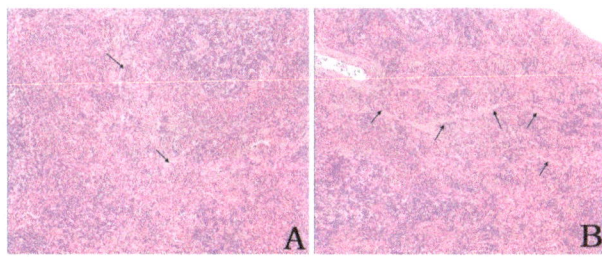

Figure 4. Delayed toxicity-associated pathological changes occurred mainly in the spleen, including fibrosis and lymphoid tissue hyperplasia. A) The group receiving deltamethrin degradation mixture; B) The control group receiving residual deltamethrin.

Pathological results

Because trypsin-catalyzed deltamethrin degradation mixture still contained 5% unconverted deltamethrin, a control group was administered with deltamethrin equal to the residual amount in the degradation mixture. No overt pathological change was observed in the nervous, cardiovascular, gastrointestinal and genitourinary systems. The skin, fur and mucosal surfaces also appeared normal. The animals did not display abnormal behaviors. During the acute phase, pathological changes were observed in the lung in the group administered with deltamethrin degradation products and the residual deltamethrin control group, with the group administered with degradation products showing more severe pathological changes. The damages occurred in the lung mainly manifested as atelectasis (Fig. 3).

Delayed toxicity-related pathological changes mainly occurred in the spleen as fibrosis and lymphoid tissue hyperplasia. There was no significant difference between the degradation products group and the residual deltamethrin control group (Fig. 4).

Discussion

In mammals, deltamethrin is subject to hydrolysis catalyzed by esterases such as carboxylesterases, with the main products being (1R,3R)-3-(2,2-dibromoethenyl)-2,2-dimethyl-cyclopropanecar-boxylic acid and 3-phenoxybenzoic acid [13]. Similar to carboxylesterases, trypsin has a catalytic triad composed of serine, histidine and aspartate, suggesting that trypsin may be able to catalyze the breakdown of the single ester bond in deltamethrin since many proteases can catalyze the hydrolysis of both amide bonds and ester bonds [14–16]. Using UV/Vis spectrum analysis and GC/MS chromatograms, we found that deltamethrin could be degraded by trypsin *in vitro*. Two major degradation products were identified by matching the MS spectra with the MS database of NIST. One product with the molecular formula of $C_9H_{12}Br_2O_2$ was predicted to be (1R,3R)-rel-3-(2,2-dibromoethenyl)-2,2-di-methyl-cyclopropanecarboxylic acid, methyl ester (CAS number 72345-91-6). Considering the molecular configuration of deltame-thrin, this product is likely (1R,3R)-3-(2,2-dibromoethenyl)-2,2-dimethyl-cyclopropanecarboxylic acid, methyl ester (CAS number 61775-87-9). The other product with the molecular formula of $C_{13}H_{10}O_2$ was predicted to be 3-phenoxybenzaldehyde (CAS number 39515-51-0). These products were similar to what would be expected from ester bond hydrolysis. However, instead of (1R,3R)-3-(2,2-dibromoethenyl)-2,2-dimethyl-cyclopropanecarboxylic acid, its methyl ester was generated in the trypsin-catalyzed reaction. The origin of the methyl group is currently unknown, since no artificial methyl esterization step was included in the reaction. Alternatively, the degradation product may not be a

methyl ester, but may be resulted from a substitution reaction of the carboxyl group in which the hydroxyl group was replaced by a nitrile group (possibly coming from the other breakdown product) which was subsequently hydrolyzed into an aldehyde group, i.e. carboxyl group converted into glycoaldehyde group. As about the other product, a possible reason that 3-phenoxybenzaldehyde was not converted into 3-phenoxybenzoic acid could be a lack of participation of oxidative enzymes that would be present *in vivo*. These results suggest that trypsin-catalyzed deltamethrin degrada-tion is more complex than promiscuous esterase activity of trypsin. More detailed mass spectrometry studies using isotope-labeled deltamethrin may provide further information about the underlying mechanism.

Because the reaction is complex, and some minor products may be generated, we used the reaction mixture to test the change in toxicity instead of purifying two major products and testing them separately or in combination. Since the degradation mixture contained unconverted deltamethrin, the observed toxicity also included that of residual deltamethrin. The LD_{50} of degradation mixture was significantly higher than that of deltamethrin, indicating the products of trypsin-catalyzed deltamethrin degra-dation had significantly reduced toxicity in mammals than their parent molecule. Assuming the degradation products and deltamethrin assert their toxic effect in mammals with similar modes of action, in which case their toxicity could be arithmet-ically additive, we used the equation (1) to calculate the toxicity of the products. The validity of such correction is contingent to the nature of toxicity of the products. Nevertheless, qualitative comparison of their toxicities is always valid.

In the pathology study, to make sure the observed toxicity reflected the effect of the products, a control group administered with residual amount of deltamethrin was included. During acute phase, only pathological changes in the lung were observed. The group administered with degradation mixture displayed more severe damages than the residual deltamethrin group. These results suggest that both deltamethrin and its degradation products are toxic to the lung. One concern is that the damages in the lung were caused by fluid leakage during gavage, i.e. deltamethrin and its degradation products were directly administered into the lung instead of being delivered to the lung by circulation. In that case, our observation would still support the conclusion that deltame-thrin and its degradation products are both toxic to the lung (when inhaled). The differences in damage were unlikely caused by different amounts of fluid leaking into the lungs because all animals in the degradation mixture group had more severe damages compared to residual deltamethrin group. On the other hand, because the rats receiving a full dose of undegraded deltamethrin displayed much more severe damages than those receiving degradation mixture, the degradation products were less toxic than deltamethrin.

Deltamethrin degradation products have much lower toxicity than deltamethrin as measured by both acute and delayed pathological damages. These results indicate that trypsin can directly degrade more non-protein molecules (organic pesticides) than previously realized. This will expand people's understanding of the functional properties of proteases.

Excessive pesticide residues in agricultural produces and environment pose a threat to human health. The fact that many pests are increasingly resistant to pesticides also aggravates the situation [17]. It is important to develop safe and effective methods to degrade pesticides as part of pollution control. Our data demonstrate that trypsin can directly degrade deltamethrin *in vitro* and significantly reduce its mammalian toxicity, suggesting that trypsin and possibly other proteases could be exploited in pesticide decontamination. Compared to some enzymes and bacterial strains that have been explored as potential tools for environmen-

tal bioremediation [18–20], trypsin has a unique advantage as it is unstable in aqueous environment around neutral pH due to its self-degradation. Our work suggests that trypsin could be selected as a promising candidate for biocatalyst to control pesticide pollution without causing secondary contamination. Alternatively, engineered microbes that secrete active trypsin may be used to control pesticide pollution in the environment.

Acknowledgments

We thank Dr. Xuewei Liao, Professors Yuyin Feng and Jianwei Zhou for their helpful suggestions.

Author Contributions

Conceived and designed the experiments: D. Zhang, CZ. Performed the experiments: CX. Analyzed the data: CX, FF, LC, QY, JH, D. Zhang. Contributed reagents/materials/analysis tools: CZ. Wrote the paper: CX, D. Zhang. Discussed the manuscript: D. Zhou, BS, LM, YS.

References

1. Krem MM, Di Cera E (2001) Molecular markers of serine protease evolution. EMBO J 20: 3036–3045.
2. Rawlings ND (1998b) Introduction: clan SC containing peptidases with the α/β hydrolase fold. In Barrett,A.J., Rawlings,N.D. and Woessner,J.F. (eds), Handbook of Proteolytic Enzymes. Academic Press, San Diego, CA, pp. 369–372.
3. Freeman HJ, Kim YS (1978) Digestion and absorption of protein. Annu Rev Med 29:99–116.
4. Erickson RH, Kim YS (1990) Digestion and absorption of dietary protein. Annu Rev Med 41:133–139.
5. Muhlia-Almazan A, Sanchez-Paz A, Garcia-Carreno FL (2008) Invertebrate trypsins: a review. J Comp Physiol B 178: 655–672.
6. Wu HW, Tian HS, Wu GL, Langdon G, Kurtis J, et al. (2004) Culex pipiens pallens:identification of genes differentially expressed in deltamethrin-resistant and -susceptible strains. Pestic Biochem Phys 79(3): 75–83.
7. Gong MQ, Shen B, Gu Y, Tian HS, Ma L, et al. (2005) Serine proteinase over-expression in relation to deltamethrin resistance in Culex pipiens pallens. Arch Biochem Biophys 438(1): 53–62.
8. Leng G, Gries W (2005) Simultaneous determination of pyrethroid and pyrethrin metabolites in human urine by gas chromatography-high resolution mass spectrometry. J Chromatogr B 814:285–94.
9. McLaren AD, Estermann EF (1957) Influence of pH on the activity of chymotrypsin at a solid-liquid interface. Arch Biochem Biophys 68(1): 157–60.
10. Fredholt K, Savolainen J, Friis GJ (1999) alpha-Chymotrypsin-catalyzed degradation of desmopressin (dDAVP):influence of pH,concentration and various cyclodextrins. Int J Pharm 178(2): 223–9.
11. Ramesh A, Ravi PE (2004) Electron ionization gas chromatography-mass spectrometric determination of residues of thirteen pyrethroid pesticides in whole blood. J Chromatogr B 802(2): 371–6.
12. Modi SJ, LaCourse WR (2006) Monitoring carbohydrate enzymatic reactions by quantitative in vitro microdialysis. J chromatogr A 1118(1): 125–33.
13. Leng G, Gries W (2005) Simultaneous determination of pyrethroid and pyrethrin metabolites in human urine by gas chromatography-high resolution mass spectrometry. J Chromatogr B Analyt Technol Biomed Life Sci 814: 285–294
14. Ross MK, Edwards CC, Potter PM (2006) Hydrolytic metabolism of pyrethroids by human and other mammalian carboxylesterases. Biochem pharmacol 71(5): 657–69.
15. Kundu N, Roy S (1970) Esterase activity of chymotrypsin on O-DNP-L-tyrosine ethyl ester substrates. Nat 226: 1171.
16. Kushida K, Kato T, Hojo H (2001) High-performance liquid chromatographic-fluorimetric assay of chymotrypsin-like esterase activity. J Chromatogr B 762(2): 137–45.
17. WHO (1992) Vector resistance to pesticides.Fifteenth Report of the WHO Expert Committee on Vector Biology and Control. World Health Organ Tech Rep ser 818: 1–62.
18. Yang QG, Sun L, Zhang DH, Qian J, Sun Y, et al. (2008) Partial characterization of deltamethrin metabolism catalyzed by chymotrypsin. Toxicol In Vitro 22(6): 1528–33.
19. Chen S, Lai K, Li Y, Hu M, Zhang Y, et al. (2011) Biodegradation of deltamethrin and its hydrolysis product 3-phenoxybenzaldehyde by a newly isolated Streptomyces aureus strain HP-S-01. Appl Microbiol Biotechnol 90(4): 1471–83.
20. Zheng YZ, Lan WS, Qiao CL, Mulchandani A, Chen W (2007) Decontamination of vegetables sprayed with organophosphate pesticides by organophosphorus hydrolase and carboxylesterase (B1). Appl Biochem Biotechnol 136: 233–241.

Four Common Pesticides, Their Mixtures and a Formulation Solvent in the Hive Environment Have High Oral Toxicity to Honey Bee Larvae

Wanyi Zhu[1]*, **Daniel R. Schmehl**[2], **Christopher A. Mullin**[1], **James L. Frazier**[1]

1 Department of Entomology, Center for Pollinator Research, The Pennsylvania State University, University Park, Pennsylvania, United States of America, **2** Honey Bee Research and Extension Laboratory, Department of Entomology and Nematology, University of Florida, Gainesville, Florida, United States of America

Abstract

Recently, the widespread distribution of pesticides detected in the hive has raised serious concerns about pesticide exposure on honey bee (*Apis mellifera* L.) health. A larval rearing method was adapted to assess the chronic oral toxicity to honey bee larvae of the four most common pesticides detected in pollen and wax - fluvalinate, coumaphos, chlorothalonil, and chloropyrifos - tested alone and in all combinations. All pesticides at hive-residue levels triggered a significant increase in larval mortality compared to untreated larvae by over two fold, with a strong increase after 3 days of exposure. Among these four pesticides, honey bee larvae were most sensitive to chlorothalonil compared to adults. Synergistic toxicity was observed in the binary mixture of chlorothalonil with fluvalinate at the concentrations of 34 mg/L and 3 mg/L, respectively; whereas, when diluted by 10 fold, the interaction switched to antagonism. Chlorothalonil at 34 mg/L was also found to synergize the miticide coumaphos at 8 mg/L. The addition of coumaphos significantly reduced the toxicity of the fluvalinate and chlorothalonil mixture, the only significant non-additive effect in all tested ternary mixtures. We also tested the common 'inert' ingredient N-methyl-2-pyrrolidone at seven concentrations, and documented its high toxicity to larval bees. We have shown that chronic dietary exposure to a fungicide, pesticide mixtures, and a formulation solvent have the potential to impact honey bee populations, and warrants further investigation. We suggest that pesticide mixtures in pollen be evaluated by adding their toxicities together, until complete data on interactions can be accumulated.

Editor: Wolfgang Blenau, Goethe University Frankfurt, Germany

Funding: Funding for this work was provided by the North American Pollinator Protection Campaign (NAPPC) and the United States Department of Agriculture (USDA)-The National Institute of Food and Agriculture (NIFA) Managed Pollinator Coordinated Agricultural Project (CAP) program. The funders had no role in study design, data collection and analysis, decision to publish, or preparation of the manuscript.

Competing Interests: The authors have declared that no competing interests exist.

* E-mail: wanyizhupsu@gmail.com

Introduction

Recently, one hundred and twenty one different pesticides and metabolites were identified in the hive with an average of seven pesticides per pollen sample, including miticides, insecticides, fungicides, herbicides, and insect growth regulators [1,2]. Feeding on pollen and nectar in the larval diet directly exposes honey bee larvae transdermally, orally and internally [3]; therefore, the potential for chronic toxicity and synergistic interactions at the brood stage seems likely to occur, especially considering the fact that early life stages might be much more sensitive to certain contaminants relative to the adult stage. Several studies have demonstrated that insecticides ranging from insect growth regulators and encapsulated organophosphate formulations to systemic insecticides are more toxic to larvae than to adult bees [4–8]. Moreover, because beebread serves as an absolute requirement for developing bee larvae, pesticide disruption of the beneficial mycofloral community in the colony may thwart the processing of pollen into beebread and allow undesirable pathogens to thrive, therefore indirectly impacting the brood health [9,10]. Indeed, chronic exposure to pesticides during the early life stage of honey bees may thus contribute to inadequate nutrition and/or direct poisoning with a resulting impact on the survival and development

of bee brood [11]. Conceivably, these impacts on the larval phase could lead to weakening of the colony structure over time. To date, only a few peer-reviewed pesticide toxicity studies assess the risks of oral toxicity of pesticides to honey bee larvae. Therefore, a goal of our study was to assess the chronic and mixture effects of common pesticides at realistic exposure concentrations on larval honey bee survival. In order to mimic realistic exposure scenarios of honey bee larvae to contaminated pollen food, we chose the four most frequently detected pesticides in the hive - fluvalinate, coumaphos, chlorothalonil, and chlorpyrifos, and tested them alone and in all combinations via chronic dietary exposure, at concentrations found in pollen and beebread.

The pyrethroid *tau*-fluvalinate and the organophosphate coumaphos have been used widely for Varroa mite control, and found highly persistent in the hive with an estimated half-life in beeswax of about 5 years [12]. These compounds have shown evidence of synergistic toxicity on adult honey bees at the level of cytochrome P450-mediated detoxification [13]. Chlorothalonil, a broad-spectrum agricultural fungicide with an unclear mode of action [14], is often applied to crops in bloom when honey bees are present for pollination, because it is currently deemed safe to bees. However, some fungicides have shown direct toxicity to honey bees or solitary bees at field use rates [15] and fungicides in stored pollen are known to inhibit the growth of beneficial fungi thereby reducing the nutritional value of the pollen to bees [10]. Chlorpyrifos is a widely employed organophosphate in crop management [16] and its residues were frequently found in honey, propolis and dead bees. These in-hive (beekeeper applied) varroacides and out-of-hive (farmer applied) insecticides and fungicides may act alone or in concert, in ways currently unknown, to create a toxic environment for honey bee growth and development.

Another goal of this study was to examine the effect of an 'inert' ingredient on brood survival. Little data exist concerning the toxicity of 'inert' ingredients on honey bees, likely because bee toxicity information for pesticide formulations is not currently required by the U.S. Environmental Protection Agency as part of the pesticide registration process in contrast to the European Union where toxicity for representative formulations is mandatory [17]. Pesticide risk assessment is largely stymied by lack of public access to product-specific information of 'inerts' or co-formulants [18]. Some 'inert' ingredients such as those in formulations of the herbicide glyphosate are more toxic than active ingredients when tested on aquatic organisms [19]. That 'inert' more than active ingredients dominate pesticide formulations and spray tank adjuvants so to increase efficacy and stability of the pesticide makes it important to examine the role of 'inerts' on honey bee toxicity. Here, we studied the chronic toxicity of N-methyl-2-pyrrolidone (NMP, CAS 872-50-4) to bee brood development. The co-solvent NMP is used extensively in chemical processing and agricultural chemical formulations [20,21]. The NMP tested alone or in formulations has demonstrated developmental toxicity in rats by various routes of administration [22] and also has shown high toxicity potential for aquatic invertebrates [23]. There is presently no information in the published literature regarding toxic effects of NMP to honey bees. Our study will be the first to test if this common 'inert' ingredient is toxic to honey bee larvae by continuous dietary exposure, and will serve as a foundation for future studies exploring 'inert' toxicity.

Specific objectives of the present study using the standardized *in vitro* larval feeding method developed by Aupinel et al. [24] are to: (i) assess possible toxic effects of single pesticides on the survival of individual *A. mellifera* larva during a 6-d continuous feeding with contaminated diet; (ii) compare the sensitivity difference between larval and adult bees to the same pesticide exposure; (iii) determine whether the selected pesticides in all combinations at realistic concentrations have any synergistic effects; and (iv) examine the toxicity of environmentally realistic levels of the formulation ingredient NMP on larval survival. Measurable impacts on larvae should demonstrate the need to extend pesticide risk assessment for honey bees from primarily acute effects on adults to chronic impacts on brood survival and development, and of the need to consider both active and 'inert' ingredients in formulations, so that more informed decisions can be made by governments, beekeepers and growers about pesticide application inside and outside the hive.

Materials and Methods

Acquisition of 1st instar larvae

Honey bee (*A. mellifera*) 1st instar larvae were collected from two colonies of *A. m. ligustica* strain reared in our experimental apiary (GPS Coordinates: 40°49′20″N, 77°51′33″W). In order to collect newly emerged larvae, a honey bee queen was confined in the queen excluder cage and placed in the 2nd super from the bottom of the hive and positioned in the center of the super to allow for proper incubation of the newly laid eggs. After being caged for

30 h, the queen was released from the cage and eggs were incubated in the hive for 3.5 days. Frames of newly-hatched 1st instar larvae were taken to the laboratory in a pre-warmed chamber (~35°C).

Diet preparation

Honey bee larval diet (adaptation of [24]) was prepared using 50% royal jelly (Beenatura.com), 12% D-glucose (Fischer Chemical, Fair Lawn, NJ, USA), 12% D-fructose (Fischer Chemical, Fair Lawn, NJ, USA), 2% yeast extract (Bacto™, Sparks, MD, USA), and distilled water (24%). Royal jelly was preserved at −80°C until use. Ingredients minus royal jelly were completely dissolved and filtered through a 0.2 μm membrane (Corning) to remove particulate matter and bacteria. This solution was poured onto royal jelly that was free of wax particles, and mixed thoroughly at room temperature using a spatula. Diet was stored at 4°C for a maximum of three days prior to use.

Pesticide application

The concentrations of applied pesticides were selected based on our previous laboratory findings of commonly found pesticides in pollen [1]. According to the survey of pesticide residues conducted on bee-related product samples from migratory and other beekeepers during the 2007–08 growing seasons, the most prevalent detections at 95th percentile values (levels at which only 5% of detections are higher) in trapped pollen samples were 0.3 mg/L (0.3 ppm) fluvalinate, 0.8 mg/L coumaphos, 0.15 mg/L chlorpyrifos, and 3.4 mg/L chlorothalonil (unpublished data up to 2009). Foraging bees may avoid and dilute contaminated pollen with that from alternative hosts; therefore, the level of contamination found in the trapped pollen pellets varies in relation to the foraging environment of the colony [1,2,25]. We have observed that apple pollen contributes approximately 10% of overall trapped pollen samples from hives placed in apple orchards during a 10-d pollination event (unpublished data). In addition, these pesticides have also been detected in other hive products at even higher levels including beebread, wax comb, foundation, and more rarely in bees. Developing bees are exposed to pesticide residues by contact with the wax, beebread and contaminated bees, so the level found in trapped pollen or royal jelly is not fully representative of actual exposure of larval bees to pesticides. For example, pollen residues of fluvalinate and coumaphos primarily originate by transfer from the contaminated comb wax, which contains much higher levels (e.g. 100-times) of these miticide residues [1,2]. Therefore, in the absence of exact measures of pollen residues in larval foods, we chose to test at 10 times the levels of these four pesticides found in pollen samples. We mixed fluvalinate (purity, 95%), coumaphos (purity, 99%), chlorpyrifos (purity, 99%), and chlorothalonil (purity, 98%) purchased from Chem Service (West Chester, PA, USA) in the larval diet at nominal concentrations of 3, 8, 1.5, and 34 mg/L, respectively. Our calculated concentrations are in accordance with the maximal levels of pesticides detected in both trapped pollen and beebread samples and within the range of 95 percentile values of four selected pesticides detected in hive samples [1]. Therefore, we believe that applying a factor of 10 can give a rough but realistic estimation of the actual exposure of larval bees through contaminated diet or direct transfer from much higher residues in the comb.

Pesticide treatments included four pesticides tested alone and in two, three, and four-component mixtures. To prepare stock solutions, each technical grade pesticide was individually dissolved in acetone and methanol, respectively. Each test solution was mixed thoroughly into the artificial diet at specific concentrations and stored in 2 ml sterile glass vials (Corning, USA). We monitored three control groups in the study: untreated diet, one solvent-treated diet containing 1% methanol and another solvent control containing 1% acetone. We also tested the dietary toxicity of a range of N-methyl-2-pyrrolidone concentrations on larval survival. NMP can be used to 100% of the solvent in pesticide formulations [26]. Table S1 lists the percentage of the solvent NMP in some pesticide formulations that disclose it in MSDS. Here, we tested seven nominal concentrations including 0.01% (100 mg/L in diet), 0.02%, 0.05%, 0.1%, 0.2%, 0.5% and 1% (10,000 mg/L).

Each experiment was repeated twice including control (3 groups), single (6 treatment groups), mixture (binary mixtures: 6 treatment groups; ternary mixtures: 6 treatment groups; four-component mixtures: 2 treatment groups), and 'inert' toxicity tests (seven concentrations of NMP). Sample size for each treatment starting from the same experimental day is 3 replicates with 24 larvae per replicate.

In vitro larval rearing technique

Newly hatched 1st instar larvae were transferred from hive frames into sterile, 48-well culture plates (Corning, USA) for the in vitro rearing technique with 24 larvae per plate. Larval transfers were done in the lab without the use of a sterile hood. The sterile, push-in queen cups (B&B Honey Farm, USA) were placed in every other well. Diet was warmed to ~34°C in a heating block prior to larval transfer. Using an Eppendorf 10–100 µl variable volume pipette, 10 µl of each diet treatment was placed per queen cup. A 00 camel hair paintbrush was used to transfer each larva from the cell on the frame to the cup. The paintbrush was dipped into distilled water between each larval transfer to aid in a smooth transfer, and was sanitized by dipping in 95% ethanol after every four to five transferred larvae. Larvae were placed directly on top of the diet and inspected for mobility to ensure a quality transfer. Four additional queen cups were equally spaced in four of the remaining open wells before placing the lid on the culture plate, allowing for adequate ventilation of the larvae throughout the experiment. Each plate was placed in a humidity chamber and kept at 95% relative humidity with a 10% aqueous solution of sulfuric acid being used at the base of the chamber to maintain humidity. Humidity chambers were placed in an incubator at 34°C in the dark and were not disturbed throughout the experiment, except when replacing the diet for ~15 min/d.

For this study, only the survivorship of honey bees during the larval stage was monitored to evaluate the impacts of selected pesticides. Larval mortality was recorded daily by probing the larvae with sanitized forceps. The dead larvae were removed daily. Diet for each larval bee was replaced daily. Old diet was removed using a glass disposable pipette and new diet was immediately placed in each queen cup according to the following schedule to account for larval growth: day 1- 10 µl, day 2- 10 µl, day 3- 20 µl, day 4- 30 µl, day 5- 40 µl, and day 6- 50 µl.

Kaplan-Meier survival analysis

The 6-d larval survival data were segregated by pesticide treatment and analyzed using Kaplan-Meier survival analysis [27]. This estimate generally assumes independence among the individual death events and randomization within the treatment group. The hazard rate h(t) is the conditional probability of failure or death in a small time period given that the subject has survived up until a specified time t. The greater the value of the hazard rate, the greater the probability of impending death. The null hypothesis of no difference between survival curves of treatment and control groups was tested by the Log-rank test that weights

each death by the square root of the total number of individuals at risk per time interval, placing less emphasis on deaths occurring later in the experiment. All the survival analyses were implemented in SAS survival program (SAS/STAT® 9.2 User's Guide).

Comparison between adult and larval sensitivity

The difference in sensitivity to the same pesticide between adult bees and larvae can be quantitatively evaluated by comparing the actual larval mortality per day from the in vitro test with the predicted mortality for adult bees if exposed to the same concentrations of pesticides. The larval mortality data were corrected with Abbott's formula beforehand. Here, the impacts of pesticide treatments on adult bees were estimated from the adult acute topical LD_{50} data converted to whole-bee LC_{50} values [1], because neither the chronic nor acute oral toxicity data of adult bees are currently available for all pesticides selected for this study. Predicted adult toxicity can be estimated as a function of the magnitude of toxicant exposure and the individual's sensitivity to a toxicant, which is generally characterized by the probit model [28]. The predicted proportion of insects killed ($\hat{p}$), in probit transformed units, calculated as $\hat{p} = a + bx$ where a = intercept and b = slope from the regression of the transformed data and x is the log-transformed concentration or time. Results of probit analyses are reported typically as a concentration or time required to kill a certain proportion of the test insects (e.g., LC_{50}). Table 1 shows the average LC_{50} values from the literature [1] and probit slopes from other sources [28]. One exception is chlorothalonil, which is estimated using the default probit slope of 4.5 because its mortality levels under topical or oral applications to honey bees are found to be insufficient to establish a dose-response relationship. Therefore, the probit function for each pesticide to adult honey bees can be inferred from the LC_{50} values (x), probit mortality ($\hat{p} = 5$) and probit slope (b) [13,28]. Then, the probit model can be extrapolated to predict the probability of an impact of each pesticide on adult bee survival for a specified concentration. Using the Probit program in SAS 9.2 (SAS/STAT® 9.2 User's Guide), the predicted probit-type mortality can be transformed to the original percent units and compared with the actual larval percent mortality data. Using the compilation of acute data from different sources may complicate the accurate estimation of the adult toxicity because of the heterogeneity introduced by differences among the studies; however, given the limitations we felt this was a reasonable approach to obtain a first approximation of the differences in adult and larval sensitivity to the same pesticide exposure.

Pesticide interaction determination

We used significant departures from additive toxicity to define antagonistic and synergistic interactions between pesticides in mixtures [29]. The expected additive toxicity for the chemical mixture is the sum of each chemical's toxicity to larval survival, calculated as f chemical components in the pesticide mixture and h_i is the hazard rate for a specific component estimated from the laboratory bioassay data. The sum of the responses (Eh_n) to the individual components is estimated based on the assumption that the selected pesticide mixtures are the combination of substances with independent modes of action or similar modes of action. The mixture toxicity can be predicted as follows: *Additive interactions*– Simultaneous action of components in which the observed response of honey bee larvae to a mixture (h_n) is equal to the sum of the responses (Eh_n) to the individual components; *Synergistic interactions*–Simultaneous action of components in which h_n is significantly higher than Eh_n; *Antagonistic interactions*–Simultaneous action of components in which h_n is significantly less than Eh_n.

Table 1. Comparison between the predicted adult mortality rate (PM, %) for each tested concentration (Conc., mg/L) of four pesticides using a probabilistic toxicity model and the observed brood mortality rate (AOM, %) for bee larva from the 6-d *in-vitro* rearing experiments.

Pesticide	Adult honey bee				Honey bee larva						
	Inverse probit prediction				*In-vitro* brood test						
	β[a]	LC$_{50}$[b]	Conc.	PM[c]	1-d[d]	2-d[d]	3-d[d]	4-d[d]	5-d[d]	6-d[d]	AOM[e]
Fluvalinate	2.5	15.86	3	3.6	3.13*	8.06	12.28	10.00	11.11	68.85**	11.72
Coumaphos	2.9	46.3	8	1.4	6.25*	1.67	8.47	5.56	3.92	53.73**	8.60
Chlorothalonil	4.5	1110	34	4 E-10	0.00	8.93	7.84	12.77	7.32	56.60**	9.82
Chlorpyrifos	10	1.22	1.5	82	0.00	4.17	8.70	33.33**	32.14**	0.00	10.07

[a]β is the slope of the probit function for different pesticides [13,28].
[b]LC$_{50}$ is the median lethal concentrations of each pesticide to adult honeybees [1].
[c]PM = predicted adult mortality rate (%) for each pesticide at the tested concentrations using inverse prediction of the probit function.
[d]1,2,3,4,5,6-d is the observed conditional mortality rate (%) for larval bees at each age (in day) in the *in vitro* rearing process.
[e]AOM = average daily mortality rate (%) for larval bees in the *in vitro* rearing process.
*Significant at $p < 0.05$;
**significant at $p < 0.001$. (Statistical differences in larval survival were assessed between pesticide-treated and solvent control groups.)

We did not test different concentrations of each pesticide component and of the combinations to fit dose-response curves. Neither food intake nor concentrations of pesticides consumed by each larva were measured during the oral feeding. Therefore, this method does not allow exact quantification of the level of interaction but makes only an initial qualitative assessment of synergism or antagonism.

Results

Control toxicity

No significant differences in larval mortality were observed when larvae were reared on untreated artificial diet or diet mixed with 1% methanol or 1% acetone (Log-rank test, $p > 0.05$) (data not shown). These three control groups showed an accumulative 6-d percent mortality of approximately 17.2% (Fig. 1), which is within the normal range observed for control mortality using the *in-vitro* larval rearing protocol [24,30]. Because control mortality exceeds 10%, the larval mortality data from treatment groups were corrected with Abbott's formula.

Single pesticide toxicity

Chronic exposure of bee larvae to each of the four pesticides at tested concentrations showed significant toxic effects on larval survival (Log-rank test, $p < 0.0001$), resulting in an overall 2- to 4-fold reduction in the total 6-d percentage survival compared to the control mortality (Fig. 1A). Based on age-specific toxicity data, mortality rates for each pesticide were uneven across different larval stages (Fig. 1B). For 1-day-old larvae, 8 mg/L coumaphos and 3 mg/L fluvalinate were more toxic than the other two pesticides. The 2 and 3-day-old larvae showed similar sensitivity to different pesticide exposures, approximately 10% mortality per day. The 4 and 5-day-old larvae were most sensitive to 1.5 mg/L chlorpyrifos, causing more than 32% larval death each day (Table 1). A dramatic increase in larval mortality for 6–day-old larvae was observed in 34 mg/L chlorothalonil and the two miticide groups, ranging from 53.73% to 68.85%. Using the probit model, notable differences were found in pesticide sensitivity between the adult bee and larvae (Table 1). Among the four pesticides tested, 1.5 mg/L chlorpyrifos was the only treatment that adult bees were more susceptible to than the larvae. For the

other pesticides, the larvae showed increased sensitivity over that of adult bees. Notably, chlorothalonil at the sublethal concentration of 34 mg/L was least toxic to adult bees, however most toxic to larvae followed by 8 mg/L coumaphos and 3 mg/L fluvalinate. On average, coumaphos was the least toxic to larval bees among the four pesticides.

Synergistic interactions

I. Chronic toxicity of chlorothalonil and coumaphos. The effects of chlorothalonil (34 mg/L), coumaphos (8 mg/L), and their mixture on larval survival through the 6-d development are shown in Fig. 2A. In the first 3 days of larval rearing, these three groups exhibited similar survival curves ($p = 0.1988$, Log-rank test). Subsequently, the larvae reared on the diet contaminated with the chlorothalonil/coumaphos mixture died most quickly. The risk of 4-day-old larvae being killed by the mixture was higher than for the other stages of larvae and the single pesticide groups. The hazard rate of the combination group (hn(4) = 0.523) was 9-times higher than the coumaphos group (hCM(4) = 0.057) and 3-times higher than the chlorothalonil group (hCL(4) = 0.136). The conditional probability of 4-day-old larvae being killed by the mixture treatment was 5-times higher than that of expected additive toxicity (Fig. 2B, Ehn(4) = 0.0965, p<0.0001, Mann–Whitney test). Therefore, the pairing of chlorothalonil and coumaphos produced a significant synergism on mortality of larvae older than 4 days.

II. Chronic toxicity of chlorothalonil and fluvalinate. For the 4-day-old larvae, the hazard rate of the mixture (hn(4) = 0.78) was the highest during the 6-d larval development, which was 7-times higher than the fluvalinate (3 mg/L) group (hFlu(4) = 0.105) and 5-times higher than the chlorothalonil (34 mg/L) group (hCL(4) = 0.136) (Fig. 2C). The chlorothalonil/fluvalinate mixture at the tested concentrations gave a synergistic interaction, which significantly magnified the hazard rate by 7 fold over the sum of the individual effects (Fig. 2D, Ehn(4) = 0.121, p<0.0001, Mann–Whitney test).

Additive interactions

I. Chronic toxicity of fluvalinate and chlorpyrifos. Larval survival on fluvalinate (3 mg/L) and chlorpyrifos (1.5 mg/L) declined the fastest among pesticide mixture treatments, ranging

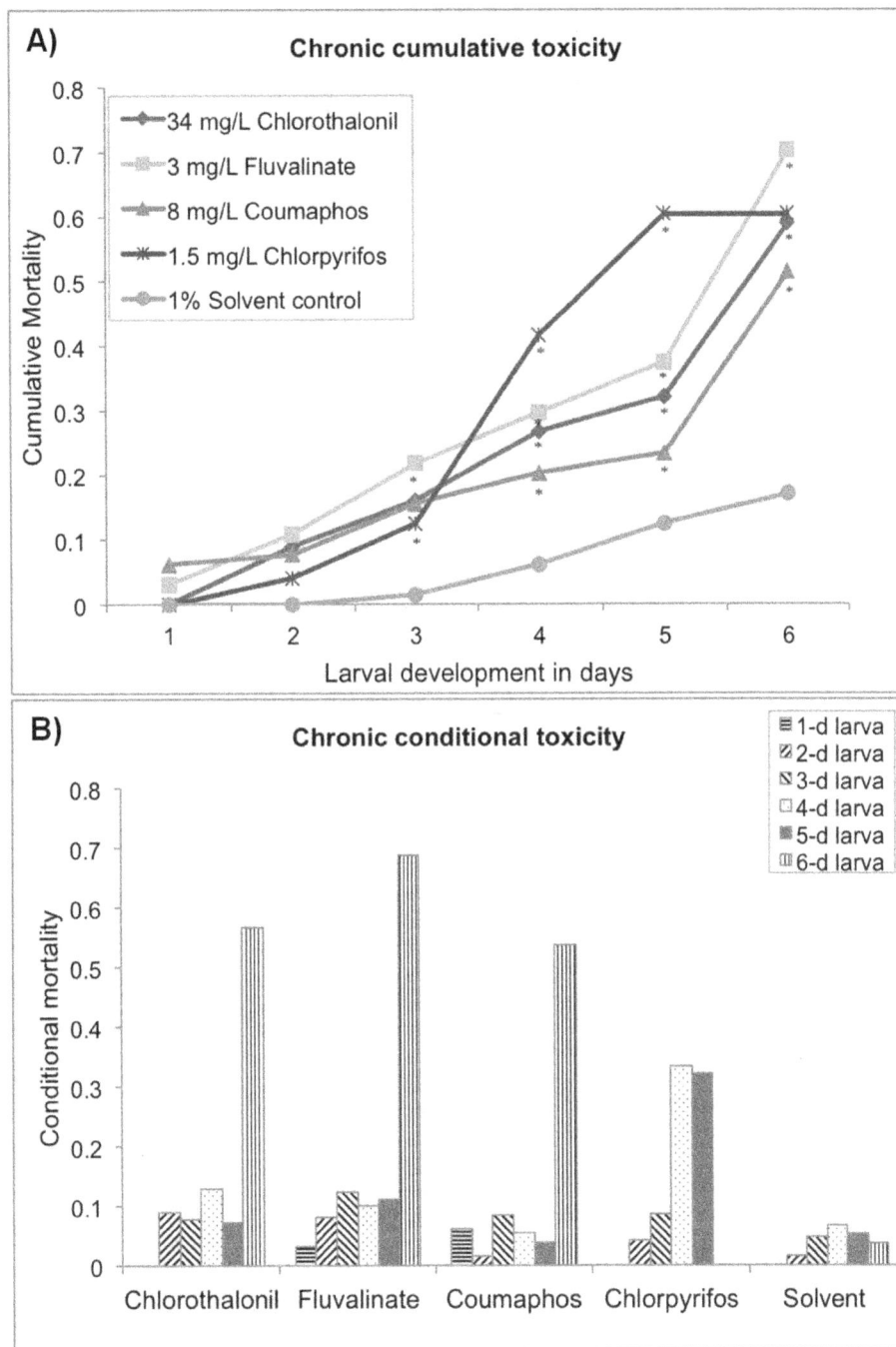

Figure 1. Larval survival during the 6-d development stage reared on artificial diet contaminated with four pesticides at the selected concentrations and a 1% solvent control. (A) shows the cumulative mortality of honey bee larvae through 6-d development continually exposed to 34 mg/L Chlorothalonil, 3 mg/L Fluvalinate, 8 mg/L Coumaphos, 1.5 mg/L Chlorpyrifos and 1% solvent; (B) illustrates the conditional mortality for different development stages of bee larva. Asterisks denote significant difference from the respective solvent controls (analysis of variance, Log-rank test, $p < 0.0001$).

from 4.17% to 70.83% (Fig. 3). No significant differences were found in larval survival between single component groups through the 6-d development (Fig. 3A, Log-rank test, p = 0.1711). This binary combination produced additive toxicity. The 6-d cumulative percent mortality caused by this mixture (hn = 71%) was slightly higher than the sum of the response to single components, but not at a significant level (Fig. 3B, Ehn = 48.96%, p = 0.171, Mann–Whitney test).

II. Chronic toxicity of chlorpyrifos and coumaphos. The larval chronic toxicity of this combination treatment was the highest among tested pesticide mixtures causing from 10.4% to 79.2% mortality during the 6 days. Survival was least affected by the diet with 8 mg/L coumaphos (Fig. 3C). The interaction between these pesticides showed an additive effect. The 6-d cumulative percent mortality of larvae reared on the mixture

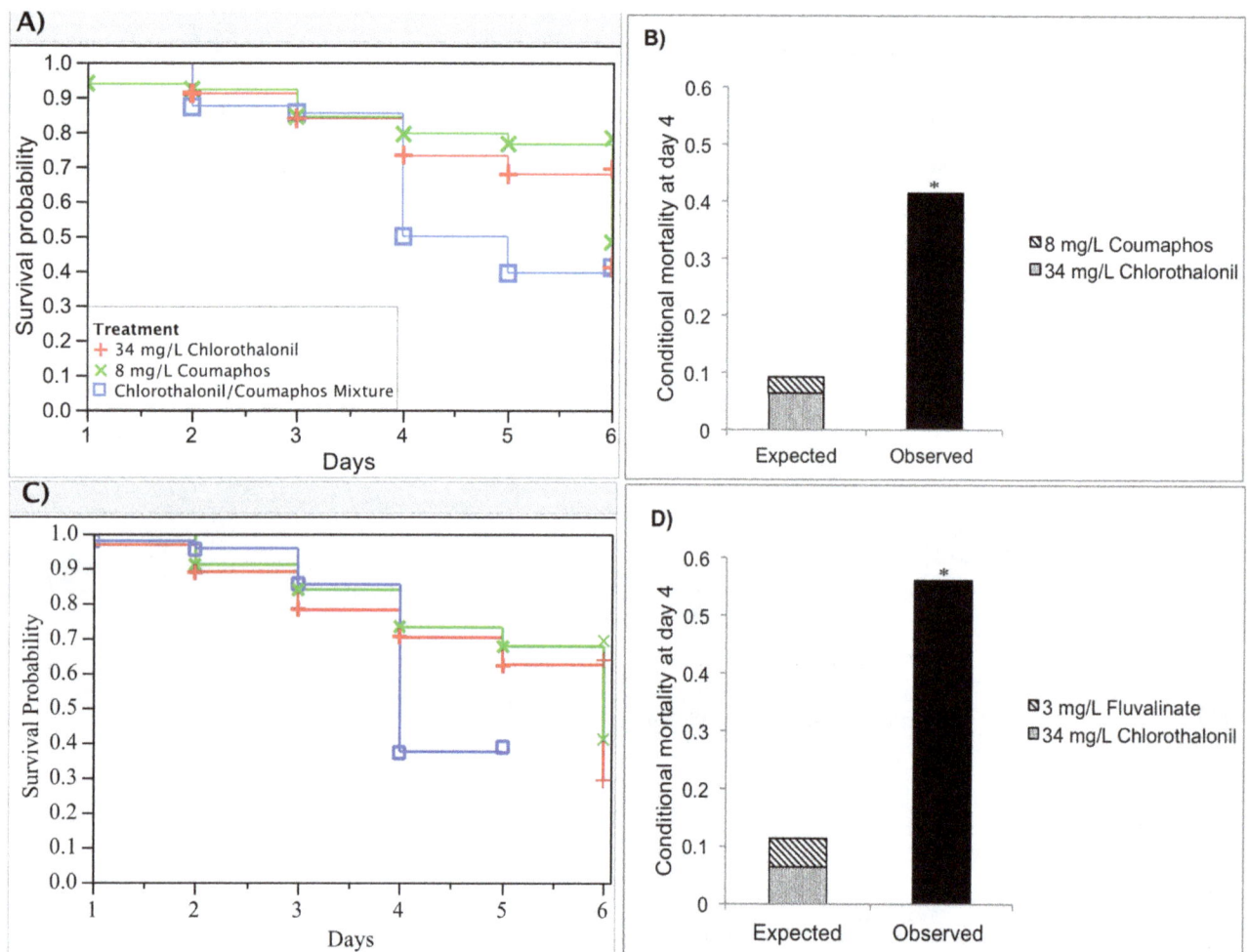

Figure 2. Synergistic interactions for two pairs of pesticide mixtures: 8 mg/L Coumaphos, 34 mg/L Chlorothalonil and the mixture; 3 mg/L Fluvalinate, 34 mg/L Chlorothalonil and the mixture. (A) and (C) show the respective Kaplan-Meier survival plots for honey bee larvae reared for each pair of pesticide mixture; (B) and (D) illustrate the interaction determination based on the deviation of observed mixture toxicity (black bar) from the expected additive toxicity (stacked bar). Asterisks denote significant difference from the expected additive toxicity (Mann–Whitney test, $p < 0.0001$).

(hn = 79.2%) did not differ significantly from expected additive toxicity (Fig. 3D, Ehn = 56%, p = 0.558, Mann–Whitney test).

III. Chronic toxicity of fluvalinate and coumaphos. The survivorship of larval bees on the combination and fluvalinate alone treatments exhibited a similar gradual declining trend, achieving the highest cumulative mortality at the end of the 6-d development (Fig. 3E). Both showed more toxicity to larval bees than coumaphos alone (Fig. 3E, p = 0.0425, Log-rank test). Fluvalinate and coumaphos, mixed at 3 mg/L and 8 mg/L respectively, showed an additive effect. The accumulative percent mortality in the mixture group (hn = 68.75%) did not vary significantly from the expected additive toxicity (Fig. 3F, Ehn = 60.94%, p = 0.052, Mann–Whitney test).

Antagonistic interactions

I. Chronic toxicity of fluvalinate and chlorothalonil at low concentrations. The 3.4 mg/L chlorothalonil and 0.3 mg/L fluvalinate mixture showed the least toxicity to larval development among pesticide combinations tested (Fig. 4A). Especially, for the 4-day-old larva, the hazard rate of individual component groups (hCL(4) = 0.214, hFlu(4) = 0.259) was greater than twice the mixture treatment (hn(4) = 0.088). This mixture showed antago-

nistic interaction, significantly reducing the hazard rate of 4-day-old larvae by three-fold from the expected additive toxicity (Fig. 4B, Ehn(4) = 0.2365, p < 0.0001, Mann-Whitney Test).

Three-component mixture toxicity

All six possible pairings were selected to determine the toxicity for three-component mixtures including chlorothalonil/fluvalinate/coumaphos and fluvalinate/coumaphos/chlorpyrifos. The only significant difference found was when coumaphos (8 mg/L) was added to the two-component mixture of fluvalinate (3 mg/L) and chlorothalonil (34 mg/L), giving a 3% reduction in the 6-d accumulative larval mortality ($h_n = 38\%$) from the expected additive effect (Fig. 4C and 4D; $Eh_n = 41.41\%$, $p = 0.006$, Mann-Whitney Test). The other five pairings did not yield significant changes in larval survival when adding one component into the existing binary mixtures.

Four-component mixture toxicity

Two pairings of mixtures including chlorothalonil added to fluvalinate/coumaphos/chlorpyrifos and chlorpyrifos added to chlorothalonil/fluvalinate/coumaphos were tested at the same concentrations as before to determine toxicity interactions in going

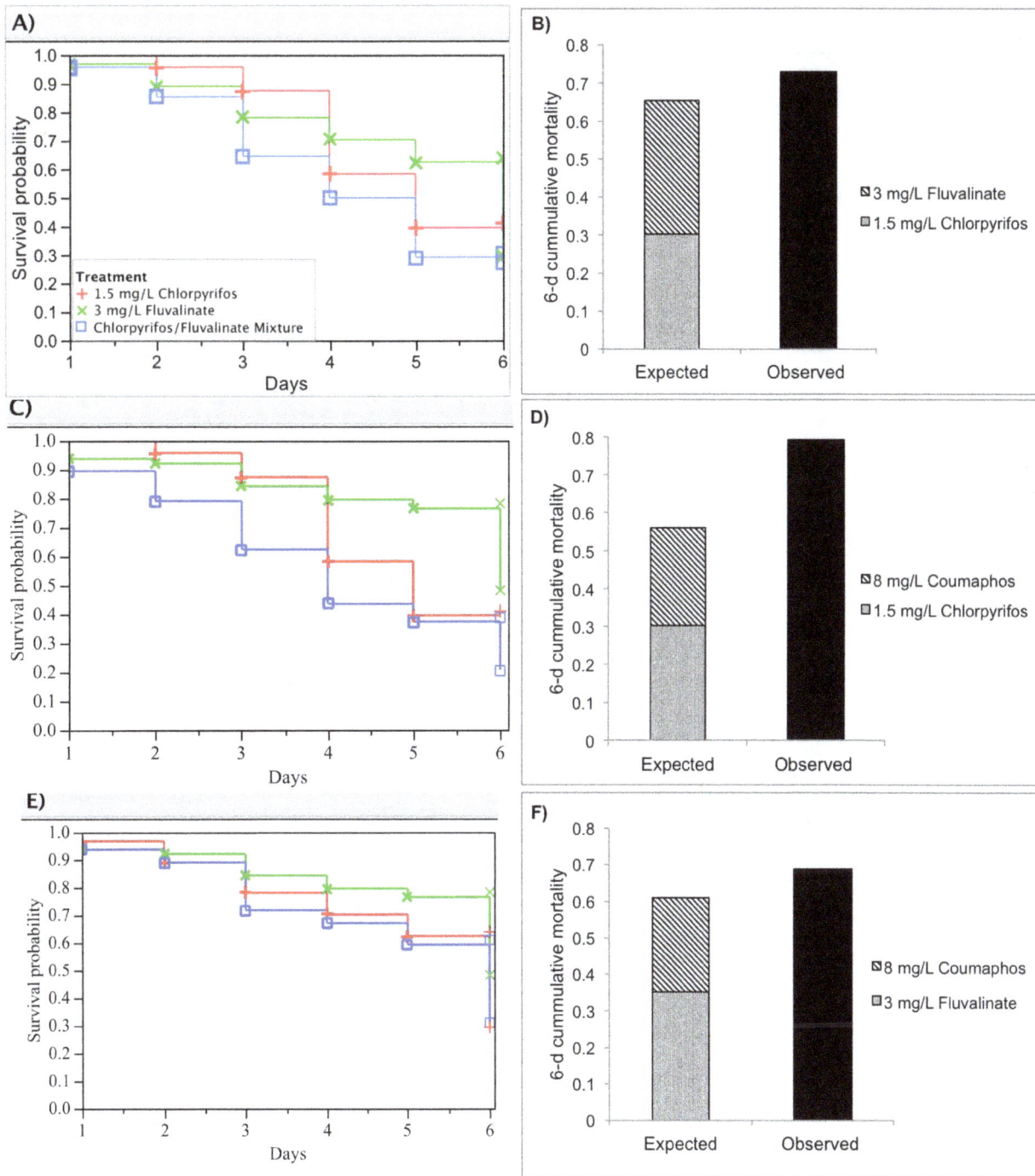

Figure 3. Additive effects for three pairs of pesticide mixtures: 3 mg/L Fluvalinate, 1.5 mg/L Chlorpyrifos and the mixture; 8 mg/L Coumaphos, 1.5 mg/L Chlorpyrifos and the mixture; 8 mg/L Coumaphos, 3 mg/L Fluvalinate and the mixture. (A), (C) and (E) show the respective Kaplan-Meier survival plots for honey bee larvae reared for each pair of pesticide mixture; (B), (D) and (F) illustrate the interaction determination based on the deviation of observed mixture toxicity (black bar) from the expected additive toxicity (stacked bar).

from three- to four-component mixtures. There were no significant changes in larval survival when integrating a fourth component into these three-component mixtures. The four-component mixture caused 54.17% larval mortality at the end of the 6-d larval development.

'Inert' ingredient toxicity

Chronic exposure of bee larvae to the 'inert' ingredient NMP at seven different concentrations ranging from 0.01% to 1% greatly impacted larval survival (Fig. 5). Increasing amounts of NMP correspondingly increased larval mortality. A 1% concentration (10,000 mg/L) of NMP was the most acutely toxic, generating 100% mortality within 24 h after treatment. Even for the lowest

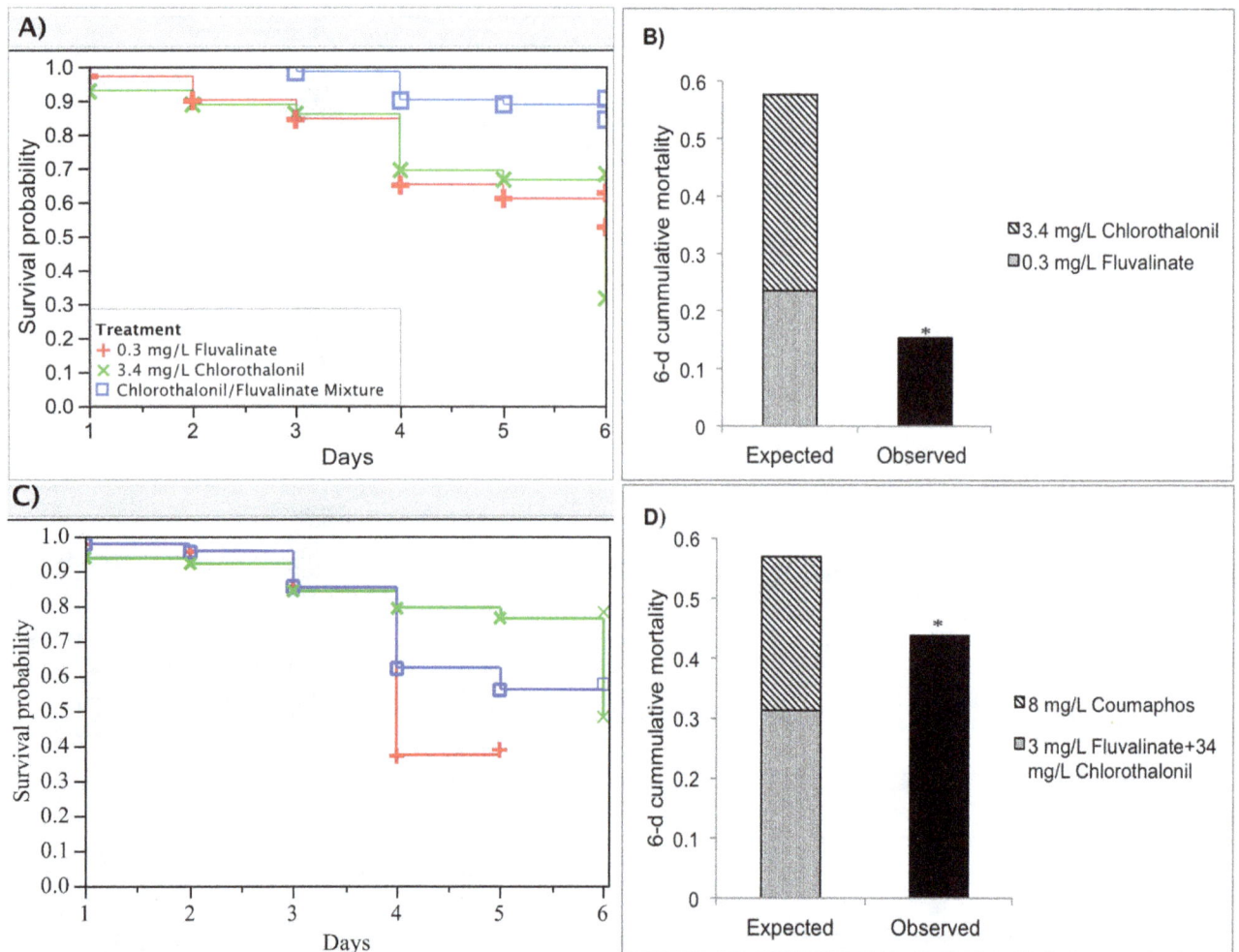

Figure 4. Antagonistic interactions for two pairs of pesticide mixtures: 0.3 mg/L Fluvalinate, 3.4 mg/L Chlorothalonil and the mixture; 3 mg/L Fluvalinate+34 mg/L Chlorothalonil mixture, 8 mg/L Coumaphos and the three-component mixture. (A) and (C) show the respective Kaplan-Meier survival plots for honey bee larvae reared for each pair of pesticide mixture; (B) and (D) illustrate the interaction determination based on the deviation of observed mixture toxicity (black bar) from the expected additive toxicity (stacked bar). Asterisks denote significant difference from the expected additive toxicity (Mann–Whitney test, $p < 0.0001$).

concentration of 0.01% (100 mg/L), the estimated time to cause 50% larval mortality was 4 days.

Discussion

Chronic toxicity

Our findings suggest that chronic dietary feeding at hive levels of common pesticide ingredients including the fungicide chlorothalonil, miticides fluvalinate and coumaphos, and insecticide chloropyrifos, individually or in mixtures, have statistically significant impacts on honey bee larval survivorship. A significant increase in larval mortality was found at or beyond 4-d of feeding. This is the first study to report serious toxic effects on developing honey bee larvae of dietary pesticides at measured hive residue concentrations. The maximum concentrations of fluvalinate, coumaphos, chlorothalonil, and chlorpyrifos found in our hive samples are 204 mg/L, 94.1 mg/L, 98.9 mg/L, and 0.9 mg/L, respectively (Table S2), which are much higher for the miticides and fungicide, or similar for the insecticide, to those levels tested here (Table 1). This chronic (6-d) toxicity is likely to be undetected in a conventional acute (24/48 h) toxicity study, resulting in

potential underestimation of pesticidal effects. The lethal effects on honey bee larvae appearing after 4-d continuous exposure to pesticides at low concentrations are also observed in adult honey bees. The accumulated dose of the organophosphorus insecticides acephate, methamidophos or dimethoate resulting in 50% adult bee mortality was over 100-fold lower than the respective acute 24 h oral LD_{50} [31]. For these organophosphates and also the pyrethroids tested, their toxicity to worker bees was significantly increased by continuous versus single ingestion of the contaminated food. At low doses of imidacloprid, adult bee mortality was observed only 72 h after onset of feeding in contrast to immediate effects at much higher doses [32].

The causes for chronic larval bee toxicity for 6-d dietary subacute pesticide exposures remain unknown. It may be associated with the extended time needed to accumulate sufficient insecticide concentrations internally to exert nerve action at central target sites, which is consistent with the pharmacological receptor theory; or may reflect variation in honey bee detoxification capacities from the more peripheral to internal tissue sites. For instance, the results of high toxicity of low doses of all imidacloprid metabolites suggest the existence of binding sites with different

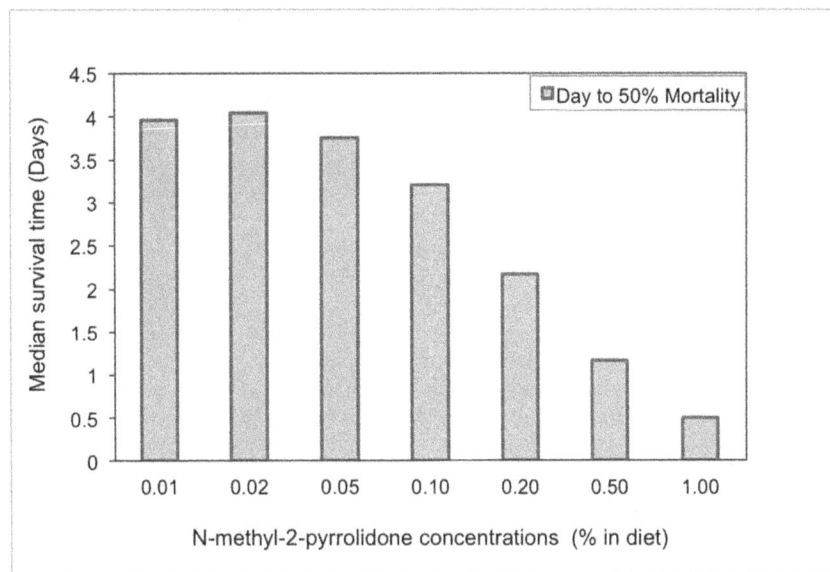

Figure 5. The estimated time to cause 50% larval mortality by seven nominal concentrations of N-methyl-2-pyrrolidone mixed in larval diet.

affinities in honey bees [32]. Another explanation may be that honey bee detoxification mechanisms are not induced by chronic exposure of low concentrations of active substances, but require higher more acute concentrations to impact honey bee susceptibilities. In the former case, bee mortality would be latent due to the time needed for pesticide bioaccumulation, further favored by the more lipophilic pesticides fluvalinate, coumaphos, chlorpyrifos and chlorothalonil tested here. The latter case of acute higher concentrations driving induction of detoxification enzymes can result in both antagonistic and synergistic effects on the target-effective insecticide concentration depending on if the induced cytochrome P450 first activates (e.g., chlorpyrifos, coumaphos to respective oxons) or detoxifies (e.g., fluvalinate) the insecticide [33,34]. Other induced enzymes (e.g., hydrolases, glutathione transferases) will further degrade and detoxify the primary metabolites.

It is also plausible that more general stress mechanisms (e.g., altered feeding, suppressed growth) dominate the chronic response. For example, exposures of some repellent pesticides such as pyrethroids at sublethal levels have been demonstrated to impair feeding behaviors of honey bees and bumble bees [3,8]. In the case of honey bee larvae, they retain internally all metabolic wastes throughout the larval stage up to the pupal molt after which they defecate a waste pellet called the meconium [25]. Concentrations of pesticides and metabolites within brood tissues may result in continuous pesticide stress [35], which differs from the adult honey bee and most other insects where excretion of toxic wastes regularly occurs. Little information is available on the distribution of fluvalinate [36] and coumaphos [37] and their degradates in honey bee adults and brood. Further studies to examine the distribution and accumulation of fluvalinate, coumaphos, chlorpyrifos and chlorothalonil and their metabolites, in honey bees at different developmental stages are needed. Meanwhile, how honey bees at different life stages withstand chronic exposure need more detailed study of metabolic regulation in this social insect.

Remarkably, among the four pesticides tested in the present study, immature honey bees are highly vulnerable to the common fungicide chlorothalonil (Figs. 1 and 2). Dietary chlorothalonil killed more than 50% of larvae in 6 days at a level of 34 mg/L, a nontoxic dose to adult bees in acute bioassays (Table 1). This difference in larval to adult susceptibility was the largest among the four pesticides tested. It is unclear why, larval bees exhibited much greater sensitivity to chlorothalonil compared to adult bees; however, the present results demonstrate that investigating fungicide impacts on honey bees is particularly necessary for a realistic evaluation of pesticide impacts on colony health, given the frequent detections of chlorothalonil in pollen and wax samples. Hence, considering that honey bees are experiencing a diverse array of agrochemicals in the hive, the chronic toxicity test may better assess pesticide exposure for a honey bee colony.

Mixture toxicity

Currently, studies of mixture toxicity between different classes of pesticides at concentrations of environmental relevance are rarely available for honey bees [34]. The present study of four pesticides in all combinations is the first study to investigate the potential synergism of common pesticides at realistic exposure levels to larval bees. The present results showed interactions between binary combinations of synthetic pesticides tested were mostly additive, which can be attributed to the same or independent mode of actions of the pesticides involved [33,34]. For instance, additivity of the coumaphos/chlorpyrifos mixture may be explained by their identical action as organophosphate inhibitors of acetylcholinesterase. The additive toxicity of the pyrethroid fluvalinate with either coumaphos or chlorpyrifos is probably due to the independent primary action of the former on nerve sodium channels. Our result with larvae is not consistent with the adult honey bee study of Johnson et al., where the combination of fluvalinate and coumaphos was synergistic [13]. This discrepancy may be explained by the different life stage, lower insecticide concentration levels, and longer length of exposure used here.

The three and four component mixtures of tested pesticides have mostly demonstrated additive effects in larval bees. This finding is in general agreement with the Funnel Hypothesis [38], which states that the toxicity will tend towards concentration additivity as the number of components in equitoxic mixtures increases. One exception was the significantly less than additive response when coumaphos was integrated into the fluvalinate and chlorothalonil mixture. That coumaphos antagonizes the synergistic effect of fluvalinate and chlorothalonil may be related to its possible induction of the detoxification of one or both of the other pesticides. This anomaly may be related to the observation that elevated coumaphos levels in brood had the highest discriminatory value with regard to healthy bee colonies whereas higher levels of this miticide in the pollen food correlated with colony collapse [39], again indicating that pesticide susceptibilities differ across honey bee developmental stages.

Remarkably, binary mixtures of chlorothalonil with the miticides fluvalinate or coumaphos were synergistically toxic to 4-day-old bee larvae. This is the first demonstration for honey bee brood of a synergistic interaction between dominant in-hive miticides and the frequently-encountered fungicide chlorothalonil at environmentally relevant concentrations. Synergism with chlorothalonil and fluvalinate but not coumaphos for adult honey bee toxicity has been noted previously [40,41].

Surprisingly, a significant antagonism was found for larval toxicity from the fluvalinate-chlorothalonil combination at one-tenth of the concentrations (Fig. 4) that otherwise exhibited a five-fold synergism (Fig. 2). One rationale behind this latter interaction, beyond the fact that the very diverging pyrethroid-multi-site chlorothalonil mechanisms of action may alone elicit synergistic effects, is that the high concentrations may directly inhibit detoxification enzymes. For example, the competitive inhibition of cytochrome P450 monooxygenase enzymes has been suggested to explain the synergistic interactions among pesticides for adult honey bees such as pyrethroid insecticides or mixtures of organophosphate insecticides and ergosterol biosynthesis inhibiting fungicides [42,43]. Also, synergism between chlorothalonil and the herbicide atrazine has been documented in aquatic species [44]. Modes of action for chlorothalonil range from inhibiting glutathione and other thiol-dependent enzymes or protein receptors, to disrupting or degrading cell membranes causing lysis that can enhance penetration of other pesticides [14]. The tendency toward antagonism of brood toxicity at the lower dietary chlorothalonil-fluvalinate concentration may be associated with alternative peripheral mechanisms such as gut microbial detoxification that may be overwhelmed at higher dosage where more internal neurotoxic effects of the pyrethroid can prevail. The consequence is that biphasic low and high dose response relationships may result depending on the extent of multiple peripheral and internal sites of action that diverge in sensitivity to the toxicants as well as to the available detoxification pathways that differ in a tissue-dependent manner to the concentrations required for their induction.

While the mechanisms of interactions among pesticides with diverse modes of action and their dynamics in the developing honey bee larvae are not known, application of the concentration-addition model combined with chronic feeding tests represents a starting point for investigation of mixture effects at realistic levels and their risks for this pollinator. Considering that the diverse arrays of chemicals [1,2,45] and general additivity exist in the hive environment, examining the toxicity of chemical mixtures in addition to single toxicants is critical for a realistic assessment of pesticide hazards experienced by honey bees and other non-target organisms. In today's agriculture dominated by mass monocultures, adults and larvae of *A. mellifera* are inevitably exposed to transgenic material via pollen consumption of GM-crops [46], which might be another confounding factor for bee health. Although minor evidence showed adverse effects of Bt-crops on *A.*

mellifera, the risk assessment of combined effects of Bt-crops and pesticides are completely lacking [47–49]. Hence, the dose dependency of the synergy, the multitude of compounds, the differences in adult bees and larvae, the possibility of continuous exposures, and the interaction with GM pollen should be taken into account in the environmental risk assessment.

'Inert' toxicity

Another important health issue that involves pesticide formulations and bees is the consequence of the additives or so-called non-active ingredients. The commonly-used 'inert' solvent N-methyl-2-pyrrolidone was found here to be highly toxic to larval honey bees (Fig. 5). Unfortunately, despite the potential toxicity of 'inert' ingredients and their widespread use in pesticide products, their testing and risk assessment seems to be inadequate. There is a growing body of research that has reported a wide range of adverse effects of 'inert' ingredients to human health, including enhancing pesticide toxicities across the nervous, cardio-vascular, respiratory, and hormonal systems [18,50,51]. However, limited data exist on the potential impacts of 'inerts' on non-target pollinators, although recent studies implicate formulation additives or adjuvants as key risk factors [52]. As one example, the toxicity of the fungicide captan to honey bee brood development was attributed to formulation ingredients other than the active ingredient alone [53]. The lack of detailed information of the usage of formulation ingredients greatly impedes appropriate risk assessment of 'inert' ingredient toxicity; therefore, label disclosure of the composition of pesticide formulations would facilitate this much-needed evaluation.

Conclusions

The current study demonstrates the chronic oral and mixture toxicity of common pesticides at hive levels to honey bees at the larval stage. Most notable are the chronic larval toxicities of the fungicide chlorothalonil and its synergistic combinations with frequently used in-hive miticides, and the unexpected high toxicity of the formulation ingredient N-methyl-2-pyrrolidone. Considering the extensive detection of chlorothalonil and its coexistence with other pesticides in diverse combinations especially in hive pollen and wax, and its substantial larval toxicity alone and in mixtures shown here, the application of this and other fungicides during crop bloom cannot be presumed innocuous to pollinating honey bees. Given the critical sensitivity of larvae to chlorothalonil and its complex interactions with other pesticides, the potential impacts of fungicides on colony survival and development need further investigation. In the more complex milieu of this social insect and its aging hive environment, pesticides, formulation additives and their resulting mixtures may have greater long-term impacts on colony health than previously considered. Consequently, the scope of pesticide risk assessment for non-target honey bees should be expanded from the present emphasis on acute toxicity of individual pesticides to a priority for assessment of chronic and mixture toxicities that incorporate fungicides, other pesticide pollutants and their 'inert' ingredients.

Acknowledgments

We thank Maryann Frazier, Sara Ashcraft, and Stephanie E. Mellott for their assistance with apicultural duties.

Author Contributions

Conceived and designed the experiments: WYZ DRS CAM JLF. Performed the experiments: WYZ DRS. Analyzed the data: WYZ CAM. Contributed reagents/materials/analysis tools: WYZ DRS CAM JLF. Wrote the paper: WYZ. Provided suggestions and comments on the manuscript: CAM.

References

1. Mullin CA, Frazier M, Frazier JL, Ashcraft S, Simonds R, et al. (2010) High levels of miticides and agrochemicals in North American apiaries: Implications for honey bee health. PloS ONE 5(3): e9754.
2. Johnson RM, Ellis MD, Mullin CA, Frazier M (2010) Pesticides and honey bee toxicity - USA. Apidologie 41(3): 312–331.
3. Desneux N, Decourtye A, Delpuech JM (2007) The sublethal effects of pesticides on beneficial arthropods. Annu Rev Entomol 52(1): 81–106.
4. Atkins EL, Kellum D (1986) Comparative morphogenic and toxicity studies on the effect of pesticides on honeybee brood. J Apicult Res 25(4): 242–255.
5. Davis A (1989) The study of insecticide poisoning of honey bee brood. Bee World 70(1): 163–174.
6. Heylen K, Gobin B, Arckens L, Huybrechts R, Billen J (2011) The effects of four crop protection products on the morphology and ultrastructure of the hypopharyngeal gland of the European honeybee, Apis mellifera. Apidologie 42(1): 103–116.
7. Tasei J-Nl (2001) Effects of insect growth regulators on honey bees and non-Apis bees. A review. Apidologie 32(6): 527–545.
8. Rortais A, Arnold G, Halm M-P, Touffet-Briens F (2005) Modes of honeybees exposure to systemic insecticides: estimated amounts of contaminated pollen and nectar consumed by different categories of bees. Apidologie 36(1): 71–83.
9. Babendreier D, Joller D, Romeis J, Bigler F, Widmer F (2007) Bacterial community structures in honeybee intestines and their response to two insecticidal proteins. FEMS Microbiol Ecol 59(3): 600–610.
10. DeGrandi-Hoffman G, Sammataro D, Alarcon R (2009) The importance of microbes in nutrition and health of honey bee colonies Part II of three parts. Am Bee J 149(7): 667–669.
11. Becher MA, Hildenbrandt H, Hemelrijk CK, Moritz RFA (2010) Brood temperature, task division and colony survival in honeybees: A model. Ecol Model 221(5): 769–776.
12. Bogdanov S (2006) Contaminants of bee products. Apidologie 37(1): 1–18.
13. Johnson RM, Pollock HS, Berenbaum MR (2009) Synergistic interactions between in-hive miticides in Apis mellifera. J Econ Entomol 102(2): 474–479.
14. Caux PY, Kent RA, Fan GT, Stephenson GL (1996) Environmental fate and effects of chlorothalonil: A Canadian perspective. Crit Rev Environ Sci Technol 26(1): 45–93.
15. Ladurner E, Bosch J, Kemp WP, Maini S (2005) Assessing delayed and acute toxicity of five formulated fungicides to Osmia lignaria Say and Apis mellifera. Apidologie 36(3): 449–460.
16. Donovan Y (2006) Reregistration Eligibility Decision for Chlorpyrifos. U.S. Environmental Protection Agency, Office of Pesticide Programs.
17. European Commission (2009) Council Directive 91/414/EEC (Plant Protection Products) as repealed under Regulation (EC) No 1107/2009. In: EC, editor Official Journal of the European Union. p. 1–50.
18. Cox C, Surgan M (2006) Unidentified inert ingredients in pesticides: Implications for human and environmental health. Environ Health Persp 114(12): 1803–1806.
19. Kudsk P, Mathiassen SK (2004) Joint action of amino acid biosynthesis-inhibiting herbicides. Weed Res 44(4): 313–322.
20. Health and Safety Executive (1997) N-Methyl-2-pyrrolidone (Risk assessment document) EH 72/10. London, UK.
21. Jouyban A, Fakhree MA, Shayanfar A (2010) Review of pharmaceutical applications of N-methyl-2-pyrrolidone. J Pharm Pharmac Sci 13(4): 524–535.
22. Saillenfait AM, Gallissot F, Sabaté JP (2007) Developmental toxic effects of N-ethyl-2-pyrrolidone administered orally to rats. J Appl Toxicol 27(5): 491–497.
23. Lan CH, Peng CY, Lin TS (2004) Acute aquatic toxicity of N-methyl-2-pyrrolidinone to Daphnia magna. Bull Environ Contam Toxicol 73(2): 392–397.
24. Aupinel P, Fortini D, Michaud B, Marolleau F, Tasei JN, et al. (2007) Toxicity of dimethoate and fenoxycarb to honey bee brood (Apis mellifera), using a new in vitro standardized feeding method. Pest Manag Sci 63(11): 1090–1094.
25. Winston ML (1987) The biology of the honey bee. Cambridge, Mass.: Harvard University Press. viii, 281 p.
26. Hung Y, Meier K (2005) Acute ®Confidor (imidacloprid-N-methyl pyrrolidone) insecticides intoxication with mimicking cholinergic syndrome. Toxicol Ind Health 21(1): 137–140.
27. Kalbfleisch J, Prentice R (1980) The statistical analysis of failure time data. John Wiley, New York, NY, USA.
28. Atkins E, Kellum D, Atkins K (1981) Reducing pesticide hazards to honey bees: mortality prediction techniques and integrated management strategies. Leafl. 2883 Division of Agriculture, University of California, Riverside.
29. Hertzberg RC, MacDonell MM (2002) Synergy and other ineffective mixture risk definitions. Sci Total Environ 288(1–2): 31–42.
30. Crailsheim K, Brodschneider R, Aupinel P, Behrens D, Genersch E, et al. (2013) Standard methods for artificial rearing of Apis mellifera larvae. J Apicul Res 52(1): 15 p.
31. Fiedler L (1987) Assessment of chronic toxicity of selected insecticides to honeybees. J Apicult Res 26(2): 115–122.
32. Suchail S, Guez D, Belzunces LP (2001) Discrepancy between acute and chronic toxicity induced by imidacloprid and its metabolites in Apis mellifera. Environ Toxicol Chem 20(11): 2482–2486.
33. Yu SJ (2008) The toxicology and biochemistry of insecticides. Boca Raton: CRC Press/Taylor & Francis. xi, 276 p.
34. Johnson RM, Dahlgren L, Siegfried BD, Ellis MD (2013) Acaricide, fungicide and drug interactions in honey bees (Apis mellifera). PloS one 8(1): e54092.
35. Wu JY, Anelli CM, Sheppard WS (2011) Sub-lethal effects of pesticide residues in brood comb on worker honey bee (Apis mellifera) development and longevity. PloS one 6(2): e14720.
36. Bonzini S, Tremolada P, Bernardinelli I, Colombo M, Vighi M (2011) Predicting pesticide fate in the hive (part 1): experimentally determined tau-fluvalinate residues in bees, honey and wax. Apidologie 42(3): 378–390.
37. Vanburen NWM, Marien AGH, Oudejans RCHM, Velthuis HHW (1992) Perizin, an acaricide to combat the mite Varroa Jacobsoni - its distribution in and influence on the honeybee Apis mellifera. Physiol Entomol 17(3): 288–296.
38. Warne MS, Hawker DW (1995) The number of components in a mixture determines whether synergistic and antagonistic or additive toxicity predominate: the funnel hypothesis. Ecotox Environ Safe 31(1): 23–28.
39. vanEngelsdorp D, Speybroeck N, Evans JD, Nguyen BK, Mullin C, et al. (2010) Weighing risk factors associated with bee colony collapse disorder by classification and regression tree analysis. J Econ Entomol 103(5): 1517–1523.
40. Johnson RM (2011) Managed pollinator CAP coordinated agricultural project miticide and fungicide interactions. Am Bee J 151(10): 975–977.
41. Thompson HM, Wilkins S (2003) Assessment of the synergy and repellency of pyrethroid/fungicide mixtures. B Insectol 56(1): 131–134.
42. Pilling ED, Bromleychallenor KAC, Walker CH, Jepson PC (1995) Mechanism of synergism between the pyrethroid insecticide Lambda-Cyhalothrin and the imidazole fungicide prochloraz, in the honeybee (Apis mellifera L). Pestic Biochem Phys 51(1): 1–11.
43. Johnson RM, Wen Z, Schuler MA, Berenbaum MR (2006) Mediation of pyrethroid insecticide toxicity to honey bees (Hymenoptera: Apidae) by cytochrome P450 monooxygenases. J Econ Entomol 99(4): 1046–1050.
44. DeLorenzo ME, Serrano L (2003) Individual and mixture toxicity of three pesticides: Atrazine, Chlorpyrifos, and Chlorothalonil to the marine phytoplankton species Dunaliella tertiolecta. J Environ Sci Health, Part B 38(5): 529–538.
45. Chauzat M-P, Martel A-C, Cougoule N, Porta P, Lachaize J, et al. (2011) An assessment of honeybee colony matrices, Apis mellifera (Hymenoptera: Apidae) to monitor pesticide presence in continental France. Environ Toxicol Chem 30(1): 103–111.
46. Babendreier D, Kalberer N, Romeis J, Fluri P, Bigler F (2004) Pollen consumption in honey bee larvae: a step forward in the risk assessment of transgenic plants. Apidologie 35(3): 293–300.
47. Hendriksma HP, Hartel S, Steffan-Dewenter I (2011) Testing pollen of single and stacked insect-resistant Bt-Maize on in vitro reared honey bee larvae. PloS ONE 6(12): e28174.
48. Hendriksma HP, Hartel S, Steffan-Dewenter I (2011) Honey bee risk assessment: new approaches for in vitro larvae rearing and data analyses. Methods Ecol Evol 2(5): 509–517.
49. Hendriksma HP, Hartel S, Babendreier D, von der Ohe W, Steffan-Dewenter I (2012) Effects of multiple Bt proteins and GNA lectin on in vitro-reared honey bee larvae. Apidologie 43(5): 549–560.
50. Bonisch U, Bohme A, Kohajda T, Mogel I, Schutze N, et al. (2012) Volatile organic compounds enhance allergic airway inflammation in an experimental mouse model. PloS ONE 7(7): e39817.
51. Schindler BK, Koslitz S, Meier S, Belov VN, Koch HM, et al. (2012) Quantification of four major metabolites of embryotoxic N-methyl- and N-ethyl-2-pyrrolidone in human urine by cooled-injection gas chromatography and isotope dilution mass spectrometry. Anal Chem 84(8): 3787–3794.
52. Ciarlo TJ, Mullin CA, Frazier JL, Schmehl DR (2012) Learning impairment in honey bees caused by agricultural spray adjuvants. PloS ONE 7(7): e40848.
53. Everich R, Schiller C, Whitehead J, Beavers M, Barrett K (2009) Effects of captan on Apis mellifera brood development under field conditions in California almond orchards. J Econ Entomol 102(1): 20–29.

Field Efficacy of Vectobac GR as a Mosquito Larvicide for the Control of Anopheline and Culicine Mosquitoes in Natural Habitats in Benin, West Africa

Armel Djènontin[1]*, **Cédric Pennetier**[2], **Barnabas Zogo**[2], **Koffi Bhonna Soukou**[2], **Marina Ole-Sangba**[2], **Martin Akogbéto**[3], **Fabrice Chandre**[4], **Rajpal Yadav**[5], **Vincent Corbel**[6]*

1 Faculté des Sciences et Techniques/MIVEGEC (IRD 224-CNRS 5290-UM1-UM2), Université d'Abomey Calavi/Centre de Recherche Entomologique de Cotonou (CREC), Cotonou, Bénin, 2 MIVEGEC (IRD 224-CNRS 5290-UM1-UM2), Centre de Recherche Entomologique de Cotonou (CREC), Cotonou, Bénin, 3 Faculté des Sciences et Techniques/Centre de Recherche Entomologique de Cotonou (CREC), Université d'Abomey Calavi/Centre de Recherche Entomologique de Cotonou (CREC), Cotonou, Bénin, 4 MIVEGEC (IRD 224-CNRS 5290-UM1-UM2), Laboratoire de lutte contre les Insectes Nuisibles (LIN), Montpellier, France, 5 Department of Control of Neglected Tropical Diseases, World Health Organization, Geneva, Switzerland, 6 MIVEGEC (IRD 224-CNRS 5290-UM1-UM2)/Department of Entomology, Kasetsart University, Ladyaow Chatuchak Bangkok, Thailand

Abstract

Introduction: The efficacy of Vectobac GR (potency 200 ITU/mg), a new formulation of bacterial larvicide *Bacillus thuringiensis var. israelensis* Strain AM65-52, was evaluated against *Anopheles gambiae* and *Culex quinquefasciatus* in simulated field and natural habitats in Benin.

Methods: In simulated field conditions, Vectobac GR formulation was tested at 3 dosages (0.6, 0.9, 1.2 g granules/m^2 against *An. gambiae* and 1, 1.5, 2 g granules/m^2 against *Cx. quinquefasciatus*) according to manufacturer's product label recommendations. The dosage giving optimum efficacy under simulated field conditions were evaluated in the field. The efficacy of Vectobac GR in terms of emergence inhibition in simulated field conditions and of reduction of larval and pupal densities in rice fields and urban cesspits was measured following WHO guidelines for testing and evaluation of mosquito larvicides.

Results: Vectobac GR caused emergence inhibition of ≥80% until 21 [20–22] days for *An. gambiae* at 1.2 g/m^2 dose and 28 [27–29] days for *Cx. quinquefasciatus* at 2 g/m^2 in simulated field habitats. The efficacy of Vectobac GR in natural habitats was for 2 to 3 days against larvae and up to 10 days against pupae.

Conclusions: Treatment with Vectobac GR caused complete control of immature mosquito within 2–3 days but did not show prolonged residual action. Larviciding can be an option for malaria and filariasis vector control particularly in managing pyrethroid-resistance in African malaria vectors. Since use of larvicides among several African countries is being emphasized through Economic Community of West Africa States, their epidemiological impact should be carefully investigated.

Editor: Georges Snounou, Université Pierre et Marie Curie, France

Funding: The authors have no support or funding to report.

Competing Interests: The authors have declared that no competing interests exist.

* E-mail: armeldj@yahoo.fr (AD); vincent.corbel@ird.fr (VC)

Introduction

Malaria in Sub-Saharan Africa is a major public health problem accounting for 79% of global incidence of cases and 90% of deaths [1]. Lymphatic filariasis is a widely prevalent neglected vector-borne disease in Africa [2]. While chemotherapy for malaria control and mass drug administration against filariasis have been extensively used in disease endemic countries, vector control can complement strategies for prevention and control of these diseases [3]. Complementary vector control tools targeting exophagic and exophilic vectors or targeting another stage in the mosquito's lifecycle (e.g. the aquatic stage) are then needed to achieving the Millennium Development Goals for malaria control by 2015 [4].

Larval source management is an important component of an integrated vector management approach [5] and has extensively been used for the control of anophelines since the 1950s [6]. Recent studies in rural areas of Eastern Africa demonstrated that larval control by hand application of larvicides can reduce the abundance of malaria mosquito larvae and adults and transmission by 70–90% where the majority of aquatic mosquito larval habitats are accessible and relatively limited in number and size [7]. Larval source management offers the dual benefits of reducing numbers of house-frequenting mosquitoes and those that bite outdoors.

Larviciding is a commonly used method of mosquito control in different ecological patterns mostly in urban areas or some coastal areas where breeding sites are well identified. Currently 10 formulations are recommended by WHOPES for mosquito larval control, including microbial agents [8]. These bio-pesticides offer interesting prospects for the control of malaria vectors through varied and diverse groups of micro-organisms including viruses, bacteria and fungi which constitute an important part of the active ingredient arsenal for Integrated Vector Control [5].

Bacillus thuringiensis israelensis (Bti) and *Bacillus sphaericus (Bs)* have been extensively evaluated in the laboratory against anophelines and culicines larvae and also tested in a variety of environmental settings [9]. *Bti* is unlikely to pose any hazard to humans, other vertebrates and non-target invertebrates, provided that it is free from non-*Bt* microorganisms and biologically active products other than the insecticidal crystal proteins [10]. It was recently demonstrated that long-term use of *Bacillus thuringiensis israelensis* in coastal wetlands had no influence on the temporal evolution of the taxonomic structure and taxa abundance of non-target aquatic invertebrate communities [11]. In Benin, larviciding by the use of *Bti* was recently integrated as a part of vector management for malaria prevention [12].

While various *Bti* formulations are available as mosquito larvicides today, there has always been a need to improve them for better efficacy, ease of application and acceptability. In the present study in southern Benin, a new granular formulation of *Bti*, Vectobac GR of Valent BioSciences Corp, USA, was evaluated against *Anopheles gambiae* and *Culex quinquefasciatus* in simulated field experiments and in natural breeding habitats. The experimental procedures followed the WHO guidelines for testing and evaluation of mosquito larvicides [13]. The National Ethical Committee for Medical Research of Benin (N°006) cleared the study and the work was supervised by the WHO Pesticide Evaluation Scheme.

Methods

1. Ethics Statement

Ethics clearance for the study was obtained from the National Ethical Committee for Medical Research in Benin (ethics clearance N°006 of 28th April, 2011). The trial on *An. gambiae* was conducted after having received formal agreement from the president of local farmers named Lokossou Nestor. Concerning the trial on *Cx. quinquefasciatus*, permission from each owner of houses where cesspits were located was obtained before the trial was conducted.

2. Test Material

Vectobac GR is a new granular formulation of *Bacillus thuringiensis, subsp. israelensis*, strain AM65-52 developed by Valent BioSciences Corp., USA. The bio potency of this larvicide is 200 International Toxic Units (ITU)/mg product. Bio potency of products based on *Bti* is compared with a lyophilized reference powder (IPS82, strain1884) of this bacterial species using early fourth instar larvae of *A. aegypti* (strain Bora Bora). The potency of IPS82 has been arbitrarily designated as 15 000 ITU/mg powder against this strain of mosquito larva.

According to the manufacturer's Material Safety Data Sheet, Vectobac GR is non-toxic by ingestion, skin contact or inhalation. It has no adverse effect on birds, earthworms, fish, or numerous other non-target aquatic invertebrates.

3. Mosquito Species

The Kisumu strain of *An. gambiae* and the F1 progeny *of* local population of *Culex quinquefasciatus* were used for the simulated field trial. Kisumu strain of *An. gambiae* is a reference strain maintained

at the insectary of the Centre de Recherche Entomologique de Cotonou (CREC) and is free of any resistance mechanism.

4. Study Area

The simulated field trial was carried out in the Centre de Recherche Entomologique de Cotonou (CREC). The field trial with *An. gambiae* was conducted in a rice field located in Lélé, Cové district located in Department of Zou (7°13′ 8″ N, 2°20′ 22″ E). Concerning *Cx. quinquefasciatus*, the field trial was conducted in Cotonou, Department of Littoral (6°23N–2°25E).

5. Study Design

5.1. Simulated field studies. The main objective of simulated field studies were to test and determine the optimum field application dosage of Vectobac GR. Vectobac GR was tested at 3 dosages against *An. gambiae* (0.6, 0.9, 1.2 g granules/m^2) and *Cx. quinquefasciatus* (1, 1.5, 2 g granules/m^2) according to manufacturer's product label recommendations. Four replicates of the experiments were run for both treatments and control.

Experimental set up: Artificial cement containers (i.e. rectangular pits of 60 cm long × 30 cm width × 30 cm depth) were used to study the Vectobac GR dose-efficacy relation. Containers were half-filled with water and covered with a mosquito netting piece to prevent oviposition by wild female mosquitoes and the deposit of debris, and were placed under a shelter to prevent direct exposure of rain and sunlight.

Bti application: At t0, measured quantity of Vectobac GR was dispensed manually taking necessary safety precautions using gloves and facial masks.

Cohort monitoring: Larvicidal activity might last longer than the developmental period. In this context, cohort of 30 to 50 second instars larvae of *An. gambiae* or *Cx. qinquefasciatus* were released in each container every 7 to 10 days, depending on the larval development time frame. Each *An. gambiae* and *Cx. quinquefasciatus* larvae cohort was fed with 0.5 g and 1 g of cat food respectively when released. Each day after treatment, pupae were counted and removed from the containers and placed in plastic cups with water and covered with a netting piece. Temperature and pH of water in the containers were recorded daily and meteorological data were obtained from the National Meteorology Department. The studies were conducted between 8 June and 21 July 2011 with *An. gambiae* and between 10 August and 21 September 2011 with *Cx. quinquefasciatus*.

5.2. Field trials. The field trial was launched with a formal agreement with the president of local rice field farmers. For the study, thirty ponds of 8 m^2 (2 m×4 m) each were delimited with natural barrier made of local mud. Thirty cesspits with a surface ranging from 0.14 to 3.46 m^2 housing *Cx. quinquefasciatus* larvae were selected in Cotonou and geo-referenced. All breeding sites were checked to confirm the presence of larvae before applying Vectobac GR.

One dose that provided the optimum efficacy in the simulated field studies was tested in natural habitats. Fifteen breeding sites of each type were treated with Vectobac while the remaining ones were left untreated and served as controls. Vectobac GR was uniformly applied manually on the water surface. Three replicates were run for each treatment or control corresponding to 45 treated habitats and 45 of untreated habitats.

Before treatment, each breeding site was sampled twice to determine the density of mosquito larvae and pupae. After treatment, sampling was done on days 1, 2, 3, and 7, and thereafter every third day until the density of larvae in the treated habitats reached to that of the control. The larval sampling method consisted of 3 dips using a ladle (350 ml). Sampling was

Table 1. Emergence and Emergence Inhibition Rates (EIR) of *An. gambiae* larvae according to treatments.

N day post treatment		Control	0.6 g/m²	0.9 g/m²	1.2 g/m²
11	N	200	200	200	200
	NE	179	2	1	1
	ER (%) [95%CI]	90 [86–94]	1 [0–2]	1 [0–2]	1 [0–1]
	EIR (%) [95%CI]	–	**99**	**99**	**99**
19	N	200	200	200	200
	NE	180	31	27	17
	ER (%) [95%CI]	90 [86–94]	16 [11–21]	14 [9–19]	9 [5–13]
	EIR (%) [95%CI]	–	**83**	**85**	**91**
26	N	200	200	200	200
	NE	191	107	86	71
	ER (%) [95%CI]	96 [93–99]	54 [47–61]	43 [36–50]	36 [29–43]
	EIR (%) [95%CI]	–	**44**	**55**	**63**
35	N	120	120	120	120
	NE	114	105	104	72
	ER (%) [95%CI]	95 [91–99]	88 [82–94]	87 [81–94]	60 [51–69]
	EIR (%) [95%CI]	–	**8**	**9**	**37**
43	N	120	120	120	120
	NE	112	107	105	87
	ER (%) [95%CI]	93 [89–97]	89 [83–95]	88 [82–94]	73 [65–81]
	EIR (%) [95%CI]	–	**5**	**6**	**22**

N = Number of larvae; NE = Number of larvae emerged; ER = Emergence Rate; EIR = Emergence Inhibition Rate.

Table 2. Emergence and Emergence Inhibition Rates (EIR) of *Cx. quinquefasciatus* larvae according to the treatments.

N days post treatment		Control	1 g/m²	1.5 g/m²	2 g/m²
11	N	200	200	200	200
	NE	191	0	0	0
	ER [95%CI]%	96 [93–99]	0	0	0
	EIR (%) [95%CI]%	–	**100**	**100**	**100**
19	N	200	200	200	200
	NE	198	51	30	3
	ER [95%CI]%	99 [98–100]	26 [20–32]	15 [10–20]	02 [0–4]
	EIR (%) [95%CI]%	–	**74**	**85**	**99**
26	N	200	200	200	200
	NE	180	96	80	28
	ER [95%CI]%	90 [86–94]	48 [41–55]	40 [33–47]	14 [9–19]
	EIR (%) [95%CI]%	–	**47**	**56**	**84**
34	N	160	160	160	160
	NE	152	93	80	65
	ER [95%CI]%	95[92–98]	58[50–66]	50[42–58]	41[33–49]
	EIR (%) [95%CI]%		**38**	**47**	**57**
42	N	180	180	180	180
	NE	171	157	154	134
	ER [95%CI]%	95[92–98]	87[82–92]	86[81–91]	74[68–80]
	EIR (%) [95%CI]%		**8**	**10**	**22**

N = Number of larvae; NE = Number of larvae emerged; ER = Emergence Rate; EIR = Emergence Inhibition Rate.

Table 3. Mean number of larvae and pupae per dip and density reduction (DR) after treatment Vectobac in natural habitats.

N day post treatment	Parameters	An. gambiae Control				Treatment (1.2 g/m²)				Cx. quinquefasciatus Control				Treatment (2 g/m²)			
		L1+L2	L3+L4	Pupae	Total	L1+L2	L3+L4	Pupae	Total	L1+L2	L3+L4	Pupae	Total	L1+L2	L3+L4	Pupae	Total
0	N larvae/dip	2.93	0.56	0.07	3.56	3.02	0.73	0.04	3.80	7.2	8.2	0.7	16.1	16.6	7.8	2.2	26.6
1	N larvae/dip	1.69	0.70	0.02	2.41	1.21	0.04	0.02	1.27	8.0	7.2	1.3	16.5	1.4	2.2	0.5	4.1
	DR (%)					31	95	0	53					92	68	88	85
2	N larvae/dip	3.36	0.90	0.08	4.33	2.0	0.2	0.00	2.3	6.5	7.4	0.9	14.8	1.9	1.2	0.3	3.4
	DR (%)					42	82	100	53					88	83	88	86
3	N larvae/dip	4.00	1.38	0.05	5.43	1.93	0.51	0.00	2.45	6.2	5.8	0.8	12.8	2.8	1.3	0.3	4.4
	DR (%)					54	73	100	60					81	76	90	79
7	N larvae/dip	3.02	2.12	0.06	5.20	3.68	3.03	0.17	6.89	5.5	6.6	0.7	12.8	4.3	2.6	0.2	7.1
	DR (%)					0	0	0	0					66	59	90	66
10	N larvae/dip	2.54	2.09	0.29	4.92	3.77	3.26	0.31	7.34	5.2	4.7	1.6	11.5	4.7	3.4	0.8	8.9
	DR (%)					0	0	0	0					61	26	83	53
13	N larvae/dip	–	–	–	–	–	–	–	–	5.9	5.4	1.5	12.8	7.7	5.4	1.7	14.9
	DR (%)					–	–	–	–					44	0	62	29
16	N larvae/dip	–	–	–	–	–	–	–	–	4.3	4.9	1.4	10.5	10.8	6.9	1.8	19.5
	DR (%)					–	–	–	–					0	0	57	0

RD = Density reduction.

done by the same operator. The larval instars as well as pupae were counted separately. Temperature and pH were monitored at each sampling day in each mosquito breeding sites. The field trials were conducted between 10 November and 23 December 2011 with *An. gambiae* and between 14 September and 5 November 2011 with *Cx. quinquefasciatus*.

6. Data Analysis

The data analyses were performed using R software (version 2.11.1). Data from the simulated field trial were used to estimate the Emergence Inhibition Rates (% EIR) for each treatment according to the following formula:

$$\% \text{ EIR} = ((CT)/C) \times 100$$

where C is the emergence rate in the control and T is the emergence rate in the treated containers at the same period of time [13].

A logistic regression model with a logit link was fitted to the data to investigate the effect of the treatments on the emergence rate. The influence explanatory covariables on the emergence rate was investigated by including in the models the dose, the number of day post-treatment and the replicates. The number of day after which the emergence rates significantly increased to more than 20% with 95% Confidence Intervals was estimated for each treatment based on the logistic regression model.

Concerning data from the field trials, the mean number of larvae and pupae collected (i.e. density) per sampling day was calculated for both treated and control groups. The first and second instars larvae (L1+L2) were pooled as early instars and the third and fourth instars (L3+L4) as late instars. Density Reduction (DR) of early and late instars larvae as well as pupae was estimated post-treatment using Mulla's formula as follows:

$$DR = 100 - (C_1/T_1) \times (T_2/C_2) \times 100$$

where C_1 is the average number of larvae or pupae in control breeding sites prior to treatment and C_2 is the average number of larvae in control breeding sites at each day of sampling. T_1 is the average number of larvae or pupae in breeding sites to be treated with Vectobac GR and T_2 is the average number of larvae or pupae in treated breeding sites for each sampling day [13]. When DR was negative (i.e. densities were higher in the treated group than the control group), the value was taken as zero.

Then, a linear regression model was fitted to the data to investigate the effect of the treatment on the density reduction. The influence of the time as explanatory covariable on the density reduction was investigated by including the time in the models. The number of day after which the density reduction reached 80% and 50% was then estimated.

Results

1. Simulated Field Studies

The average temperature recorded in containers during trials with *An. gambiae* was 26.5°C (ranging from 24.0°C to 27.8°C) and 26.5°C (25.0°C to 27.7°C) with *Cx. quinquefasciatus*. The water pH was 7.5 (6.8 to 8.7) and 6.9 (6.6 to 8.0) for *An. gambiae* and *Cx. quinquefasciatus* containers, respectively.

Three thousands three hundred and sixty (3,360) larvae of *An. gambiae* were released in the containers for the trial. Emergence rates (ER) and Emergence Inhibition Rates (EIR) for each treatment are shown in Table 1. Emergence rates in the control ranged from 90% [86–94] to 96% [93–99]. The EIR were >80% for all dosages up to day 19 post-treatment. After day 26, the EIR was 44%, 55% and 63% at 0.6 g/m², 0.9 g/m² and 1.2 g/m² doses of Vectobac GR, respectively. According to the logistic

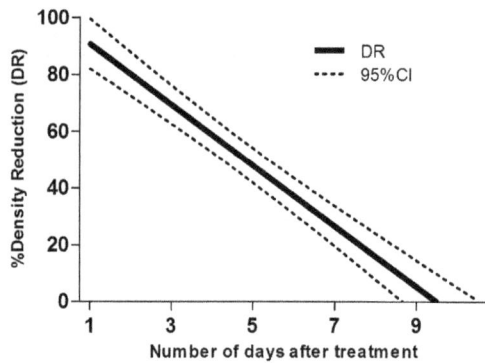

Figure 1. Density reduction (DR) of *An. gambiae* old instars larvae estimated by the regression model according to the number of days after treatment.

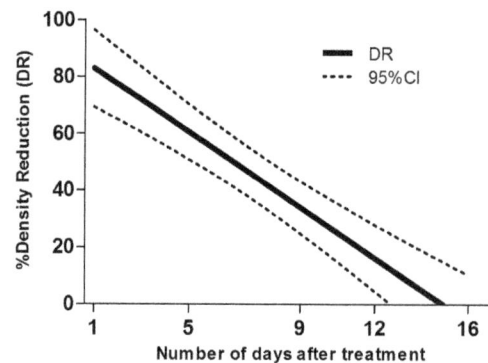

Figure 3. Density reduction (DR) of *Cx. quinquefasciatus* old instars larvae estimated by the regression model according to the number of days after treatment.

regression model, the estimated period of effectiveness (i.e. emergence rates <20%) was 15 days [14–17], 17 days [16–18] and 21 days [20–22] at the doses of 0.6, 0.9 and 1.2 g granule/m^2, respectively.

Three thousands seven hundred and sixty (3,760) larvae of *Cx. quinquefasciatus* were released in the containers during the simulated studies. Emergence and EIR for each treatment are shown in Table 2. Emergence rates in the control ranged from 90% [86–94] to 99% [98–100]. The EIR were 100% at day 11 regardless of the doses, and then decreased to 80% after 19 days of treatment with 1 g/m^2 dose, 26 days at 1.5 g/m^2 and 34 days at 2 g/m^2. According to the logistic regression model, the estimated period of effectiveness (i.e. emergence rates <20%) was 19 days [18–20], 22 days [20–23] and 28 days [27–29] days at 1, 1.5 and 2 g granules/m^2, respectively.

2. Field Trials

Based on the results of simulated studies, doses that showed highest efficacies and residual activities against *An. gambiae* (1.2 g granules/m^2) and *Cx. quinquefasciatus* (2 g granules/m^2) were selected for the field trials.

The average temperature recorded in the breeding sites through the trial was 35.1°C (ranging from 28°C to 41.7°C) and 27.1°C (ranging from 25.1°C to 32.2°C) for *An. gambiae* and *Cx. quinquefasciatus*, respectively. The water pH was 6.6 (ranging from 5.1 to 8.8) and 6.8 (ranging from 5.7 to 8.1) in habitats with *An.*

gambiae and *Cx. quinquefasciatus*, respectively. No rain was recorded during the trial on *An. gambiae* as the study was conducted during the dry season. The aerial average temperature recorded in the rice fields was 27.8°C, (23.2°C to 35.0°C). During the trial on *Cx. quinquefasciatus*, there was 345.90 mm rainfall while the average temperature was 27.3°C (24.6°C to 30.0°C).

The mean number of *An. gambiae* larvae sampled per dip and the density reduction (DR) at each sampling day are shown in Table 3. Before treatment, mosquito larvae densities in the control ponds were 2.93 per dip, 0.56 per dip and 0.07 per dip for early instars larvae, late instars larvae and pupae respectively. In the ponds to be treated these densities were 3.02 per dip, 0.73 per dip and 0.04 per dip for early instars larvae, late instars larvae and pupae respectively. The highest efficacy of Vectobac GR in terms of reduction of early instars larvae of *An. gambiae* was observed three days post-treatment but was below 60% reduction. Vectobac GR reduced late instars larvae density by >80% up to 2 days post-treatment. The DR decreased to 73% after 3 days and to nil after day 7. The number of pupae was too low to make any comparisons between control and treated ponds. According to the logistic regression model, the estimated period for which the density of late instars larvae would be reduced by 80% (DR$_{80}$) and 50% (DR$_{50}$) was 2 days (1–3) and 5 days (4–6), respectively (Figure 1). The highest DR induced by Vectobac GR against early instars larvae was about 50%. Consequently The DR$_{80}$ and DR$_{50}$ values could not be estimated.

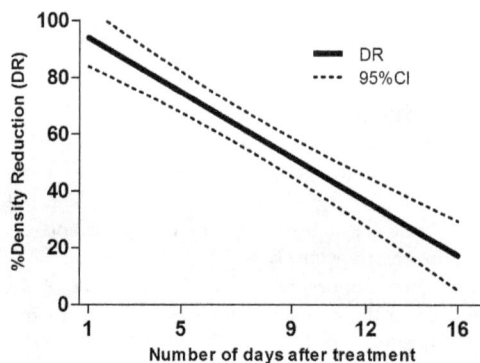

Figure 2. Density reduction (DR) of *Cx. quinquefasciatus* young instars larvae estimated by the regression model according to the number of days after treatment.

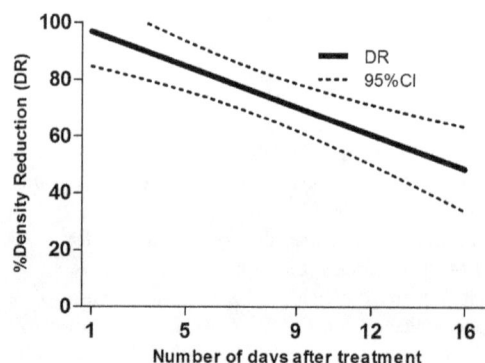

Figure 4. Density reduction (DR) of *Cx. quinquefasciatus* pupae estimated by the regression model according to the number of days after treatment.

Field Efficacy of Vectobac GR as a Mosquito Larvicide for the Control of Anopheline and Culicine Mosquitoes...

125

The densities of early instars, late instars and pupae of *Cx. quinquefasciatus* in control cesspits before treatment were 7.2 per dip, 8.2 per dip and 0.7 per dip respectively. In the cesspits to be treated the densities were 16.6 per dip, 7.8 per dip and 2.2 per dip respectively. After treatment with Vectobac GR, >80% reduction was observed until day 3 in early instars, until 2 days in late instars and until day 10 in pupae (Table 3). According to the linear regression model, the estimated numbers of days after which the density of late instars larvae would be reduced by 80% (DR_{80}) and 50% (DR_{50}) were 2 days (0–4) and 7 days (5–8), respectively. DR_{80} and DR_{50} were 4 days (2–5) and 10 days (8–11), respectively for early instars larvae and 6 days (4–9) and 16 days (11–20), respectively for pupae (Figures 2, 3 and 4).

Discussion and Conclusions

The efficacy of Vectobac GR, a new formulation of *Bacillus thuringiensis var. israelensis* Strain AM65-52, was evaluated against *An. gambiae* and *Cx. quinquefasciatus* in both simulated and natural conditions.

Under simulated field conditions, Vectobac GR caused emergence inhibition of ≥80% up to 21 days (20–22) post-treatment for *An. gambiae* (at 1.2 g/m^2) and 28 days (27–29) for *Cx. quinquefasciatus* (at 2 g/m^2). The longer efficacy of Vectobac GR against *Cx. quinquefasciatus* can be explained by the higher dosage of Vectobac GR used during the trial (as per manufacturer's recommendation) and/or by a higher susceptibility of *Cx. quinquefasciatus* larvae to *Bti* as reported elsewhere [14,15].

In the field, Vectobac GR formulation, designed for deep penetration of overgrown vegetation after application, induced a ~50% reduction of *An. gambiae* density in rice fields 3 days after application. The reduction of *Cx. quinquefasciatus* larvae was about 80% in urban cesspits 3 days after application. The short residual efficacy against both *Anopheles* and *Culex* mosquitoes in open water bodies may be due to its low ITU content (200 ITU/mg compared with previous *Bti* formulations with 3000 ITU/mg) or a faster degradation or sequestration of *Bti* toxins in natural habitats as previously reported [16].This short residual efficacy of Vectobac GR in natural habitats inevitably rose the question about the bioavailability of the *Bti* toxins in the rice field ponds and highly polluted habitats such as cesspits. *Bti* toxins are known to sediment in the breeding sites and this is also true with the Vectobac GR (granules were found at the bottom of the cement containers in simulated field conditions). In rice fields, the targeted species population (*An. gambiae s.s.*) was exclusively made of the M molecular form (recently renamed *Anopheles coluzzii*) [17]. Compared to the S molecular (now *Anopheles gambiae*) form, the M form larvae of are known to spend significantly more time at the bottom of the water column in breeding sites to collect food than the S form larvae [18]. Consequently, it is unlikely that the sedimen-

tation of the Vectobac GR toxins might have caused lower control of M form *An. gambiae*. In contrast, we observed that in the water ponds, the granules were sometimes found buried in the mud. Between the granule sp, thus were not fully available for mosquito control.

In addition, it is likely that the direct sunlight exposure of the larval habitats contributed to reduce the residual efficacy of Vectobac GR. Regarding *Culex quinquefasciatus*, results obtained in this present study are consistent with previous trials conducted in polluted (stagnant) waters in Africa [19] and India [20]. The presence of debris and heavy load of organic materials in the cesspits are known to absorb the *Bt* toxins and hence reduce the performance of *Bti*-based products.

With the development and rapid spread of insecticide resistance in malaria vectors [21] and increased proportion of malaria vectors that feed outdoors in response to the implementation of vector control intervention such treated nets and indoor residual spraying [22,23], there is a urgent need for complementary vector control strategies that could better impact vector density and malaria transmission. As suggested by Corbel et al. [24], the use of larvicide products could be complementary tool in the context of an integrated vector management for malaria transmission reduction and vector resistance management.

The use of larvicidal products for malaria control has long history in Africa with however more or less success [6]. In Gambia, hand application of water-dispersible granular formulations of *Bti* (Valent BioSciences, USA) to water bodies was associated with a 88% reduction in larval densities but had no effect on adult mosquito density and clinical malaria [25]. It is essential to better assess the impact of larviciding on mosquito density, malaria transmission and malaria morbidity. The opportunity to reinforce the use of larviciding in public health is currently under the spotlights among African countries through Economic Community of West Africa States (ECOWAS) including Benin [26]. This can also be an option to manage the spread of pyrethroid-resistance in African malaria vectors, as well as complement control of lymphatic filariasis in Africa south of Sahara. Nevertheless this complementary tool is highly dependent of the larval breeding site dynamics. The cost-effectiveness of such vector control strategy should be also carefully investigated. The present results emphasize the crucial need to improve basic knowledge on mosquito ecology as well as precise identification, mapping and monitoring of larval habitats in order to enhance the public health benefit to implement larval control programs.

Author Contributions

Conceived and designed the experiments: VC AD FC RY MA. Performed the experiments: BZ MO AD. Analyzed the data: KBS AD VC. Contributed reagents/materials/analysis tools: AD VC CP RY.

References

1. WHO (2012) World Malaria Report.
2. WHO (2013) Sustaining the drive to overcome the global impact of neglected tropical diseases. Second WHO report on neglected tropical diseases. WHO/HTM/NTD/2013.1. Geneva, World Health Organization.
3. Townson H, Nathan MB, Zaim M, Guillet P, Manga L, et al. (2005) Exploiting the potential of vector control for disease prevention. Bull World Health Organ 83: 942–947.
4. Matthew TB, Godfray HCJ, Read AF, Berg Hvd, Tabashnik BE, et al. (2012) Lessons from Agriculture for the Sustainable Management of Malaria Vectors. PLoS Med 9.
5. WHO (2008) WHO position statement on integrated vector management. Weekly epidemiological record 20: 177–184.
6. Fillinger U, Lindsay SW (2011) Larval source management for malaria control in Africa: myths and reality. Malaria J 10: 353.

7. Geissbühler Y, Kannady K, Chaki P, Emidi B, Govella N, et al. (2009) Microbial Larvicide Application by a Large-Scale, Community-Based Program Reduces Malaria Infection Prevalence in Urban Dar Es Salaam, Tanzania. PLoS ONE 4.
8. WHO (2011) WHO position statement on integrated vector management to control malaria and lymphatic filariasis. *Weekly Epidemiological Record* Number 13: 120–127.
9. Lacey LA (2007) *Bacillus thuringiensis* serovariety *israelensis* and *Bacillus sphaericus* for mosquito control. Amer Mosq Cont Assoc 23, Supplement to No.2: 133–163.
10. WHO (2006) Pesticides and their application for the control of vectors and pests of public health importance. Sixth edition. 1–125 p.
11. Lagadic L, Roucaute M, Caquet T (2013) *Bti* sprays do not adversely affect non-target aquatic invertebrates in French Atlantic coastal wetlands. Journal of Applied Ecology doi: 10.1111/1365-2664.12165.

12. Kinde-Gazard D, Baglo T (2012) Assessment of microbial larvicide spraying with *Bacillus thuringiensis israelensis*, for the prevention of malaria. Med Mal Infect 42: 114–118.

13. WHO (2005) Guidelines for laboratory and field testing of mosquito larvicides. WHO/CDS/WHOPES/GCDPP/200513.

14. Porter A, Davidson E, Liu J (1993) Mosquitocidal toxins of bacilli and their genetic manipulation for effective biological control of mosquitoes. Microbiological Reviews 57: 838–861.

15. Charles JF, Nielsen-LeRoux C (2000) Mosquitocidal bacterial toxins : diversity, mode of action and resistance phenomena. Mem Inst Oswaldo Cruz 95 Suppl 1: 201–206.

16. Madliger M, Gasser CA, Schwarzenbach RP, Sander M (2010) Adsorption of transgenic insecticidal Cry1Ab protein to silica particles. Effects on transport and bioactivity. Environ Sci Technol 45: 4377–4384.

17. Coetzee M, Hunt RH, Wilkerson R, Della Torre A, Mamadou B, et al. (2013) *Anopheles coluzzii* and *Anopheles amharicus*, new members of the *Anopheles gambiae* complex. *Zootaxa* 3619 246–274.

18. Gimonneau G, Pombi M, Dabiré RK, Diabaté A, Morand S, et al. (2012) Behavioural responses of *Anopheles gambiae* sensu stricto M and S molecular form larvae to an aquatic predator in Burkina Faso. Parasites & Vectors 5: 65.

19. Skovmand O, Sanogo E (1999) Experimental formulations of *Bacillus sphaericus* and *B. thuringiensis israelensis* against *Culex quinquefasciatus* and *Anopheles gambiae* (Diptera: Culicidae) in Burkina Faso. Journal of Med Entomol 36: 62–67.

20. Haq S, Bhatt R, Vaishnav K, Yadav R (2004) Field evaluation of biolarvicides in Surat city, India. J Vector Borne Dis 41: 61–66.

21. Ranson H, N'Guessan R, Lines J, Moiroux N, Nkuni Z, et al. (2011) Pyrethroid resistance in African anopheline mosquitoes: what are the implications for malaria control? Trends Parasitol 27: 91–98.

22. Okumu F, Moore SJ (2011) Combining indoor residual spraying and insecticide-treated nets for malaria control in Africa: a review of possible outcomes and an outline of suggestions for the future. *Malar J* 10: 208.

23. Moiroux N, Gomez MB, Pennetier C, Elanga E, Djènontin A, et al. (2012) Changes in Anopheles funestus Biting Behavior Following Universal Coverage of Long-Lasting Insecticidal Nets in Benin. J Infect Dis 206: 1622–1629.

24. Corbel V, Akogbéto M, Damien GB, Djènontin A, Chandre F, et al. (2012) Combination of malaria vector control interventions in pyrethroid resistance area in Benin: a cluster randomised controlled trial. Lancet Infect Dis 12: 617–626.

25. Majambere S, Lindsay SW, Green C, Kandeh B, Fillinger U (2007) Microbial larvicides for malaria control in The Gambia. Malar J 6: 76.

26. Economic Community of West African States (ECOWAS) (2011) ECOWAS Report of technical meeting on malaria vector control in ECOWAS region, Cotonou, Republic of Benin 28–30 nov. 2011.

Prenatal Exposure to Organophosphate Pesticides and Neurobehavioral Development of Neonates: A Birth Cohort Study in Shenyang, China

Ying Zhang[1], Song Han[1]*, Duohong Liang[1], Xinzhu Shi[1], Fengzhi Wang[1], Wei Liu[2], Li Zhang[2], Lixin Chen[2], Yingzi Gu[2], Ying Tian[3]

1 Department of Epidemiology, Public Health School, Shenyang Medical College, Shenyang, China, **2** Department of Obstetrics, Central Hospital Affiliated to Shenyang Medical College, Shenyang, China, **3** School of Medicine, Shanghai JiaoTong University, Shanghai, China

Abstract

Background: A large amount of organophosphate pesticides (OPs) are used in agriculture in China every year, contributing to exposure of OPs through dietary consumption among the general population. However, the level of exposure to OPs in China is still uncertain.

Objective: To investigate the effect of the exposure to OPs on the neonatal neurodevelopment during pregnancy in Shenyang, China.

Methods: 249 pregnant women enrolled in the Central Hospital Affiliated to Shenyang Medical College from February 2011 to August 2012. A cohort of the mothers and their neonates participated in the study and information on each subject was obtained by questionnaire. Dialkyl phosphate (DAP) metabolites were detected in the urine of mothers during pregnancy to evaluate the exposure level to OPs. Neonate neurobehavioral developmental levels were assessed according to the standards of the Neonatal Behavioral Neurological Assessment (NBNA). Multiple linear regressions were utilized to analyze the association between pregnancy exposure to OPs and neonatal neurobehavioral development.

Results: The geometric means (GM) of urinary metabolites for dimethyl phosphate (DMP), dimethyl thiophosphate (DMTP), diethyl phosphate (DEP), and diethyl thiophosphate (DETP) in pregnant women were 18.03, 8.53, 7.14, and 5.64 µg/L, respectively. Results from multiple linear regressions showed that prenatal OP exposure was one of the most important factors affecting NBNA scores. Prenatal total DAP concentrations were inversely associated with scores on the NBNA scales.?Additionally, a 10-fold increase in DAP concentrations was associated with a decrease of 1.78 regarding the Summary NBNA (95% CI, -2.12 to -1.45). And there was an estimated 2.11-point difference in summary NBNA scores between neonates in the highest quintile of prenatal OP exposure and the lowest quintile group.

Conclusion: The high exposure of pregnant women to OPs in Shenyang, China was the predominant risk factor for neonatal neurobehavioral development.

Editor: Aditya Bhushan Pant, Indian Institute of Toxicology Reserach, India

Funding: This study was supported by the National Natural Science Foundation of China (Grant No. 30972533). The funders had no role in study design, data collection and analysis, decision to publish, or preparation of the manuscript.

Competing Interests: The authors have declared that no competing interests exist.

* E-mail: hansong@symc.edu.cn

Introduction

As one of the largest developing agriculture countries, China must manage to maintain and increase crop yields from year to year. Because of this, pesticides are widely used in agriculture. The annual application of synthetic pesticides on food crops in China exceeds 300,000 tons and the average amount of the pesticides used in one field unit in China is more than 2.5-5 fold higher than the global average [1]. Even though the Chinese government emphasizes and promotes the reduction of pesticide use in agriculture, data from Chinese Agricultural Ministry for 2006–2010 showed that the annual pesticide consumption has not decreased. Along with the wide-spread use of highly toxic

organophosphate pesticides (OPs) (with potent toxicity to insects, relatively low costs and the decreased likelihood for pest resistance [2]), OPs account for more than 1/3 of all insecticide use in China.

As a result of the heavy use of OPs in agriculture, more than 10 percent of fruits, vegetables, and cereal grains grown in China contain pesticide residues exceeding the national safety standard [3,4]. For several middle toxicity Ops, such as chlorpyrifos and malathion, their residues mostly are detectable in vegetables [5–7]. Moreover, some highly toxic OPs such as parathion and methamidophos, which have been banned in China since 2007, have still been detected in vegetable samples through routine monitoring by the Ministry of Agriculture [8]. Besides raw foods treated by OPs, contaminated drinking water, dust, and spray drift commonly contribute to OPs exposure among the general population [9]. However, to our knowledge, it is uncertain to know the exposure level to OPs and the effect on the health of the general population in China.

OPs are also likely to be neurotoxic to humans, possibly by utilizing similar mechanisms that target the nervous system of insect pests. This concern is of particular relevance to the developing human brain, which is inherently much more vulnerable to injury caused by toxic agents than adult brains [10]. Because OPs can potentially cross the placenta, fetuses face the risk of *in utero* exposure to OPs. Together with lower levels of detoxifying enzymes deactivating OPs in fetuses than those in adults, fetus development could be severely affected by OPs [11,12].

In animals, prenatal exposure to OPs may lead to embryo toxicity and developmental toxicity, including neurotoxicity [13–15]. The neurotoxic effects of high level acute poisoning caused by the enzyme acetylcholinesterase (AChE), for example, are well established. Thus, AChE is prohibited for use because it can impair the peripheral, autonomic, and central nervous system (the cholinergic crisis) [16,17] [18]. The effects of long-term exposure to OPs resulting in acute toxicity are currently controversial. Several epidemiologic studies amongst agricultural workers found adverse neurobehavioral and neuropsychological effects, as well as other chronic occupational hazards, associated with long-term exposure to OPs [19–22]. An inner-city multiethnic cohort study, the Mount Sinai Children's Environmental Health Center (New York City) Survey, showed that prenatal exposure to OPs (through indoor pesticides use) was associated with anomalies in primary reflexes of neonates, and even had long-term adverse effects on neurodevelopment in young children [23,24]. While the studies from the cohort of mothers and children in an agricultural area of California, the Center for the Health Assessment of Mothers and Children of Salinas (CHAMACOS) cohort, showed that the prenatal exposure to OPs was adversely associated with attention span, intelligence, cognitive and other neurodevelopment problems in young children, but no associations were discovered in neonates [25,26]. We speculate that the association of exposure to OPs during neurodevelopment is dependent on the level of exposure to OPs in different regions. Moreover, the effects of OP exposure among the general population are currently unknown.

In this study, we enrolled a cohort of pregnant mothers and their neonates in Shenyang, where OPs are often detected in vegetables and fruits [8], and investigated the maternal exposure to OPs during pregnancy by measuring urinary concentrations of dialkyl phosphate (DAP), which are the metabolites of OPs. The association between the prenatal exposure to OPs and neonatal neurobehavioral development was analyzed. We provide evidence for the effects of OPs on the general population and also provide valuable information in planning policies in the management of pesticides use and public health, especially in pregnant women and children.

Methods

Participants

Healthy pregnant women were recruited to participate in this study from the obstetric wards of the Central Hospital Affiliated with Shenyang Medical College in Shenyang from February 2011 to August 2012. The pregnant women who met the inclusion criteria were enrolled in this study: living in the city for over three years; must not present with hypertension, diabetes, thyroid hypofunction, heart diseases, and other chronic diseases prior to

pregnancy; participants must be without heavy complications during gestation such as diabetes, anemia, and hypertension; no family or medical history of mental retardation, phenylketonuria, and Pompe's syndrome for pregnant women or their spouses. Women were screened for eligibility, and enrolled if they consented to participate in the study. Infants with disorders associated with adverse neurodevelopment such as traumatic brain injury, meningitis, and severe neonatal illnesses were excluded. Of 307 eligible women, 249 pregnant women and their neonates were finally enrolled as participants in the study (response rate: 81.1%).

The study was approved by the Medical Ethics Committee of Shenyang Medical College. All the participants signed written informed consents before the study.

Questionnaire

Face-to-face interviews were conducted with the women after delivery by a specially trained nurse. Information about each subject was obtained by questionnaire, which included data on demographic information, home characteristics, residential history, reproductive history, active and passive smoking, dietary habits, and alcohol and drug use. The use of pesticides during pregnancy was also questioned, including whether or not pesticides were used by pregnant mother or her family member, the types of pesticides, and frequency of use.

Urine OP Metabolites Measurements

Urine samples were collected from each mother for the analysis of dialkyl phosphate (DAP) and other pesticide-specific metabolites. Specimens were aliquoted into pre-cleaned glass containers with Teflon-lined caps, bar coded and stored at $-70°C$. Samples were then shipped on dry ice to Shanghai Center for Disease Control (CDC) for analysis. Gas chromatography with flame photometric detection (GC-FPD) was used to analyze the DAP metabolites of OPs following the method of Wu et al [37].

Briefly, one milliliter of urine was pipetted into a 10 mL screw-top glass test tube, and 250 µL of a Dibutyl phosphate (DBP, 97% purity, obtained from Fluka and used as an internal standard (IS)) solution (4.0 mg/L) were added. Subsequently, 4mL acetonitrile was added and the sample was mixed. After vigorous mechanical shaking for 5min, the test tube was centrifuged (1200×g, 5 min, 25°C). The supernatant fluid containing DAP and DBP was transferred into a clean screw-top glass test tube. Sample volume was then reduced to 70°C to a volume of 0.5 mL with a gentle nitrogen stream. Residues were re-extracted with 3mL of acetonitrile that contained 1 g of Na_2SO_4, were shaken for 10 min and then centrifuged. The resulting extract was repeatedly evaporated at 70°C to 0.1–0.2 mL under a gentle stream of nitrogen. To the final extracts, 20 mg of K_2CO_3, and 25 µL of pentafluorobenzyl bromide (PFBBr) were added and heated at 50°C for 16 h to covert the phosphate acids to their pentafluorobenzyl (PFB) esters. The PFB-DAP derivatives were dissolved in 100 µL of toluene for injection into the GC-FPD.

Capillary gas chromatography with P-specific flame photometric detection (GC-FPD) after derivatization with PFBBr was used to determinate the DAP metabolites of OPs in urine. PFB derivatives were identified by GC-MS using a GC-FPD (Shimadzu GC-14A) in the electron ionization (EI) mode. The GC operating conditions were as follows: GC column, BP-10, 25 m×0.33 mm i.d., 0.25 µm film thickness (SGE, Australia). Column temperatures, 110°C (1min) $-8°C/min$ $-210°C$ (1min) $-20°C/min$ $-280°C$ (10 min). Injection port temperature, 280°C; detector temperature, 300°C. Nitrogen gas (99.99% purity) was used as carrier gas at a head pressure of 150 kPa. The detector gases used were air at 60 kPa and hydrogen at 80 kPa. The injection volume was 1.0 μL in the splitless mode (splitless time, 1min). A GC-MS (HP5890–5973) was used for structural elucidation of PFB derivatives of DAP. Injector conditions and chromatographic conditions used were the same as for the GC-FPD. Operating conditions were as follows: carrier gas, helium gas (high purity grade) at a flow rate of 1.0 mL/min; ion-source temperature, 250°C; electron ionization, 70 eV; interface temperature, 280°C.

Five nonspecific OPs metabolites were measured in each sample, including two dimethyl (DM) phosphate metabolites (DMP,DMTP) and three diethyl (DE) phosphate metabolites (DEP, DETP and DEDTP). To provide an overall assessment of precision, accuracy and overall reliability of the method, quality control (QC) samples were used as blank samples and inserted blindly among the study samples. The limits of detection (LOD) for the five metabolites were 2.0 μg/L for DMP and 1.0 μg/L for DMTP, DEP, DETP, and DEDTP, respectively. Individual metabolite levels below the LOD was assigned a value equal to the LOD divided by the square root of two, and this value was included in each sum.

Summed concentrations of DM and DE (the two dimethyl metabolites; DMP, DMTP) and three diethyl metabolites (DEP, DETP, DEDTP) were calculated to provide summary measurements for exposure. To compare with other studies, we converted each metabolite from its untransformed concentration (μg/L) to the corresponding molar concentration (nmol/L) according to the forum described elsewhere [28].

Metabolite concentrations were adjusted using creatinine concentrations to correct for variable urine dilutions in the spot urine samples. Creatinine concentrations in urine were determined using a commercially available diagnostic enzyme method (Vitros CREA slides; Ortho Clinical Diagnostics, Raritan, NJ).

Assessment of Growth and Neurodevelopment Status of Newborns

Neonatal Behavioral Neurological Assessment (NBNA) was used as the measurement of neurodevelopment and was administered when the infants were 3 days old. NBNA was formulated by Bao et al. [29], based on the method of Brazelton and Amiel-Tison for behavioral neurological measurement in newborns as well as the experience in Chinese newborns, tested with distinct stability and reliability by several large cohort in China, and was observed to not be influenced by the geographic location and suitable for large surveys [30–32]. The NBNA assessed functional abilities, most reflexes and responses, and stability of behavioral status during the examination. It involves five scales: Behavior (six items), Passive Tone (four items), Active Tone (four items), Primary Reflexes (three items), and General Assessment (three items). Each item has three dimensions of score (0, 1 and 2). Twenty items are summarized in the Summary Score with a maximum score of 40. Neonates with Summary Score more than 37 were considered to be well developed, lower than 34 abnormal, and between those scores was the acceptable range. NBNA were conducted by two examiners who were trained in the training center of Beijing Children's Hospital affiliated with the Capital Institute of Pediatrics in Beijing, China. Examiners were blinded with regard to exposure status when NBNA were carried out.

Lead Concentration in Umbilical Cord Blood

Considering the potential confounding effects of other suspected neurotoxins such as lead, we measured the lead concentration in umbilical cord blood, which were collected at delivery. Graphite furnace atomic absorption spectrophotometer (MG2) was used to determine the concentration of lead and the LOD was 0.1 μg/L.

Statistical Analysis

Data were inputted with Epidata 3.02 and analyzed with SPSS17.0 for Windows (IBM, New York, USA). The levels of urinary OP metabolites (DAPs) were calculated with and without adjustment for creatinine. DAP levels were normalized by logarithmic transformation. All analyses were conducted on non-creatinine adjusted values and models were rerun with creatinine adjusted values (ng/g creatinine) in sensitivity analyses.

Nonparametric tests (Kruskal-Wallis tests) were used to compare the levels of urinary OPs metabolites among the pregnant women. Pearson Correlation analysis was used to determine the relationship between maternal urinary OP metabolites during pregnancy and neonatal neurobehavioral assessments scores (NBNA scores).

Multiple stepwise linear regressions were applied to examine the association of maternal OP exposure during pregnancy and the neurodevelopment of neonates. First, the scores of five Scales and the Summary of NBNA were used as dependent variables respectively, and the concentrations (logarithmic transformation) of OPs urinary metabolites, total DAPs (or DMs and DEs) and as well as the potential prenatal risk factors, including maternal age, education, gestational age, prenatal BMI, neonatal bodyweight and the lead concentration in umbilical cord blood were used as independent variables. The statistical significance level was set at 0.10 for inclusion and 0.20 for exclusion of the independent variables in the stepwise process. Secondly, the potential dose-response relationships between maternal OP exposure and neonatal neurodevelopment outcomes were assessed. Subjects were grouped into five equally sized exposure groups with quintiles of urinary DAP concentrations, from the lowest exposure group (as reference group) to the highest OP exposure. Mean NBNA scores among each exposure group (second to fifth categories) were calculated by the multiple regression formula and compared with those in the reference group.

Ethical Statement

The study protocol was approved by the Medical Ethics Committee of Shenyang Medical College. Written informed consents were obtained from all the mothers for present study.

Results

General Characteristics of the Participants

The characteristics of the participants are listed in Table 1. The average age of the pregnant women and their average prenatal BMI was 28.9 ± 3.2 years old and 22.5 ± 4.5 kg/m^2, respectively. About 51% of the subjects had university or higher education background and about 86% was primiparous. Over 68% of the pregnant women consumed vegetables or fruit every day. The average gestation of the women at delivery was 38.9 ± 1.0 weeks. The average birth weight and height of the neonates were 3500.0 ± 462.2 g and 50.9 ± 1.9 cm, respectively. Mean Summary Score of NBNA of the newborns was 37.6 ± 1.6 and the individuals over 37 scores occupy 96% of the whole new babies.

Organophosphate Pesticide Metabolites Level in Urine

OP urinary metabolite levels of the study sample both adjusted and not adjusted for creatinine are shown in Table 2. The OP metabolite concentrations higher than the LOD ranged from a low of 6.8% for DEDTP, to a high of 95.58% for DEP. The maximum value without creatinine adjustment was 334.02 μg/L for DMP and 167.06 μg/L for DEP, respectively. The maximum value with creatinine adjustment was 453.04 μg/g for DMP and 305.92 μg/g for DMTP, respectively. The geometric mean (GM) values without creatinine adjustment for DMP, DMTP, DEP and DETP levels

Table 1. Demographic and Exposure Characteristics for 249 pairs of the pregnant women and neonates in Shenyang, China.

		N	Percentage
Age(year)			
18–24		23	9.2
25–29		127	51.0
30–34		91	36.6
≥35		8	3.2
Parity			
0		207	83.1
>1		42	16.9
Education			
Junior and Secondary school		35	14.1
High school		64	25.7
College and postgraduates		150	60.2
Resident			
urban		217	87.1
rural		32	12.9
Occupational exposure to pesticides before pregnancy			
yes		2	0.8
no		247	99.2
Occupational exposure to pesticides during pregnancy			
yes		1	99.6
no		248	0.4
Using insecticides in household during pregnancy			
no		124	49.8
rarely		76	30.5
often		49	19.7
Passive smoking during pregnancy			
rarely		190	76.3
often		32	12.9
Always		17	6.8
Vegetables consumed weekly during pregnancy			
1–3times		23	9.2
4–6times		68	27.3
daily		158	63.5
Fruit consumed weekly during pregnancy			
1–3times		22	8.8
4–6times		70	28.1
daily		157	63.1
Monthly household income (RMB)			
<1000		76	30.5
1000~3000		94	37.8
3001~5000		61	24.5
>5000		18	7.2
Newborns sex			
Male		138	55.4
Female		111	44.6
Birth weight(g), mean(sd)		3500.44 (462.24)	
Body length(cm), mean(sd)		50.94 (1.92)	
Length of gestation(week), mean (sd)		38.7 (0.9)	
NBNA total score	<37	10	4.0
	≥37	239	96.0

were 18.03 μg/L, 8.53 μg/L, 7.14 μg/L and 5.64 μg/L, respectively. The GM for DEDTP levels was not calculated because of the low detection frequency. The creatinine-adjusted GMs for DMP, DMTP, DEP, DETP and DEDTP levels were 24.02 μg/g, 11.29 μg/g, 9.49 μg/g, 7.58 μg/g and 0.46 μg/g, respectively.

The median of urinary metabolites of OPs among different characteristic of the pregnant women were compared (Table 3). The median concentrations of DMs, DEs and DAPs of the pregnant women aged more than 30 years were significantly higher than those under 30 years (*P<0.05*); the median concentrations of DMs, DEs and DAPs of the pregnant women living in rural areas were slightly higher than those living in urban regions, but this did not reach statistical significance (*P>0.05*). Pregnant women with a prenatal BMI higher than 28 had higher concentrations of DMs, DEs and DAPs than those with prenatal BMIs less than 28 (*P>0.05*); the median of DMs, Des, and DAPs in the group of high level of passive smoking were higher than others, while there were no significant differences (*P>0.05*).

Neurodevelopment of Newborns

In the study, 96% of neonates were considered well-developed from a neurodevelopmental level (NBNA Summary Scores≥37) and less than 1% abnormal (NBNA Summary Scores<34). The median NBNA Summary scores for male and female neonates did not differ and equaled 38 (interquartile range: 37–39). NBNA total scores, behavior scores, passive tone scores, and primary reflexes scores between male and female newborns were not significantly different (Z = −0.377, Z = −0.406, Z = −0.705, Z = −0.543, Z = −0.524, respectively; P>0.05). Almost all neonates (>99%) scored full marks in the scale of General Assessment, therefore this scale was not further analyzed in the study.

Relationship between Prenatal OP Exposure and Neurodevelopment of Newborns

First, Pearson correlation analyses were used to explore the relationship between maternal urinary OP metabolites (logarithmic transformation) during pregnancy and neonatal neurobehavioral assessments scores (NBNA scores). There were significantly negative correlations between summary NBNA scores and the concentration of maternal DMs, DEs, DAPs during pregnancy (r = −0.510, r = −0.494, r = −0.561, *P<0.01*).

Multiple stepwise linear regression analyses were then used to examine the association between neonatal neurobehavioral development and maternal OP exposure (total DAP as the index) during pregnancy. Maternal urinary DAP concentrations measured during pregnancy revealed significant associations with poorer NBNA scores (without other prenatal confounding influences). The adjusted coefficients (β) (95% CIs) for NBNA scores compared to OP metabolite levels (DAP, DM, DE) are represented in Table 4. From the results, higher prenatal DAP concentrations were associated with lower scores in all NBNA scales, especially the Behavior Scale (β for a 10-fold increase in concentration = −0.65, 95% CI, −0.85 to −0.45, *P<0.01*). Moreover, a 10-fold increase in total DAP concentration was associated with a decrease of 1.78 in NBNA Summary scores (95% CI, −2.12 to −1.45). Urinary DM concentrations during pregnancy was also associated with poorer NBNA scores in Passive Tone, Active Tone, Primary Reflex scales and the Summary (*P< 0.05*), although point estimates were slightly lower than for total DAP concentrations. While urinary DE concentrations were associated with poorer NBNA scores in Behavior scale and the Summary (*P<0.05*), they were not as highly correlated compared to total DAP and DM concentrations.

The NBNA scores in different maternal DAP concentration groups are shown in Figure 1. No evidence of departure from linearity in the relation between maternal DAP concentrations and NBNA scores was observed. There was a 2.11-point (in Summary) to 0.27-point (in Passive tone Scale) difference of estimate NBNA scores between neonates in the highest quintile of maternal prenatal DAP levels and those in the lowest quintile. Moreover, there is no difference between male and female neonates in that NBNA scores were negatively associated with the concentrations

Table 2. Detection Frequency, Creatinine Unadjusted and Adjusted Geometric Mean, Range and Percentile of organophosphate pesticides urinary metabolites in 249 pregnant women in Shenyang, China.

	DM		DE		
	DMP	DMTP	DEP	DETP	DEDTP
Detection rate (%)	94.78	83.94	95.58	88.76	6.8
Unadjusted					
GM	18.03	8.53	7.14	5.64	-
25th	7.83	3.4	3.54	2.34	LOD
50th	24.04	11.84	5.42	7.04	LOD
75th	39.43	15.67	17.17	13.55	LOD
90th	94.18	47.48	42.32	29.15	LOD
Range	<LOD~334.02	<LOD~137.95	<LOD~167.06	<LOD~133.00	<LOD~6.61
Creatine-adjusted					
GM	24.02	11.29	9.49	7.58	0.78
25th	9.72	4.07	4.29	2.93	0.46
50th	26.86	13.39	8.58	8.2	0.77
75th	60.1	29.49	20.45	19.95	1.34
90th	151.25	61.83	53.02	44.74	2.75
Range	0.19~453.04	0.10~305.92	0.36~125.09	0.32~102.21	0.03~15.41

Table 3. Comparisons of urinary metabolites of OPs medians by population characteristic of pregnant women in Shenyang, China*.

characteristic		n	Metabolites		
			DMs	DEs	DAP
Age (year)	<30	150	29.61	214.34	45.80
	≥30	99	38.88	18.23	58.19
	P value		0.004	0.04	0.013
Education	Junior and secondary school	35	31.31	14.85	46.58
	High School	64	34.97	18.73	60.56
	College and postgraduates	150	35.11	14.85	50.49
	P value		0.503	0.564	0.823
Residues	Urban	217	32.75	14.85	47.19
	Rural	32	38.89	27.13	74.09
	P value		0.199	0.038	0.055
Prenatal BMI	BMI<28	69	36.99	15.78	53.35
	BMI≥28	180	55.87	24.98	78.91
	P value		0.19	0.66	0.46
Passive smoking	rarely	190	34.62	15.78	50.99
	often	32	31.21	13.07	46.91
	Always	17	67.71	21.95	108.50
	P value		0.517	0.675	0.467
Household income (Yuan/Month)	≤3000	154	34.33	15.78	51.99
	>3000	95	34.85	14.85	48.85
	P value		0.09	0.79	0.24
Gestational Age (week)	<37	8	25.76	14.85	40.61
	≥37	241	34.27	14.99	50.99
	P value		0.97	0.82	0.95

*compared by Kruskal-Wallis test.

of maternal urinary OP metabolites (total DAPs) during pregnancy. For male neonates, a 10-fold increase in maternal pregnant urinary DAP concentrations was associated with a decrease of 1.47 points of summary NBNA scores. For female neonates, maternal pregnant urinary DAP concentrations were associated with poorer summary NBNA scores ($\beta = -2.03$, $P = 0.01$). There were similar associations between urinary DM, DE concentrations and poor NBNA scores among male and female neonates, with slightly stronger estimates in girl neonates (Table 4).

The standard coefficients of the multiple stepwise linear analyses for NBNA scores and prenatal OPs metabolites (DAP) and other potential prenatal risk factors among the neonates are shown in Table 5. Prenatal OP exposure was evidently the strongest risk factor for lower NBNA scores, with the largest standardized regression coefficients (in absolute value) in all of the NBNA scales both in boys and girls ($P<0.05$). High levels of lead concentration in the umbilical cord blood and prenatal BMI were associated with poor NBNA scores ($P<0.05$). Maternal age, education, and neonatal bodyweight had no associations with NBNA scores in the present study as determined by multiple linear regression analysis ($P>0.05$).

Point estimates for creatinine-adjusted DAP concentrations were similar to those for non-creatinine adjusted concentrations (data not shown).

Discussion

Fetal exposure to OPs occurs because OPs can cross the placenta [33,34]. Thus, fetuses are more vulnerable to OPs [35]. Exposure to low level of OPs can influence emotional behaviors [36] and neuronal cell development [37] through a variety of noncholinergic mechanisms such as disruption of various cellular processes [38], up regulation of serotonin neurotransmitters [39] and oxidative stress [40]. Since many OPs are lipophilic and rapidly metabolized in the human body by hydrolysis or oxidative desulfurization, dialkyl phosphate (DAP) metabolites in urine are often used as biomarkers to reflect the cumulative exposure to OPs in humans [41,42]. In this study, we measured five non-specific

Table 4. Adjusted Coefficients (β) (95% CIs) on the Behavior, Passive tone, Active tone, Primary reflexes And Summary scores of NBNA for a Log10 Unit Increase in OPs Urinary Metabolites among the neonates in Shenyang, China.

	Summary	Behavior	Passive Tone	Active Tone	Primary Reflexes
Total neonates					
DAPs	−1.78 (−2.12,−1.45)	−0.65 (−0.85,−0.45)	−0.22 (−0.34,−0.10)	−0.48 (−0.66,−0.30)	−0.36 (−0.51,−0.21)
DM	−0.96 (−1.35,−0.57)		−0.22 (−0.33,−0.11)	−0.41 (−0.57,−0.29)	−0.30 (−0.44,−0.17)
DE	−0.88 (−1.30,−0.47)	−0.59 (−0.79,−0.40)			
Boy neonates					
DAP	−1.47 (−1.93,−1.01)	−0.50 (−0.76,−0.23)	−0.21 (−0.36,−0.02)	−0.46 (−0.72,−0.21)	−0.34 (−0.55,−0.13)
DM	−0.93 (−1.45,−0.40)		−0.19 (−0.35,−0.07)	−0.34 (−0.58,−0.11)	
DE	−0.61 (−1.15,−0.07)	−0.42 (−0.67,−0.17)			−0.28 (−0.48,−0.09)
Girl neonates					
DAP	−2.03 (−2.55,−1.52)	−0.84 (−1.15,−0.52)	−0.21 (−0.40,−0.02)	−0.51 (−0.76,−0.25)	−0.39 (−0.61,−0.17)
DM	−1.22 (−1.89,−0.55)		−0.18 (−0.35,−0.01)	−0.41 (−0.65,−0.18)	−0.34 (−0.54,−0.14)
DE	−0.98 (−1.58,−0.39)	−0.83 (−1.15,−0.53)			

Estimates were adjusted for maternal age, education, gestational age, prenatal BMI and the lead concentration in umbilical cord blood.
Variable coding: Gestational age (<37week = 1, ≥37week = 2); Maternal Age (<30 = 1, ≥30 = 2); Prenatal BMI (BMI<28 = 1, BMI≥28 = 2); Passive smoking (rarely = 1, often = 2, always = 3); Neonatal bodyweight (<3000 = 1,3000−3500 = 2; >3500 = 3).

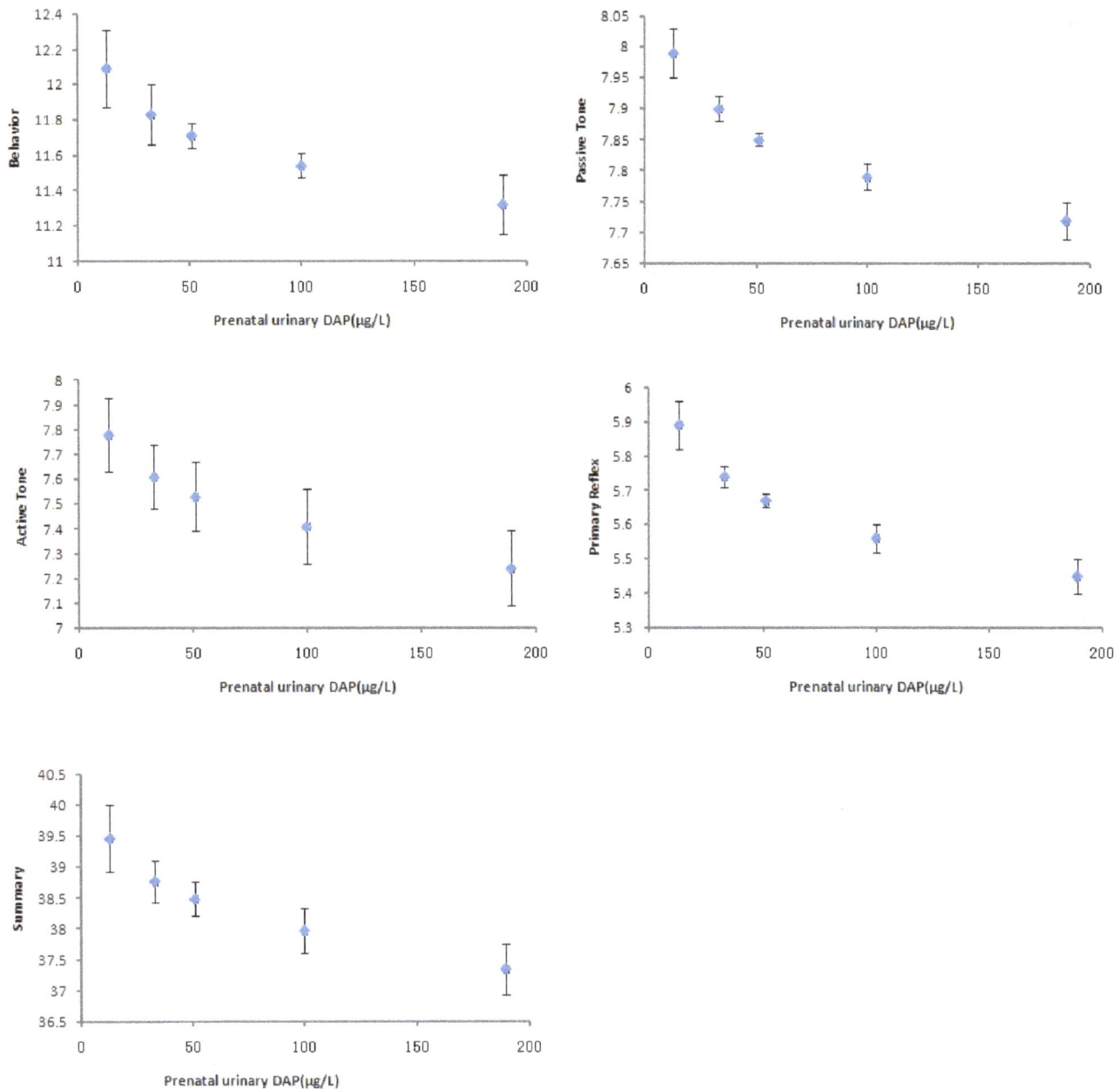

Figure 1. NBNA scores per quintile of prenatal urinary DAP concentration (Mean ± SD): Behavior, Passive Tone, Active Tone, Primary Reflex, and Summary. The medians (ranges) for DAP quintiles (µg/L) are as follows: first quintile, 13 (3–21); second, 33 (>21–42); third, 51 (>42–65); fourth, 100(>65–140); fifth, 189 (>140).

metabolites of OPs in the urine, DMP, DEP, DMTP, DETP and DEDTP (referred to as DAPs) to assess the exposure to OPs.

In our study, the GM levels (nmol/L) of DAPs in pregnant women were 167.14 for DMP, 56.62 for DMTP, 44.68 for DEP, and 40.76 for DETP. The concentrations (GM) of DM and DE were 283.66 and 107.39 nmol/L, respectively. In a previous study from the Netherlands, the DAPs concentrations from 100 pregnant women were 79.9 for DMP, 60.9 for DMTP, 13.0 for DEP and 4.7 for DETP [43]. Data from the National Health and Nutrition Examination Survey (NHANES, 1999–2000) in the U.S showed that the median level of urinary DAP metabolites in pregnant women [44] was 72 nmol/L. The results from the Center for Health Assessment of Mothers and Children of Salinas (CHAMACOS) cohort study indicated that the average GM for prenatal maternal urinary DAPs, DM and DE are 109.0, 76.8 and 17.7 nmol/L [25]. One investigation in Shanghai also revealed high levels of urinary DAPs in pregnant women, similar to those found in our study [45]. The results above indicate that GM levels (nmol/L) of DAPs in China are significantly higher than those in the Western Counties, suggesting that people are exposed to higher levels of OPs in developing counties like China compared to those in developed countries. The reason for the high level of OP exposure in China is mainly due to the heavy use of OPs in agriculture, leading to high residues in food, especially in vegetables and fruits [27]. The pesticide (most of them are OPs) residues of vegetables and fruits in Chinese markets are easily

Table 5. Standardized regression coefficients from the Stepwise Linear Regressions for the NBNA scores and prenatal risk factors among the neonates in Shenyang, China (n = 249).

	Summary	Behavior	Passive Tone	Active Tone	Primary Reflexes
Total neonates					
DAP [a] [#]	−0.615**	−0.403**	−0.253**	−0.364**	−0.329**
Gestational Age	0.127*	0.170*			
Maternal Age					
Prenatal BMI		−0.122			
Passive smoking					
Blood Lead	−0.192**			−0.217**	
Neonatal bodyweight					
Boy neonates					
DAPs [a] [#]	−0.555**	−0.342**	−0.250*	−0.338**	−0.303**
Gestational Age	0.165	0.266**			
Maternal Age					
Prenatal BMI					
Passive smoking					
Blood Lead	−0.178*		−0.177	−0.198*	
Neonatal bodyweight					
Girl					
DAPs [a] [#]	−0.672**	−0.503**	−0.236*	−0.394**	−0.361**
Gestational Age					
Maternal Age					
Prenatal BMI			−0.231*		
Passive smoking					
Blood Lead	−0.158			−0.240*	
Neonatal bodyweight					

Note: ** $P<0.01$. * $P<0.05$.
[a] the concentration of maternal OPs urinary metabolites during pregnancy.
[b] the concentration of lead in umbilical cord blood of neonates.
[#] the value in regression was changed by the log10 transition.
Variable coding: Gestational age (<37week = 1, ≥37week = 2); Maternal Age (<30 = 1, ≥30 = 2); Prenatal BMI (BMI<28 = 1, BMI≥28 = 2); Passive smoking (rarely = 1, often = 2, always = 3); Neonatal bodyweight (<3000 = 1,3000–3500 = 2; >3500 = 3).

detectable. Some of these fruits and vegetables show high levels of residues exceeding the national safe standard [6], even though the Chinese government announced in 2005 that the high toxic OPs such as dichlorvos, dimethoate, Parathion, methamidophos, monocrotophos and phosphamidon, are forbidden in use in agriculture. The mid-toxic pesticides such as chlorpyrifos and diazinon are always detected because they are commonly used for the control of pests in vegetables and fruits.

More than 68% of the pregnant subjects in this study consumed fresh vegetables and fruits every day during the pregnancy, suggesting that diet might be the primary source of OP exposure. One exception is that of one woman who was occupationally exposed to OPs (she is a pesticide production worker) before pregnancy. Investigations in the general American population [9] and in children [46] also concluded that diet exposure provided a vehicle for OPs to affect children as well as the general population. Compared to average individuals, pregnant women tend eat more food (especially more fresh vegetables and fruits) and drink more water than usual to obtain much more nutrition [47], therefore they face a higher risk of OP exposure.

Interestingly, we found a significant inverse association between maternal urinary metabolites of OPs (DM, DE and DAPs) during

pregnancy and the NBNA scores of neonates. Among several potential prenatal risk factors for neonatal neurobehavioral development, including OPs exposure, maternal age, education, passive smoking, gestational age, prenatal BMI, blood lead concentration, maternal OP exposure (DAP concentrations) during pregnancy was the predominant factors to these scores. The estimate NBNA scores of neonates in the high-exposure group for the four scales and the sum of the scores were 0.27–0.77 points and 2.11 points higher than those in the low-exposure group, respectively. These results suggest that maternal exposure to OPs during pregnancy could influence neonatal neurobehavioral development. Our results are consistent with those from a pregnancy cohort study in New York City (the Mount Sinai Children's Environmental Health Center), which concluded that prenatal OP metabolites in urine (primary DEs) are associated with an increasing number of abnormal primitive reflexes in neonates as evaluated by the Brazelton Neonatal Behavioral Assessment Scale [23]. Another longitudinal cohort in California, USA, from a cohort of the Center for the Health Assessment of Mothers and Children of Salinas (CHAMACOS) suggested that maternal pesticide exposure during pregnancy and not postnatal exposure is associated with poorer neonatal reflexes and long-term effects on children's mental development at 2 years at age, poor attention skills at 5-years-old, as well as poor intellectual development in 7-year-olds are present [25,26,48,49]. Although there are some differences between our study and the studies mentioned above, such as different measurements of neonatal neurodevelopment, similar conclusions were found: that the concentrations of maternal OP metabolites during pregnancy are inversely associated with the neonatal neurodevelopment.

The primary target of organophosphate insecticides is the enzyme acetylcholinesterase (AChE), which hydrolyses the neurotransmitter acetylcholine in both the peripheral and the central nervous system. In this study, we applied NBNA as the measurement of neonatal neurodevelopment. Based on the method of Brazelton and Amiel-Tison for behavioral neurological measurement in newborns, NBNA was utilized according to the condition of Chinese newborns [29]. NBNA was tested with distinct stability and reliability by several large cohorts in China, and was observed not to be influenced by the geographic location and suitable for large surveys [30–32]. In our study, NBNA was administered to the neonates when they were 3 days old to ensure the uniformity of the tests.

This study was not without limitations. For example, we measured the OP urinary metabolites at a single time point only prior to delivery. Urinary DAPs have a half-life of <24 h in humans, so the urinary concentrations reflect only recent exposure [50,51]. The combination of the short elimination half-life and the episodic nature of exposures to OP pesticides results in a high degree of variability in urinary DAP concentrations. Secondly, urinary DAP concentrations cannot quantify exposure to a particular pesticide. Even though the serum OPs can evaluate the actual exposure to a particular pesticide, they are often undetectable. At present, urinary organophosphate metabolites and DAPs are still used as biomarkers of organophosphate pesticide exposure in many studies [25,26,43,52,53]. Nevertheless, our study provides important information on the adverse health effects of OPs on pregnant women and neonates. Further longitudinal studies with larger sample size and more representative samples will confirm the association of prenatal OPs and the neurodevelopment of neonates and young children.

Conclusion

In summary, our study reveals that there exist high levels of exposure to OPs among pregnant women in Shenyang, China. This maternal exposure to OPs during pregnancy strongly associated with adverse neonatal neurobehavioral development. This study helps the local government to legislate against the abuse of OPs to improve the human health, especially the health of pregnant women and children.

Acknowledgments

We are extremely grateful to all the mothers who took part and to the midwives in the Central Hospital Affiliated to Shenyang Medical College, for their cooperation and help. The whole study team comprised interviewers, laboratory technicians, and volunteers whose efforts made the study possible.

Author Contributions

Conceived and designed the experiments: YZ SH. Performed the experiments: DL XS FW YT WL LZ LC YG. Analyzed the data: YZ HS DL. Wrote the paper: YZ SH.

References

1. Agriculture Information Network. (2006) Analysis of pesticides demand in China. (Chinese) Plant Doctor 19: 16–16.
2. Karalliedde et al. (2001) Organophosphates and Health. London: Imperial College Press.
3. Chen C, Qian Y, Chen Q, Tao C, Li C, et al. (2011) Evaluation of pesticide residues in fruits and vegetables from Xiamen, China. Food Control 22: 1114–1120.
4. Wang L, Liang Y, Jiang X (2008) Analysis of eight organophosphorus pesticide residues in fresh vegetables retailed in agricultural product markets of Nanjing, China. Bull Environ Contam Toxicol 81: 377–382.
5. Chen C, Qian Y, Liu X, Tao C, Liang Y, et al. (2012) Risk assessment of chlorpyrifos on rice and cabbage in China. Regul Toxicol Pharmacol 62: 125–130.
6. Wang S, Wang Z, Zhang Y, Wang J, Guo R (2013) Pesticide residues in market foods in Shaanxi Province of China in 2010. Food Chem 138: 2016–2025.
7. Yu Y, Tao S, Liu W, Lu X, Wang X, et al. (2009) Dietary intake and human milk residues of hexachlorocyclohexane isomers in two Chinese cities. Environ Sci Technol 43: 4830–4835.
8. Jiang L, Zhang Y, He R, Pan W, Jiang B, et al. (2011) Analysis of Pesticide Residues in Vegetables from Shenyang, China. International Conference on Intelligent Computation Technology and Automation DOI: 10.1109/ICICTA.2011.489.
9. Barr DB, Bravo R, Weerasekera G, Caltabiano LM, Whitehead RD, Jr., et al. (2004) Concentrations of dialkyl phosphate metabolites of organophosphorus pesticides in the U.S. population. Environ Health Perspect 112: 186–200.
10. Rodier PM (1995) Developing brain as a target of toxicity. Environ Health Perspect 103 Suppl 6: 73–76.
11. Whyatt RM, Rauh V, Barr DB, Camann DE, Andrews HF, et al. (2004) Prenatal insecticide exposures and birth weight and length among an urban minority cohort. Environ Health Perspect 112: 1125–1132.
12. Eskenazi B, Harley K, Bradman A, Weltzien E, Jewell NP, et al. (2004) Association of in utero organophosphate pesticide exposure and fetal growth and length of gestation in an agricultural population. Environ Health Perspect 112: 1116–1124.
13. Venerosi A, Ricceri L, Scattoni ML, Calamandrei G (2009) Prenatal chlorpyrifos exposure alters motor behavior and ultrasonic vocalization in CD-1 mouse pups. Environ Health 8: 12.
14. Lazarini CA, Lima RY, Guedes AP, Bernardi MM (2004) Prenatal exposure to dichlorvos: physical and behavioral effects on rat offspring. Neurotoxicol Teratol 26: 607–614.
15. Qiao D, Seidler FJ, Tate CA, Cousins MM, Slotkin TA (2003) Fetal chlorpyrifos exposure: adverse effects on brain cell development and cholinergic biomarkers emerge postnatally and continue into adolescence and adulthood. Environ Health Perspect 111: 536–544.
16. Slotkin TA, Seidler FJ (2007) Comparative developmental neurotoxicity of organophosphates in vivo: transcriptional responses of pathways for brain cell development, cell signaling, cytotoxicity and neurotransmitter systems. Brain Res Bull 72: 232–274.
17. Kamel F, Engel LS, Gladen BC, Hoppin JA, Alavanja MC, et al. (2007) Neurologic symptoms in licensed pesticide applicators in the Agricultural Health Study. Hum Exp Toxicol 26: 243–250.
18. Sultatos LG (1994) Mammalian toxicology of organophosphorus pesticides. J Toxicol Environ Health 43: 271–289.
19. Rohlman DS, Anger WK, Lein PJ (2011) Correlating neurobehavioral performance with biomarkers of organophosphorous pesticide exposure. Neurotoxicology 32: 268–276.
20. Rohlman DS, Lasarev M, Anger WK, Scherer J, Stupfel J, et al. (2007) Neurobehavioral performance of adult and adolescent agricultural workers. Neurotoxicology 28: 374–380.
21. Rothlein J, Rohlman D, Lasarev M, Phillips J, Muniz J, et al. (2006) Organophosphate pesticide exposure and neurobehavioral performance in agricultural and non-agricultural Hispanic workers. Environ Health Perspect 114: 691–696.
22. Roldan-Tapia L, Parron T, Sanchez-Santed F (2005) Neuropsychological effects of long-term exposure to organophosphate pesticides. Neurotoxicol Teratol 27: 259–266.
23. Engel SM, Berkowitz GS, Barr DB, Teitelbaum SL, Siskind J, et al. (2007) Prenatal organophosphate metabolite and organochlorine levels and performance on the Brazelton Neonatal Behavioral Assessment Scale in a multiethnic pregnancy cohort. Am J Epidemiol 165: 1397–1404.
24. Engel SM, Wetmur J, Chen J, Zhu C, Barr DB, et al. (2011) Prenatal exposure to organophosphates, paraoxonase 1, and cognitive development in childhood. Environ Health Perspect 119: 1182–1188.
25. Eskenazi B, Marks AR, Bradman A, Harley K, Barr DB, et al. (2007) Organophosphate pesticide exposure and neurodevelopment in young Mexican-American children. Environ Health Perspect 115: 792–798.
26. Bouchard MF, Chevrier J, Harley KG, Kogut K, Vedar M, et al. (2011) Prenatal exposure to organophosphate pesticides and IQ in 7-year-old children. Environ Health Perspect 119: 1189–1195.

27. Wu C, Liu P, Zheng L, Chen J, Zhou Z (2010) GC-FPD measurement of urinary dialkylphosphate metabolites of organophosphorous pesticides as pentafluorobenzyl derivatives in occupationally exposed workers and in a general population in Shanghai (China). J Chromatogr B Analyt Technol Biomed Life Sci 878: 2575–2581.

28. Arcury TA, Grzywacz JG, Davis SW, Barr DB, Quandt SA (2006) Organophosphorus pesticide urinary metabolite levels of children in farmworker households in eastern North Carolina. Am J Ind Med 49: 751–760.

29. Bao XL, Yu RJ, Li ZS, Zhang BL (1991) Twenty-item behavioral neurological assessment for normal newborns in 12 cities of China. Chin Med J (Engl) 104: 742–746.

30. Bao XL, Yu RJ, Li ZS (1993) 20-item neonatal behavioral neurological assessment used in predicting prognosis of asphyxiated newborn. Chin Med J (Engl) 106: 211–215.

31. Yu XD, Yan CH, Shen XM, Tian Y, Cao LL, et al. (2011) Prenatal exposure to multiple toxic heavy metals and neonatal neurobehavioral development in Shanghai, China. Neurotoxicol Teratol 33: 437–443.

32. Gao Y, Yan CH, Tian Y, Wang Y, Xie HF, et al. (2007) Prenatal exposure to mercury and neurobehavioral development of neonates in Zhoushan City, China. Environ Res 105: 390–399.

33. Rauh VA, Garfinkel R, Perera FP, Andrews HF, Hoepner L, et al. (2006) Impact of prenatal chlorpyrifos exposure on neurodevelopment in the first 3 years of life among inner-city children. Pediatrics 118: e1845–1859.

34. Whyatt RM, Garfinkel R, Hoepner LA, Andrews H, Holmes D, et al. (2009) A biomarker validation study of prenatal chlorpyrifos exposure within an inner-city cohort during pregnancy. Environ Health Perspect 117: 559–567.

35. Tau GZ, Peterson BS (2010) Normal development of brain circuits. Neuropsychopharmacology 35: 147–168.

36. Roegge CS, Timofeeva OA, Seidler FJ, Slotkin TA, Levin ED (2008) Developmental diazinon neurotoxicity in rats: later effects on emotional response. Brain Res Bull 75: 166–172.

37. Slotkin TA, Bodwell BE, Levin ED, Seidler FJ (2008) Neonatal exposure to low doses of diazinon: long-term effects on neural cell development and acetylcholine systems. Environ Health Perspect 116: 340–348.

38. Howard AS, Bucelli R, Jett DA, Bruun D, Yang D, et al. (2005) Chlorpyrifos exerts opposing effects on axonal and dendritic growth in primary neuronal cultures. Toxicol Appl Pharmacol 207: 112–124.

39. Slotkin TA, Tate CA, Ryde IT, Levin ED, Seidler FJ (2006) Organophosphate insecticides target the serotonergic system in developing rat brain regions: disparate effects of diazinon and parathion at doses spanning the threshold for cholinesterase inhibition. Environ Health Perspect 114: 1542–1546.

40. Slotkin TA, Seidler FJ (2009) Oxidative and excitatory mechanisms of developmental neurotoxicity: transcriptional profiles for chlorpyrifos, diazinon, dieldrin, and divalent nickel in PC12 cells. Environ Health Perspect 117: 587–596.

41. Margariti MG, Tsakalof AK, Tsatsakis AM (2007) Analytical methods of biological monitoring for exposure to pesticides: recent update. Ther Drug Monit 29: 150–163.

42. Margariti MG, Tsatsakis AM (2009) Analysis of dialkyl phosphate metabolites in hair using gas chromatography-mass spectrometry: a biomarker of chronic exposure to organophosphate pesticides. Biomarkers 14: 137–147.

43. Ye X, Pierik FH, Hauser R, Duty S, Angerer J, et al. (2008) Urinary metabolite concentrations of organophosphorous pesticides, bisphenol A, and phthalates among pregnant women in Rotterdam, the Netherlands: the Generation R study. Environ Res 108: 260–267.

44. Bradman A, Eskenazi B, Barr DB, Bravo R, Castorina R, et al. (2005) Organophosphate urinary metabolite levels during pregnancy and after delivery in women living in an agricultural community. Environ Health Perspect 113: 1802–1807.

45. Wang P, Tian Y, Wang XJ, Gao Y, Shi R, et al. (2012) Organophosphate pesticide exposure and perinatal outcomes in Shanghai, China. Environ Int 42: 100–104.

46. Lu C, Bravo R, Caltabiano LM, Irish RM, Weerasekera G, et al. (2005) The presence of dialkylphosphates in fresh fruit juices: implication for organophosphorus pesticide exposure and risk assessments. J Toxicol Environ Health A 68: 209–227.

47. Han Y, Li L, Shu L, Yu T, Bo Q, et al. (2011) [Evaluation on the dietary quality of pregnant women with an adjusted dietary balance index]. Wei Sheng Yan Jiu 40: 454–456, 460.

48. Young JG, Eskenazi B, Gladstone EA, Bradman A, Pedersen L, et al. (2005) Association between in utero organophosphate pesticide exposure and abnormal reflexes in neonates. Neurotoxicology 26: 199–209.

49. Marks AR, Harley K, Bradman A, Kogut K, Barr DB, et al. (2010) Organophosphate pesticide exposure and attention in young Mexican-American children: the CHAMACOS study. Environ Health Perspect 118: 1768–1774.

50. Wessels D, Barr DB, Mendola P (2003) Use of biomarkers to indicate exposure of children to organophosphate pesticides: implications for a longitudinal study of children's environmental health. Environ Health Perspect 111: 1939–1946.

51. Barr DB, Angerer J (2006) Potential uses of biomonitoring data: a case study using the organophosphorus pesticides chlorpyrifos and malathion. Environ Health Perspect 114: 1763–1769.

52. Grandjean P, Harari R, Barr DB, Debes F (2006) Pesticide exposure and stunting as independent predictors of neurobehavioral deficits in Ecuadorian school children. Pediatrics 117: e546–556.

53. Lacasana M, Lopez-Flores I, Rodriguez-Barranco M, Aguilar-Garduno C, Blanco-Munoz J, et al. (2010) Association between organophosphate pesticides exposure and thyroid hormones in floriculture workers. Toxicol Appl Pharmacol 243: 19–26.

Habitat Availability Is a More Plausible Explanation than Insecticide Acute Toxicity for U.S. Grassland Bird Species Declines

Jason M. Hill[1]*, J. Franklin Egan[2], Glenn E. Stauffer[1], Duane R. Diefenbach[3]

1 Pennsylvania Cooperative Fish and Wildlife Research Unit, Pennsylvania State University, University Park, Pennsylvania, United States of America, **2** USDA-ARS Pasture Systems and Watershed Management Research Unit, University Park, Pennsylvania, United States of America, **3** U.S. Geological Survey, Pennsylvania Cooperative Fish and Wildlife Research Unit, Pennsylvania State University, University Park, Pennsylvania, United States of America

Abstract

Grassland bird species have experienced substantial declines in North America. These declines have been largely attributed to habitat loss and degradation, especially from agricultural practices and intensification (the habitat-availability hypothesis). A recent analysis of North American Breeding Bird Survey (BBS) "grassland breeding" bird trends reported the surprising conclusion that insecticide acute toxicity was a better correlate of grassland bird declines in North America from 1980–2003 (the insecticide-acute-toxicity hypothesis) than was habitat loss through agricultural intensification. In this paper we reached the opposite conclusion. We used an alternative statistical approach with additional habitat covariates to analyze the same grassland bird trends over the same time frame. Grassland bird trends were positively associated with increases in area of Conservation Reserve Program (CRP) lands and cropland used as pasture, whereas the effect of insecticide acute toxicity on bird trends was uncertain. Our models suggested that acute insecticide risk potentially has a detrimental effect on grassland bird trends, but models representing the habitat-availability hypothesis were 1.3–21.0 times better supported than models representing the insecticide-acute-toxicity hypothesis. Based on point estimates of effect sizes, CRP area and agricultural intensification had approximately 3.6 and 1.6 times more effect on grassland bird trends than lethal insecticide risk, respectively. Our findings suggest that preserving remaining grasslands is crucial to conserving grassland bird populations. The amount of grassland that has been lost in North America since 1980 is well documented, continuing, and staggering whereas insecticide use greatly declined prior to the 1990s. Grassland birds will likely benefit from the de-intensification of agricultural practices and the interspersion of pastures, Conservation Reserve Program lands, rangelands and other grassland habitats into existing agricultural landscapes.

Editor: R. Mark Brigham, University of Regina, Canada

Funding: The authors have no support or funding to report.

Competing Interests: The authors have declared that no competing interests exist.

* E-mail: jmh656@psu.edu

Introduction

The U. S. Geological Survey (USGS) North American Breeding Bird Survey (BBS) has documented that grassland birds declined faster than any other habitat guild of birds in North America over recent decades, including our study period of 1980–2003 [1,2]. These declines have been largely attributed to habitat loss and degradation, particularly from an increasing scope and intensity of agricultural practices (the habitat-availability hypothesis) [3–5]. Total grassland losses from agriculture and other land uses have been dramatic – e.g., of >60 million ha of tallgrass prairie historically present in North America less than 14% remains in many areas [6]. While substantial acreage of grasslands were lost prior to the 1980s, native grassland in the tallgrass prairie region and elsewhere is still being lost to agriculture [6,7], succession [8], and other land use changes [9]. Indeed, between 1982 and 1997 (the approximate timeframe of our study) the United States lost approximately 97,000 km^2 of grasslands–primarily due to agricultural practices [6]. Intensive agricultural practices can also degrade habitat and adversely affect grassland bird populations via the overgrazing of grasslands, harvesting of forage crops earlier and more frequently, and through the conversion of heterogeneous rural landscapes into intensively-managed crop monocultures [10].

The influence of insecticides on grassland bird populations has received less attention than habitat loss, but researchers have long expressed concerns about excessive and widespread insecticide use [11,12]. Pesticide use, especially insecticides, can poison and kill birds and reduce reproductive success (e.g., through eggshell thinning) and food supplies [10,13,14]. Recently, Mineau and Whiteside [15] evaluated possible correlates, including several measures of insecticide exposure, of declines in BBS trends of North American grassland birds. They reported that exposure to insecticides was more strongly correlated with grassland bird declines (the insecticide-acute-toxicity hypothesis) in the conterminous U.S. from 1980–2003 than several other indices of agricultural practices and habitat availability, such as herbicide use and changes in permanent pasture area. The study was highlighted in the *New York Times* [16] and other popular media, which demonstrates public concern for the declines of grassland birds and the impacts of insecticides on the environment. Mineau and Whiteside [15] concluded that toxic insecticides "offer a more plausible explanation for overall declines than does the oft-cited 'habitat loss through agricultural intensification.'"

The conclusions of Mineau and Whiteside [15] have important implications for understanding grassland bird population dynamics, but these results are surprising for several reasons. While grassland birds have experienced substantial declines [1,2] insecticide use in the U.S. has been substantially reduced over the past thirty years [17]. Alternative chemistries that replaced the most toxic organophosphates were available starting in the mid-1970s [18], and most of the reduction in insecticide use occurred prior to the 1990s [17]. Acute poisoning events have also declined over the last 20 years [19]. Meanwhile, substantial grassland loss to agricultural use has continued to occur over that same timeframe [7]. The resulting smaller and fragmented grasslands are more likely to harbor smaller and less diverse grassland bird communities because many grassland birds are area- and edge-sensitive species [4,5,20] that respond strongly to the configuration and amount of grasslands in agricultural landscapes [21,22].

By suggesting that insecticide exposure is a more important correlate of grassland bird declines than changes in landscape structure and habitat availability, Mineau and Whiteside's [15] analysis could have substantial implications for agricultural policy and practice. Therefore, we reassessed whether habitat availability or insecticide acute toxicity offers a more plausible explanation of grassland bird declines in North America. Our analysis differed in two important ways from that of Mineau and Whiteside [15]. Firstly, Mineau and Whiteside [15] reduced individual grassland species' trends in each state to a dichotomous variable (1 = declining; 0 = increasing) and examined only two measures of habitat availability for grassland birds in agricultural landscapes. In contrast, we used the overall trend reported for the "grassland breeding" bird guild for each state [2], which accounts for the uncertainty among individual species' trends within a state [23]. Secondly, we identified and evaluated several additional covariates representing habitat availability and reexamined the correlation between U.S. grassland bird declines (1980–2003), habitat availability, and insecticide use and acute toxicity.

Methods

North American Grassland Bird Data

The USGS North American Breeding Bird Survey is the only long-term continental monitoring program of North American breeding bird species. Each year volunteers across North America count breeding birds at 50 locations along 34.9-km roadside routes. We used the BBS "grassland breeding" species trends produced for each of 45 states between 1980 and 2003 (http://www.mbr-pwrc.usgs.gov/bbs/trend/guild03.html) - the same time period examined by Mineau and Whiteside [15]. These composite statewide trends incorporate any grassland bird species detected on ≥14 survey routes within a state, and account for uncertainty among individual species' trends by adjusting them toward the mean trend of that state [23]. Additional details of BBS statistical procedure can be found in Link and Sauer [24,25]. Some states with few grassland birds detected during BBS surveys (Alaska, Delaware, Hawai'i, Iowa, and Maryland) did not have state-wide trend estimates and hence were not included in the analysis.

Agricultural Data and Covariates

We used publically available data primarily from the United States Department of Agriculture [USDA] Economic Research Service [26–28], the 1992 National Land Cover Data (NLCD [29]), the Natural Resources Conservation Service (NRCS [30]), and Mineau and Whiteside [15] to construct nine covariates

describing agricultural intensity and landscape composition that we hypothesized would be relevant to grassland bird trends in the United States. Six of these covariates were used by Mineau and Whiteside [15], including the change in permanent pasture from 1978–2002, change in cropland for pasture from 1978–2002, agricultural intensity in 1992, agricultural herbicide use in 1992, agricultural insecticide use in 1992, and lethal insecticide risk to grassland birds in 1992. Change in permanent pasture or cropland for pasture, for each state, was calculated as the number of hectares in 2002 minus the number of hectares in 1978 divided by the number of hectares in 1978 [26,28]. Permanent pasture was defined as land not used for arable crops and composed of primarily introduced forage plants managed for intensive grazing, as defined by the U.S. Department of Agriculture [USDA] Economic Research Service [26,28]. In contrast, cropland for pasture (referred to as "cropped pasture" by Mineau and Whiteside [15]) is defined as land that was potentially suitable for arable crop production without improvements (e.g., adding tile drains or grading) but was used for short- or long-term livestock grazing rather than crop production in 1978 or 2002, respectively.

We indexed agricultural intensity as the percentage of all agricultural lands within a state allocated to active cropping (i.e., "cropland used for crops") in 1992; data were obtained from the USDA Economic Research Service [27]. For pesticide use, we used the herbicide and insecticide indices presented in Mineau and Whiteside [15], which were calculated with data obtained from the National Pesticide Use Database maintained by the National Center for Food and Agricultural Policy (http://www.ncfap.org/database/state.php). This database provides estimates of areas of crops treated with individual pesticides. The insecticide and herbicide use indices were created with summed areas of crops within a state treated with each insecticide or herbicide, respectively, divided by the total areas of farmland within that state in 1992 [15]. In some states the herbicide index greatly exceeded 1.0, because many crops were treated with multiple herbicides. Unlike Mineau and Whiteside [15], we did not restrain herbicide use values at 1.0. Such an approach would have both underestimated herbicide use estimates and assumed equal herbicide exposure across 14 states. We used the lethal insecticide risk covariate presented in Mineau and Whiteside [15]. Briefly, lethal insecticide risk was taken to be the percentage of insecticide-treated farmland within a state over which logistic exposure models predicted some avian mortality for birds [18,31]. State-specific estimates of pesticide use were only available in 1992 and 1997. Therefore, following Mineau and Whiteside [15], we used the lethal insecticide risk, agricultural intensity, herbicide use, and insecticide use data from only 1992 (the mid-point of the avian trend estimates). Thus our models, and those of Mineau and Whiteside [15], made the implicit assumption that these covariate values in 1992 were predictably related to other covariate values, such that states with high lethal insecticide risk in 1992 would have high insecticide risk in other years from 1980–2003.

Other herbaceous land cover types besides the two habitat types (permanent pasture and cropland for pasture) considered by Mineau and Whiteside [15] can benefit grassland birds [32,33]. We defined three such covariates: statewide grassland coverage in 1992, Conservation Reserve Program (CRP) area in 1992, and change in rangeland from 1982–2002. We calculated the percentage of grassland coverage within a state from the "grassland" land cover classification from the 1992 NLCD [29], which excluded all grass or herbaceous cover within a state subjected to intensive management (e.g., residential lawns, golf courses, and crop fields). The NLCD categorization of grasslands includes grasslands (e.g., reclaimed surface mine grasslands and other non-grazed grasslands) not included in our other covariate categories. While percent change values for state-wide grassland variables could be more informative than a static value for 1992 grassland coverage, there are unfortunately no other national land cover estimates that cover the timeframe of our study.

The CRP covariate was expressed as the area (km^2) of CRP lands within a state obtained from the NASS [34]. Change in rangeland was calculated as the rangeland area in 2002 minus the rangeland area in 1982 and expressed as a proportion of the 1982 area for each state using data from the NRCS [30]. Rangelands are extensively managed lands where the natural vegetation are mixtures of native grasses, forbs, and shrubs suitable for grazing as well as introduced forage species that have become naturalized [35].

Statistical Analysis

We used an information theoretic approach to compare linear regression models with Akaike's Information Criterion adjusted for sample size (AIC$_c$). We separated our analyses into two suites of regression models to reduce model complexity and to enable a more direct comparison of our results with Mineau and Whiteside [15]. In the first suite of models we compared models created with the six covariates used by Mineau and Whiteside [15]. In the second suite of models we compared models created with lethal insecticide risk and the three additional covariates not considered by Mineau and Whiteside [15]. Our linear regression models took the general form of

$$Y_i = \alpha + \beta * \text{covariate}_i + \varepsilon_i$$

where Y_i is the overall grassland bird trend (1980–2003) for state i, α is the global intercept for all states, β is the slope of the relationship for each covariate, and ε_i is the residual where $\varepsilon_i \sim N(0,\sigma^2)$.

Plots of the grassland bird trends and each covariate did not suggest any non-linear relationships, but at the suggestion of a reviewer we evaluated models with quadratic covariate terms. These models with quadratic terms yielded extraordinary large confidence intervals for coefficients (e.g., lethal insecticide risk) and hence we did not further consider such models. We examined residual plots from univariate linear regression models to determine if transformations of covariates were necessary. Based on these plots and the subsequent elimination of residual patterns, we square-root transformed the covariates representing agricultural intensity, lethal insecticide risk, insecticide use, herbicide use, and natural log (x+1) transformed the CRP area. We scaled and centered (i.e., standardized) all covariates to allow direct comparison of effect sizes among coefficients. Therefore, positive coefficients indicated a positive association with grassland bird trends and larger (absolute value) regression coefficients indicate a larger effect on grassland bird trends than coefficients closer to zero.

To reduce multicollinearity impacts [36], we excluded models from consideration (from all subsets of models) where combinations of predictors produced variance inflation factor (VIF) scores >5.0. In the first suite of models this criterion resulted in only models with ≤4 covariates and excluded models that contained both lethal insecticide risk and insecticide use. Lethal insecticide risk and insecticide use were correlated ($r = 0.95$, $P = <0.001$). At the suggestion of a reviewer we did evaluate models that contained both insecticide risk and insecticide use as predictors. Resulting coefficient estimates for these two covariates were highly uncertain with model-averaged 95% confidence intervals that were >4 times

the width of the confidence intervals of any other covariate. Thus, correcting for multicollinearity allowed us to much more precisely estimate the effects of lethal insecticide risk. Consequently, we did not further consider models that included both insecticide use and insecticide risk. VIF scores did not exceed 5.0 for any of the models in the second model suite.

In the first model suite we considered 45 models (of all 56 possible models) that did not contain combinations of lethal insecticide risk and insecticide use; we considered all 16 possible models in the second set. When comparing models with AIC is it not uncommon for researchers to mistakenly assign importance to non-competitive models [37]. This often occurs when comparing a model with K+1 parameters to a better (i.e., lower AIC) nested model with only K parameters. The K+1 model may fall within 2 AIC units (ΔAIC≤2) of a simpler model, but such models "should not be interpreted as having any ecological effect" [37] unless they also substantially lower the model deviance [37,38]. Therefore, after running both suites of models we then removed any nested model with one additional parameter (K+1) with an AIC$_c$ score greater than the simpler model with K parameters and a log-likelihood value <0.5 units smaller than the simpler model. We model-averaged coefficient estimates from the confidence model set (<4.0 ΔAIC$_c$ units from the top model [38]) and produced unconditional standard errors and 95% confidence intervals with the MuMIn package [39] in program R [40]. We model-averaged coefficient estimates using only the coefficient estimates from models that included that parameter of interest in the confidence sets.

Results

First Model Suite: Reevaluation of Mineau and Whiteside

Our model selection procedure from the first model suite resulted in 10 competitive models, three of which were included in the final confidence set (Table 1). However, only change in cropland for pasture had an estimated effect for which the confidence interval did not include zero (Figure 1). Change in cropland for pasture occurred in two of the three models in the confidence set (Table 1) and was positively associated with increasing bird trends (Figure 1). Only considering the best-performing models representing both hypotheses, habitat availability had 21.0 times more support (ratio of AIC$_c$ weights = 0.42/ 0.02) than lethal insecticide risk as the most parsimonious explanation of the data (Table 1).

Second Model Suite: Lethal Insecticide Risk & Additional Habitat Covariates

For the second model suite, our model selection procedure resulted in nine competitive models, six of which were included in the confidence set. All four possible covariates were included in the confidence set (Table 2), but again, confidence intervals for the estimated effect of most covariates included zero. CRP area had the largest effect size, was positively associated with grassland bird trends, and was the only covariate that had an estimated effect for which the confidence interval did not include zero (Figure 1). We had originally expressed the CRP covariate as the change in CRP area over time within a state, but we changed it to its current form at the request of a reviewer–our results and conclusions did not change. Although included in the confidence set, change in rangeland, percent grass, and lethal insecticide risk all had uncertain effects on grassland bird trends (Table 2, Figure 1). The most parsimonious model included only CRP area (Table 2), and this model had >1.3 times more support (ratio of AIC$_c$ weights = 0.28/0.22) than the highest ranking model containing

Agricultural intensity

Lethal insecticide risk

Change in rangeland

Percent grassland

Change in permanent pasture

Change in cropland for pasture

CRP area

-2   -1   0   1   2

Standardized coefficients

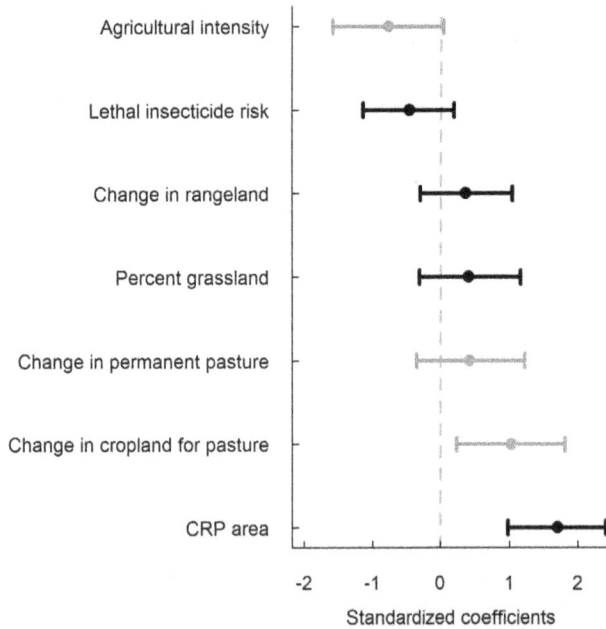

Figure 1. Model-averaged coefficients of covariates used to predict grassland breeding bird trends (1980–2003). Model-averaged coefficients (95% confidence intervals) occurring in the confidence sets of model suites one (gray bars) and two (black bars) that describe agricultural practices and habitat availability used to predict grassland breeding bird trends from 1980 to 2003 in the U.S.

lethal insecticide risk (Table 2). Comparing only univariate models, the CRP model had >2300 times more support than the lethal insecticide risk model. CRP had approximately 3.06 times the effect (ratio of standardized regression coefficients = 1.71/0.47) on grassland bird populations than did lethal insecticide risk.

Discussion

Our results indicate that population trends of grassland birds in the U.S. are primarily associated with habitat availability, rather than insecticide use or insecticide acute toxicity. Thus, in direct contrast to Mineau and Whiteside [15], our results do not support the insecticide-acute-toxicity hypothesis because we found stronger connections with declines in grassland bird abundance using habitat-based covariates. Pesticides may negatively influence grassland birds, but our results strongly support habitat associations over insecticide-acute-toxicity as the more plausible explanation for observed BBS trends.

We suggest three reasons why we arrived at different conclusions than Mineau and Whiteside [15]. First, Mineau and Whiteside's [15] use of a dichotomous response variable based on the point estimate of a species' trend means that species with steeply negative or precisely estimated trends are treated identically to species with near-zero or poorly estimated trends. In contrast, we used a single state trend for grassland bird species which explicitly accounted for the similarity and uncertainty among all grassland bird trends within that state [23]. Second, Mineau and Whiteside [15] analyzed individual grassland species' trends (1 = decreasing, 0 = increasing) without accounting for the lack of independence among trends from within a state (e.g., by treating state identity as a random effect). Trends of grassland birds from within a state are unlikely to be independent of one

Table 1. Linear models from the first model suite with standardized coefficients[a] that described the relationship between U.S. state grassland bird trends (1980–2003) and agriculture-related covariates (see Methods).

Change in cropland for pasture	Change in permanent pasture	Agricultural intensity	Herbicide use	Insecticide use	Lethal insecticide risk	K[b]	log(L)	AIC_c	ΔAIC_c	w_i[c]
1.02						2	−106.87	220.32	0.00[d]	0.42
1.06	0.43	−0.77				3	−106.27	221.54	1.21[d]	0.23
	0.43	−0.82				2	−108.25	223.08	2.76[d]	0.11
		−1.36	0.71			3	−107.68	224.37	4.05	0.06
	0.43	−1.36				3	−107.74	224.49	4.16	0.05
				−0.49		2	−109.32	225.22	4.90	0.04
			0.41			2	−109.53	225.64	5.32	0.03
		−1.53	1.27	−0.63		4	−107.06	225.66	5.34	0.03
	0.33					2	−109.69	225.97	5.64	0.03
					−0.27	2	−109.79	226.16	5.83	0.02

[a] Model-averaged coefficients with unconditional standard errors: change in cropland for pasture (β = 1.03, SE = 0.40), change in permanent pasture (β = 0.43, SE = 0.40), and agricultural intensity (β = −0.77, SE = 0.41).
[b] No. parameters.
[c] Akaike model weights.
[d] Models appearing in the confidence set.

Table 2. Linear models with standardized coefficients[a] that appeared in the second model suite that described the relationship between U.S. state grassland bird trends (1980–2003), lethal insecticide risk, and three additional habitat covariates not considered by Mineau and Whiteside (2013).

CRP area	Percent grassland	Change in rangeland	Lethal insecticide risk	K^b	$\log(L)$	AIC_c	AIC_c	w_i^c
1.71				2	−99.71	206.01	0.00^d	0.28
1.76			−0.47	3	−98.73	206.46	0.45^d	0.22
1.52	0.42			4	−99.08	207.15	1.14^d	0.16
1.78		0.36		3	−99.15	207.30	1.28^d	0.15
1.84	0.42	0.39	−0.49	4	−98.05	207.64	1.63^d	0.12
1.59	1.09	0.36		4	−98.49	208.53	2.51^d	0.08
			−0.27	2	−106.34	219.28	13.26	<0.01
				2	−109.79	226.16	20.15	<0.01
		0.02		2	−110.00	226.58	20.57	<0.01

^{a}Model-averaged coefficients with unconditional standard errors: CRP area ($\beta = 1.71$, SE = 0.38), change in rangeland ($\beta = 0.42$, SE = 0.35), percent grassland ($\beta = 1.71$, SE = 0.37), and lethal insecticide risk ($\beta = −0.47$, SE = 0.34).
^{b}No. parameters.
^{c}Akaike model weights.
^{d}Models appearing in the confidence set.

another due to similar land management, agricultural policies, pesticide exposure, weather conditions, and habitat availability that affect all species within that state. Failure to account for this lack of independence would not necessarily favor insecticide acute toxicity in the modeling process, but the violation of basic linear model assumptions can result in spurious results [41]. Third, we observed evidence of the effects of multicollinearity in Mineau and Whiteside's [15] analysis. For example, when insecticide use and lethal insecticide risk occurred within the same model (16 such occurrences in Mineau and Whiteside [15]) the coefficient for insecticide use was always negative, whereas it was always positive when it occurred in models without lethal insecticide risk (16 occurrences) (JMH unpublished data). Multicollinearity can complicate model selection, result in large variances, and cause coefficient estimates to reverse sign when another co-linear variable is included in the model [36,42–44]. We reduced multicollinearity effects by limiting model selection to combinations of linear predictors where VIF scores were <5.0.

Consistent with the prediction of the habitat-availability hypothesis we found that grassland bird trends were more strongly associated with (i.e., larger effect sizes of standardized coefficients) CRP area, agricultural intensity, and trends in cropland for pasture than they were for lethal insecticide risk (Tables 1 and 2). These results are consistent with the continuing decline of grassland birds over the study period [2] and the dramatic conversion of rangelands, pasturelands, and native grasslands to intensive croplands and developed land over the last 35 years [7,30]. For example, cropland for pasture area in the U.S. declined by 20% (6.3 million ha) from 1978 to 2002 (approximately the time frame of this study) and declined by a further 41% (10 million ha) from 2002 to 2007 alone [30,45]. Meanwhile overall pesticide use in the U.S. has likely remained relatively stable and insecticide use has likely declined since the 1980s [17]. Additionally, several researchers have compared the influence of habitat availability relative to the prevalence of organic (i.e. pesticide free) cropland. These studies have consistently found that the extent and configuration of grassland habitats has stronger effects on the richness and abundance of most grassland bird species than the amount of organic acreage [46–48]. As grassland area declines in agricultural landscapes the potential for negative impacts to grassland bird populations increases because many grassland bird species are sensitive to fragmentation and area and edge effects [4,5,20,49,50].

Our models suggest that grassland birds respond negatively to agricultural intensification [32,51] (i.e., increases in row-crop agriculture), but less-intensive agricultural practices may benefit some grassland bird species [52]. For example, light grazing pressure, and rangelands in general, have been positively linked to abundances of many species of grassland birds such as grasshopper sparrows (*Ammodramus savannarum*), Henslow's sparrows (*A. henslowii*), and upland sandpipers (*Bartramia longicauda*) [9,53–55]. Some grassland birds breed in hayfields and pasturelands [32,56], and adult bobolink (*Dolichonyx oryzivorus*) and savannah sparrow (*Passerculus sandwichensis*) apparent survival is negatively related to the intensity of agricultural practices in these hayfields [57]. The value of hayfields to breeding grassland birds, however, may be dependent upon the timing of mowing activities [58,59]. The point estimate for the effect of permanent pasture on grassland bird trends was positive, but the effect was uncertain and smaller than the effect of change in cropland for pasture area (Figure 1). This result may be related to the relatively small amount of permanent pasture changes observed across states from 1978 to 2002 (mean = −647 ha) compared to cropland for pasture changes (mean = −120,880 ha) over the same time period [26,28].

We found that CRP area was the strongest predictor (i.e., largest standardized coefficient; Figure 1) of grassland bird trends in the U.S. Other researchers have also documented positive effects of CRP area on some species of grassland birds [33,60,61], but see Pabian et al. [62]. We found nearly identical results when we expressed the CRP variable as the change in percent of acreage over time within a state (JMH unpub.). The Conservation Reserve Program is a type of agricultural de-intensification where farmers are paid to remove environmentally sensitive lands from agricultural production. However, as farmland values have risen sharply over the last decade [63], CRP enrolled hectares have dropped by 29% from 2006 to 2013 [64,65]. Our results suggest that CRP program enrollment may be an important component of grassland bird conservation strategy in the U.S. However, while CRP and other restored or non-native grasslands may benefit some species, these areas may not provide optimal habitat for some grassland bird species (e.g., Sprague's pipt [*Anthus spragueii*], and short-eared owl [*Asio flammeus*]) that are relatively intolerant to anthropogenic disturbance ([66,67]). The best conservation strategy for such species may be to protect remaining native grasslands from conversation to agricultural lands [5,68,69].

Our results are consistent with the scarcity of data linking pesticides to vertebrate population changes [19], but depend on several important assumptions. Following Mineau and Whiteside [15] we developed several covariates based on data from 1992 (the midpoint of our study period), because data for these covariates was not available circa 1980; thus our models assume that these 1992 covariate values were predictably related to other years in our analysis. Insecticides have direct and indirect negative effects on many groups of birds [12,14,70,71], but the phase-out of especially toxic active pesticide ingredients (e.g., organophosphates and cholinesterase-inhibiting carbamates) combined with the increased use of Bt crops, and reduced application rates of insecticides [17,18] may have lowered the risk of lethal avian exposures to North American grassland bird species over the past 20 years. If overall insecticide risk to grassland birds declined non-linearly between 1992 and 2003 then both we and Mineau and Whiteside [15] may have imprecisely estimated the risk of insecticides to grassland birds. Our models suggested that

insecticide acute toxicity was negatively related to grassland bird trends, but insecticide acute toxicity was a relatively poor predictor of grassland bird trends on its own (Table 1). Our analyses could be improved if finer scale (temporal and spatial resolution) land use, pesticide application, and pesticide exposure data were available for the conterminous United States from 1980–2003.

Our results and those of others [15,32,33] that are based on broad-scale correlations of BBS and agricultural data clearly cannot prove or disprove the insecticide-acute-toxicity or habitat-availability hypotheses. However, the strong majority of the grassland bird research [5,20,61,72–76] suggests that grassland bird populations in the U.S. are most strongly related to grassland configuration and amount (the habitat-availability-hypothesis). Further grassland bird population declines are foreseeable given the continued loss of U.S. grasslands [7,30,45]. Building multi-functional landscapes that provide an abundant food supply while also conserving biodiversity and supplying ecosystem services, has become a central challenge for agriculture in the 21st century [77,78]. Farmers, conservation organizations, and governments have taken many approaches to build multifunctional working landscapes [79–82]. Our results, and those of others [5,61,72] suggest that an approach most effective for grassland bird conservation would include a focus on habitat creation and preservation, encouraging CRP acreages, pastures, and low-intensity grazing land within intensive agriculture landscapes, and strongly disincentivizing the conversion of remaining native grasslands to agricultural lands.

Acknowledgments

We thank three anonymous reviewers, Margaret Brittingham, and Pierre Mineau for editorial comments, and we thank Pierre Mineau and Mélanie Whiteside for sharing their data. Any use of trade, firm, or product names is for descriptive purposes only and does not imply endorsement by the U.S. Government.

Author Contributions

Conceived and designed the experiments: JMH JFE GES. Analyzed the data: JMH. Wrote the paper: JMH JFE GES DRD.

References

1. North American Bird Conservation Initiative, U.S. Committee (2011) The State of the Birds 2011 Report on Public Lands and Waters: United States of America. US Fish Wildl Publ: 48.

2. Sauer JR, Hines JE, Fallon JE, Pardieck KL, Ziolkowski DJ Jr, et al. (2012) The North American Breeding Bird Survey, Results and Analysis 1966–2011. Laurel, MD: USGS Patuxent Wildlife Research Center.

3. Herkert JR, Sample DW, Warner RE (1996) Management of Midwestern grassland landscapes for the conservation of migratory birds. St. Paul, MN: U.S. Dept. of Agriculture, Forest Service, North Central Forest Experiment Station.

4. Vickery PD, Tubaro PL, Cardoso da Silva JM, Peterjohn B, Herkert JR, et al. (1999) Conservation of grassland birds in the western hemisphere. Stud Avian Biol: 2–26.

5. Askins RA, Chávez-Ramírez F, Dale BC, Haas CA, Herkert JR, et al. (2007) Conservation of Grassland Birds in North America: Understanding Ecological Processes in Different Regions: "Report of the AOU Committee on Conservation." Ornithol Monogr: iii–46. doi:10.2307/40166905.

6. Samson F, Knopf FL, Ostlie W (2004) Great Plains Ecosystems: Past, Present, and Future. USGS Staff – Publ Res. Available: http://digitalcommons.unl.edu/usgsstaffpub/45.

7. Wright CK, Wimberly MC (2013) Recent land use change in the Western Corn Belt threatens grasslands and wetlands. Proc Natl Acad Sci 110: 4134–4139. doi:10.1073/pnas.1215404110.

8. Briggs JM, Knapp AK, Blair JM, Heisler JL, Hoch GA, et al. (2005) An Ecosystem in Transition: Causes and Consequences of the Conversion of Mesic Grassland to Shrubland. BioScience 55: 243. doi:10.1641/0006-3568(2005)055 [0243:AEITCA]2.0.CO;2.

9. Powell AFLA (2006) Effects of Prescribed Burns and Bison (Bos bison) Grazing on Breeding Bird Abundances in Tallgrass Prairie. The Auk 123: 183–197. doi:10.2307/4090640.

10. Newton I (2004) The recent declines of farmland bird populations in Britain: an appraisal of causal factors and conservation actions. Ibis 146: 579–600. doi:10.1111/j.1474-919X.2004.00375.x.

11. Carson R (2002) Silent spring. Boston: Houghton Mifflin. 400 p.

12. Risebrough RW (1986) Pesticides and Bird Populations. In: Johnston RF, editor. Current Ornithology. Springer US. 397–427. Available: http://link.springer.com/chapter/10.1007/978-1-4615-6784-4_9. Accessed 30 September 2013.

13. Mineau P, Fletcher MR, Glaser LC, Thomas NJ, Brassard C, et al. (1999) Poisoning of raptors with organophosphorus and carbamate pesticides with emphasis on Canada, U.S. and U.K. United States Geological Survey. Available: http://pubs.er.usgs.gov/publication/70021842. Accessed 30 June 2013.

14. Bright JA, Morris AJ, Winspear R (2008) A Review of indirect effects of pesticides on birds and mitigating land-management practices. Research. Royal Society for the Protection of Birds. Available: http://www.rspb.org.uk/Images/bright_morris_winspear_tcm9-192457.pdf.

15. Mineau P, Whiteside M (2013) Pesticide acute toxicity is a better correlate of U.S. grassland bird declines than agricultural intensification. PLoS ONE 8: e57457. doi:10.1371/journal.pone.0057457.

16. Annonymous (2013) Toxic Threats to Grassland Birds. N Y Times. Available: http://www.nytimes.com/2013/03/12/opinion/toxic-threats-to-grassland-birds.html. Accessed 30 September 2013.

17. Osteen CD, Fernandez-Cornejo J (2013) Economic and policy issues of U.S. agricultural pesticide use trends. Pest Manag Sci 69: 1001–1025. doi:10.1002/ps.3529.

18. Mineau P, Whiteside M (2006) Lethal risk to birds from insecticide use in the United States–a spatial and temporal analysis. Environ Toxicol Chem 25: 1214–1222. doi:10.1897/05-035R.1.

19. Köhler H-R, Triebskorn R (2013) Wildlife Ecotoxicology of Pesticides: Can We Track Effects to the Population Level and Beyond? Science 341: 759–765. doi:10.1126/science.1237591.

20. Helzer CJ, Jelinski DE (1999) The relative importance of patch area and perimeter-area ratio to grassland breeding birds. Ecol Appl 9: 1448–1458. doi:10.1890/1051-0761(1999)009[1448:TRIOPA]2.0.CO;2.

21. Virkkala R, Luoto M, Rainio K (2004) Effects of landscape composition on farmland and red-listed birds in boreal agricultural-forest mosaics. Ecography 27: 273–284. doi:10.1111/j.0906-7590.2004.03810.x.

22. Wretenberg J, Pärt T, Berg Å (2010) Changes in local species richness of farmland birds in relation to land-use changes and landscape structure. Biol Conserv 143: 375–381. doi:10.1016/j.biocon.2009.11.001.

23. Link WA, Sauer JR (1996) Extremes in Ecology: Avoiding the Misleading Effects of Sampling Variation in Summary Analyses. Ecology 77: 1633–1640. doi:10.2307/2265557.

24. Link WA, Sauer JR (1997) Estimation of Population Trajectories from Count Data. Biometrics 53: 488–497. doi:10.2307/2533952.

25. Link WA, Sauer JR (1998) Estimating Population Change from Count Data: Application to the North American Breeding Bird Survey. Ecol Appl 8: 258–268. doi:10.1890/1051-0761(1998)008[0258:EPCFCD]2.0.CO;2.

26. Frey HT (1982) Major Uses of Land in the United States: 1978. Natural Resources and Environment Division, Economic Research Service, U.S. Department of Agriculture.

27. Daugherty AB (1995) Major Uses of Land in the United States: 1992. Agricultural Economic Report. Natural Resources and Environment Division, Economic Research Service, U.S. Department of Agriculture. Available: http://naldc.nal.usda.gov/download/CAT10825179/PDF.

28. Lubowski RN, Vesterby M, Bucholtz S, Baez A, Roberts MJ (2006) Major Uses of Land in the United States: 2002. Natural Resources and Environment Division, Economic Research Service, U.S. Department of Agriculture.

29. Fry J, Xian G, Jin S, Dewitz J, Homer C, et al. (2011) Completion of the 2006 National Land Cover Database for the Conterminous United States. Photogramm Eng Remote Sens 77: 858–864.

30. U.S. Department of Agriculture (2009) Summary Report: 2007 National Resources Inventory. Iowa State University, Ames, IA: Natural Resources Conservation Service and the Center for Survey Statistics and Methodology.

31. Mineau P (2002) Estimating the probability of bird mortality from pesticide sprays on the basis of the field study record. Environ Toxicol Chem 21: 1497–1506.

32. Murphy MT (2003) Avian population trends within the evolving agricultural landscape of Eastern and Central United States. The Auk 120: 20–34. doi:10.2307/4090137.

33. Riffell S, Scognamillo D, Burger L, Wes J, Bucholtz S (2010) Broad-scale relations between Conservation Reserve Program and grassland birds: do cover type, configuration and contract age matter? Open Ornithol J 3: 112–123. doi:10.2174/1874453201003010112.

34. U.S. Department of Agriculture (2004) 2002 Census of Agriculture. United States Summary and State Data. National Agricultural Statistics Service.

35. Nickerson C, Ebel R, Borchers A, Carriazo F (2011) Major Land Uses in the United States, 2007. USDA Economic Research Service.

36. Graham MH (2003) Confronting multicollinearity in ecological multiple regression. Ecology 84: 2809–2815. doi:10.1890/02-3114.

37. Arnold TW (2010) Uninformative parameters and Model selection using Akaike's information criterion. J Wildl Manag 74: 1175–1178. doi:10.2307/40801110.

38. Burnham KP, Anderson DR (2002) Model selection and multimodel inference: a practical information-theoretic approach. New York: Springer. 488 p.

39. Barton K. (2013) MuMIn: Multi-model inference. Available: http://cran.r-project.org/web/packages/MuMIn/MuMIn.pdf.

40. R Core Team (2013) R: A language and environment for statistical computing. Vienna, Austria: R Foundation for Statistical Computing.

41. Lazic SE (2010) The problem of pseudoreplication in neuroscientific studies: is it affecting your analysis? BMC Neurosci 11: 5. doi:10.1186/1471-2202-11-5.

42. Mansfield ER, Helms BP (1982) Detecting multicollinearity. Am Stat 36: 158–160. doi:10.2307/2683167.

43. Smith AC, Koper N, Francis CM, Fahrig L (2009) Confronting collinearity: comparing methods for disentangling the effects of habitat loss and fragmentation. Landsc Ecol 24: 1271–1285. doi:10.1007/s10980-009-9383-3.

44. Mela C, Kopalle P (2002) The impact of collinearity on regression analysis: the asymmetric effect of negative and positive correlations. Appl Econ 34: 667–677.

45. U.S. Department of Commerce (1981) 1978 Census of Agriculture. United States Summary and State Data. Bureau of the Census.

46. Piha M, Tiainen J, Holopainen J, Vepsäläinen V (2007) Effects of land-use and landscape characteristics on avian diversity and abundance in a boreal agricultural landscape with organic and conventional farms. Biol Conserv 140: 50–61. doi:10.1016/j.biocon.2007.07.021.

47. Smith HG, Dänhardt J, Lindström Å, Rundlöf M (2010) Consequences of organic farming and landscape heterogeneity for species richness and abundance of farmland birds. Oecologia 162: 1071–1079. doi:10.1007/s00442-010-1588-2.

48. Hiron M, Berg Å, Eggers S, Josefsson J, Pärt T (2013) Bird diversity relates to agri-environment schemes at local and landscape level in intensive farmland. Agric Ecosyst Environ 176: 9–16. doi:10.1016/j.agee.2013.05.013.

49. Renfrew RB, Ribic CA, Nack JL, Bollinger EK (2005) Edge Avoidance by Nesting Grassland Birds: a futile strategy in a fragmented landscape. The Auk 122: 618–636. doi:10.1642/0004-8038(2005)122[0618:EABNGB]2.0.CO;2.

50. Davis SK, Brigham RM, Shaffer TL, James PC, Stouffer PC (2006) Mixed-grass prairie passerines exhibit weak and variable responses to patch size. The Auk 123: 807–821. doi:10.1642/0004-8038(2006)123[807:MPPEWA]2.0.CO;2.

51. Geiger F, de Snoo GR, Berendse F, Guerrero I, Morales MB, et al. (2010) Landscape composition influences farm management effects on farmland birds in winter: A pan-European approach. Agric Ecosyst Environ 139: 571–577. doi:10.1016/j.agee.2010.09.018.

52. Best LB, Freemark KE, Dinsmore JJ, Camp M (1995) A review and synthesis of habitat use by breeding birds in agricultural landscapes of Iowa. Am Midl Nat 134: 1–29. doi:10.2307/2426479.

53. Smith CR (1997) Use of public grazing lands by Henslow's Sparrows, Grasshopper Sparrows, and associated grassland birds in New York State. In: Vickery PD, Dunwiddie PW, editors. Grasslands of Northeastern North America: Ecology and Conservation of Native and Agricultural Landscapes. Lincoln, MA: Massachusetts Audubon Society. 171–186.

54. Walk JW, Warner RE (2000) Grassland management for the conservation of songbirds in the Midwestern USA. Biol Conserv 94: 165–172. doi:10.1016/S0006-3207(99)00182-2.

55. Brennan LA, Kulesky WP (2005) Invited Paper: North American grassland birds: an unfolding conservation crisis? J Wildl Manag 69: 1–13. doi:10.2193/0022-541X(2005)069<0001:NAGBAU>2.0.CO;2.

56. Perlut NG, Strong AM, Donovan TM, Buckley NJ (2006) Grassland songbirds in a dynamic management landscape: behavioral responses and management strategies. Ecol Appl 16: 2235–2247. doi:10.1890/1051-0761(2006)016[2235:GSIADM]2.0.CO;2.

57. Perlut NG, Strong AM, Donovan TM, Buckley NJ (2008) Grassland Songbird Survival and Recruitment in Agricultural Landscapes: Implications for Source–Sink Demography. Ecology 89: 1941–1952. doi:10.1890/07-0900.1.

58. Perlut NG, Strong AM, Donovan TM, Buckley NJ (2008) Regional population viability of grassland songbirds: Effects of agricultural management. Biol Conserv 141: 3139–3151. doi:10.1016/j.biocon.2008.09.011.

59. Dale BC, Martin PA, Taylor PS (1997) Effects of Hay Management on Grassland Songbirds in Saskatchewan. Wildl Soc Bull 25: 616–626.

60. Johnson DH, Igl LD (1995) Contributions of the Conservation Reserve Program to populations of breeding birds in North Dakota. Wilson Bull 107: 709–718.

61. Herkert JR (2009) Response of Bird Populations to Farmland Set-Aside Programs. Conserv Biol 23: 1036–1040. doi:10.1111/j.1523-1739.2009.01234.x.

62. Pabian SE, Wilson AM, Brittingham MC (2013) Mixed responses of farmland birds to the Conservation Reserve Enhancement Program in Pennsylvania. J Wildl Manag 77: 616–625. doi:10.1002/jwmg.514.

63. Nickerson C, Morehart M, Kuethe T, Beckman J, Ifft J, et al. (2012) Trends in U.S. Farmland Vluaes and Ownership. USDA, Economic Research Service. Available: http://www.ers.usda.gov/media/377487/eib92_2_.pdf. Accessed 1 January 2014.

64. USDA Farm Service Agency (2007) Conservation Reserve Program summary and enrollment statistics FY 2006. Available: http://www.fsa.usda.gov/Internet/FSA_File/06rpt.pdf.

65. USDA Farm Service Agency (2013) Conservation Reserve Program November 2013 Monthly Report. Available: http://www.fsa.usda.gov/FSA/webapp?area=home&subject=copr&topic=rns-css.

66. Dechant JA, Sondreal ML, Johnson DH, Igl LD, Goldade CM, et al. (2003) Effects of managmment practices on grassland birds: short-eared owl. Northern Prairie Wildlife Research Center, Jamestown, ND. Available: http://www.npwrc.usgs.gov/resource/literatr/grasbird/seow/seow.htm. Accessed 15 March 2014.

67. Dechant JA, Sondreal ML, Johnson DH, Igl LD, Goldade CM, et al. (2003) Effects of management practices on grassland birds: Sprague's pipit. Northern Prairie Wildlife Research Center, Jamestown, ND. Available: http://www.npwrc.usgs.gov/resource/literatr/grasbird/sppi/sppi.htm. Accessed 15 March 2014.

68. Johnson DH (1996) Management of northern prairies and wetlands for the conservation of neotropical migratory birds. In: Thompson FRI, editor. Management of Midwestern Landscapes for the Conservation of Neotropical Migratory Birds. North Central Forest Experiment Station, St. Paul, MN: Northern Prairie Wildlife Research Center, Jamestown, ND. 53–67. Available: http://www.npwrc.usgs.gov/resource/habitat/neobird/index.htm.

69. Koford RR, Best LB (1996) Management of agricultural landscapes for the conservation of neotropical migratory birds. In: Thompson FRI, editor. Management of Midwestern Landscapes for the Conservation of Neotropical Migratory Birds. Nort: Northern Prairie Wildlife Research Center, Jamestown, ND. 68–88. Available: http://www.nrs.fs.fed.us/pubs/gtr/other/gtr-nc187/Page%2068%20RR%20Koford,%20NC-GTR-187.pdf.

70. Mineau P, Downes CM, Kirk DA, Bayne E, Csizy M (2005) Patterns of bird species abundance in relation to granular insecticide use in the Canadian prairies. Ecoscience 12: 267–278.

71. McEwen LC, Knittle CE, Richmond ML (1972) Wildlife Effects from Grasshopper Insecticides Sprayed on Short-Grass Range. J Range Manag 25: 188. doi:10.2307/3897053.

72. Veech JA (2006) A comparison of landscapes occupied by increasing and decreasing populations of grassland birds. Conserv Biol 20: 1422–1432. doi:10.1111/j.1523-1739.2006.00487.x.

73. Vickery PD, Hunter ML, Melvin SM (1994) Effects of Habitat Area on the Distribution of Grassland Birds in Maine. Conserv Biol 8: 1087–1097. doi:10.1046/j.1523-1739.1994.08041087.x.

74. Coppedge BR, Engle DM, Masters RE, Gregory MS (2001) Avian Response To Landscape Change In Fragmented Southern Great Plains Grasslands. Ecol Appl 11: 47–59. doi:10.1890/1051-0761(2001)011[0047:ARTLCI]2.0.CO;2.

75. Davis SK, Brittingham M (2004) Area Sensitivity In Grassland Passerines: Effects Of Patch Size, Patch Shape, And Vegetation Structure On Bird Abundance And Occurrence In Southern Saskatchewan. The Auk 121: 1130–1145. doi:10.1642/0004-8038(2004)121[1130:ASIGPE]2.0.CO;2.

76. Perlut NG (2014) Grassland birds and dairy farms in the northeastern United States. Wildl Soc Bull: n/a–n/a. doi:10.1002/wsb.415.

77. Boody G, Vondracek B, Andow DA, Krinke M, Westra J, et al. (2005) Multifunctional agriculture in the United States. BioScience 55: 27–38. doi:10.1641/0006-3568(2005)055[0027:MAITUS]2.0.CO;2.

78. Foley JA, DeFries R, Asner GP, Barford C, Bonan G, et al. (2005) Global Consequences of Land Use. Science 309: 570–574. doi:10.1126/science. 1111772.

79. Jordan N, Warner KD (2010) Enhancing the Multifunctionality of US Agriculture. BioScience 60: 60–66. doi:10.1525/bio.2010.60.1.10.

80. Scherr SJ, McNeely JA (2003) Making space for wildlife in agricultural landscapes. In: EcoAgriculture: strategies to feed the world and save wild biodiversity. In: Harvest F, editor. Ecoagriculture: Strategies to Feed the World and Save Wild Biodiversity. Washington, D.C.: Island Press. 115–148.

81. Atwell RC, Schulte LA, Westphal LM (2011) Tweak, adapt, or transform: Policy scenarios in response to emerging bioenergy markets in the U.S. corn belt. Ecol Soc 16: 15.

82. Robertson GP, Gross KL, Hamilton SK, Landis DA, Schmidt TM, et al. (2014) Farming for Ecosystem Services: An Ecological Approach to Production Agriculture. BioScience: biu037. doi:10.1093/biosci/biu037.

Self-Reported Parental Exposure to Pesticide during Pregnancy and Birth Outcomes: The MecoExpo Cohort Study

Flora Mayhoub[1,6]**, Thierry Berton**[1,2]**, Véronique Bach**[1]**, Karine Tack**[2]**, Caroline Deguines**[1,3]**, Adeline Floch-Barneaud**[1,4]**, Sophie Desmots**[1,5]**, Erwan Stéphan-Blanchard**[1]**, Karen Chardon**[1]*

1 Laboratoire PériTox, Unité mixte Université – INERIS (EA 4285-UMI 01), Université de Picardie Jules Verne, Amiens, France, 2 Unité NOVA, Institut National de l'Environnement Industriel et des Risques, Verneuil en Halatte, France, 3 Médecine Néonatale, Pôle Femme-Couple-Enfant, Centre Hospitalier Universitaire d'Amiens, Amiens, France, 4 Unité ISAE, Institut National de l'Environnement Industriel et des Risques, Verneuil en Halatte, France, 5 Unité TOXI, Institut National de l'Environnement Industriel et des Risques, Verneuil en Halatte, France, 6 Faculty of Medicine, Tishreen University, Latakia, Syria

Abstract

The MecoExpo study was performed in the Picardy region of northern France, in order to investigate the putative relationship between parental exposures to pesticides (as reported by the mother) on one hand and neonatal parameters on the other. The cohort comprised 993 mother-newborn pairs. Each mother completed a questionnaire that probed occupational, domestic, environmental and dietary sources of parental exposure to pesticides during her pregnancy. Multivariate regression analyses were then used to test for associations between the characteristics of parental pesticide exposure during pregnancy and the corresponding birth outcomes. Maternal occupational exposure was associated with an elevated risk of low birth weight (odds ratio (OR) [95% confidence interval]: 4.2 [1.2, 15.4]). Paternal occupational exposure to pesticides was associated with a lower than average gestational age at birth (-0.7 weeks; $p = 0.0002$) and an elevated risk of prematurity (OR: 3.7 [1.4, 9.7]). Levels of domestic exposure to veterinary antiparasitics and to pesticides for indoor plants were both associated with a low birth weight (-70 g; $p = 0.02$ and -160 g; $p = 0.005$, respectively). Babies born to women living in urban areas had a lower birth length and a higher risk of low birth length (-0.4 cm, $p = 0.006$ and OR: 2.9 [1.5, 5.5], respectively). The present study results mainly demonstrate a negative correlation between fetal development on one hand and parental occupational and domestic exposure to pesticides on the other. Our study highlights the need to perform a global and detailed screening of all potential physiological effects when assessing in utero exposure to pesticides.

Editor: Olga Y. Gorlova, Geisel School of Medicine at Dartmouth College, United States of America

Funding: This work was funded by the Picardy Regional Council. F. Mayhoub was funded by Tishreen University (Latakia, Syria) and T. Berton was funded by the European Regional Development Fund. The funders had no role in study design, data collection and analysis, decision to publish, or preparation of the manuscript.

Competing Interests: The authors have declared that no competing interests exist.

* Email: karen.chardon@u-picardie.fr

Introduction

Human exposure to pesticides is a very complex phenomenon, since it involves many different compounds, sources of exposure and exposure pathways (i.e. respiratory, cutaneous and intestinal pathways). Once pesticides have been applied, the primary compounds and their degradation products are dispersed into the air, water and soil. Human exposure to pesticides can be occupational (through agriculture, floristry, municipal mainte-nance, etc.), dietary (through the consumption of food contami-nated by pesticide residues), domestic (through the spraying of houseplants or garden plants, the eradication of domestic insect pests (such as mosquitoes, flies, etc.) and the use of antiparasitics in humans or in domestic pets) or environmental (i.e. the inhalation of volatilized pesticides of agricultural or non-agricultural origin) [1–2]. Although occupational exposure has been extensively investigated, there are few studies of domestic exposure [2].

Pesticide exposure during pregnancy is becoming an increas-ingly important public health issue because it may affect the development of the exposed fetus. The association between

pesticide exposure in pregnant women and fetal growth has [11,14]. However, this topic is still subject to debate because the various studies did not reach consistent conclusions - probably because of differences in location, exposure assessment methods and the type and number of compounds investigated [3–12]. Furthermore, most of these epidemiological studies focused on a very specific population, such as farmers or other populations with high levels of pesticide exposure (e.g. those living near to crop-farming areas or other areas with intensive pesticide use). To the best of our knowledge, very few studies [9,10,13,14] have studied the relationship between "routine" domestic/dietary parental exposure on one hand and birth outcomes on the other. There are few data on the need for prevention of these types of exposure, and these data can only be gathered in general population cohorts.

The primary objective of the present MecoExpo study was to assess the different types of prenatal exposure to pesticides in the Picardy region of northern France (a region which is characterized by a high birth rate (13.1‰, according to the French National Institute of Statistics and Economic Studies (*Institut National de la*

Statistique et des Études Économiques), relative to the national average. A secondary objective was to investigate the relationship between the different modes of exposure on one hand and birth outcomes on the other. Prenatal exposure to pesticides was assessed via a self-questionnaire filled out by the mother; this is the only method that can simultaneously gather information on all the various sources of intrauterine exposure to pesticides (i.e. occupational, domestic, environmental and dietary exposure) in a sample of the general population.

Methods

2.1 Study participants

The MecoExpo cohort (comprising 993 mother-newborn pairs) was recruited between January 2011 and January 2012. Eleven of Picardy's 16 maternity clinics agreed to participate in the study. Unfortunately, the region's neonatal intensive care units (which treat newborns suffering from a severe neonatal disease or with a gestational age <32 weeks at birth) did not participate.

To be included in the MecoExpo cohort, the mother had to be had to be (i) 18 years of age or older and (ii) fluent enough in French to understand the study's objectives and procedures. Mothers aged under 18 and mothers who did not have full parental authority over their child (such as incarcerated persons) were excluded from the study. Multiple births were also excluded from the present study.

The study and both maternal and neonatal data collection (questionnaire, medical records) were approved by the local investigational review board (*Comité de Protection des Personnes dans la Recherche Biomédicale de Picardie*). Potential participants were given a verbal presentation of the study by their attending pediatrician in the maternity clinic. All mothers signed an informed consent giving us permission to enroll them and their infants in the study. A study information sheet was given to the women prior to their completion of the study questionnaire. During completion of the self-administred questionnaire, an investigating physician was always on hand to provide explanations but did not influence the answers. In view of the many different aspects of possible exposure probed by the questionnaire and in order to shorten and facilitate the questionnaire's completion, the respondee was instructed not to answer a given question if in any doubt. Most of the questions were "yes"/"no" or multiple choice questions. Open questions were used to gather additional information, when necessary. None of the data collected during the study enabled the direct identification of participants.

2.2 Characteristics of intrauterine exposure to pesticides in a questionnaire-based survey

Mothers from the cohort completed the questionnaire during their stay in the maternity clinic (4 to 5 days after giving birth). The questionnaire addressed occupational, domestic, environmental and dietary sources of pesticide exposure. Prior to use in the study, the questionnaire was validated in a sample of 25 women (aged between 25 and 45 and from various socio-professional categories). It was then modified by taking account of the women's answers, comments and questions. The final questionnaire comprised 44 items and took respondees about 15 minutes to read and complete.

2.2.1 Occupational exposure to pesticides. The mother was asked to state (i) whether she had ever worked during the pregnancy (and, if so, the duration), (ii) whether her occupation and/or that of the father involved exposure to pesticides (e.g. in agriculture, animal husbandry, gardening, etc.) and (iii) the modalities of maternal exposure to pesticides (the duration and location of exposure, the type of compound and the use or not of protective equipment).

2.2.2 Domestic exposure to pesticides. The questionnaire probed possible uses of pesticides in the mother's home during her pregnancy (whether applied directly by the mother or indirectly by another person). The mother was asked to report use of (i) antiparasitics for human administration (to treat lice, scabies, ticks, etc.), (ii) antiparasitics for domestic pets (to treat fleas, ticks, etc.) and the number and type of treated animals, (iii) insecticides, herbicides, fungicides and plant growth regulators for treating houseplants or garden plants and (iv) pesticides for eradicating domestic insect pests (such as flies, mosquitoes, etc.).

2.2.3 Environmental exposure to pesticides. The mother was asked to state whether she lived within 1 km of (i) green areas (a park, a sports field, etc.), (ii) a crop field, (iii) a highway, a railway or an airport. The mother was also asked to state the locality in which she had lived during her pregnancy. We then queried a database (produced by the French National Institute of Statistics and Economic Studies) to obtain the population density for each locality. The localities were then arbitrarily divided into rural localities (<2 000 inhabitants/km^2) and urban localities (≥2 000 inhabitants/km^2).

2.2.4 Questions on dietary exposure. The mother was asked to state (i) her total dietary intake of the main food categories (fruits and vegetables, cereals, milk products and meat) during pregnancy (possible answers were "never or almost never", "not every day", "every day", "several times a day") and (ii) the frequency with which the consumed fruits and vegetables, milk products and meat were organic foods (with possible answers of "never", "sometimes", "most of the time ").

2.3 Birth outcomes

Neonatal data - gestational age (weeks of amenorrhea), birth weight (g), birth length (cm) and head circumference at birth (cm) - were extracted from medical records by the maternity clinic's pediatrician. Prematurity was defined as birth before 37 weeks of amenorrhea. Low birth weight or length was defined as a birth weight or length (adjusted for the mother's age, weight, and height, the rank of pregnancy, the gestational age and the baby's gender) below the 5th percentile (relative to normative data for France) [15–16]. Small head circumference was defined as a measurement (adjusted for gestational age and the baby's gender) below the 5th percentile (relative to normative data for France) [15]. Infants with missing neonatal or maternal data on adverse birth outcomes were not considered in these analyses.

2.4 Confounding factors

Factors known or suspected (on the basis of the scientific literature) to have an effect on fetal development were considered as confounding factors, i.e. the mother's age and body mass index (BMI), parity, diabetes, hypertension, tobacco use, alcohol use, drug abuse, socioprofessional category (educational level and type of work), and the baby's gender.

2.5 Data processing

The study data (questionnaire data, birth outcomes and covariates) were recorded using Sphinx plus^2 software (version 5.1.0.6, Le Sphinx Développement, Chavanod, France). Some questionnaire answers were combined, in order to obtain new, composite variables that were more relevant than those obtained from single questions. Thus, the composite variables used in the assessment of intrauterine exposure to pesticides included maternal or paternal occupational exposure, maternal domestic exposure to human antiparasitics, veterinary antiparasitics, and

Table 1. Maternal characteristics in the MecoExpo cohort (n = 993).

Variables	
Age (year)	Mean (SD)
	29 (5.2)
Parity:	n(%)
Primiparous	409 (41.7)
Multiparous	573 (58.3)
Missing data	*11*
Maternal diabetes during pregnancy:	n(%)
Yes	62 (6.2)
No	931 (93.8)
Maternal hypertension during pregnancy:	n(%)
Yes	47 (4.7)
No	946 (95.3)
Maternal smoking during pregnancy:	n(%)
Yes	293 (29.5)
No	700 (70.5)
Maternal educational level:	n(%)
≤High school completed	504 (53.7)
>High school completed	434 (46.3)
Missing data	*55*
Working during pregnancy:	n(%)
Yes	671 (67.6)
No	322 (32.4)
Place of residence during pregnancy:	n(%)
Rural locality	774 (85.9)
Urban locality	127 (14.1)
Missing data	*92*

Table 2. Newborn characteristics in the MecoExpo cohort (n = 993).

Variables	Mean (SD)	n (%)
Gestational age (weeks)	39.3 (1.3)	924 (93.1)
Missing data		*69 (6.9)*
Birth weight (g)	3340 (492)	932 (93.9)
Missing data		*61 (6.1)*
Birth length (cm)	49.5 (2.1)	922 (92.8)
Missing data		*71 (7.2)*
Head circumference at birth (cm)	34.4 (1.4)	911 (91.7)
Missing data		*82 (8.3)*
Gender:		
male		508 (51.2)
female		485 (48.8)
Prematurity:		
Yes		24 (2.6)
No		900 (97.4)
Missing data		*69*
Low birth weight:		
Yes		788 (94.9)
No		42 (5.1)
Missing data		*163*
Low birth length:		
Yes		56 (6.8)
No		764 (93.2)
Missing data		*173*
Small head circumference:		
Yes		34 (4.0)
No		822 (96.0)
Missing data		*137*

domestic pesticides for houseplants or garden plants, environmental exposure (the home's proximity to a crop fields, green areas, highways, railways or airports) and dietary exposure (consumption of fruits and vegetables, cereals, milk products and meat; consumption of organic fruits and vegetables, etc.).

2.6 Statistical analyses

The MecoExpo cohort's demographics and intrauterine exposure to pesticides were first characterized in a descriptive analysis. Quantitative parameters (gestational age, birth weight, etc.) were expressed as the mean and the standard deviation (SD). Qualitative parameters (prematurity, low birth weight, maternal occupational exposure to pesticides, etc.) were expressed as the number and the percentage of the study population.

Multivariate linear regression and logistic regression analyses were used to study the putative associations between birth outcomes and the characteristics of *in utero* exposure to pesticides. Using forward selection, covariates with a p-value<0.20 in a univariate analysis were fed into in the multivariate analyses. The risk of an adverse birth outcome (prematurity, small head circumference, and low birth weight and low birth length) was expressed as an odds ratio (OR) [95% confidence interval (CI)]. The 95% CI was calculated according to Woolf's method, with an

alpha risk of 0.05. All statistical analyses were performed with SPSS software (V.20.0, Chicago, IL, USA).

Results

3.1 Characteristics of the study population

A total of 993 mother-newborn pairs were included in the MecoExpo cohort. Given that some questionnaire answers or clinical values were missing, the total n for some variables was below 993 and so the corresponding missing data rates are also reported.

The characteristics of the MecoExpo study population are summarized in Table 1 and Table 2. The mean (SD) maternal age was 29 (5.2) and the mean pre-pregnancy BMI was 24.4 (5.5) kg/m^2. Six percent of the mothers were diabetic and less than 5% had arterial hypertension during pregnancy. Forty-two percent of the women were primiparous, 46% had completed high school and 68% had been working during their pregnancy. About 30% of the mothers stated smoking during pregnancy; the corresponding values for drinking alcohol and illicit drug use were 3% and 1%, respectively. Overall, 86% of the women lived in rural localities (<2 000 inhabitants/km^2).

Table 3. Characteristics of prenatal exposure to pesticides (n = 993).

Variables	n (%)
Occupational exposure to pesticides	
Mother:	
Yes	43 (4.3)
No	950 (95.7)
Father:	
Yes	86 (8.7)
No	907 (91.3)
Maternal domestic exposure to pesticides	
Human antiparasitics:	
Yes	328 (33.0)
No	665 (67.0)
Veterinary antiparasitics:	
Yes	215 (21.7)
No	776 (78.3)
Missing data	*2*
Pesticides against insects:	
Yes	127 (13.0)
No	850 (87.0)
Missing data	*16*
Pesticides for indoor plants:	
Yes	37 (3.7)
No	956 (96.3)
Pesticides for outdoor plants:	
Yes	166 (17.0)
No	813 (83.0)
Missing data	*14*
Maternal environmental exposure to pesticides	
Proximity to a crop field (<1 km):	
Yes	527 (58.2)
No	378 (41.8)
Missing data	*88*
Proximity to a green area:	
Yes	707 (72.7)
No	266 (27.3)
Missing data	*20*
Proximity to a highway, railway or airport:	
Yes	366 (37.4)
No	612 (62.6)
Missing data	*15*

Table 4. Characteristics of prenatal exposure to pesticides (n = 993) (continued).

Variables	n (%)
Maternal dietary exposure to pesticides (total food intake)	
Fruits and vegetables:	
Never or almost never	18 (1.8)
Not every day	319 (32.3)
Every day	362 (36.6)
Several times a day	289 (29.3)
Missing data	*5*
Cereals:	
Never or almost never	4 (0.4)
Not every day	173 (17.6)
Every day	532 (54.1)
Several times a day	274 (27.9)
Missing data	10
Dairy products:	
Never or almost never	8 (0.8)
Not every day	111 (11.3)
Every day	430 (43.7)
Several times a day	436 (44.3)
Missing data	*8*
Meat:	
Never or almost never	23 (2.4)
Not every day	365 (37.5)
Every day	489 (50.2)
Several times a day	97 (10.0)
Missing data	*19*
Maternal dietary exposure to pesticides (organic food intake)	
Fruits and vegetables:	
Never	518 (52.4)
Sometimes	375 (38.0)
Most of the time	95 (9.6)
Missing data	*5*
Dairy products:	
Never	654 (66.3)
Sometimes	265 (26.9)
Most of the time	68 (6.9)
Missing data	*6*
Meat:	
Never	827 (84.0)
Sometimes	137 (13.9)
Most of the time	20 (2.0)
Missing data	*9*

3.2 Characteristics of in utero exposure to pesticides

The main characteristics of prenatal exposure to pesticides (as assessed by the mothers' questionnaire data) are summarized in Table 3 and 4.

3.2.1 Occupational exposure to pesticides. 43 mothers and 86 fathers (4.3% and 8.7% of the total sample, respectively) worked in an occupation in which there was potential for exposure to pesticides. Fifteen mothers and 42 fathers (1.5% and 4.2% of the total sample, respectively) had agricultural occupations.

3.2.2 Domestic exposure to pesticides. 33% of the mothers were exposed to human antiparasitics, with 22% exposed to veterinary antiparasitics, 13% exposed to domestic insecticides, 4% exposed to pesticides for houseplants and 17% exposed to pesticides for garden plants.

Table 5. Multivariate associations between factors related to *in utero* exposure to pesticides and gestational age and prematurity.

Variable	Gestational age (weeks)		Prematurity	
	n	Mean (SD)	Cases/ controls	OR [95% CI]
Total sample	924	-	24/900	-
Occupational exposure				
Mother:				
Yes	21	39.2 (1.4)	1/20	NA
No	903	39.5 (1.3)	23/880	Reference
Father:				
Yes	80	38.8 (1.7)**	6/74	3.7 [1.4, 9.7]*
No	844	39.5 (1.3)	18/826	Reference
Maternal domestic exposure				
Human antiparasitics:				
Yes	307	39.6 (1.2)	7/300	0.8 [0.3, 2.0]
No	617	39.4 (1.4)	17/600	Reference
Veterinary antiparasitics:				
Yes	201	39.4 (1.4)	6/195	1.2 [0.5, 3.1]
Non	721	39.5 (1.3)	18/703	Reference
Missing data	2		0/2	-
Pesticides against insects:				
Yes	115	39.6 (1.2)	3/112	1.0 [0.3, 3.4]
No	794	39.4 (1.3)	20/774	Reference
Missing data	15	-	1/14	-
Pesticides for indoor plants:				
Yes	35	39.4 (1.2)	1/34	NA
No	889	39.5 (1.3)	23/866	Reference
Pesticides for outdoor plants:				
Yes	156	39.5 (1.2)	2/154	NA
No	754	39.5 (1.3)	21/733	Reference
Missing	14	-	1/13	-
Maternal environmental exposure				
Proximity to a crop field:				
Yes	490	39.4 (1.4)	17/473	2.1 [0.8, 5.4]
No	356	39.6 (1.3)	6/350	Reference
Missing data	78	-	1/77	-
Proximity to a green area:				
Yes	660	39.5 (1.3)	19/641	1.8 [0.6, 5.3]
No	244	39.5 (1.3)	4/240	Reference
Missing data	20	-	1/19	-
Proximity to a highway, railway or airport:				
Yes	339	39.5 (1.3)	8/331	0.9 [0.4, 2.2]
No	571	39.5 (1.3)	15/556	Reference
Missing data	14	-	1/13	-

3.2.3 Environmental exposure to pesticides. 58% of the mothers lived near to a crop field crop, 73% lived near to a green area and 37% lived near to a highway, railroad line or airport.

3.2.4 Dietary exposure to pesticides. 29% of mothers consumed fruits and vegetables "several times" a day and 44% consumed milk products "several times a day". Fifty percent consumed cereals "every day" and 50% consumed meat "every day".

Only 10% of the mothers consumed organic fruits and vegetables "most of the time", whereas the corresponding proportions organic milk products and organic meat were 7% and 2%, respectively.

3.3 Multivariate associations between fetal growth and estimated in utero exposure to pesticides

3.3.1 Gestational age and prematurity (Table 5, 6). Estimated paternal occupational exposure to pesticides was significantly associated with lower gestational age (-0.7 weeks, relative to infants whose parents were not occupationally exposed to pesticides; p = 0.006) and a higher risk of prematurity (OR [95% CI]: 3.7 [1.4, 9.7]; p = 0.02). Maternal exposure (whether occupational, domestic, environmental or dietary) was not significantly associated with either gestational age or the risk of prematurity.

3.3.2 Birth weight (as a continuous variable) and low birth (Table 7, 8). Estimated maternal occupational exposure was associated with higher risk of low birth weight (OR [95% CI]: 4.2 [1.2, 15.4]; p = 0.01). Exposure to human and veterinary antiparasitics were both significantly associated with a lower birth weight (-70 g; p = 0.02 and -85 g; p = 0.04, respectively). Exposure to pesticides for houseplants was also associated with lower birth weight (-160 g; p = 0.03). No association with dietary exposure was observed.

3.3.3 Birth length (as a continuous variable) and low birth length (Table 9, 10). The risk of low of low birth length was significantly associated with the mother's residence in an urban area (OR [95% CI]: 2.9 [1.5, 5.5]; p = 0.01). No associations with parental occupational exposure or dietary exposure were observed.

3.3.4 Birth head circumference at birth (as a continuous variable) and small head circumference (Table 11, 12). In multivariate statistical models, no significant association was found between head circumference at birth or small head circumference on one hand and the variables describing the *in utero* exposure to pesticides on the other.

Discussion

To the best of our knowledge, this study is the first one to simultaneously examine the association between all sources of *in utero* exposure to pesticides (i.e. occupational, domestic, environmental and dietary sources) and four common descriptors of fetal development (gestational age, birth weight, birth length and head circumference at birth). The prevalence of fetal growth restriction with respect to these parameters (i.e. prematurity, low birth weight, low birth length and small head circumference) was also considered.

Fetal growth is conditioned by many different environmental, genetic, metabolic, nutritional and placental factors. It is well known that adverse fetal growth is a good predictor of neonatal mortality and morbidity [17–19]. Fetal growth restriction may therefore be a determining factor for some diseases of childhood (since infants with poor head growth appear to run an increased risk of cerebral palsy, cognitive impairment and behavioral disorders) [20,21] (ii) diseases of adolescence (since extremely low birth weight may be associated with a higher prevalence of developmental delay, neurosensory impairments (seizures, visual

Table 6. Multivariate associations between factors related to *in utero* exposure to pesticides and gestational age and prematurity (continued).

Place of residence during pregnancy:				
Rural locality	714	39.4 (1.4)	21/693	Reference
Urban locality	123	39.6 (1.3)	1/122	NA
Missing data	*87*	-	*2/85*	-
Maternal dietary exposure (total food intake)				
Fruits and vegetables:				
Never or almost never	16	39.3 (1.4)	0/16	NA
Not every day	298	39.5 (1.3)	8/290	Reference
Every day	333	39.4 (1.3)	10/323	1.1 [0.4, 2.9]
Several times a day	272	39.6 (1.3)	6/266	0.8 [0.3, 2.4]
Missing data	*5*	-	*0/5*	-
Cereals:				
Never or almost never	4	39.7 (0.5)	0/3	NA
Not every day	162	39.4 (1.2)	3/159	Reference
Every day	488	39.5 (1.4)	15/473	1.7 [0.5, 5.9]
Several times a day	260	39.5 (1.3)	6/254	1.3 [0.3, 5.1]
Missing data	*79*	-	*0/79*	-
Dairy products:				
Never or almost never	7	39.6 (0.7)	0/7	NA
Not every day	105	39.4 (1.1)	2/103	Reference
Every day	399	39.4 (1.3)	10/389	1.3 [0.3, 6.1]
Several times a day	405	39.5 (1.4)	12/393	1.6 [0.4, 7.1]
Missing data	*8*	-	*0/8*	-
Meat:				
Never or almost never	22	39.5 (2.0)	1/21	NA
Not every day	338	39.5 (1.2)	4/334	Reference
Every day	458	39.4 (1.4)	16/442	3.0 [1.0, 9.1]
Several times a day	88	39.5 (1.6)	3/85	3.0 [0.7, 13.4]
Missing	*87*	-	*0/87*	-
Maternal dietary exposure (organic food intake)				
Fruits and vegetables:				
Never	481	39.4 (1.4)	13/468	0.6 [0.2, 1.9]
Sometimes	347	39.6 (1.2)	7/340	0.5 [0.1, 1.6]
Most of the time	91	39.3 (1.5)	4/87	Reference
Missing data	*5*	-	*0/5*	-
Dairy products:				
Never	607	39.5 (1.3)	17/590	1.8 [0.2, 13.9]
Sometimes	247	39.5 (1.3)	6/241	1.6 [0.2, 13.3]
Most of the time	64	39.5 (1.3)	1/63	Reference
Missing data	*6*	-	*0/6*	
Meat:				
Never	773	39.5 (1.3)	21/752	Reference
Sometimes	124	39.5 (1.3)	2/122	NA
Most of the time	18	39.2 (1.4)	1/17	NA
Missing data	*78*	-	*0/78*	-

*p = 0.05;
**p = 0.01;
***p = 0.001.
NA: an odds ratio was not available because too few cases were observed (n<3).

Table 7. Multivariate associations between (i) variables related to *in utero* exposure to pesticides, (ii) birth weight (BW, in g) and low birth weight (LBW).

Variables	BW (g)		LBW	
	n	Mean (SD)	Cases/controls	OR [95% CI]
Total sample	932		42/786	
Occupational exposure				
Mother:				
Yes	19	3177 (521)	3/14	4.2 [1.2, 15.4]**
No	913	3343 (491)	39/772	Reference
Father:				
Yes	83	3213 (517)	1/96	NA
No	849	3353 (488)	41/690	Reference
Domestic exposure				
Humans antiparasitic:				
Yes	309	3397 (477)*	11/241	0.8 [0.4, 1.6]
No	623	3312 (497)	31/545	Reference
Veterinary antiparasitic:				
Yes	206	3286 (520)*	10/177	1.1 [0.5, 2.3]
No	724	3358 (482)	31/608	Reference
Missing data	*2*	*-*	*1/1*	*-*
Pesticides against insects:				
Yes	120	3373 (472)	4/95	0.8 [0.3, 2.2]
No	796	3338 (493)	37/680	Reference
Missing data	*16*	*-*	*1/11*	*-*
Pesticides for indoor plants:				
Yes	33	3186 (464)**	2/26	NA
No	899	3346 (492)	40/760	Reference
Pesticides for outdoor plants:				
Yes	156	3369 (479)	2/136	NA
No	762	3337 (492)	40/638	Reference
Missing data	*14*	*-*	*0/12*	*-*
Environmental exposure				
Proximity to a crop field:				
Yes	490	3328 (491)	25/411	1.2 [0.6, 2.4]
No	359	3352 (486)	15/304	Reference
Missing data	*83*	*-*	*2/71*	*-*
Proximity to a green area:				
Yes	663	3365 (498)	26/564	0.6 [0.3, 1.1]
No	250	3275 (470)	16/204	Reference
Missing data	*19*	*-*	*0/18*	*-*
Proximity to a highway, railway or airport:				
Yes	344	3380 (523)	15/259	0.9 [0.5, 1.8]
No	573	3317 (470)	27/484	Reference
Missing data	*15*	*-*	*0/13*	*-*

problems), learning disabilities and hyperactivity) and diseases of adulthood (hypertension, diabetes and hyperlipidemia) [20,21]. Unfortunately, these factors are only seldom analyzed in the literature.

The inconsistency of the literature results may be explained (at least in part) by the fact that studies evaluating the association between pesticide exposure and fetal growth have used several different definitions of low birth weight. In fact, studies considering the proportion of small-for-gestational-age infants generally consider the 10th percentile and include only the gestational age (plus, in some cases, the baby's gender) as an adjustment parameter. In contrast, our study low birth weight or length as being below the 5th percentile after adjustment for various maternal and neonatal characteristics (the mother's age, weight,

Table 8. Multivariate associations between (i) variables related to *in utero* exposure to pesticides, (ii) birth weight (BW, in g) and low birth weight (LBW) (continued).

Dietary exposure (Total food intake)				
Fruits and vegetables:				
Never or almost never	17	3152 (501)	2/13	Reference
Not every day	298	3316 (481)	16/250	0.4 [0.1, 2.0]
Every day	342	3329 (483)	11/289	0.3 [0.1, 1.3]
Several times a day	270	3388 (511)	13/230	0.4 [0.1, 1.8]
Missing data	5	-	0/4	-
Cereals:				
Never or almost never	4	3525 (527)	0/4	NA
Not every day	164	3360 (466)	7/141	Reference
Every day	495	3321 (489)	22/409	0.6 [0.3, 1.5]
Several times a day	259	3360 (508)	12/227	1.1 [0.4, 2.8]
Missing data	71	-	1/5	-
Dairy products:				
Never or almost never	8	3098 (431)	1/5	NA
Not every day	103	3251 (473)	8/86	Reference
Every day	408	3345 (475)	17/341	0.5 [0.1, 4.5]
Several times a day	405	3357 (510)	16/348	0.3 [0.0, 2.3]
Missing data	8		0/6	
Meat:				
Never or almost never	22	3275 (520)	1/20	NA
Not every day	340	3329 (465)	19/288	Reference
Every day	458	3332 (494)	19/388	0.7 [0.4, 1.4]
Several times a day	93	3393 (557)	3/78	0.6 [0.2, 2.0]
Missing data	80	-	0/12	-
Dietary exposure (organic food intake)				
Fruits and vegetables:				
Never	483	3304 (509)	23/399	1.6 [0.5, 5.4]
Sometimes	354	3399 (472)	16/300	1.5 [0.4, 5.2]
Most of the time	90	3283 (453)	3/83	Reference
Missing data	5	-	0/5	-
Dairy products:				
Never	617	3327 (505)	32/508	1.4 [0.7, 2.8]
Sometimes	245	3367 (469)	10/216	Reference
Most of the time	64	3346 (439)	0/4	NA
Missing data	6	-	0/6	-
Meat:				
Never	777	3329 (490)	39/657	Reference
Sometimes	126	3422 (492)	2/108	NA
Most of the time	20	3238 (518)	1/16	NA
Missing data	70	-	0/5	-

*P<0.05;
**P<0.01.
BW: birth weight; LBW: Low birth weight; NA: an odds ratio was not available because too few cases were observed (n<3).

and height, the rank of birth, the gestational age and the baby's gender), as recommended by the AUDIPOG study [15–16].

In the present study, we considered almost all the confounding factors known or suspected in the literature to have an effect on fetal development. However, some of these factors (such as the consumption of alcohol and/or illicit drugs during pregnancy) could not be analyzed further because of the low number of affirmative replies. For ethical reasons, we chose not to probe a number of other factors (such as the parent's financial situation and ethnic origin); this might constitute a source of study bias. Moreover, the fact that pesticide levels were not measured in this study might constitute another source of bias; however, data

Table 9. Multivariate associations between (i) variables related to *in utero* exposure to pesticides, (ii) birth length (BL, in cm) and low birth length).

Variable	BL (cm)		LBL	
	n	Mean (SD)	Cases/controls	OR [95% CI]
Total sample	922		56/764	
Occupational exposure				
Mother:				
Yes	19	49.5 (1.9)	2/15	NA
No	903	49.5 (2.1)	54/749	Reference
Father:				
Yes	79	49.3 (2.0)	1/67	NA
No	843	49.5 (2.1)	55/697	Reference
Domestic exposure				
Humans antiparasitics:				
Yes	309	49.8 (2.0)	19/252	1.0 [0.6, 1.9]
No	613	49.3 (2.1)	37/512	Reference
Veterinary antiparasitics:				
Yes	201	49.3 (2.1)	9/173	0.7 [0.3, 1.4]
No	719	49.5 (2.1)	46/590	Reference
Missing data	*2*	*-*	*1/1*	*-*
Pesticides against insects:				
Yes	115	49.5 (1.9)	8/88	1.3 [0.6, 2.9]
No	793	49.5 (2.1)	46/666	Reference
Missing data	*14*	*-*	*2/10*	*-*
Pesticides for indoor plants:				
Yes	35	49.2 (2.0)	4/26	2.2 [0.7, 6.5]
No	887	49.5 (2.1)	52/738	Reference
Pesticides for outdoor plants:				
Yes	157	49.7 (2.2)	9/130	0.9 [0.4, 1.9]
No	752	49.4 (2.1)	47/623	Reference
Missing data	*13*	*-*	*0/11*	*-*
Environmental exposure to pesticides				
Proximity to a crop field (<1 km):				
Yes	491	49.6 (2.1)	32/404	1.0 [0.6, 1.8]
No	350	49.3 (2.0)	22/289	Reference
Missing data	*81*	*-*	*2/71*	*-*
Proximity to a green area:				
Yes	655	49.6 (2.2)	42/541	1.1 [0.6, 2.1]
No	249	49.2 (1.9)	14/206	Reference
Missing data	*18*	*-*	*0/17*	*-*
Proximity to a highway, railway or airport:				
Yes	339	49.6 (2.1)	23/276	1.2 [0.7, 2.9]
No	570	49.4 (2.1)	33/477	Reference
Missing data	*13*	*-*	*0/11*	*-*
Level of urbanization:				
Rural	726	49.5 (2.1)	34/606	Reference
Urban	119	49.1 (2.0)	15/93	2.9 [1.5, 5.5]*
Missing data	*77*	*-*	*7/65*	*-*

Table 10. Multivariate associations between (i) variables related to *in utero* exposure to pesticides, (ii) birth length (BL, in cm) and low birth length) (continued).

Dietary exposure (total food intake)			
Fruits and vegetables:			
Never or almost never	18	49.3 (2.2)	1/15	Reference
Not every day	301	49.4 (2.0)	19/253	1.1 [0.1, 9.0]
Every day	329	49.4 (2.2)	19/268	1.1 [0.1, 8.5]
Several times a day	269	49.7 (2.1)	17/224	1.1 [0.1, 9.2]
Missing data	5	-	0/4	-
Cereals:			
Never or almost never	4	50.2 (1.7)	0/4	NA
Not every day	159	49.4 (2.1)	11/132	Reference
Every day	493	49.4 (2.1)	27/402	0.8 [0.4, 1.7]
Several times a day	256	49.6 (2.2)	18/218	1.0 [0.5, 2.2]
Missing data	81	-	0/8	-
Dairy products:			
Never or almost never	7	48.1 (1.2)	0/5	NA
Not every day	105	49.3 (2.1)	8/87	Reference
Every day	396	49.4 (2.0)	21/328	0.7 [0.3, 1.6]
Several times a day	407	49.6 (2.2)	27/339	0.9 [0.4, 2.0]
Missing data	7	-	0/5	-
Meat:			
Never or almost never	22	49.3 (2.2)	0/20	NA
Not every day	338	49.4 (2.1)	25/279	Reference
Every day	457	49.5 (2.1)	24/382	0.7 [0.4, 1.3]
Several times a day	87	49.6 (2.3)	6/71	0.9 [0.4, 2.4]
Missing data	89	-	1/12	-
Dietary exposure (organic food intake)			
Fruits and vegetables:			
Never	484	49.3 (2.2)	36/388	1.5 [0.6, 4.0]
Sometimes	345	49.7 (1.9)	15/330	0.6 [0.3, 2.1]
Most of the time	88	49.4 (2.1)	5/83	Reference
Missing data	5	-	0/4	-
Dairy products:			
Never	611	49.4 (2.1)	39/498	1.1 [0.6, 2.0]
Sometimes	242	49.6 (2.0)	15/207	2.0 [0.4, 9.0]
Most of the time	62	49.8 (2.0)	2/55	Reference
Missing data	3	-	0/4	-
Meat:			
Never	769	49.4 (2.1)	47/642	Reference
Sometimes	126	49.6 (1.9)	7/102	0.9 [0.4, 2.1]
Most of the time	18	49.0 (2.1)	1/14	NA
Missing data	80	-	1/6	-

*P<0.05;
**P<0.01.
BL: birth length, LBL: low birth length. NA: an odds ratio was not available because too few cases were observed (n<3).

obtained from maternal questionnaires were used as a proxy for pesticide exposure.

4.1 Occupational exposure to pesticides

Our results revealed that 4.3% of the mothers and 8.7% of the fathers had an occupation that potentially involved pesticide exposure. We found that 1.5% of the mothers and 4.2% of the fathers worked in agriculture (i.e. 2.9% of all parents). The literature studies differ in terms of their locations, study populations, exposure assessment methods and fetal parameters - making it difficult to compare the respective results. Our results highlighted an association between self-reported maternal occu-

Table 11. Multivariate associations between (i) variables related to *in utero* exposure to pesticides, (ii) head circumference at birth (HCB, in cm) and (iii) small head circumference at birth.

Variable	HCB (cm)		SHCB	
	N	Mean (SD)	Cases/controls	OR [95% CI]
Total sample	911		34/822	
Occupational exposure				
Mother:				
Yes	21	34.5 (1.7)	2/36	NA
No	890	34.4 (1.4)	32/786	Reference
Father:				
Yes	79	34.2 (1.6)	4/70	1.3 [0.5, 4.2]
No	832	34.4 (1.4)	30/752	Reference
Domestic exposure				
Human antiparasitics:				
Yes	305	34.5 (1.4)	9/276	0.7 [0.3, 1.5]
No	606	34.3 (1.4)	25/546	Reference
Veterinary antiparasitics:				
Yes	193	34.3 (1.5)	8/175	1.1 [0.5, 2.6]
No	717	34.4 (1.4)	26/646	Reference
Missing data	*1*	*-*	*0/1*	*-*
Pesticides against insects:				
Yes	115	34.4 (1.3)	7/99	1.9 [0.8, 4.4]
Non	781	34.4 (1.4)	27/709	Reference
Missing data	*15*	*-*	*0/14*	*-*
Pesticides for indoor plants:				
Yes	33	34.2 (1.4)	2/29	NA
No	878	34.4 (1.4)	32/793	Reference
Pesticides for outdoor plants:				
Yes	149	34.6 (1.4)	8/135	1.5 [0.7, 3.5]
No	748	34.4 (1.4)	26/674	Reference
Missing data	*14*	*-*	*0/13*	*-*
Environmental exposure				
Proximity to a crop field:				
Yes	487	34.4 (1.4)	20/437	1.3 [0.6, 2.8]
No	345	34.4 (1.4)	11/314	Reference
Missing data	*79*	*-*	*3/71*	*-*
Proximity to a green area:				
Yes	644	34.5 (1.4)	21/586	0.7 [0.3, 1.3]
No	249	34.2 (1.4)	12/219	Reference
Missing data	*18*	*-*	*1/17*	*-*
Proximity to a highway, railway or airport:				
Yes	332	34.4 (1.4)	12/301	0.9 [0.5, 1.9]
No	565	34.5 (1.4)	22/508	Reference
Missing data	*14*	*-*	*0/13*	*-*

pational exposure to pesticides during pregnancy and the risk of low birth weight. We also found an association between estimated paternal occupational exposure to pesticides on one hand and low gestational age and the risk of the prematurity on the other. However, it is important to note that our cohort was recruited from the general population, in which only some of the parents were occupationally exposed to pesticides. This situation contrasts with literature studies of specifically exposed cohorts of farmers or other agricultural workers. The retrospective study of the general population performed in the USA by Savitz et al. [3] found an association between small-for-gestational-age status and self-reported occupational parental exposure to pesticides. In Colombia, the study by Restrepo et al. [5] revealed a moderately strong relationship between an increased risk of prematurity and reported

Table 12. Multivariate associations between (i) variables related to *in utero* exposure to pesticides, (ii) head circumference at birth (HCB, in cm) and (iii) small head circumference at birth (continued).

Place of residence during pregnancy:				
Rural locality	722	34.4 (1.4)	24/649	Reference
Urban locality	114	34.4 (1.5)	9/102	2.4 [1.1, 5.3]
Missing data	*75*	-	*1/71*	-
Dietary exposure (total food intake)				
Fruits and vegetables:				]
Never or almost never	18	34.0 (1.3)	0/16	NA
Not every day	292	34.4 (1.4)	16/263	Reference
Every day	334	34.4 (1.4)	11/297	0.6 [0.3, 1.3]
Several times a day	262	34.5 (1.5)	7/241	0.5 [0.2, 1.2]
Missing data	*5*	-	*0/5*	-
Cereals:				
Never or almost never	4	34.7 (2.2)	0/4	NA
Not every day	162	34.4 (1.4)	5/148	Reference
Every day	488	34.4 (1.4)	24/428	2.6 [1.0, 6.9]
Several times a day	247	34.5 (1.4)	5/232	0.6 [0.2, 2.3]
Missing data	*92*	-	*0/10*	-
Dairy products:				
Never or almost never	7	33.1 (1.1)	2/4	Reference
Not every day	104	34.3 (1.3)	5/94	0.1 [0.0, 0.8]
Every day	395	34.4 (1.4)	14/356	0.1 [0.0, 0.5]
Several times a day	397	34.4 (1.5)	13/360	0.1 [0.0, 0.4]
Missing data	*8*	-	*0/8*	-
Meat:				
Never or almost never	22	34.0 (1.4)	1/20	NA
Not every day	336	34.5 (1.4)	11/302	Reference
Every day	449	34.4 (1.4)	16/409	1.1 [0.5, 2.3]
Several times a day	85	34.4 (1.6)	6/73	2.3 [0.8, 6.3]
Missing	*101*	-	*0/18*	-
Dietary exposure (organic food intake)				
Fruits and vegetables:				
Never	481	34.2 (1.4)	25/425	2.6 [1.1, 6.1]
Sometimes	340	34.6 (1.4)	7/312	Reference
Most of the time	85	34.4 (1.3)	2/80	NA
Missing data	*5*	-	*0/5*	-
Dairy products:				
Never	612	34.3 (1.4)	25/549	1.3 [0.3, 5.6]
Sometimes	231	34.6 (1.5)	7/210	1.0 [0.2, 4.7]
Most of the time	62	34.5 (1.3)	2/57	Reference
Missing data	*2*	-	*0/6*	-
Meat:				
Never	762	34.4 (1.4)	31/688	Reference
Sometimes	123	34.7 (1.5)	3/110	0.6 [0.2, 2.0]
Most of the time	17	34.1 (1.2)	0/15	NA
Missing data	*91*	-	*0/9*	-

HCB: head circumference at birth. SHCB: small head circumference at birth. NA: an odds ratio was not available because too few cases were observed (n<3).

occupational exposure to pesticides among female workers and the wives of male workers in floriculture. In contrast, the study in agricultural areas in Norway by Kristensen et al. [4] found that the rates of prematurity and small-for-gestational-age status were lower for farmers than for non-farmers. In the latter study, pesticide exposure was assessed by means of national census data on indicators such as farm size, the number and types of livestock, and so on.

Despite some small discrepancies between the above-mentioned results, there appears to be a general consensus in which maternal and/or paternal occupational exposure to pesticides is associated with adverse effects on fetal development and adverse birth outcomes.

4.2 Domestic exposure to pesticides

Our study also considered domestic pesticides but did not focus on any one substance or group of substances in particular (in contrast to several literature studies). We found that self-reported maternal exposure to pesticides for houseplants was correlated with lower birth weight. In contrast, we did not find any association between maternal exposure to pesticides for outdoor plants and poor fetal growth. Our findings are consistent with Petit et al.'s [14] results on insecticide use on indoor plants in the Brittany region of western France but not with their data on use on garden plants. In fact, Petit et al.'s retrospective study found that higher self-reported *in utero* exposure to household insecticides for use on plants (and especially outdoor plants) was related to lower birth weight and head circumference [14].

We did not find any association between fetal growth and maternal exposure to domestic insecticides. Our findings are consistent with those of Petit et al. [14] regarding the association between residential use of insect control on one hand and birth weight and head circumference at birth on the other hand. Similarly, our results are in line with the reports by Berkowitz et al. [8] and Whyatt et al. [9] in the USA, which did not find any association between self-reported use of domestic pesticides, prenatal personal ambient air or umbilical cord blood levels of some domestic insecticides on one hand and birth weight, birth length or head circumference at birth on the other. The questionnaire used in these two studies of the same cohort probed domestic use of pesticides against three groups of pests only (cockroaches, rodents and others). In the retrospective study performed by Savitz et al. [3] in the USA, self-reported exposure to household pesticides was associated with the risk of small-for-gestational-age status but not with prematurity. The questionnaire used in Savitz et al.'s study [1] probed exposure very generally via a single question on whether or not the mother and/or the father had been exposed to household pesticides.

In the present study, we found that maternal exposure to veterinary antiparasitics was associated with a lower birth weight. To the best of our knowledge, our study is the first to have assessed the relationship between maternal exposure to veterinary antiparasitics and fetal growth. In contrast, we did not find any association between fetal growth and maternal exposure to human antiparasitics.

4.3 Environmental exposure to pesticides

In our questionnaire, environmental exposure was assessed as self-reported proximity (within 1 km) of the mother's home during pregnancy to areas in which pesticides are usually widely applied (crop fields, green areas and transportation networks (highways, railroads and airports)). We did not observed any association between fetal growth and this proximity, in agreement with the results of the study by Petit et al. [14] in France; the latter researchers did not find any association between agricultural activities in the mother's place of residence during early pregnancy (based on data on agricultural activities from the national census) and birth weight (as a continuous variable) or the risk of low birth weight. In contrast, Xiang et al.'s [6] study in the USA (based on remote sensing and a geographic information system (GIS)) found an association between low birth weight and total crop production area within a 300 m buffer zone around the mother's residence. However, Xiang et al. [6] did not examine fetal growth parameters other than low birth weight. However, it is possible that the mothers in the present MecoExpo study (especially those living on the outskirts of cities) may have over-estimated the distance between their residence and crop fields.

4.4 Dietary exposure to pesticides

We did not find a significant association between fetal growth on one hand and overall or organic food intake on the other. Our questionnaire asked the mother about her eating behavior during a period (pregnancy) in which the diet is very likely to change. We cannot rule out the presence of recall bias, since the questionnaire was completed after delivery.

In conclusion, our present results demonstrate that both maternal and paternal occupational exposure to pesticides during pregnancy is associated with low gestational age and a greater risk of prematurity and low birth weight. Maternal domestic exposure to pesticides for houseplants and to veterinary antiparasitics was associated with low birth weight. Our questionnaire served as a wide-ranging tool for characterizing all the possible sources of prenatal exposure to pesticides. In the future, we could better characterize the exposure pathways and the nature of the compounds involved by complementing our questionnaire-based data with data generated by tools such as GISs and environmental or biological assays of pesticide residues.

Acknowledgments

We thank the parents who participated in this study, the staff at the maternity clinics and the staff from Picardy's perinatal networks. We also wish to thank the *Biobanque de Picardie* for technical assistance. Lastly, we thank Dr. David Fraser for advice on the manuscript's English.

This work was funded by the Picardy Regional Council and the Grenelle Environment. F. Mayhoub was funded by Tishreen University (Latakia, Syria) and T. Berton was funded by the European Regional Development Fund.

Author Contributions

Conceived and designed the experiments: FM VB KT AFB SD ESB KC. Performed the experiments: FM TB CD KC. Contributed reagents/materials/analysis tools: TB KT AFB. Wrote the paper: FM VB ESB KC.

References

1. Bedos C, Cellier P, Calvet R, Barriuso E, Gabrielle B (2002) Mass transfer of pesticides into the atmosphere by volatilization from soils and plants: overview. Agronomie 22 21–33

2. Whyatt RM, Barr DB, Camann DE, Kinney PL, Barr JR, et al. (2003) Contemporary-use pesticides in personal air samples during pregnancy and blood samples at delivery among urban minority mothers and newborns. Environ Health Perspect 111(5):749–56.

3. Savitz DA, Whelan EA, Kleckner RC (1989) Self-reported exposure to pesticides and radiation related to pregnancy outcome-results from National Natality and Fetal Mortality Surveys. Public Health Rep 104(5): 473–477.

4. Kristensen P, Irgens LM, Andersen A, Bye AS, Sundheim L (1997) Gestational age, birth weight, and perinatal death among births to Norwegian farmers, 1967–1991. Am J of Epidemiol 146(4): 329–338.

5. Restrepo M, Munoz N, Day NE, Parra JE, de Romero L, et al. (1990) Prevalence of adverse reproductive outcomes in a population occupationally exposed to pesticides in Colombia. Scand J Work Environ Health 16(4): 232–238.

6. Xiang H, Nuckols JR, Stallones L (2000) A geographic information assessment of birth weight and crop production patterns around mother's residence. Environ Res 82(2): 160–167.

7. Eskenazi B, Harley K, Bradman A, Weltzien E, Jewell NP, et al. (2004) Association of in utero organophosphate pesticide exposure and fetal growth and length of gestation in an agricultural population. Environ Health Perspect 112(10): 1116–1124.

8. Berkowitz GS, Wetmur JG, Birman-Deych E, Obel J, Lapinski RH, et al. (2004) In utero pesticide exposure, maternal paraoxonase activity, and head circumference. Environ Health Perspect 112(3): 388–391.

9. Whyatt RM, Camann D, Perera FP, Rauh VA, Tang D, et al. (2005) Biomarkers in assessing residential insecticide exposures during pregnancy and effects on fetal growth. Toxicol Appl Pharmacol 206(2): 246–254.

10. Sathyanarayana S, Basso O, Karr CJ, Lozano P, Alavanja M, et al. (2010) Maternal pesticide use and birth weight in the agricultural health study. J Agromedicine 15(2): 127–136.

11. Petit C, Chevrier C, Durand G, Monfort C, Rouget F, et al. (2010) Impact on fetal growth of prenatal exposure to pesticides due to agricultural activities: a prospective cohort study in Brittany, France. Environ Health 9: 71.

12. Brucker-Davis F, Wagner-Mahler K, Bornebusch L, Delattre I, Ferrari P, et al. (2010) Exposure to selected endocrine disruptors and neonatal outcome of 86 healthy boys from Nice area (France). Chemosphere 81(2): 169–176.

13. Wohlfahrt-Veje C, Main KM, Schmidt IM, Boas M, Jensen TK, et al. (2011) Lower birth weight and increased body fat at school age in children prenatally exposed to modern pesticides: a prospective study. Environ Health 10: 79.

14. Petit C, Blangiardo M, Richardson S, Coquet F, Chevrier C, et al. (2012) Association of environmental insecticide exposure and fetal growth with a Bayesian model including multiple exposure sources: the PELAGIE mother-child cohort. Am J Epidemiol 175(11): 1182–1190.

15. Association des Utilisateurs de Dossiers Informatisés en Pédiatrie, Obstétrique et Gynécologie. Étude sur la croissance fœtale et infantile. www.audipog.net.

16. Mamelle N, Boniol M, Rivière O, Joly MO, Mellier G, et al. (2006) Identification of newborns with Fetal Growth Restriction (FGR) in weight and/or length based on constitutional growth potential. Eur J Pediatr 165: 717–725.

17. McCormick MC (1985) The contribution of low birth weight to infant mortality and childhood morbidity. N Engl J Med 312(2): 82–90.

18. Kramer MS, Olivier M, McLean FH, Willis DM, Usher RH (1990) Impact of intrauterine growth retardation and body proportionality on fetal and neonatal outcome. Pediatrics 86(5): 707–713.

19. McIntire DD, Bloom SL, Casey BM, Leveno KJ (1999) Birth weight in relation to morbidity and mortality among newborn infants. N Engl J Med 340(16): 1234–1238.

20. Yanney M, Marlow N (2004) Paediatric consequences of fetal growth restriction. Semin Fetal Neonatal Med 9(5): 411–418.

21. Strauss RS, Dietz WH (1997) Effects of intrauterine growth retardation in premature infants on early childhood growth. J Pediat 130(1): 95–102.

Identification of Genomic Features in Environmentally Induced Epigenetic Transgenerational Inherited Sperm Epimutations

Carlos Guerrero-Bosagna[1,2], Shelby Weeks[1], Michael K. Skinner[1]*

1 Center for Reproductive Biology, School of Biological Sciences, Washington State University, Pullman, Washington, United States of America, **2** Department of Physics, Biology and Chemistry, Linköping University, Linköping, Sweden

Abstract

A variety of environmental toxicants have been shown to induce the epigenetic transgenerational inheritance of disease and phenotypic variation. The process involves exposure of a gestating female and the developing fetus to environmental factors that promote permanent alterations in the epigenetic programming of the germline. The molecular aspects of the phenomenon involve epigenetic modifications (epimutations) in the germline (e.g. sperm) that are transmitted to subsequent generations. The current study integrates previously described experimental epigenomic transgenerational data and web-based bioinformatic analyses to identify genomic features associated with these transgenerationally transmitted epimutations. A previously identified genomic feature associated with these epimutations is a low CpG density (<12/100bp). The current observations suggest the transgenerational differential DNA methylation regions (DMR) in sperm contain unique consensus DNA sequence motifs, zinc finger motifs and G-quadruplex sequences. Interaction of molecular factors with these sequences could alter chromatin structure and accessibility of proteins with DNA methyltransferases to alter de novo DNA methylation patterns. G-quadruplex regions can promote the opening of the chromatin that may influence the action of DNA methyltransferases, or factors interacting with them, for the establishment of epigenetic marks. Zinc finger binding factors can also promote this chromatin remodeling and influence the expression of non-coding RNA. The current study identified genomic features associated with sperm epimutations that may explain in part how these sites become susceptible for transgenerational programming.

Editor: W. Steven Ward, University of Hawaii at Manoa, John A. Burns School of Medicine, United States of America

Funding: The study was supported by an NIH NIEHS grant to MKS. The funders had no role in study design, data collection and analysis, decision to publish, or preparation of the manuscript.

Competing Interests: The authors have declared that no competing interests exist.

* E-mail: skinner@wsu.edu

Introduction

A number of environmental factors have been shown to induce the epigenetic transgenerational inheritance of disease and phenotypic variation [1,2,3,4,5,6]. The initiation of this transgenerational inheritance process involves exposure of a gestating female and the developing fetus during gonadal sex determination to environmental factors (e.g. toxicants). The exposures promote alterations in the epigenetic programming of the germline that are transmitted to subsequent generations [3,6,7]. A variety of environmental toxicants have been shown to induce the epigenetic transgenerational inheritance of disease including the fungicide vinclozolin [1,3,4], dioxin [2,6], pesticides [5,6], jet fuel hydrocarbons [8] and platicizers (i.e. bisphenol A (BPA) and phthalates) [6]. Environmentally-induced epigenetic modifications in the germline have been shown to involve DNA methylation changes that are transmitted transgenerationally [6]. These germline epigenetic modifications also induce epigenetic alterations in somatic tissues which correlate with transgenerational transcriptome changes [9] and phenotypic abnormalities [10].

Germline epigenetic transgenerational inheritance has been described in several different organisms including plants, flies, worms, rodents, and humans [3,6,11,12,13,14,15]. The role of the germline in the transgenerational process is crucial since it is the only cell that transmits genetic material and stable epigenetic marks (e.g. imprinted genes) to subsequent generations. The initiation of germline development involves a major epigenetic reprogramming through alterations in DNA methylation [16,17,18]. DNA methylation erasure takes place during the migration of primordial germ cells to the genital ridge (before colonization of the gonads), while re-methylation is initiated during gonadal sex determination in a sex specific manner [19,20]. This reprogramming of DNA methylation and the occurrence of other major epigenetic events during primordial germ cell development [21] represents a critical window of exposure for environmental factors [22]. Environmental exposures [23,24] and epigenetic alterations [25] in this developmental window have been shown to promote the epigenetic transgenerational inheritance of disease and phenotypic variation.

Previous studies have shown that different exposures produce distinct sets of transgenerationally altered differential DNA methylation regions (DMR) in male germ cells, termed epimutations [6]. Interestingly, the transgenerationally altered sperm epimutations among these different exposure groups were found to have minimal overlap [6]. The methylation status of these DMR appears to be transmitted transgenerationally in similar ways to DNA methylation transmission of imprinted genes (imprinted-like mechanism). The DMR identified in these previous studies were found to be exposure specific suggesting potential genomic features among these distinct DMR may exist. The current study was designed to elucidate the potential molecular mechanisms involved in the susceptibility of these epimutations to escape the DNA methylation erasure following fertilization and become transgenerationally programmed.

Recent studies have investigated a variety of DNA-protein interactions that are involved in the establishment of DNA methylation. For example, the presence of protein binding factors in CpG-poor regulatory regions is one feature that would influence DNA methylation [26]. The configuration of specific protein/DNA complexes, formed for example by CTCF or Sp1, can prevent local actions of *de novo* methyltransferases in the genome [27]. The presence of repeat elements has also been reported to influence the establishment of DNA methylation. For example, the presence of CTG/CAG repeats act as a DNA methylation sensitive insulator [28]. Features such as repeat element composition in genomic domains [29] or the composition of nucleotides flanking CpG sites [30] are shown to influence the susceptibility of methylation by Dnmt3 methytransferases. Chromatin marks of histone binding are also shown to correlate with genomic domains in which DNA methylation changes occur [31,32]. The current study used a bioinformatic analysis of patterns of DNA sequence (motifs) in previously identified exposure specific sets of transgenerational DMR. Observations provide insights into the DNA sequence motifs potentially involved in the establishment of these transgenerational sperm epimutations.

A variety of approaches have been used to identify DNA sequence patterns (motifs) that could be functionally relevant from the perspective of gene expression regulation [33,34]. These DNA motifs are known to serve as binding sites for transcription factors and other regulatory factors [33]. Methods to identify DNA motifs have evolved from the visual alignment of a few sequences to the use of complex algorithms and computer programs [34]. Identification of consensus sequences or position weight matrices in genomic regions characterize these DNA motifs [34]. Recently, several algorithms have been developed to identify DNA motifs in a given set of sequences and to determine if they are over-represented compared to that expected by chance [33]. The integration of these computational analyses with experimental techniques is becoming fundamental to identify genome-scale regulatory elements [35,36,37]. Examples of recent studies using motif analysis at a genomic scale include genome-wide identification of estrogen receptor binding sites [38], identification of CTCF-binding sites in the human genome [39] and identification of motifs associated with aberrant CpG island methylation [40]. The current study integrates previously described experimental epigenomic transgenerational data and web-based bioinformatic analyses to identify DNA motifs to help elucidate the molecular mechanisms involved in environmentally induced transgenerational inheritance of sperm epimutations.

Results

The main goal of the current study was to identify genomic features associated with the environmentally induced epigenetic transgenerational inherited sperm epimutations. Previously described transgenerational differential DNA methylation regions (DMR) in the rat sperm were investigated [3,6]. These DMR were identified using a methylated DNA immunoprecipitation (MeDIP) followed by genome wide promoter tiling array (Chip) for an MeDIP-Chip protocol previously described [3,6]. The web-based bioinformatics tool GLAM2 (Gapped Local Alignment of Motifs) [41], which is part of MEME suite [42], was used to identify DNA motifs associated with these transgenerational sperm epimutations. DNA motifs were built from different sets of germline transgenerational DMR derived from different environmental exposures and compared. These were then grouped based on similarities using a familial binding analysis available in the web-based tool for motif analysis termed STAMP [43]. The association tree produced showed two groups of motifs. One branch of the tree was represented only by the "environmentally induced DNA methylation motif 1″ (EDM1) (vinclozolin) previously identified [3], while another branch included motifs from the other exposures (plastics, pesticides, dioxin and jet fuel [6]). A familial binding motif representing this branch was named "environmentally induced DNA methylation motif 2″ (EDM2) (Figure 1). Interestingly, EDM1 is A/T rich while EDM2 is G rich.

The presence of EDM1 was tested in a variety of transgenerationally altered DMR from sperm and from somatic Sertoli and granulosa cells [44,45]. These transgenerational F3 generation DMR sets included vinclozolin (52 DMR) [3] dioxin (50 DMR) [2], hydrocarbons, jet fuel (33 DMR) [8], pesticide, permethrin and DEET (367 DMR) [5], plasticizers, BPA and phthalates (198 DMR) [46]. The somatic Sertoli cells and granulosa cells were obtained from F3 generation vinclozolin lineage animals. The DMR were identified with a comparative hybridization MeDIP-Chip analysis on F3 generation control versus exposures lineage cells. A subset of the vinclozolin F3 generation sperm DMR that were confirmed with bisulfite-mass spectrometry were also examined separately and termed "confirmed". A computer generated random set of DNA sequences using the same genetic features of size and promoter association was created to act as a control for the comparisons (random occurrence). The presence of EDM1 was found to be significantly increased in the vinclozolin DMR set (52 sequences) and in the "confirmed" subset of 16 sequences [3] when compared to a random occurrence set of computer-generated sequences (Figure 2). Interestingly, EDM1 incidence tended to be decreased in sperm DMR from non-vinclozolin exposures or in sets from somatic cells when compared with its occurrence in a random set of sequences. Significant decreases are observed for the sperm plastics and pesticides DMR groups and for the somatic group of Sertoli cells from the vinclozolin exposure lineage (Figure 2A). The presence of EDM2 motif was also tested against the distinct sets of DMR sequences (Figure 3). It was found that EDM2 was significantly increased in the promoter associated sperm DMR sets of dioxin, plastics and pesticides lineages, and in vinclozolin lineage Sertoli cells when compared with its occurrence in a random set of sequences. The most significant increases were in the plastics and pesticides DMR groups with an over two-fold increase in EDM2 incidence. Therefore, two different motifs were identified with EDM1 being primarily associated with vinclozolin lineage DMR and EDM2 being predominant in a number of the other exposures.

The two DNA motifs identified from the analysis of the DMR groups were then compared to a database of eukaryotic transcription factor binding sites. The top five similarities of known transcription factors binding sites for each motif are shown in Table 1. The presence of motifs of these transcription factor binding sites was then tested against the different exposure lineage sets of DMR sequences (Figure 4). Observations indicate the KROX, SP1, UF1H3-beta and ZNF219 were consistently increased in dioxin, plastics and pesticides groups. However, their incidence in the jet fuel group was variable, with significant increases of only KROX in this group. RREB1 was observed in

FAMILIAL BINDING ANALYSIS OF GLAM2 MOTIFS FROM DMR SETS

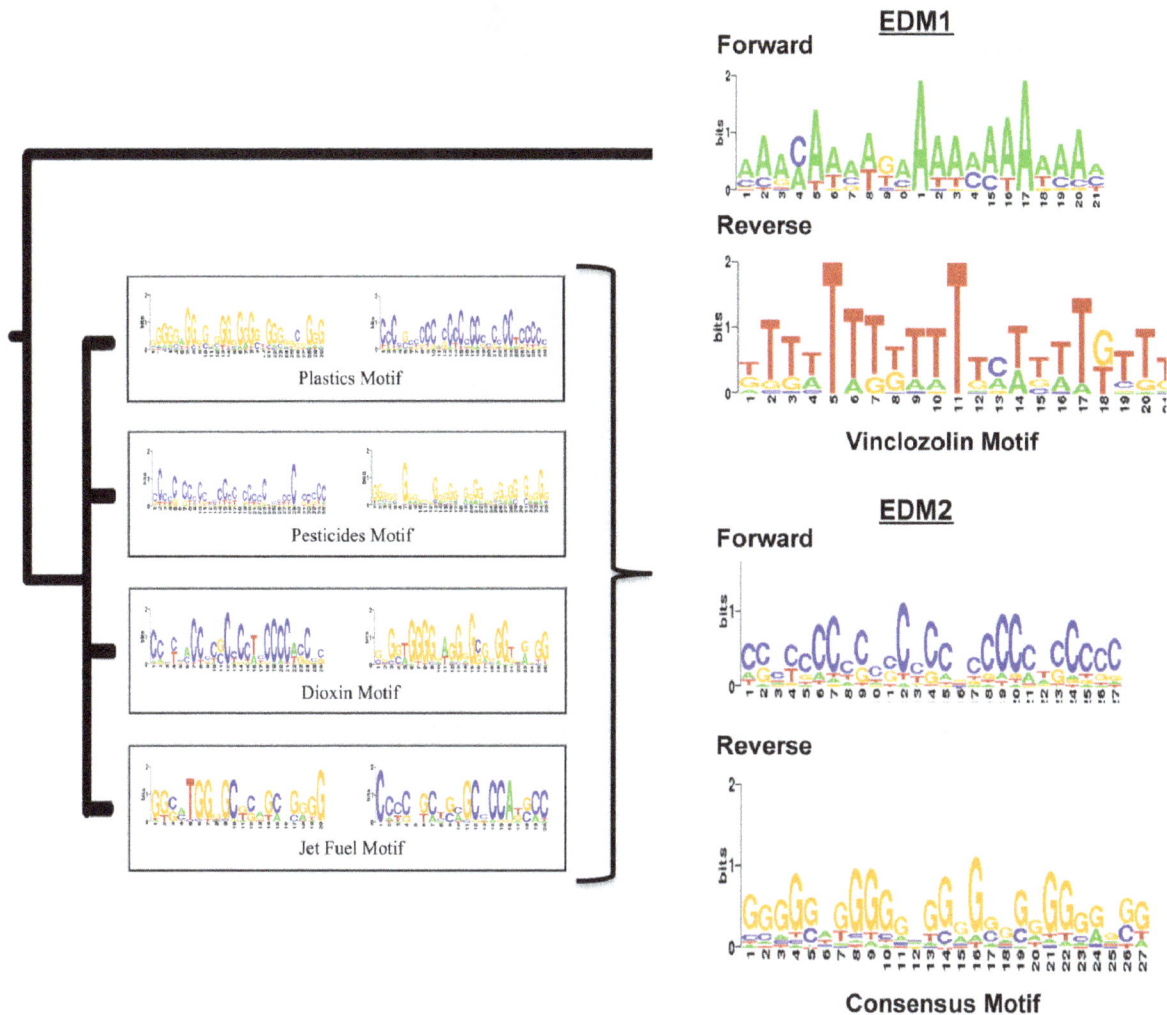

Figure 1. Exposure specific DMR set DNA sequence motifs using GLAM2 and the familial binding tree. The forward (left) and reverse (right) sequences for each motif are presented. The consensus EDM2 motif is presented.

the jet fuel, plastics and pesticides groups. With the exception of UF1H3-beta, all these transcription factors are zinc fingers. FOXP1 was significantly decreased only in the pesticides group. Alfin1 was significantly increased in the plastics and pesticides group, and had a tendency to increase in the dioxin group. Therefore, zinc finger binding sites are apparently associated with the sperm transgenerational epimutations.

The possibility that an altered density of EDM1 or EDM2 motifs could be observed in DMR sets versus the random set was examined. No significant changes were observed for the density of EDM1 or EDM2 in DMR from the exposure groups in reference to the random set (Figure 5). Given the composition of EDM1 in terms of being an A/T rich sequence and the reported role of A/T rich sequences as a recognition site for de novo DNA methylation [47], the density distribution of this feature across the different sets of DMR was also analyzed. Differences in the density distribution of A/T strings were found between the plastics, pesticides, dioxin and jet fuel groups and equivalent random sets of sequences. There is an overall reduction in the density of A/T strings in DMRs from these groups in comparison to the random set

(Figure 6 A–E; p<0.01). However, A/T string density in the vinclozolin group is similar to the random distribution (Figure 6F).

EDM2 was observed to be a G/C rich sequence. Interestingly, previous reports show that G quadruplexes associate with zinc finger binding sites [48] and have a role in restricting DNA methylation [49] to influence chromatin dependent epigenetic instability [50]. Therefore, the distribution of G-quadruplexes across the different sets of DMR was analyzed. Interesting differences were also found in the distribution of G-quadruplexes in the exposure lineage DMR sets versus the random set of sequences. In the plastics, pesticides, jet fuel and dioxin groups an overall increase in G-quadruplex density regarding the random group was observed (Figure 7 A–E; p<0.01). The vinclozolin group had a distribution comparable with the random set (Figure 7F).

Schematic visualization of the features analyzed in selected sets of transgenerational DMRs previously confirmed for the vinclozolin and other exposure DMR groups are shown in Figure 8A,B. The locations of EDM1 and EDM2 in selected sequences are shown in Figure 8. Detailed schematic representations of the locations of these features in selected DMR from the vinclozolin

A   EDM1 INCIDENCE IN TRANSGENERATIONAL SETS OF DMR

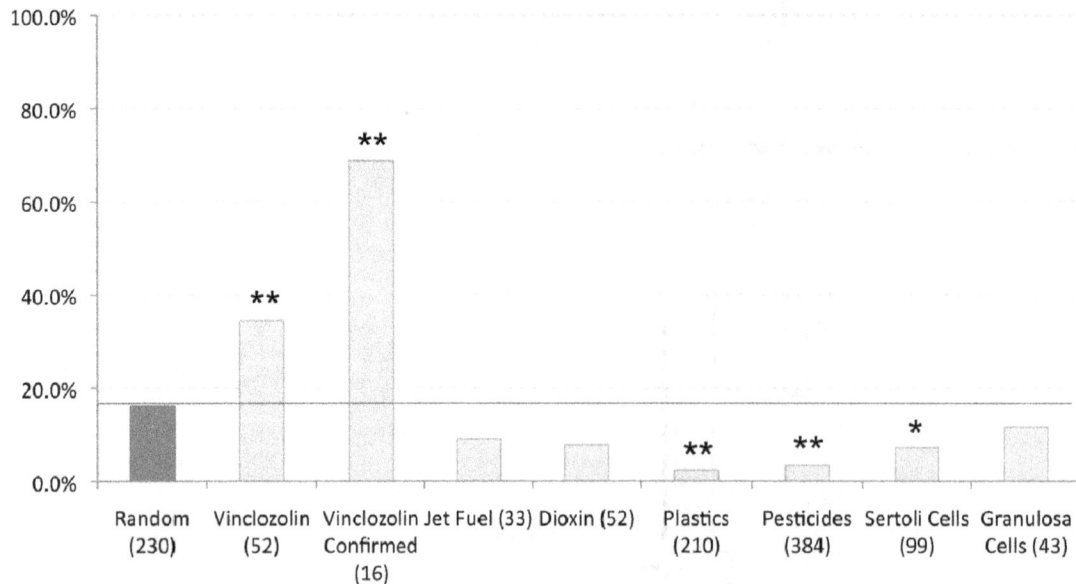

B   EDM1 RELATIVE CHANGE IN TRANSGENERATIONAL SETS OF DMR

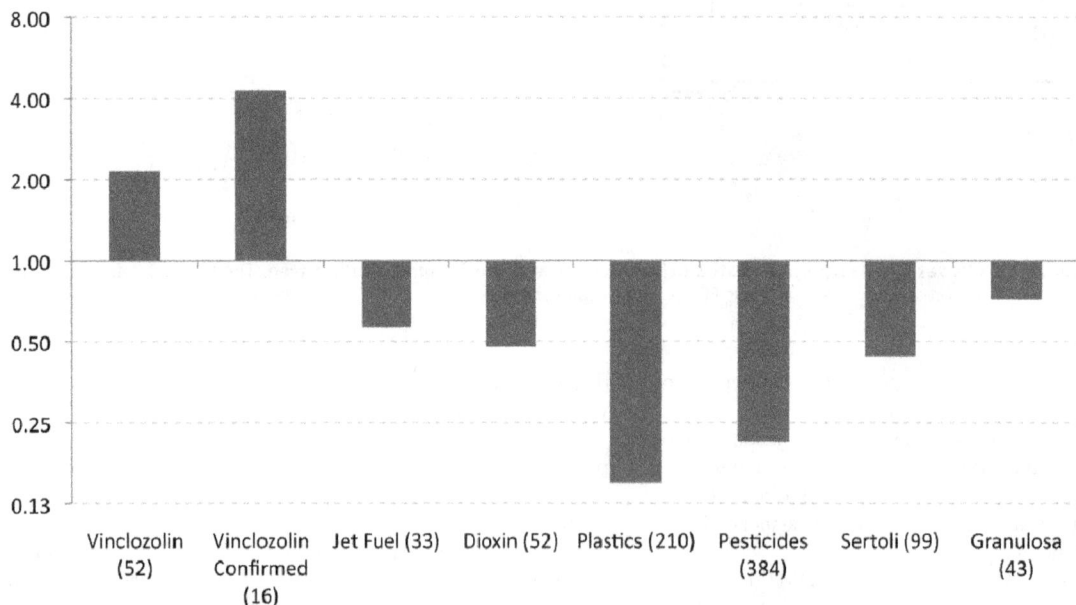

Figure 2. EDM1 incidence in exposure specific epimutation data sets. (A) Individual occurrence (percentage) of EDM1 in a variety of sets with transgenerational DMR and (B) relative change of EDM1 in these DMR. Columns with ** represent significant change with p<0.01, while columns with * represent significant change with p<0.05.

lineage exposure are shown in Figure 9. Therefore, a number of genomic features were identified and appeared to be associated with the transgenerational sperm epimutations investigated.

A follow up experiment was done to help confirm the observations regarding the genomic features associated with the transgenerational DMR (i.e. epimutations). A more recently

developed dichlorodiphenyltrichloroethane (DDT) induced trans-generational set of DMR in F3 generation sperm was investigated [51]. This DDT transgenerational DMR set was not used in the development of the EDM1 or EDM2 sequences, nor the other genomic feature identification. A comparison of the DDT DMR set with the random set of sequences demonstrated a 12.8%

A   EDM2 INCIDENCE IN TRANSGENERATIONAL SETS OF DMR

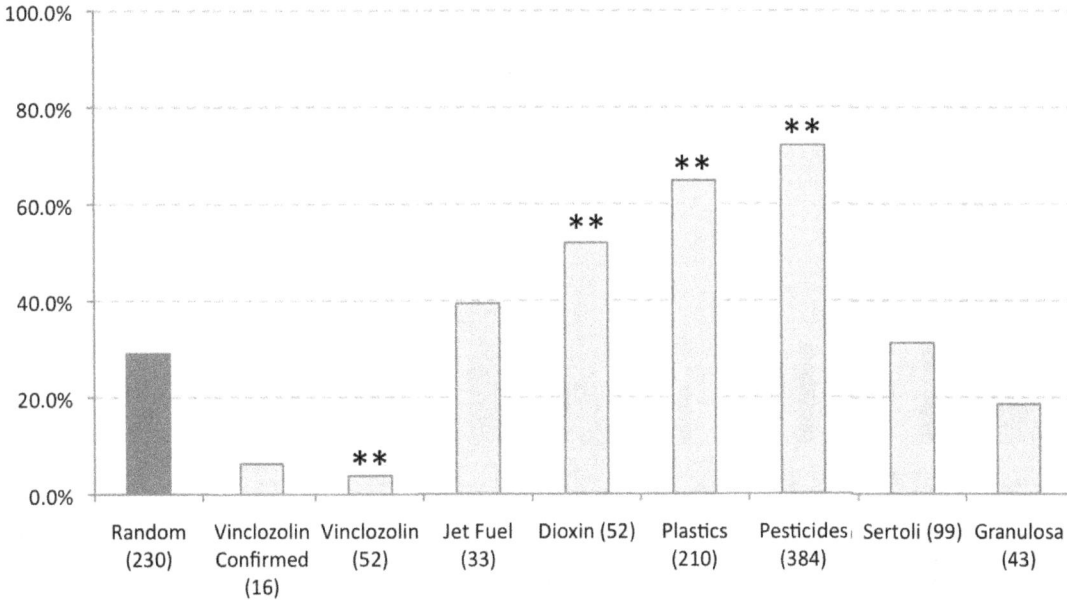

B   EDM2 RELATIVE CHANGE IN TRANSGENERATIONAL SETS OF DMR

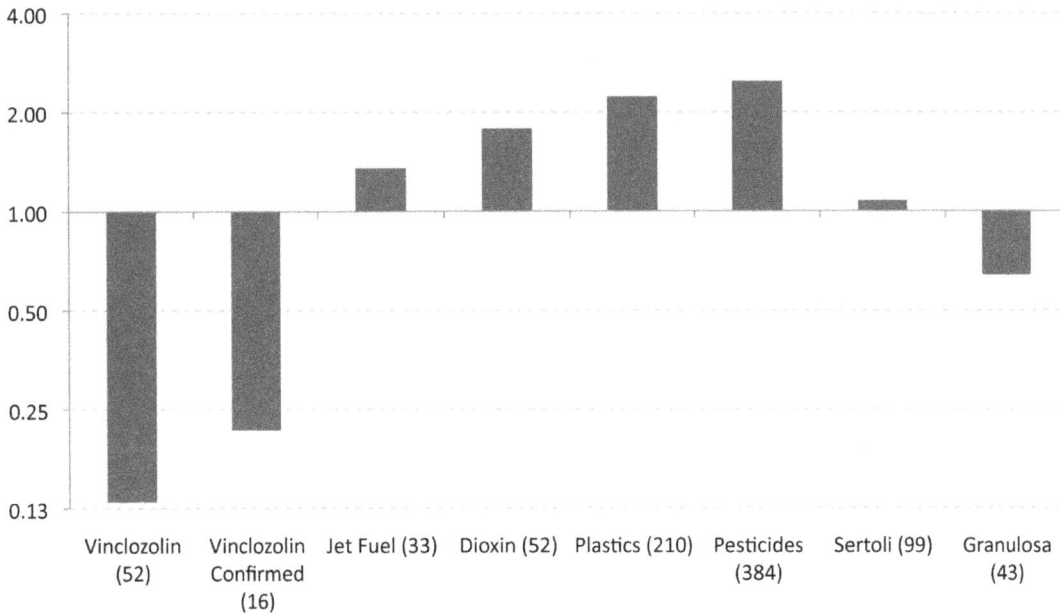

Figure 3. EDM2 incidence in exposure epimutation data sets. (A) Individual occurrence (percentage) of EDM2 in a variety of sets with transgenerational DMR and (B) relative change of EDM2 in these DMR. Columns with ** represent significant change with p<0.01, while columns with * represent significant change with p<0.05.

incidence in the presence of EDM1, which is similar to the random sequence incidence. A 64.1% incidence of EDM2 is observed, which represents a statistically significant increase (p< 0.01) compared to the random sequence occurrence. The density distribution of the A/T strings decreased significantly (p<0.01) and G-quadruplexes increased significantly (p<0.01) in the DDT DMR compared to random sequences (Figure S1). Therefore, the

patterns of incidence of EDM1, EDM2, G-quadruplexes and A/T strings in the DDT set are similar to the plastic, pesticide, jet fuel and dioxin DMR sets. Therefore, many of the same genomic features were also present in the transgenerational DDT sperm DMR [51] compared to the random sequences. Although further experiments are required to address the functional importance of these genomic features, this analysis helps confirm the presence of

INCIDENCE OF TRANSCRIPTION FACTOR MOTIFS ASSOCIATED WITH EPIMUTATIONS

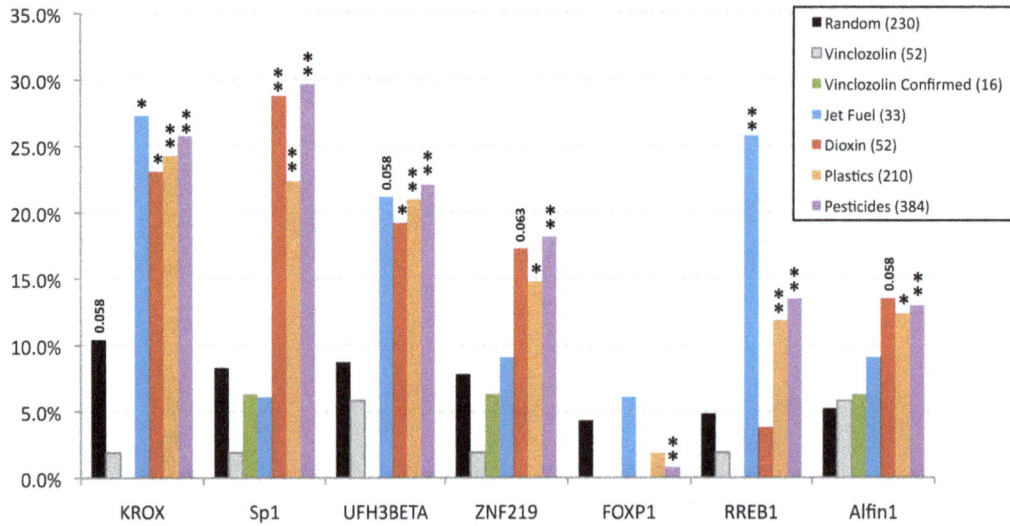

Figure 4. Incidence of consensus motifs of known transcription factor binding sites in DMR from a variety of F3 generation exposure lineage sperm DNA. Columns with (**) represent significant change with p<0.01, while columns with (*) represent significant change with p<0.05. For nearly significant changes p-values are shown in the respective column. Colored legend for specific exposure lineage DMR sets with total number indicated. The percentage incidence is indicated for the different transition factor sites.

Discussion

The phenomenon of environmentally induced epigenetic transgenerational inheritance is a germline mediated process [1,2,3,4,5,6]. Germline epigenetic marks are altered during early fetal development and these environmentally induced epigenetic modifications (epimutations) can be transmitted to subsequent generations. Although the initial genomic feature associated with all the epimutations previously identified was a low CpG density (<12 GpG/100bp), other genomic features are anticipated. The current study is designed to perform a bioinformatic analysis and

these genomic features in the environmentally induced epigenetic transgenerational inheritance of the sperm epimutations.

identify patterns of DNA sequences (motifs) in sperm and somatic cell DMR. Observations are anticipated to provide insights into the potential molecular mechanisms involved in establishing these epigenetic marks.

Identification of DNA sequence motif incidences was performed in different sets of previously identified sperm and somatic cell DMR sequences. Two motifs were identified that are associated with the DMR sets from different environmental exposures. Two different motifs were identified, EDM1, which is an A/T rich motif that is present in the transgenerational vinclozolin DMR set, and EDM2 that is a G/C rich motif present in the other germline transgenerational DMR sets investigated (jet fuel, pesticides, plastics and dioxin). The incidence of EDM1 is over-represented only in the vinclozolin lineage sperm DMRs, but is not over-

Table 1. Similarities of GLAM2 created motifs with known transcription factor for binding sites.

Sequence sets used to build the Glam2 Motif	Best Motif similarities in STAMP database with respective E-values				
	1st	**2nd**	**3rd**	**4th**	**5th**
Vinclozolin	AZF1	FOXP1	HMG-IY	STE11	BR-C
	1.44E-14	1.53E-10	5.09E-08	4.95E-07	2.11E-02
Dioxin	UF1H3BETA	KROX	RREB1	ZNF219	PAX
	3.41E-14	2.68E-08	2.01E-07	6.39E-07	49.56E-07
Jet Fuel	CDC5	KROX	Dde	IME1	HEN
	2.07E-05	4.26E-05	8.58E-05	9.53E-05	11.05E-04
Pesticides	ZNF219	PAX	KROX	UF1H3BETA	MAZ
	3.04E-10	41.50E-09	3.27E-08	1.74E-08	1.61E-07
Plastics	KROX	RREB1	PAX	ZNF219	UF1H3BETA
	1.25E-12	1.57E-11	41.39E-10	1.89E-09	8.94E-09

Similarity between motifs created from a variety of F3 generation exposure lineage sperm DNA and known transcription factor binding site matrices. The top five similarities are shown for each created motif with their respective statistically significant E-values.

A   DENSITY OF EDM1 IN DMR

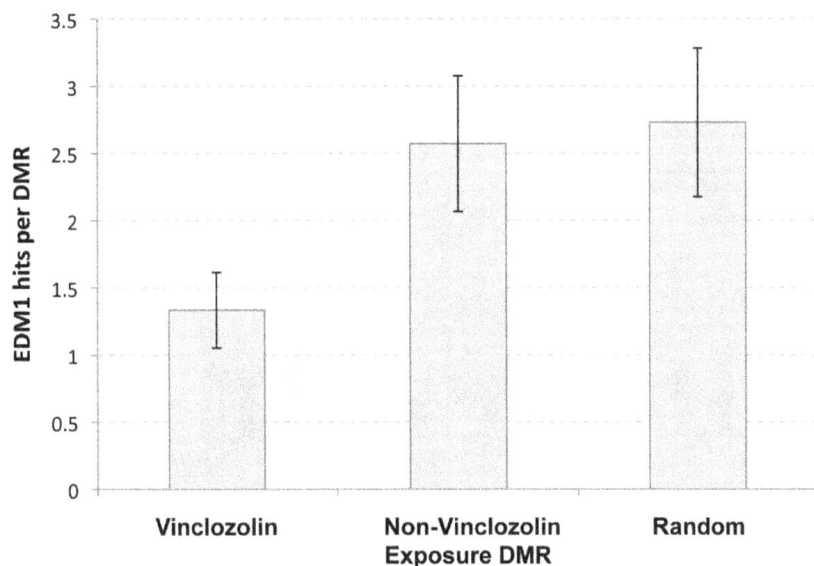

B   DENSITY OF EDM2 IN DMR

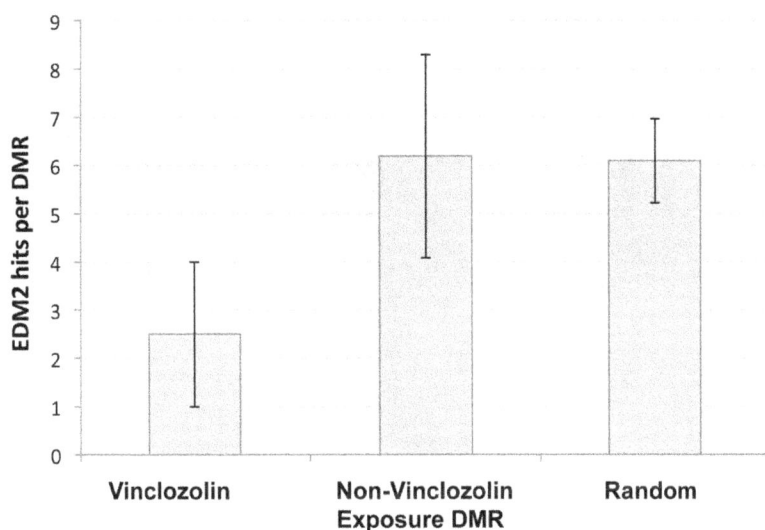

Figure 5. Density of incidences of (A) EDM1 and (B) EDM2 in sets of transgenerational DMR. The vinclozolin DMR, combination of plastics (BIP), pesticide (PIP), jet fuel (JIP) and dioxin (HIP) DMR, and a random set of genomic sites were investigated. The number of EDM1 or EDM2 sites per DMR is presented with the mean± SEM.

represented in the somatic cell vinclozolin lineage DMR sets tested. These observations suggest the somatic cell epigenetic modifications are secondary to the germline epigenetic modifications and probably occur through alternate developmental mechanisms. EDM2 shows the opposite pattern of EDM1, being over-represented in the DMRs of all the exposure lineages, except for vinclozolin.

An analysis of known motifs with a database of transcription factor motifs shows that zinc finger motifs are associated with the sperm epimutations. Further analysis of the presence of these transcription factor binding sites in promoter DMR regions of the exposure sets shows that zinc finger transcriptome factor consensus

DNA binding motifs for Krox, Sp1, Znf219 and Rreb1 are over-represented in the majority of the transgenerational sperm DMR sets investigated. These observations suggest that zinc finger containing DNA binding factors may have a role in the molecular mechanism of epigenetic transgenerational inheritance of sperm epimutations. UFH3Beta might also be relevant, since it follows the same patterns as these zinc finger binding factors. Interestingly, previous studies have highlighted the role of zinc fingers in epigenetic reprogramming. For example, the zinc-finger protein UHRF1 has recently been shown to have a role in maintaining DNA methylation in specific genomic regions in mammals [52]. Other studies have shown that the zinc-finger ZBTB4 preferen-

A/T STRING (WWWW) FREQUENCY DISTRIBUTION IN DMR

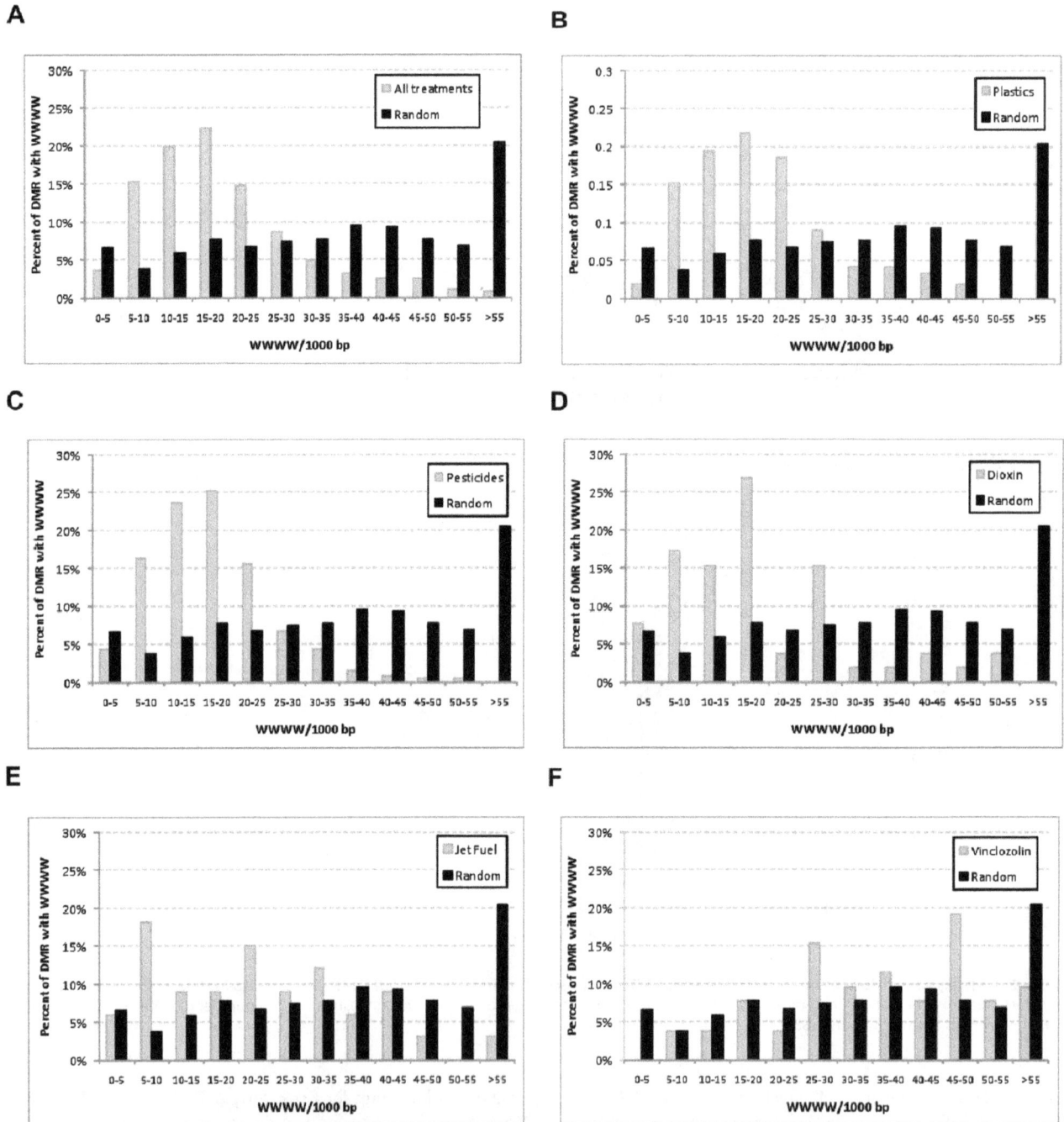

Figure 6. Distribution of A/T string (WWWW) incidence across the transgenerational DMR sets. The percent of DMR with A/T string sequences for all DMR (A), plastics (B), pesticides (C), dioxin (D), jet fuel (E), and vinclozolin (F) are presented compared to the random sequence data set.

tially bind to methylated DNA [53]. The potential that zinc finger binding regions may be targets for DNA methylation changes that will maintain DNA methylation alterations transgenerationally needs to be further investigated. DNA methylation changes in zinc finger rich regions may also preferentially occur in the germline.

Another correlation of interest is FoxP1, which is reduced in all but one treatment (jet fuel). Interestingly, FoxP1 expression has been shown to be altered by the hypomethylating agent 5-azacytidine and by micro RNA expression neighboring the FoxP1 gene in human hepatocellular carcinoma cell lines [54]. The

G QUADRUPLEX (GGGG) FREQUENCY DISTRIBUTION IN DMR

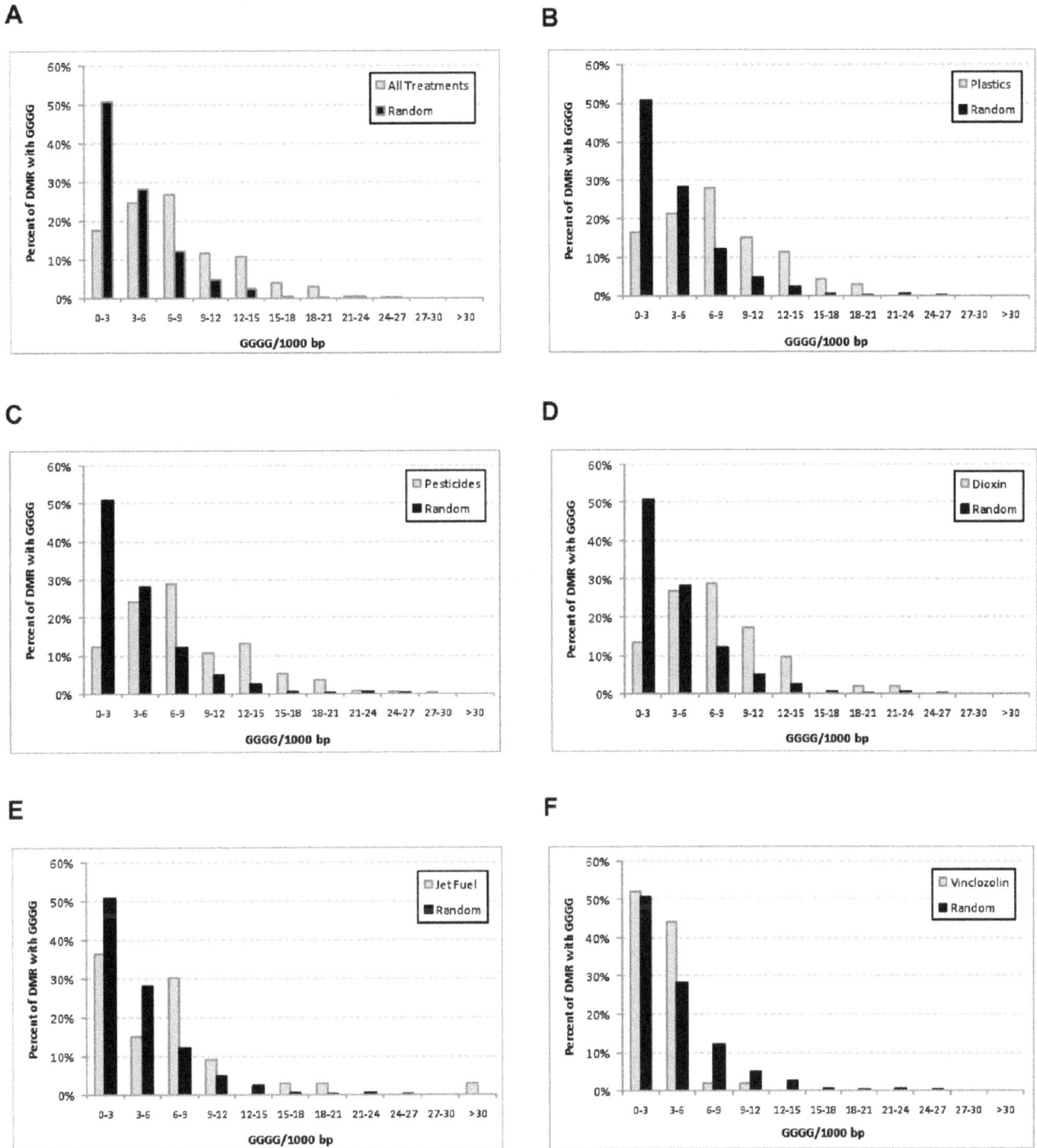

Figure 7. Distribution of G quadruplexes (GGGG) incidence across the transgenerational DMR sets. The percent of DMR with G-quadruplexes of sequences for all DMR (A), plastics (B), pesticides (C), dioxin (D), jet fuel (E), and vinclozolin (F) are presented compared to the random sequence data set.

reduced incidence of FoxP1 sequences in the transgenerational DMR suggests the potential absence of epigenetic mechanisms that may correct the epigenetic defect induced. This may allow for these modifications in DNA methylation to be permanently transmitted to subsequent generations. These potential mechanisms need to be further investigated.

Another genomic feature analyzed was the density of the incidence of EDM1 or EDM2 within the DMR. Because of the

A) INCIDENCE OF EDM1, EDM2 IN A SET OF CONFIRMED DMR FROM VINCLOZOLIN LINEAGE

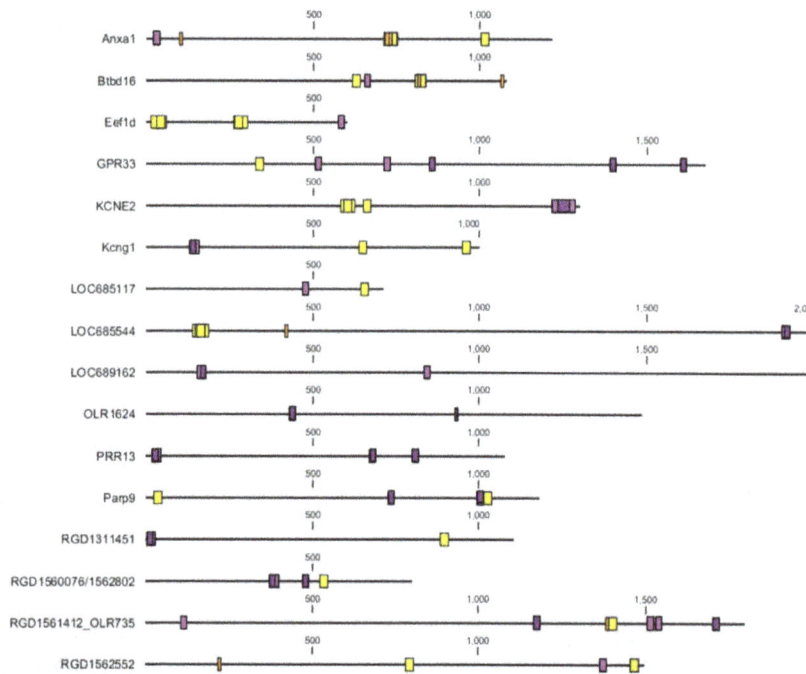

B) INCIDENCE OF EDM1 AND EDM2 IN A SET OF CONFIRMED DMR FROM LINEAGE ORIGINATED FROM A VARIETY OF TREATMENTS

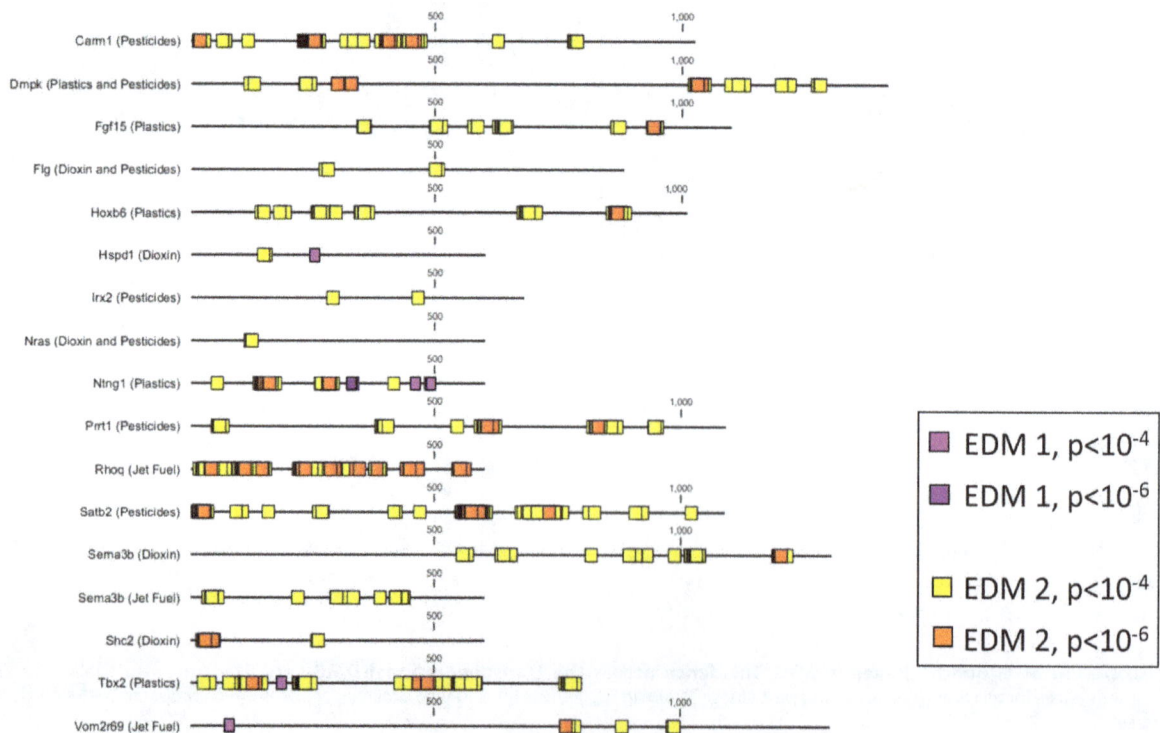

Figure 8. Visualization of DNA motifs associated with epigenetic transgenerational inheritance in selected sets of sequences: (A) incidence of EDM1 and EDM2 in a set of confirmed DMR from vinclozolin-lineage and (B) incidence of EDM1 and EDM2 in a set of confirmed DMR from lineage originated from a variety of exposures listed. The colored legend for EDM1 versus EDM2 motifs are presented.

LOCATION OF EDM1, EDM2, A/T STRINGS, G QUADRUPLEXES AND CPG SITES IN SELECTED DMR

A KCNE2

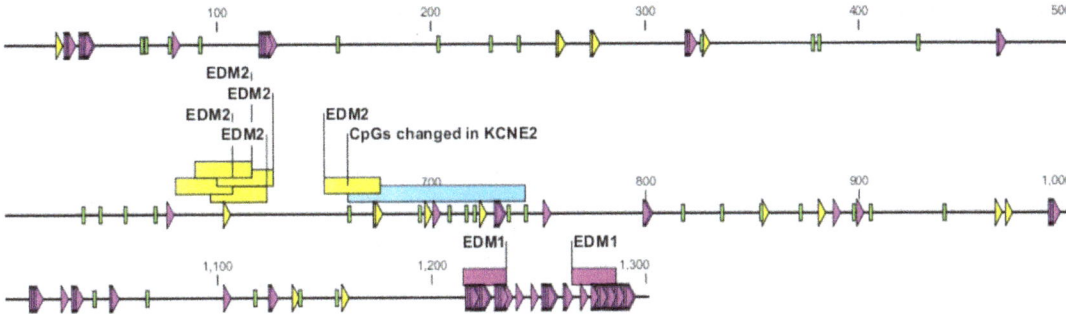

B OLR 735

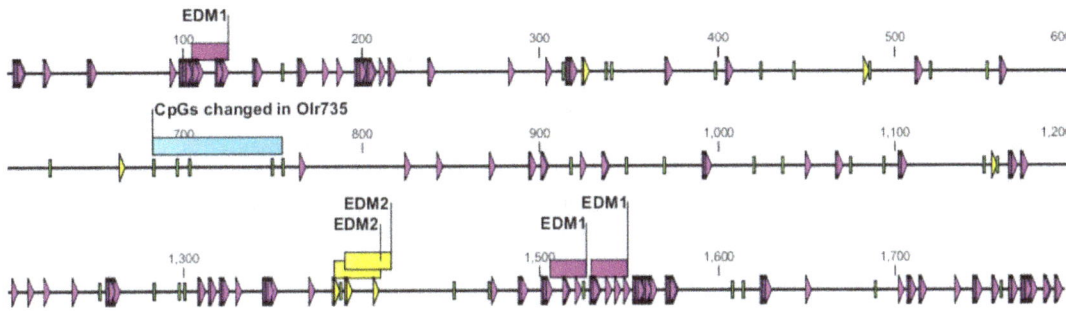

C GPR 33

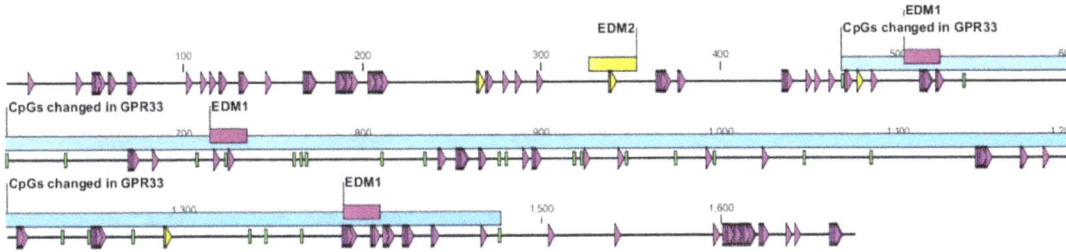

	CpG
	GGGG
	A/T(4) string

Figure 9. Location of EDM1, EDM2, G quadruplexes, A/T strings and CpG sites in selected DMR from the vinclozolin set: (A) KCNE2; (B) OLR 735; (C) GPR 33. The colored legend for CpG, GGGG sequence, and A/T string is presented, with the blue box being the DMR, yellow box EDM2, and purple box EDM1.

possibility that even if over-representation of individual matches does not occur, a cluster incidence of these motifs might occur. The density of EDM1 or EDM2 in the DMR was determined and no difference was found between any of the exposure DMR groups and the random sequence set (Figure 5). Since EDM1 is A/T rich, the frequency of short A/T strings was also tested in the DMR sets. Interestingly, the incidence of A/T strings (WWWW) is

less frequent than in the random sequence set for all the exposures but vinclozolin. In *Neurospora crassa* A/T-rich sequences are shown to be a fundamental recognition site for *de novo* DNA methylation [47]. A/T strings adjacent to CpGs seem to be a requirement for binding of some DNA binding proteins such as MeCP2 [55]. Therefore, the presence of the A/T string is a genomic feature that contributes to the susceptibility of the epimutations to develop and/or be transmitted.

The other DNA motif obtained was called EDM2 and it was found to have guanine rich regions. Previous studies have shown that CpGs with high methylation are generally not present in G-quadruplexes (GGGG), which suggests that DNA methylation is restricted when G-quadruplex features exist [49]. G-quadruplex unwinding is a conserved mechanism which prevents G-quadruplex-induced damages such as genetic and epigenetic changes [56]. Interestingly, the observations show that the incidence of G-quadruplexes is more frequent in the random sequence group than in all the exposure DMR groups, but vinclozolin. This G-quadruplex conformation forms pockets of accessibility that could open during specific times during development, allowing for epigenetic modifications to be established. Indeed the formation of G-quadruplex structures depends of the RAV1 factor, which when absent alters the incorporation of histones [50]. As mentioned above, zinc finger binding sites are enriched in the transgenerational sperm DMR. Interestingly, previous reports show that G-quadruplexes associate with zinc fingers [48]. Therefore, the presence of G-quadruplexes, zinc fingers and/or chromatin remodeling proteins appear to be associated with the transgenerational sperm epimutations.

The motif associated with the vinclozolin DMR (EDM1) was found to be distinct from the motif associated to the other exposure DMR (EDM2). One speculative mechanism to explain the difference is the variable signaling mechanisms of the compounds generating the transgenerational germline epimutations. While vinclozolin is a known anti-androgenic compound [57], several of the other compounds investigated are associated with estrogenic effects. The estrogenic effects of bisphenol A (BPA) and the phthalates have been established [58]. The action of both permethrin [59] and dioxin [60] are also reported to have estrogenic effects. Jet fuel (JP8) has been reported to reduce LH levels in women [61] by disruption of testosterone conversion to estradiol by aromatase [62]. Although the actions of the toxicants are on the F1 generation fetus, the altered epigenetic programming may be in part different due to the distinct signaling. This potential differential signaling effects on the germline epimutations needs to be further investigated.

An initial experiment to help validate the presence of these genomic features in the transgenerational sperm epimutations used a recent DDT sperm DMR set for analysis [51]. This DDT DMR set was not used in the identification of the genomic features. A number of the genomic features were also found to be present in the DDT sperm epimutations. The distribution of the genomic features in the DDT DMR group has a similar pattern to the plastics, pesticides, jet fuel and dioxin DMR groups. The mechanism of action of DDT is primarily to act as an estrogenic compound [63], which is generally similar to the actions of the other compounds, and distinct from the anti-androgenic actions of vinclozolin. This initial validation helps confirm the presence of the features, however, the functional role of these features remains to be elucidated.

Observations lead to the speculation that the mechanism of the transgenerational epigenetic programming of germline epimutations may be in part based on the action of zinc finger motifs and G-quadruplex sequences that can alter chromatin structure and accessibility to proteins. This alteration may allow an opening of DNA that alters the action of DNA methyltransferases or interacting factors. G-quadruplex regions would be more prone for this opening to epigenetic marks to occur. The zinc finger factors may interact with other proteins to promote this chromatin remodeling and/or altered expression of non-coding RNA. The

current study identified a number of motifs and genomic features potentially associated with the DMR involved in the environmentally induced epigenetic transgenerational inheritance of sperm epimutations. Future studies are now needed to further investigate the specific proteins involved and developmental aspects of these epimutations.

Methods

DMR Sequence Sets

Exposure sets of DMR sequences used to perform the bioinformatic analyses were obtained from previous studies from our group showing transgenerational epigenetic changes in the F3 generation sperm and somatic cells [3,6,10,45]. The DMR sequences used were from the published data sets, using a p-value cut-off of 10^{-7} instead of the p-value cut-off of 10^{-5} used in these publications. The DMR data sets were reduced in size, based on the above statistics, to allow for creation of the DNA motifs by the web-based tools used, which have size limitations. The GEO accession number (GSE57693) for these previous publications and additional information on the data access and bioinformatics can be found at www.skinner.wsu.edu/arrays.

Motif Analyses

Two main computational methods exist to identify shared motifs in sets of sequences: (i) application of *ab initio* motif discovery algorithms, which search for recurring patterns in a set of DNA sequences and (ii) assessment of statistical over-representation of previously characterized motifs (from transcription factor binding site databases) in sequences [64]. In the present study the GLAM2 algorithm (Gapped Local Alignment of Motifs, available online on MEME suite) was used for *ab initio* motif discovery. GLAM2 considers insertions or deletions that are a variable and not incorporated by other algorithms [41]. The DMR sets were uploaded to the GLAM2 website and the best motif produced with the default settings was chosen for each sequence dataset. Previously, motifs were identified using GLAM2 in sets of sequences with vinclozolin-induced transgenerational changes in DNA methylation in the male germline [3,4]. The web-based tool FIMO (Find Individual Motif Occurrences, available online on MEME suite) is a general-purpose web-based tool for identifying candidate binding sites [65]. FIMO assigns scores (p and q values) to individual motif occurrences in a defined set of sequences. FIMO was used in the present study to interrogate whether the motifs previously created with GLAM2 would be overrepresented in the tested sets of sequences versus a random set. FIMO was set with a p-value of 10^{-6} and the scan was on both strands. The sequences with matches of motifs were counted and the percentage of the sequences with matches was calculated in for each DMR dataset. The number of motif matches per sequence was also counted to determine the density of motif matches per DMR dataset. In addition to these analyses, comparison of created motifs with known motifs of transcription factors was performed with STAMP. STAMP is a web-based bioinformatic toolbox used to detect similarities of input motifs to motifs representing transcription factor binding sites, which are present in a database that include binding sites information from several organisms [43]. GLAM2 built motifs were compared to the 'selected eukaryotic' database of transcription factor motifs in STAMP, using the default settings. The top five matches (default setting) were selected (Table 1). These selected motifs were then scanned against the DMR sequence datasets using FIMO, as described above. Another feature of STAMP is that it groups motifs based on similarities using a familial binding analysis. This analysis was performed to contruct a tree of similarities between the GLAM2 built motifs (Figure 1). The incidence of G quadruplexes and A/T strings in DMR sets were computed with R (R Development CoreTeam (2010), R: A language for statistical computing, R Foun-dation for Statistical Computing, Vienna, Austria. ISBN 3-900051-07-0, URL http://www.R-project.org). For this, matches for GGGG and WWWW motifs were interrogated, respectively. The number of matches per sequences was then obtained and the density of matches per sequence sets was calculated. Figures showing visualization of all motif incidences (GLAM2 built, GGGG and WWWW) were created using CLC Workbench (Cambridge, MA).

Statistical Analyses

Incidences of motifs between the DMR from different treatments and random sets were tested with Fisher's test. Comparison of density of motif incidences between treatment sets of DMR and random sets were performed with Student's t-test. The average of the distributions of G quadruplexes and A/T strings was also compared with Student's t-test between the treatment DMRs and random sets.

Supporting Information

Figure S1 Validation with DDT DMR data set. (A) Distribution of A/T string (WWWW) incidence in the DDT DMR data set. The percent of DMR with A/T string sequences are presented compared to the random sequence data set. (B) Distribution of G quadruplexes (GGGG) incidence in the DDT DMR data set. The percent of DMR with G-quadruplexes are presented compared to a random sequence data set.

Acknowledgments

We thank the assistance and advice of Mr. Md Haque and Dr. Eric Nilsson, and the assistance of Ms. Heather Johnson in the preparation of the manuscript. Current address for Dr. Carlos Guerrero-Bosagna: Avian Group, Department of Physics, Biology and Chemistry (IFM), Linköping University, Linköping, Sweden.

Author Contributions

Conceived and designed the experiments: MKS CGB. Performed the experiments: CGB SW. Analyzed the data: MKS CGB SW. Wrote the paper: MKS CGB. Edited the Manuscript: MKS CGB SW.

References

1. Anway MD, Cupp AS, Uzumcu M, Skinner MK (2005) Epigenetic transgenerational actions of endocrine disruptors and male fertility. Science 308: 1466–1469.
2. Manikkam M, Tracey R, Guerrero-Bosagna C, Skinner MK (2012) Dioxin (TCDD) induces epigenetic transgenerational inheritance of adult onset disease and sperm epimutations. PLoS One 7: e46249.
3. Guerrero-Bosagna C, Settles M, Lucker B, Skinner M (2010) Epigenetic transgenerational actions of vinclozolin on promoter regions of the sperm epigenome. PLoS ONE 5: e13100.
4. Guerrero-Bosagna C, Covert T, Haque MM, Settles M, Nilsson EE, et al. (2012) Epigenetic Transgenerational Inheritance of Vinclozolin Induced Mouse Adult Onset Disease and Associated Sperm Epigenome Biomarkers. Reproductive Toxicology 34: 694–707.
5. Manikkam M, Tracey R, Guerrero-Bosagna C, Skinner M (2012) Pesticide and Insect Repellent Mixture (Permethrin and DEET) Induces Epigenetic Transgenerational Inheritance of Disease and Sperm Epimutations. Reproductive Toxicology 34: 708–719.
6. Manikkam M, Guerrero-Bosagna C, Tracey R, Haque MM, Skinner MK (2012) Transgenerational Actions of Environmental Compounds on Reproductive Disease and Epigenetic Biomarkers of Ancestral Exposures. PLoS ONE 7: e31901.
7. Skinner M, Guerrero-Bosagna C, Haque MM, Nilsson E, Bhandari R, et al. (2013) Environmentally Induced Transgenerational Epigenetic Reprogramming of Primordial Germ Cells and Subsequent Germline. PLoS ONE 8: e66318.
8. Tracey R, Manikkam M, Guerrero-Bosagna C, Skinner M (2013) Hydrocarbon (Jet Fuel JP-8) Induces Epigenetic Transgenerational Inheritance of Adult-Onset Disease and Sperm Epimutations. Reproductive Toxicology 36: 104–116.
9. Skinner MK, Mohan M, Haque MM, Zhang B, Savenkova MI (2012) Epigenetic transgenerational inheritance of somatic transcriptomes and epigenetic control regions. Genome Biol 13: R91.
10. Nilsson E, Larsen G, Manikkam M, Guerrero-Bosagna C, Savenkova MI, et al. (2012) Environmentally induced epigenetic transgenerational inheritance of ovarian disease. PLoS One 7: e36129.
11. Arico JK, Katz DJ, van der Vlag J, Kelly WG (2011) Epigenetic patterns maintained in early Caenorhabditis elegans embryos can be established by gene activity in the parental germ cells. PLoS Genet 7: e1001391.
12. Carone BR, Fauquier L, Habib N, Shea JM, Hart CE, et al. (2010) Paternally induced transgenerational environmental reprogramming of metabolic gene expression in mammals. Cell 143: 1084–1096.
13. Dunn GA, Bale TL (2011) Maternal high-fat diet effects on third-generation female body size via the paternal lineage. Endocrinology 152: 2228–2236.
14. Morgan CP, Bale TL (2011) Early prenatal stress epigenetically programs dysmasculinization in second-generation offspring via the paternal lineage. J Neurosci 31: 11748–11755.
15. Saze H (2012) Transgenerational inheritance of induced changes in the epigenetic state of chromatin in plants. Genes Genet Syst 87: 145–152.
16. Lees-Murdock DJ, Walsh CP (2008) DNA methylation reprogramming in the germ line. Epigenetics 3: 5–13.
17. Reik W, Dean W, Walter J (2001) Epigenetic reprogramming in mammalian development. Science 293: 1089–1093.
18. Smith ZD, Chan MM, Mikkelsen TS, Gu H, Gnirke A, et al. (2012) A unique regulatory phase of DNA methylation in the early mammalian embryo. Nature 484: 339–344.
19. Allegrucci C, Thurston A, Lucas E, Young L (2005) Epigenetics and the germline. Reproduction 129: 137–149.
20. Durcova-Hills G, Hajkova P, Sullivan S, Barton S, Surani MA, et al. (2006) Influence of sex chromosome constitution on the genomic imprinting of germ cells. Proc Natl Acad Sci U S A 103: 11184–11188.
21. Hackett JA, Zylicz JJ, Surani MA (2012) Parallel mechanisms of epigenetic reprogramming in the germline. Trends Genet 28: 164–174.
22. Skinner MK, Manikkam M, Guerrero-Bosagna C (2010) Epigenetic transgenerational actions of environmental factors in disease etiology. Trends Endocrinol Metab 21: 214–222.
23. Cowin PA, Gold E, Aleksova J, O'Bryan MK, Foster PM, et al. (2010) Vinclozolin exposure in utero induces postpubertal prostatitis and reduces sperm production via a reversible hormone-regulated mechanism. Endocrinology 151: 783–792.
24. Uzumcu M, Suzuki H, Skinner MK (2004) Effect of the anti-androgenic endocrine disruptor vinclozolin on embryonic testis cord formation and postnatal testis development and function. Reprod Toxicol 18: 765–774.
25. Webster KE, O'Bryan MK, Fletcher S, Crewther PE, Aapola U, et al. (2005) Meiotic and epigenetic defects in Dnmt3L-knockout mouse spermatogenesis. Proc Natl Acad Sci U S A 102: 4068–4073.
26. Stadler MB, Murr R, Burger L, Ivanek R, Lienert F, et al. (2011) DNA-binding factors shape the mouse methylome at distal regulatory regions. Nature 480: 490–495.
27. Molaro A, Hodges E, Fang F, Song Q, McCombie WR, et al. (2011) Sperm methylation profiles reveal features of epigenetic inheritance and evolution in primates. Cell 146: 1029–1041.
28. Filippova GN, Thienes CP, Penn BH, Cho DH, Hu YJ, et al. (2001) CTCF-binding sites flank CTG/CAG repeats and form a methylation-sensitive insulator at the DM1 locus. Nat Genet 28: 335–343.
29. Ichiyanagi K, Li Y, Watanabe T, Ichiyanagi T, Fukuda K, et al. (2011) Locus- and domain-dependent control of DNA methylation at mouse B1 retrotransposons during male germ cell development. Genome Res 21: 2058–2066.
30. Wienholz BL, Kareta MS, Moarefi AH, Gordon CA, Ginno PA, et al. (2010) DNMT3L modulates significant and distinct flanking sequence preference for DNA methylation by DNMT3A and DNMT3B in vivo. PLoS Genet 6.
31. Alabert C, Groth A (2012) Chromatin replication and epigenome maintenance. Nat Rev Mol Cell Biol 13: 153–167.
32. Wen D, Wu H, Bjornsson H, Green RD, Irizarry R, et al. (2008) Overlapping euchromatin/heterochromatin- associated marks are enriched in imprinted gene regions and predict allele-specific modification. Genome Res 18: 1806–1813.
33. Das MK, Dai HK (2007) A survey of DNA motif finding algorithms. BMC Bioinformatics 8 Suppl 7: S21.
34. Stormo GD (2000) DNA binding sites: representation and discovery. Bioinformatics 16: 16–23.
35. Elnitski L, Jin VX, Farnham PJ, Jones SJ (2006) Locating mammalian transcription factor binding sites: a survey of computational and experimental techniques. Genome Res 16: 1455–1464.

36. Nardone J, Lee DU, Ansel KM, Rao A (2004) Bioinformatics for the 'bench biologist': how to find regulatory regions in genomic DNA. Nat Immunol 5: 768–774.

37. Wasserman WW, Sandelin A (2004) Applied bioinformatics for the identification of regulatory elements. Nat Rev Genet 5: 276–287.

38. Carroll JS, Meyer CA, Song J, Li W, Geistlinger TR, et al. (2006) Genome-wide analysis of estrogen receptor binding sites. Nat Genet 38: 1289–1297.

39. Kim TH, Abdullaev ZK, Smith AD, Ching KA, Loukinov DI, et al. (2007) Analysis of the vertebrate insulator protein CTCF-binding sites in the human genome. Cell 128: 1231–1245.

40. Feltus FA, Lee EK, Costello JF, Plass C, Vertino PM (2006) DNA motifs associated with aberrant CpG island methylation. Genomics 87: 572–579.

41. Frith MC, Saunders NF, Kobe B, Bailey TL (2008) Discovering sequence motifs with arbitrary insertions and deletions. PLoS Comput Biol 4: e1000071.

42. Bailey TL, Boden M, Buske FA, Frith M, Grant CE, et al. (2009) MEME SUITE: tools for motif discovery and searching. Nucleic Acids Res 37: W202–208.

43. Mahony S, Benos PV (2007) STAMP: a web tool for exploring DNA-binding motif similarities. Nucleic Acids Res 35: W253–258.

44. Nilsson E, Larsen G, Manikkam M, Guerrero-Bosagna C, Savenkova M, et al. (2012) Environmentally Induced Epigenetic Transgenerational Inheritance of Ovarian Disease. PLoS ONE 7: e36129.

45. Guerrero-Bosagna C, Savenkova M, Haque MM, Sadler-Riggleman I, Skinner MK (2013) Environmentally Induced Epigenetic Transgenerational Inheritance of Altered Sertoli Cell Transcriptome and Epigenome: Molecular Etiology of Male Infertility. PLoS ONE 8: e59922.

46. Manikkam M, Tracey R, Guerrero-Bosagna C, Skinner M (2013) Plastics Derived Endocrine Disruptors (BPA, DEHP and DBP) Induce Epigenetic Transgenerational Inheritance of Adult-Onset Disease and Sperm Epimutations. PLoS ONE 8: e55387.

47. Tamaru H, Selker EU (2003) Synthesis of signals for de novo DNA methylation in Neurospora crassa. Mol Cell Biol 23: 2379–2394.

48. Kumar P, Yadav VK, Baral A, Kumar P, Saha D, et al. (2011) Zinc-finger transcription factors are associated with guanine quadruplex motifs in human, chimpanzee, mouse and rat promoters genome-wide. Nucleic Acids Res 39: 8005–8016.

49. Halder R, Halder K, Sharma P, Garg G, Sengupta S, et al. (2010) Guanine quadruplex DNA structure restricts methylation of CpG dinucleotides genome-wide. Mol Biosyst 6: 2439–2447.

50. Sarkies P, Reams C, Simpson LJ, Sale JE (2010) Epigenetic instability due to defective replication of structured DNA. Mol Cell 40: 703–713.

51. Skinner MK, Manikkam M, Tracey R, Nilsson E, Haque MM, et al. (2013) Ancestral DDT Exposures Promote Epigenetic Transgenerational Inheritance of Obesity BMC Medicine 11: 228.

52. Bostick M, Kim JK, Esteve PO, Clark A, Pradhan S, et al. (2007) UHRF1 plays a role in maintaining DNA methylation in mammalian cells. Science 317: 1760–1764.

53. Sasai N, Nakao M, Defossez PA (2010) Sequence-specific recognition of methylated DNA by human zinc-finger proteins. Nucleic Acids Res 38: 5015–5022.

54. Datta J, Kutay H, Nasser MW, Nuovo GJ, Wang B, et al. (2008) Methylation mediated silencing of MicroRNA-1 gene and its role in hepatocellular carcinogenesis. Cancer Res 68: 5049–5058.

55. Klose RJ, Sarraf SA, Schmiedeberg L, McDermott SM, Stancheva I, et al. (2005) DNA binding selectivity of MeCP2 due to a requirement for A/T sequences adjacent to methyl-CpG. Mol Cell 19: 667–678.

56. Paeschke K, Bochman ML, Garcia PD, Cejka P, Friedman KL, et al. (2013) Pif1 family helicases suppress genome instability at G-quadruplex motifs. Nature 497: 458–462.

57. Wong C, Kelce WR, Sar M, Wilson EM (1995) Androgen receptor antagonist versus agonist activities of the fungicide vinclozolin relative to hydroxyflutamide. J Biol Chem 270: 19998–20003.

58. Singh S, Li SS (2012) Bisphenol A and phthalates exhibit similar toxicogenomics and health effects. Gene 494: 85–91.

59. Brander SM, He G, Smalling KL, Denison MS, Cherr GN (2012) The in vivo estrogenic and in vitro anti-estrogenic activity of permethrin and bifenthrin. Environ Toxicol Chem 31: 2848–2855.

60. Tanaka J, Yonemoto J, Zaha H, Kiyama R, Sone H (2007) Estrogen-responsive genes newly found to be modified by TCDD exposure in human cell lines and mouse systems. Mol Cell Endocrinol 272: 38–49.

61. Reutman SR, LeMasters GK, Knecht EA, Shukla R, Lockey JE, et al. (2002) Evidence of reproductive endocrine effects in women with occupational fuel and solvent exposures. Environ Health Perspect 110: 805–811.

62. Pitteloud N, Dwyer AA, DeCruz S, Lee H, Boepple PA, et al. (2008) Inhibition of luteinizing hormone secretion by testosterone in men requires aromatization for its pituitary but not its hypothalamic effects: evidence from the tandem study of normal and gonadotropin-releasing hormone-deficient men. J Clin Endocrinol Metab 93: 784–791.

63. Robison AK, Schmidt WA, Stancel GM (1985) Estrogenic activity of DDT: estrogen-receptor profiles and the responses of individual uterine cell types following o,p'-DDT administration. J Toxicol Environ Health 16: 493–508.

64. Frith MC, Fu Y, Yu L, Chen JF, Hansen U, et al. (2004) Detection of functional DNA motifs via statistical over-representation. Nucleic Acids Res 32: 1372–1381.

65. Grant CE, Bailey TL, Noble WS (2011) FIMO: scanning for occurrences of a given motif. Bioinformatics 27: 1017–1018.

Paraoxonase Enzyme Protects Retinal Pigment Epithelium from Chlorpyrifos Insult

Jagan Mohan Jasna, Kannadasan Anandbabu, Subramaniam Rajesh Bharathi, Narayanasamy Angayarkanni*

R.S Mehta Jain Department of Biochemistry and Cell Biology, KBIRVO Block, Vision Research Foundation, Sankara Nethralaya, Chennai, India

Abstract

Retinal pigment epithelium (RPE) provides nourishment and protection to the eye. RPE dysfunction due to oxidative stress and inflammation is one of the major reason for many of the retinal disorders. Organophosphorus pesticides are widely used in the agricultural, industrial and household activities in India. However, their effects on the eye in the context of RPE has not been studied. In this study the defense of the ARPE19 cells exposed to Chlorpyrifos (1 nM to 100 μM) in terms of the enzyme paraoxonase (PON) was studied at 24 hr and 9 days of treatment. Chlorpyrifos was found to induce oxidative stress in the ARPE19 cells as seen by significant increase in ROS and decrease in glutathione (GSH) levels without causing cell death. Tissue resident Paraoxonase 2 (PON2) mRNA expression was elevated with chlorpyrifos exposure. The three enzymatic activities of PON namely, paraoxonase (PONase), arylesterase (PON AREase) and thiolactonase (PON HCTLase) were also found to be significantly altered to detoxify and as an antioxidant defense. Among the transcription factors regulating PON2 expression, SP1 was significantly increased with chlorpyrifos exposure. PON2 expression was found to be crucial as ARPE19 cells showed a significant loss in their ability to withstand oxidative stress when the cells were subjected to chlorpyrifos after silencing PON2 expression. Treatment with N-acetyl cysteine positively regulated the PON 2 expression, thus promoting the antioxidant defense put up by the cells in response to chlorpyrifos.

Editor: Alfred S. Lewin, University of Florida, United States of America

Funding: The research work was financially supported by the Indian Council of Medical Research (ICMR), India. Grant number is 5/8/4-18(Env)/08-NCD-1. The funders had no role in study design, data collection and analysis, decision to publish, or preparation of the manuscript.

Competing Interests: The authors have declared that no competing interests exist.

* Email: drak@snmail.org

Introduction

Retinal pigment epithelium (RPE) is a monolayer of epithelial cells between the neural retina and the choriocapillaris [1]. RPE cells act as a selective barrier in regulating the movement of nutrients and solutes from the choroid to the sub-retinal space forming the outer blood-retinal barrier [2]. Loss in the RPE function is associated with oxidative stress, inflammation, fibrosis and contribute to pathophysiological processes in age-related macular degeneration (AMD), proliferative vitreoretinopathy (PVR) and proliferative diabetic retinopathy (PDR) [3]. Tumor necrosis factor alpha (TNF-α) [2], glycated-albumin [4] and oxidized low density lipoprotein [5] are capable of inducing RPE dysfunction. Pesticides like paraquat are also reported to induce oxidative damage to the RPE [6].

Organophosphate insecticide, Chlorpyrifos (CPF; O,O-diethyl-O-(3,5,6-trichloro-2-pyridyl) phosphorothioate) is common in agricultural, industrial and household pesticide formulations [7–8]. It is classified by WHO as class II moderately hazardous compound that has an LD50 range of 20–2000 mg//kg body weight in rat [9]. Chlorpyrifos is a neurotoxicant that inhibits neuronal and blood cholinesterase leading to overstimulation of cholinergic neurotransmission [8]. Exposure to chlorpyrifos can produce ocular toxicity with long-lasting changes in retinal physiology and anatomy [10–11]. Abnormal electroretinograms were noticed in rats after administration of chlorpyrifos [12].

Chlorpyrifos is reported to cause cell apoptosis, lipid peroxidation and DNA damage in mouse retina and pretreatment with antioxidants, vitamins C and E were effective in reverting these damages [13]. Chlorpyrifos is reported to induce oxidative stress by inhibiting mammalian acetylcholine esterase. In addition it also disrupts the endocrine actions of androgenic, estrogenic, thyroid and parathyroid hormones [14].

Cytochrome P450 (CYP450) metabolically activates chlorpyrifos to chlorpyrifos oxon, which is acted upon by alpha-esterases, like paraoxonase and is further converted to diethyl phosphate and 3,5,6-trichloro-2-pyridinol in the liver by the CYP450 system [15–16]. Chlorpyrifos is absorbed rapidly with 80% excretion in urine within 48 hr as studied in rats [17]. Paraoxonase (PON) is a calcium-dependent enzyme having enzyme activities towards varied substrates. It can hydrolyze paraoxon (PONase activity) and exhibits arylesterase (PON AREase) and thiolactonase activity (PON HCTLase). PON has 3 isoforms- PON1, PON2 and PON3 [18–19]. PON1 and PON3 are associated with serum HDL while PON2 is predominantly seen in tissues [20–21]. Antioxidant properties of human PON1 prevents oxidative modifications of lipoproteins apart from hydrolyzing oxidized phospholipids, hydroperoxides and lactones [22].

Few studies report on the detrimental effects of chlorpyrifos on retina in animal models. However, the effect of chlorpyrifos on retinal pigment epithelium has not been studied so far. This study is focused on how the RPE cells respond to the toxic pesticide chlorpyrifos *in vitro* as studied at the level of antioxidant enzyme paraoxonase.

Materials and Methods

Reagents

Mouse monoclonal anti-PON2 antibody (sc373981), mouse monoclonal anti-ACTIN antibody (sc32251) and goat anti-mouse horseradish peroxidase-conjugated secondary antibody (sc2005) were purchased from Santa-Cruz, USA. DMEM F12 and fetal calf serum were procured from Invitrogen (Carlsbad, CA). Dimethyl sulfoxide (DMSO), 2',7'-dichlorodihydrofluorescein diacetate (DCFDA), ter-butyl hydroperoxide (tBH) and paraoxon (O, O-diethyl-o-p-nitro-phenylphosphate), chlorpyrifos and mithramycin was from Sigma-Aldrich (St. Louis, MO). Enhanced chemiluminescence western blotting detection reagents were from Amersham Biosciences UK, Ltd. (Little Chalfont, Buckinghamshire, UK). Redox assay kit was procured from Oxford Biomedical Research, MI, USA.

Human RPE Cell culture

ARPE19 cells (ATCC; Manassas, VA) were grown in DMEM F12, supplemented with 10% fetal bovine serum and antibiotics (100 µg/mL penicillin/streptomycin mix) in a humidified atmosphere at 37°C with 5% CO_2. When cells were 80 % confluent, they were shifted from DMEM F12 supplemented with 10% fetal bovine serum to DMEM F12, supplemented with 1% fetal bovine serum for 3 hr. The cells were then exposed to chlorpyrifos in 1% fetal bovine serum for 3 hr and 24 hr as for acute exposure and for 9 days in case of chronic exposure, with change of media and chlorpyrifos added at every 3 days interval. The exposure regimen was the same for all the experiments performed.

MTT Assay

To measure cytotoxicity, ARPE19 cells were plated in a 96-well plate at a density of 5×10^4 cells/well. After exposing the cells to chlorpyrifos for 24 hr and 9 days, medium was aspirated and 0.25 mg/ml MTT was added and incubated for 4 hr at 37 °C. The formazan crystals formed were dissolved in DMSO and absorbance at 540 nm was measured using SpectraMax M2 (Molecular Devices) with 680 nm as reference wavelength. Cell viability was defined relative to the untreated control.

Detection of reactive oxygen species (ROS)

ROS was measured by DCFDA method. 10 µM DCFDA was added to each well and incubated for 30 minutes at 37°C and DCFDA fluorescence was measured using SpectraMax M2 (Molecular Devices) in 96-well plates at an excitation wavelength of 485 nM and an emission wavelength of 530 nM. ROS production was expressed as relative fluorescence. All assays were performed in triplicates. *N*-acetylcysteine (NAC; 5 mM), which can decrease ROS production by increasing the intracellular GSH concentration was used as antioxidant positive control, while H_2O_2 (1 mM) / tBH (500 µM) was used as the pro oxidant.

Changes in Redox status

The ARPE19 cells were grown to 80% confluence in 6 well plates for the reductase assay. After treatment period, a pyridine derivative was added as a thiol-scavenger to the sample, which was then used for estimation of oxidized form of the glutathione (GSSG). Cells for GSH (without pyridine derivative) and GSSG determination were harvested in PBS. The re-suspended cells were lysed by sonication, centrifuged and the supernatant was used for the assay as per the Oxford Biomedical Kit for determination of

GSH/GSSG [23]. The change in absorbance at 412 nm was measured every minute for 10 minutes.

Determination of paraoxonase activity

The enzyme assay for estimating the paraoxon hydrolyzing activity of PON in ARPE19 cell lysates was established after modification of the existing PONase enzyme assays [24–26].

PON-ase activity was determined spectrophotometrically using 1 mM paraoxonase as the substrate and measured by increase in absorbance at 405 nm due to the formation of 4-nitrophenol for 10 min. Briefly, the activity was measured at 37°C by adding cell lysate to 300 µl of Tris-HCl buffer (100 mM at pH 8.5) containing 2 mM $CaCl_2$ and 2 M NaCl. One unit is defined as 1 nmol of para-nitrophenol formed per minute.

PON-AREase activity was measured using 1 mM phenylacetate as the substrate. The increase in phenol liberated after hydrolysis of phenyl acetate by the addition of cell lysate was measured spectrophotometrically in kinetic mode at 217 nm following an established procedure [27–28]. The assay conditions were performed in buffer containing 10 mM Tris and 1 mM CaCl2, pH 8.0. One unit was defined as the enzyme quantity that disintegrates 1 µmol phenylacetate per minute.

PON-HCTLase activity assay was measured using γ thiobutyrolactone as substrate, and the rate of hydrolysis was measured spectrophotometrically in the kinetic mode at 450 nm (main wavelength) and 546 nm (sub wavelength) using 10 µl of cell lysate at pH 7.2, using 100 mmol/L of phosphate buffer. Enzyme activity was expressed in U/L [29].

Semi quantitative PCR and Real time PCR for quantification of transcripts

After treatment with chlorpyrifos, the total RNA was isolated from the cells using Trizol reagent (Sigma, St. Louis, MO, USA) following manufacturer's instructions. One microgram of RNA was used for conversion into cDNA using iScript™ cDNA synthesis kit (Biorad). GAPDH was used as the house-keeping gene for normalization. Semi quantitative PCR was performed with 50 ng of cDNA. The PCR cycle conditions were: one cycle of 95°C for 2 min, 30 cycles at 95°C for 30 sec followed by 62°C for 30 sec, 72°C for 50 sec and a final extension of 72°C for 30 sec. The products were visualized under UV transilluminator on 2% agarose gel containing 0.5 mg/ml ethidium bromide. The bands obtained were quantified using ImageJ software (developed by Wayne Rasband, National Institutes of Health, Bethesda, MD; available at http://rsb.info.nih.gov/ij/index.html) after normalization to GAPDH.

10 ng of cDNA was then used for real-time PCR quantitation of products for human PON2, ARH2, JUN, NRF2, SP1, SREBP2, STAT5B with GAPDH serving as an internal control. Real-time PCR was done using POWER SYBR green PCR master mix on 7300 Real Time PCR System from Applied Biosystems. The specificity of PCR amplification products were checked by performing melting curve analysis. The primers used in the study are listed in table1.

Silencing the expression of PON2 using siRNA

2×10^5 cells per well were seeded in six-well plates and transfected with 10 nM of PON2 siRNA (Predesigned Flexi Tube siRNA,Qiagen) according to the manufacturer's instructions. Negative control (non silencing siRNA) incorporated in the experiment was All Stars siRNA (Qiagen) with scrambled sequence. After transfection for 24 hr, Real time PCR and western blot was done to prove the down regulation of PON2

Table 1. Primers used in PCR.

Gene	Sequence
arh forward	5' CTGCCTTTCCCACAAGATGT 3'
arh reverse	5' AGTTATCCTGGCCTCCGTTT 3'
gapdh forward	5' GAACATCATCCCTGCCTCTACTG 3'
gapdh reverse	5' CGCCTGCTTCACCACCTTC 3'
jun forward	5' ACAGAGCATGACCCTGAACC 3'
jun reverse	5' CCGTTGCTGGACTGGATTAT 3'
nrf2 forward	5' CGGTATGCAACAGGACATTG 3'
nrf2 reverse	5' GTTTGGCTTCTGGACTTGGA 3'
pparg forward	5' GCCCAGGTTTGCTGAATGTG 3'
pparg reverse	5' Tgggcgagaggcttctggca 3'
PON2 forward	5' CCACAGCTTTGCACCAGATA 3'
PON2 reverse	5' ATGCCATGTGGATTGAATGA 3'
PON2 forward	5'TGGGCGAGAGGCTTCTGGCA 3'
PON2 reverse	5'TGTGCCGGTCCAACAGCTGT 3'
sp1 forward	5' GGCTACCCCTACCTCAAAGG 3'
sp1 reverse	5' CACAACATACTGCCCACCAG 3'
srebp forward	5' GACATCATCTGTCGGTGGTG 3'
srebp reverse	5' GGGCTCTCTGTCACTTCCAG 3'
stat5b forward	5' GTTGGTGGAAATGAGCTGGT 3'
stat5b reverse	5' AGGCTCTGCAAAAGCATTGT 3'

expression. The cells post transfection with siRNA were treated with chlorpyrifos and analysed for cell viability and ROS generation.

Immunoblot Analysis

Cells were lysed in M-PER (Thermo Fisher Scientific Inc), with protease inhibitors at 4°C and centrifuged at 5000 rpm for 10 min. The supernatant was collected and the protein was quantified using BCA method. 40 µg of protein was mixed with Laemmli sample buffer containing 100 mM DTT and electrophoresed onto a discontinuous acrylamide gel having 10% resolving gel (pH 8.8) and 4% stacking gel (pH 6.8). Gels were run on a Mini Protean III vertical electrophoresis system (BioRad) at 100V. The proteins were then transferred to Hybond-P PVDF (0.45 µ, Amersham Pharmacia Biotech) in transfer buffer (2.5 mM Tris, 19 mM glycine (pH 8.3), 20 % methanol (v/v) using a Mini Transblot cell (BioRad) at constant voltage of 100 V for each membrane for 1 hr. The non-specific protein sites on the membrane were blocked using 5% nonfat milk for 2 hr at RT on rocking shaker. The membranes were then washed thrice (3×10 min) with PBST (pH 7.4, 0.1% Tween-20). The membranes were incubated in primary antibody for 2 hr at RT. After incubation with primary antibody (1:2000 dilution for PON2 and 1:4000 dilution for ACTIN), the membranes were again washed for 3×10 min with PBST followed by 1 hr incubation in horseradish peroxidase conjugated secondary antibody (1:6000 dilution) at RT. Protein bands of interest were developed using enhanced chemiluminescence system where the chemiluminescence resulting from the peroxidase-catalyzed oxidation of luminol was captured on Fluor Chem FC3 from Protein Simple. Equal protein loading was verified by immunoblotting for β ACTIN.

Statistical Analysis

Statistical results were expressed as mean ± standard deviation of the mean obtained from each independent experiment. The results of the experimental and control groups were tested for statistical significance by a one-tailed Student's t test or a two-tailed ANOVA. The level of statistical significance was set at $p < 0.05$. All experiments performed were in triplicates.

Results

Cell viability assay

The ARPE19 cells were treated with varying concentration of Chlorpyrifos (1 nM to 100 µM) for 24 hr (acute) and 9 days (chronic exposure). The viability of the cells were determined by the MTT assay. The cells were viable at all the concentrations tested and there was no significant cell death even at the highest concentration of 100 µM in both acute 24 hr treatment and 9 days chronic treatment (Figure 1.A & 1.B).

ROS generated as an index of oxidative stress

ARPE19 cells exposed to chlorpyrifos (1 nM to 100 µM) for 24 hr induced a significant increase in ROS production even at 1 nM chlorpyrifos ($p < 0.05$) (Figure 1.C). The increase in ROS generation observed, indicates that chlorpyrifos mediates its toxic effects by causing oxidative stress to the cells. In chronic exposure conditions as well, there was a significant dose dependent increase in ROS production ($p < 0.01$), (Figure 1.D). The maximal increase in ROS production after chlorpyrifos exposure for 24 hr was 87 %. Upon chronic exposure for 9 days there was 112 % increase in ROS seen at the maximal concentration of 100 µM chlorpyrifos.

N acetyl cysteine can diminish the pesticide induced cytotoxicity

Treatment with chlorpyrifos caused an increase in ROS. Therefore, ROS scavenger namely N acetyl cysteine (NAC) was treated to see whether the cells are protected from the oxidative insult given. Pretreatment with NAC caused a significant reduction in the ROS production thus rescuing the cells of the pesticide induced oxidative stress (Figure 1E). There was 49 % decrease in ROS production when ARPE19 cells were pretreated with 5 mM NAC prior to 100 µM of chlorpyrifos exposure ($p < 0.05$). The reduction in ROS generation with NAC pretreatment was observed in all the doses of chlorpyrifos exposure.

Changes in levels of intracellular antioxidants

Reduced glutathione (GSH) is a major tissue antioxidant that provides reducing equivalents for the glutathione peroxidase (GPx) catalyzed reduction. When cells are exposed to increased levels of oxidative stress, the oxidized form of GSH, GSSG will accumulate and the ratio of GSH to GSSG will decrease. Determination of the GSH/GSSG ratio is a useful indicator of balance between the pro and antioxidants in cells and tissues. The level of reduced glutathione (GSH) and oxidized glutathione (GSSG) was estimated in the ARPE19 cell lysates treated with varying concentrations of chlorpyrifos for 24 hr and 9 days. When ARPE19 cells were exposed to chlorpyrifos for 24 hr, there was a significant dose dependent decrease in levels of GSH with increasing concentration of chlorpyrifos. However, there was a significant dose dependent increase in GSSG levels, the net result being that the ratio of GSH to GSSG decrease with increasing concentration of the pesticide exposure for 24 hr ($p < 0.001$), (Figure 1G). The redox status of the cell after 9 days of chlorpyrifos exposure also showed a significant decrease (Figure 1H) ($p < 0.01$). A significant

A MTT assay - 24 hr chlorpyrifos treatment

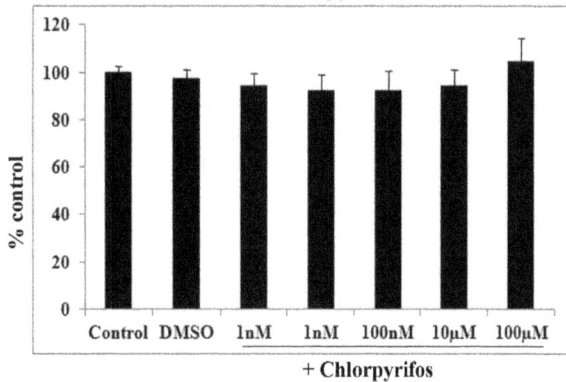

B MTT assay - 9 days of chlorpyrifos treatment

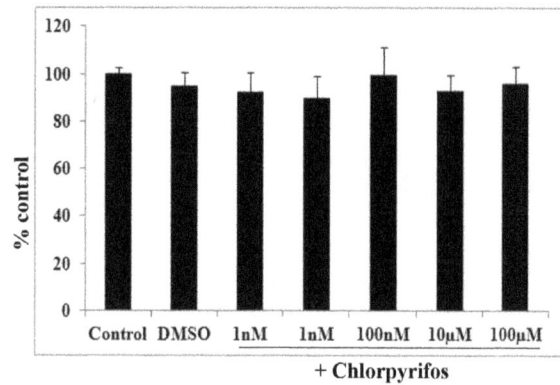

C ROS assay - 24 hr chlorpyrifos treatment

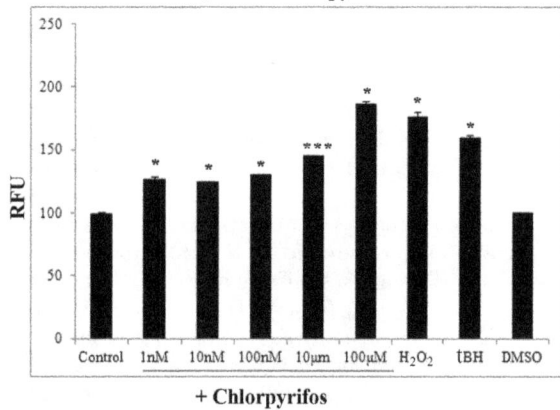

D ROS assay - 9 days of chlorpyrifos treatment

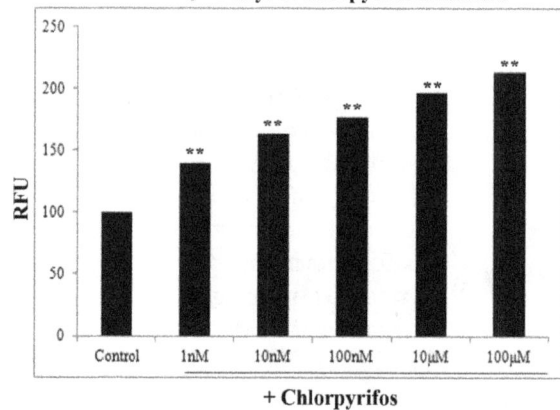

E ROS inhibited by NAC

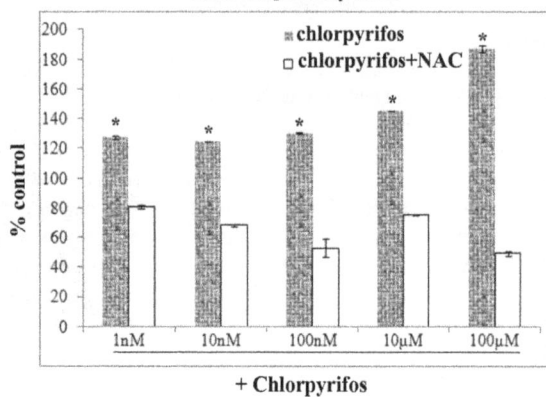

F Correlation between ROS & GSH

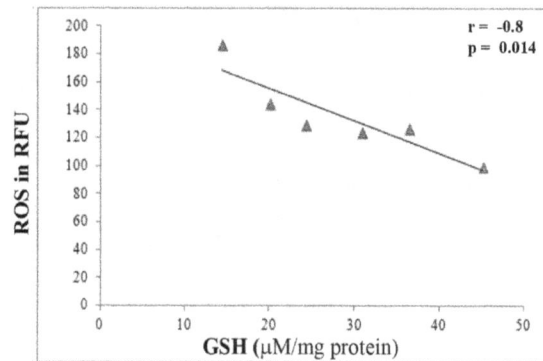

G GSH :GSSG ratio- 24 hr

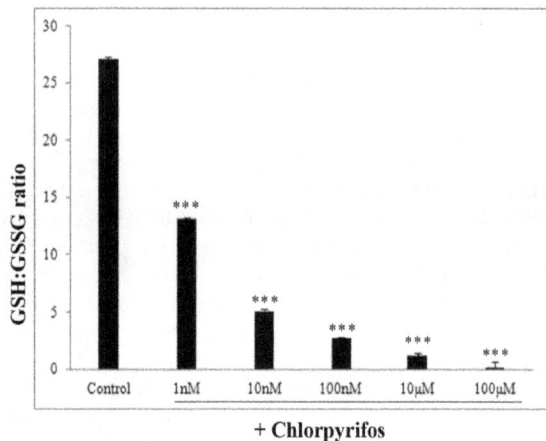

H GSH :GSSG ratio- 9days

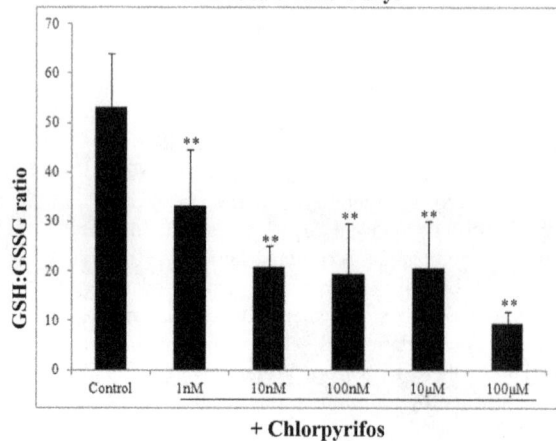

Figure 1. Effect of chlorpyrifos on cell viability and oxidative stress in ARPE19 cells. Effect of chlorpyrifos on cell viability in ARPE19 cells was assessed using MTT assays (A) after 24 hr chlorpyrifos treatment, (B) After 9 days of chlorpyrifos treatment. (C) ROS production measured by the DCFDA method in ARPE19 cells after 24 hr chlorpyrifos treatment. DMSO was the vehicle control used. tBH (500 µM) and H_2O_2 (1 mM) was the positive control for ROS generation, (D) ROS production after 9 days of chlorpyrifos treatment. (E) ROS production measured in ARPE19 cells pretreated with NAC & exposed to chlorpyrifos. (F) Negative correlation between ROS generation and the GSH level upon chlorpyrifos treatment. Dose-dependent response of the GSH/GSSG ratio to chlorpyrifos exposure (G) for 24 hr (H) after 9 days of chlorpyrifos treatment. p values are the comparison between treated control and the respective treatments. * $p < 0.05$, ** $p < 0.01$, *** $p < 0.001$. All value expressed are a mean of 3 experiments done in triplicates and the values are expressed as Mean ± SD.

negative correlation between the ROS generation and the GSH level was observed with chlorpyrifos treatment and is indicative of oxidative stress ($p = 0.014$) (Figure 1F).

Chlorpyrifos induces the expression of paraoxonase

Since paraoxonase is the enzyme that comes into action when an organism is exposed to organophosphate pesticides, we analyzed the expression of paraoxonase at the mRNA level in the RPE cell when exposed to chlorpyrifos. Semi quantitative PCR was employed to analyse the expression of PON2, which is the predominant tissue form of PON [21]. In ARPE19 cells also the predominant form expressed was PON2 followed by PON3 and PON1 (supplementary data). The expression of PON2 mRNA was significantly increased when ARPE19 cells were exposed to chlorpyrifos for 3 hr, 24 hr and 9 days. At 3 hr, exposure to 100 nM and above concentration of chlorpyrifos, increase in PON2 expression was observed ($p < 0.05$). However at 24 hr, this was 6 fold higher as seen at 100 nM ($p < 0.05$). The chronic 9 days exposure revealed an increase in the PON2 in response to all concentration of chlorpyirofs including the lowest at 1 nM. A maximum is reached at 10 nM with a 0.5 fold increase over the untreated control ($p < 0.05$) (Figure 2). Thus, the maximal increase was seen at the end of 24 hr.

Treatment with chlorpyrifos showed a 0.9 fold increase in PON2 expression by qPCR. In order to check whether this is a ROS mediated effect, H_2O_2 was used as a prooxidant and the relative PON2 expression was quantified using Real time PCR. H_2O_2 was also found to induce expression of PON2 in ARPE19 cells. The Real Time PCR showed a significant increase in PON2 expression when 100 µM H_2O_2 were added to the cells ($p < 0.05$) (Figure 2D). Pretreatment with NAC however did not abrogate the effect of H_2O_2 but in turn increased the PON2 expression.

Measurement of paraoxonase enzyme activity

The specific activity of PONase in the untreated ARPE19 cells was found to be 5 ± 1.23 nmol/mg protein. The PONase activity showed a significant dose dependent increase with increase in concentration of the chlorpyrifos treatment for 3 hr ($p < 0.001$). A maximum specific activity of 24.08 ± 2.14 nmol/mg protein was observed at 100 µM chlorpyrifos (Figure 3A). With 24 hr treatment, the specific activity was significantly lower than the control, upto 100 nM chlorpyrifos ($p < 0.01$). However, a dose dependent increase was observed, that was significantly higher at 10 µM and 100 µM chlorpyrifos ($p < 0.01$). A maximum of 5 fold increase was seen at 100 µM at the end of 3 hr of exposure and less than 2 fold increase is seen at 24 hr.

With respect to PON-AREase activity, no significant change in the activity was observed after 3 hr of chlorpyrifos treatment. However at 24 hr, a decrease in specific activity was seen at lower concentration of chlorpyrifos from 1 nM to 10 nM, while at concentrations above 100 nM a dose dependent increase was observed with a maximal activity of 67 ± 7.09 µmol/mg protein at 100 µM of chlorpyrifos exposure ($p < 0.05$) (Figure 3B).

The PON-HCTLase activity in both chlorpyrifos treated and untreated ARPE19 cells were found to be low at 3 hr and 24 hr of chlorpyrifos exposure, with specific activity > 0.1 U/mg protein (Figure 3C). However, after treatment with chlopyrifos for 9 days, there was a dose dependent increase in specific activity ranging from 4.5 ± 0.54 U/mg protein to 13 ± 1.76 U/mg protein ($p < 0.05$) (Figure 3F). The specific activity of PONase and PON-AREase was found to be lower in the 9 days grown cells (control). After 9 days of chlorpyrifos exposure, a dose dependent significant increase in PONase activity was observed compared to control (Figure 3D). The specific activity of PON-AREase was found to be significantly lowered and it showed a dose dependent decrease (Figure 3E).

Blocking PON2 expression disturbed the ARPE19 cells ability to withstand oxidative insult

Since PON2 expression was observed to be elevated upon exposure to chlorpyrifos study was done to determine the effect of PON2 silencing in ARPE19 cells exposed to chlorpyrifos. Down regulation of PON2 after siRNA silencing in ARPE19 cells was shown by western blot (Figure 4A) and qRT PCR (Figure 4B). Cell viability studies showed that upon silencing the PON2 expression in ARPE19 cells and exposing the cells to chlorpyrifos, there was a significant loss in viability as seen by MTT assay (Figure 4C). With 26 % cell death, a significant decrease in cell viability was observed when compared to the control cells ($p < 0.01$). There was a 44% increase in ROS production ($p < 0.001$) in the PON2 silenced ARPE19 cells exposed to chlorpyrifos in comparison to the non silencing siRNA (scrambled siRNA) transfected cells exposed to chlorpyrifos (Figure 4D).

Exploring the transcription factors regulating the expression of PON by qRT-PCR

In ARPE19 cells, chlorpyrifos increased the synthesis of PON2 and experiments where PON2 was silenced also proved that PON2 in the cells provide protection during stress. The expression of transcription factors that regulate transcription of PON2 was further looked into. The transcription factors studied were shortlisted based on literature on PON expression. Expression of transcription factors namely ARH, STAT5B [30–31], SREBP2 [32], NRF2 [33], JUN [34], PPARG [35] and SP1 were analysed in the chlorpyrifos exposed cells and the control ARPE19 cells. At 24 hr of chlorpyrifos exposure, there was a significant 2.5 fold increase in expression of SP1 in the chlorpyrifos exposed ARPE19 cells compared to the untreated controls ($p < 0.05$) (Figure 5A).

To further prove SP1 mediated PON2 expression, mithramycin an inhibitor that interferes with SP1 transcription factor binding was treated before chlorpyrifos exposure. Mithramycin treated cells showed 0.8 fold decrease in PON2 expression even after chlorpyrifos challenge, as per the Real Time PCR results (Figure 5B), thereby showing that the transcription factor SP1 plays a role in regulating PON2 expression.

PON2 -3hr Chlorpyrifos

A

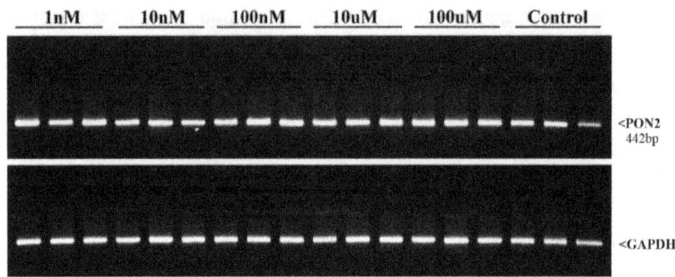

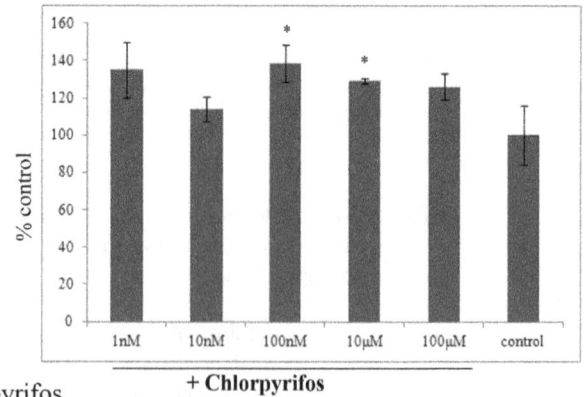

PON2 -24hr Chlorpyrifos

B

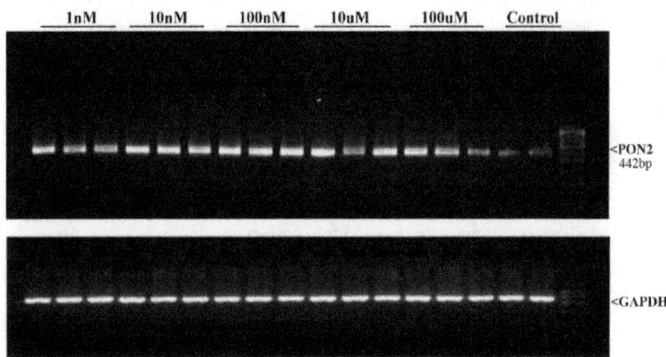

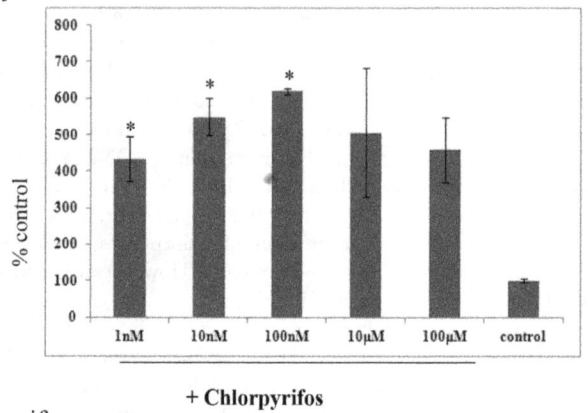

PON2- 9 Days Chlorpyrifos

C

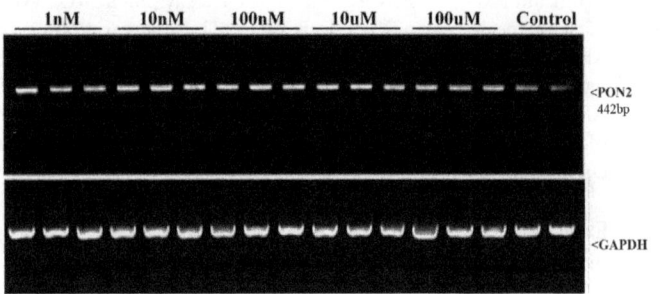

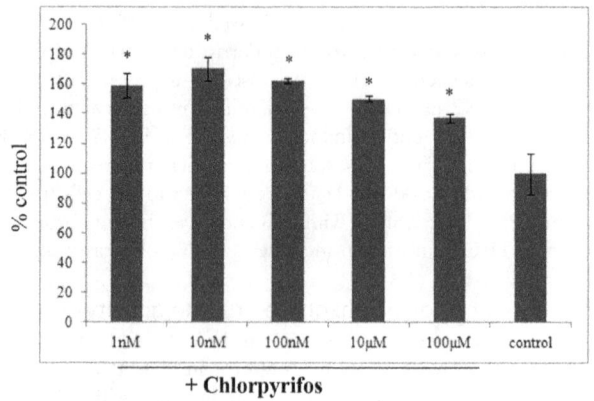

D

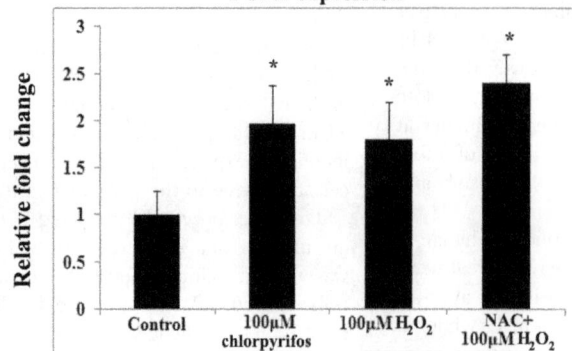

Figure 2. Expression of PON2 in ARPE19 cells upon chlorpyrifos exposure. (A) Expression of PON2 at 3 hr chlorpyrifos exposure. (B) PON2 at 24 hr chlorpyrifos exposure. (C) PON2 at 9 days of chlorpyrifos exposure. PCR product is visualized on ETBR gel and the Histogram represents the quantification of PCR band normalized to the GAPDH. (D) Relative quantification of PON2 expression after exposure to 100 µM chlorpyrifos and 100 µM of H_2O_2 with and without pretreatment with 5 mM NAC. Histogram represents the fold change after normalization to the GAPDH. All value expressed are a mean of 3 experiments done in triplicates and the values are expressed as Mean ±SD. p values are the comparison between treated control and the respective treatments.*$p < 0.05$, t test.

Discussion

Ocular damage due to organophosphate pesticides exposure was studied by the US environmental protection agency subsequent to reports from India and Japan [36–37]. A few of these studies have indicated damage in the eye at the level of retina. Acute dose of Fenthion in experimental rats caused increased expression of glial fibrillary acid protein (GFAP) in rat retina [38]. Macular damage suggestive of RPE defect was noticed in significant proportion of farmers using chlorpyrifos [39].

In this study cultured human retinal pigment epithelium was treated with sub lethal dose of chlorpyrifos ranging from 1 nM to 100 µM, and no significant cell death was observed. Earlier reports on acute and chronic pesticide toxicity in mouse, revealed markers of oxidative stress such as oxidized lipids, lowering of antioxidant enzymes and DNA damage in the retina [13]. Our study revealed that chlorpyrifos treatment to ARPE19 cells induces a dose dependent ROS generation. The antioxidant N-acetyl cysteine was able to significantly reduce the ROS generated, thus showing that organophosphorous pesticide, chlorpyrifos causes an oxidative stress to ARPE19 cells. Chlorpyrifos also led to lowering of the redox status as measured by the ratio of GSH/GSSG in the ARPE19 cells with significant reduction in glutathione levels. The activities of antioxidant enzymes superoxide dismutase, catalase and glutathione peroxidase are previously reported to be decreased in the retina of chlorpyrifos administrated mice [13]. Reduced glutathione levels associated with

chlorpyriofs treatment is reported in rat brain [40] and recently in JEG-3 cells [41].

As seen in this study, ARPE19 cells were capable of withstanding the oxidative insult in response to the chlorpyrifos exposure with no significant cell death as depicted in the cell viability assay. Concentrations up to 1 mM did not show significant cell death at 24 hours (data not shown).

Organophosphorous pesticides like chlorpyrifos are cleaved by the enzyme paraoxonase. The paraoxonase (PON) gene cluster contains three members (PON1, PON2 and PON3), located on chromosome 7q21.3–22.1. Of the three, PON2 the tissue resident form is more expressed in ARPE19 cells with low levels of PON1 and PON3 mRNA (Supplementary Figure S1). PON2 is an intracellular form that protects cells against oxidative stress and is not associated with HDL unlike PON 1 and 3 [21–42]. In this study, the expression and activity of paraoxonase were determined in ARPE19 cells exposed to chlorpyrifos. An increase in the specific activity of PONase was observed at 3 hr, which can be attributed to the paraoxon substrate availability and its detoxification. PON2 mRNA levels are not increased at 3 hr, as much as seen at 24 hr. A 4-fold increase in PON2 mRNA expression was observed at 24 hr and this gene expression change can be attributed to the pro oxidant insult mediated signaling. This is supported by the fact that chlorpyrifos exposure at 24 hr, showed increased ROS, lowered glutathione and altered redox status apart from increase in the PON-AREas antioxidant activity. Silencing of PON2 increased the ROS levels and caused significant cell death, indicating the crucial role of PON2 in RPE. Thus, the acute

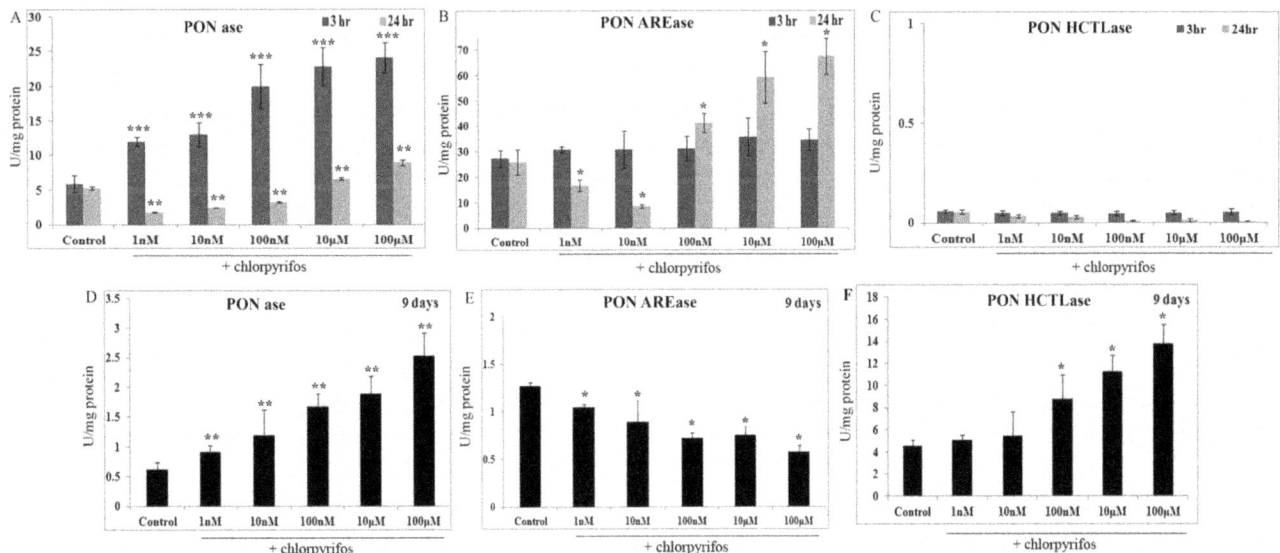

Figure 3. Effect of chlorpyrifos on paraoxonase enzyme activity. (A) PON-ase specific activity upon acute exposure to chlorpyrifos (3 hr and 24 hr). (B) PON-AREase specific activity upon acute exposure to chlorpyrifos (3 hr and 24 hr). (C) PON-HCTLase specific activity upon acute exposure to chlorpyrifos (3 hr and 24 hr). (D) PON-ase specific activity in chronic chlorpyrifos exposure (9 days). (E) PON-AREase specific activity in chronic chlorpyrifos exposure (9 days). (F) PON-HCTLase specific activity in chronic chlorpyrifos exposure (9 days). All value expressed are a mean of 3 experiments done in triplicates and the values are expressed as Mean ±SD. p values are the comparison between treated control and the respective treatments. *$p < 0.05$, **$p < 0.01$, ***< 0.001.

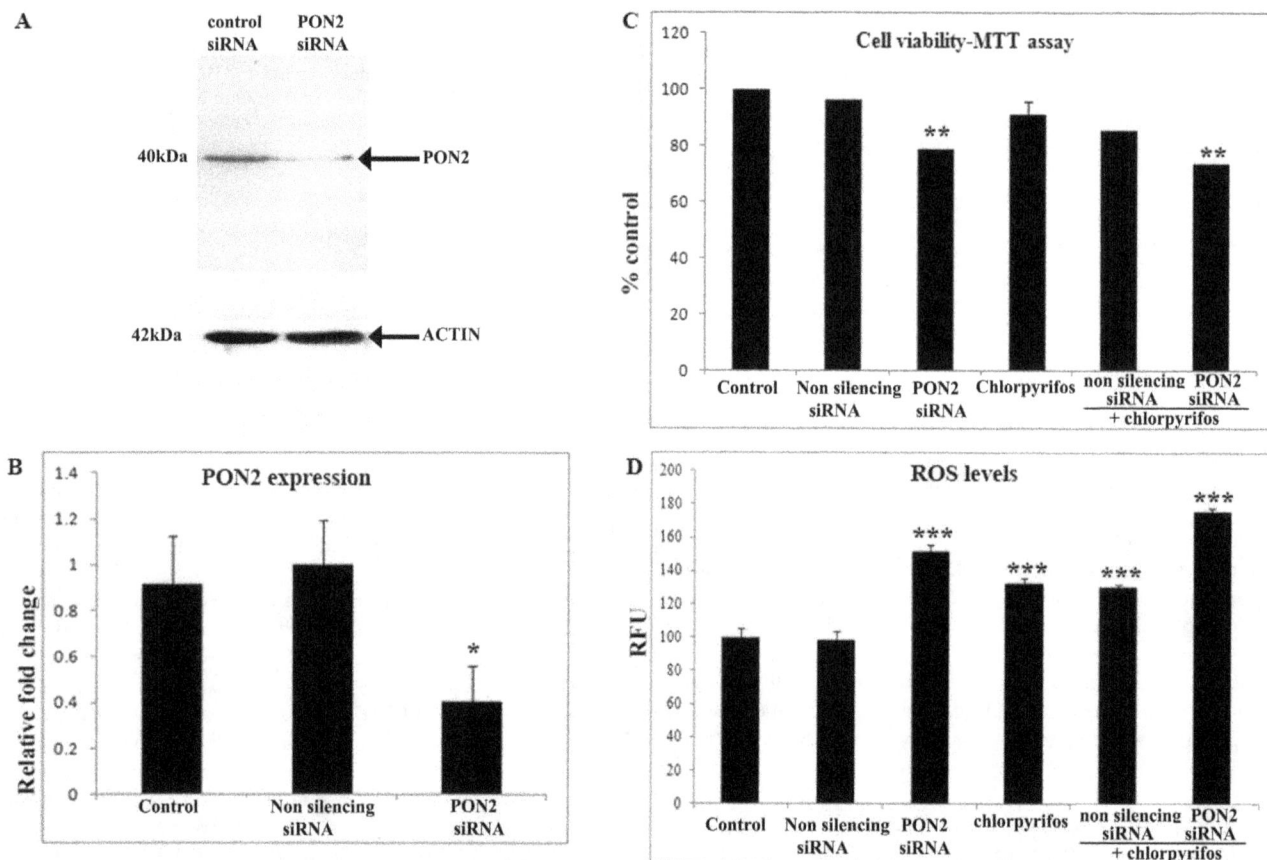

Figure 4. Silencing PON2 expression causes cell death and increases the ROS production. (A) Immunoblot analysis of PON2 from ARPE19 cells transfected with PON2 and non silencing siRNA. β ACTIN used for normalizing the protein load. (B) Histogram represents the expression of PON2 in the control and siRNA treated ARPE19 cells as fold change obtained from the qRT PCR. (C) Effect of 100 μM chlorpyrifos on cell viability in ARPE19 cells where PON2 expression was silenced using siRNA. (D) ROS production measured by the DCFDA method in ARPE19 cells transfected with PON2 siRNA or non silencing siRNA and exposed to chlorpyrifos for 24 hr. All value expressed are a mean of 3 experiments done in triplicates and the values are expressed as Mean ±SD. p values are the comparison between treated control and the respective treatments.*p<0.05, **p<0.01, ***p<0.001, t test.

exposure of chlorpyrifos to ARPE19 cells is taken care by the cellular paraoxonase activity and gene expression.

Nine days chronic exposure of chlorpyrifos at varying concentration showed a net increase in the mRNA levels of PON2 compared to the untreated control, but was less when compared to the 24 hr levels, indicating the lowering of defense through PON expression. Oxidative stress has been shown to decrease the mRNA expression of PON2 and over expression of PON2 to be protective, as studied in mice macrophages and HeLa cells respectively [21–43]. On the other hand, Shamir et al, showed that induced oxidative stress did not alter the mRNA expression of PON2 in Caco-2 cells [44]. A recent study by Chiapella et al indicates that the placental JEG-3 cells are able to attenuate the oxidative stress induced by chlorpyrios [41]. Thus, there seems to be variations in PON expression based on the tissue type. RPE cells detoxify the chlorpyrifos and show an up regulation of the PON2 expression probably as a defense in response to the chlorpyrifos induced oxidative stress.

Treatment with the proxidant such as H_2O_2 increased the PON expression, showing the ROS mediated gene expression of PON2. Interestingly this was not abrogated by pretreatment with antioxidant NAC. NAC treatment further increased the PON2 expression. This can be explained by the fact that NAC action is not restricted to just free radical scavenging or in improving the intracellular glutathione as cysteine precursor, but also in influencing the redox state of the cysteine residues of signaling molecules such as Raf-1, MEK and ERK. NAC-mediated signaling resulting in the activation of SP1 is reported [45]. Thus with NAC treatment, an increase in PON 2 expression was observed in the ARPE19 cells. Therefore, treatment with NAC, not only improves the redox status, but can also increase the antioxidant response through PON2 upregulation. The beneficial effects of NAC supplements in chlorpyrifos toxicity can be further looked into.

ARPE19 cells grown post confluent for 9 days showed an overall decrease in the activity of PONase and PON-AREase but the thiolactoanse activity showed an increase. Interestingly there was around 10 fold increase in the PON-HCTLase activity when compared to the untreated control at 9 days. The specific activity of PON-HCTLase increased dose dependently with chlorpyrifos treatment. This can also be due to the accumulation of substrates for the thiolactonase activity. However the range of physiological substrate for this enzyme activity is still unclear [42–46].

Thus it is observed that PONase activity plays an immediate role to detoxify chlorpyrifos and reduces cellular ROS that helps in the survival of the cell. The PON-AREase activity increases with

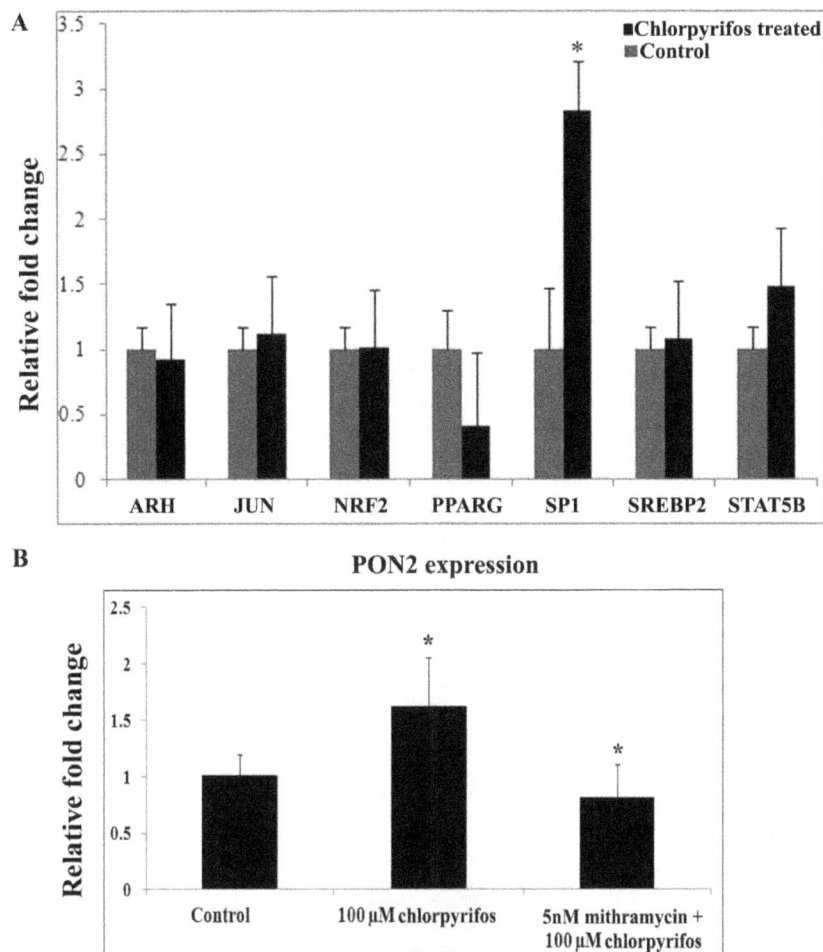

Figure 5. Transcriptional factors regulating PON2 expression. (A) Relative quantification of the expression of transcription factors regulating PON2 in ARPE19 cells in response to chlorpyrifos exposure for 24 hr. (B) Effect of mithramycin on PON2 expression. 5 nM of mithramycin was used to interfere with SP1 function. Histogram represents the fold change after normalization to the GAPDH. All value expressed are a mean of 3 experiments done in triplicates and the values are expressed as Mean ±SD. p values are the comparison between treated control and the respective treatments. * $p < 0.05$.

time to handle the accumulating oxidized substrates. However, in spite of the augmented paraoxanase activity and expression, the oxidative stress and altered redox status still seems to be predominant, clearly showing that the cells antioxidant machinery is down. Studies in animal models, on long term chlorpyrifos exposure is required for further understanding the RPE metabolism and molecular mechanism in handling the chlorpyrifos toxicity especially with chronic exposure.

Amongst the known transcriptional regulators of PON, SP1 was significantly increased in RPE cells exposed to chlorpyrifos. SP1 is reported to act as a positive regulator of PON1 transcription, mediated through PKC, a cellular sensor of the intra cellular redox changes [47–48]. SP1 activation during excess ROS generation is previously reported and the same is seen in this study [49]. Thus based on this study it is inferred that chlorpyrifos that induces ROS generation and glutathione depletion, results in increased PON2 expression through SP1 activation, as a protective response. While regulation of ROS using antioxidants may prove to be

beneficial in chlorpyrifos induced toxicity, the role of PON polymorphism in the pesticide users needs attention.

Supporting Information

Figure S1 Expression of PON1, PON2 and PON3 in ARPE19 cells and HUH cells. HUH cells are used as positive control to show that primers of PON1 and PON3 are working.

Acknowledgments

Dr.Venil Sumantran is acknowledged for her information on chlorpyrifos while initiating the study.

Author Contributions

Conceived and designed the experiments: NA JMJ. Performed the experiments: JMJ. Analyzed the data: NA JMJ KA SRB. Contributed reagents/materials/analysis tools: NA JMJ. Wrote the paper: NA JMJ.

References

1. Garcia-Ramirez M, Villarroel M, Corraliza L, Hernandez C, Simo R. (2011) Measuring permeability in human retinal epithelial cells (ARPE-19): implications for the study of diabetic retinopathy. Methods Mol.Biol 763: 179–94.

2. Shirasawa M, Sonoda S, Terasaki H, Arimura N, Otsuka H, et al. (2013) TNF-alpha disrupts morphologic and functional barrier properties of polarized retinal pigment epithelium. Exp Eye Res. 110: 59–69.

3. Bastiaans J, van Meurs JC, van Holten-Neelen C, Nijenhuis MS, Kolijn-Couwenberg MJ, et al. (2013) Factor Xa and thrombin stimulate proinflammatory and profibrotic mediator production by retinal pigment epithelial cells: a role in vitreoretinal disorders? Graefes Arch.Clin.Exp Ophthalmol. 251: 1723–33.

4. Dahrouj M, Alsarraf O, Liu Y, Crosson CE, Ablonczy Z. (2013) C-type natriuretic peptide protects the retinal pigment epithelium against advanced glycation end product-induced barrier dysfunction. J Pharmacol Exp Ther. 344: 96–102.

5. Kim JH, Lee SJ, Kim KW, Yu YS, Kim JH. (2012) Oxidized low density lipoprotein-induced senescence of retinal pigment epithelial cells is followed by outer blood-retinal barrier dysfunction. Int J Biochem Cell Biol 44: 808–14.

6. Lu L, Hackett SF, Mincey A, Lai H, Campochiaro PA. (2006) Effects of different types of oxidative stress in RPE cells. J Cell Physiol 206: 119–25.

7. Chen WQ, Ma H, Bian JM, Zhang YZ, Li J. (2012) Hyper-phosphorylation of GSK-3beta: possible roles in chlorpyrifos-induced behavioral alterations in animal model of depression. Neurosci Lett 528: 148–52.

8. Geller AM, Abdel-Rahman AA, Peiffer RL, Abou-Donia MB, Boyes WK. (1998) The organophosphate pesticide chlorpyrifos affects form deprivation myopia. Invest Ophthalmol.Vis.Sci.39: 1290–4.

9. World health statistics 2009. Available: http://www.who.int/whosis/whostat/2009. Accessed 7 June 2014 .

10. Geller AM, Sutton LD, Marshall RS, Hunter DL, Madden V, et al. (2005) Repeated spike exposure to the insecticide chlorpyrifos interferes with the recovery of visual sensitivity in rats. Doc.Ophthalmol. 110: 79–90.

11. Akhtar N, Srivastava MK, Raizada RB. (2009) Assessment of chlorpyrifos toxicity on certain organs in rat, Rattus norvegicus. J Environ.Biol 30: 1047–53.

12. Yoshikawa H, Yoshida M, Hara I. (1990) [Effect of administration with chlorpyrifos on electroretinogram in rats]. Nihon Eiseigaku Zasshi 45: 884–9.

13. Yu F, Wang Z, Ju B, Wang Y, Wang J, Bai D. (2008) Apoptotic effect of organophosphorus insecticide chlorpyrifos on mouse retina in vivo via oxidative stress and protection of combination of vitamins C and E. Exp Toxicol.Pathol. 59: 415–23.

14. Tripathi S, Suzuki N, Srivastav AK. (2013) Response of serum minerals (calcium, phosphate, and magnesium) and endocrine glands (calcitonin cells and parathyroid gland) of Wistar rat after chlorpyrifos administration. Microsc.Res. Tech. 76: 673–8.

15. Timchalk C, Nolan RJ, Mendrala AL, Dittenber DA, Brzak KA, et al. (2002) Physiologically based pharmacokinetic and pharmacodynamic (PBPK/PD) model for the organophosphate insecticide chlorpyrifos in rats and humans. Toxicol.Sci. 66: 34–53.

16. Hines CJ, Deddens JA. (2001) Determinants of chlorpyrifos exposures and urinary 3,5,6-trichloro-2-pyridinol levels among termiticide applicators. Ann Occup.Hyg. 45: 309–21.

17. European commission health and consumer protection directorate-general (2005). EC. Available http://ec.europa.eu/food/plant/protection/evaluation/existactive/list. Accessed 2 June 2014.

18. Loscalzo J. (2008) Paraoxonase and coronary heart disease risk: language misleads, linkage misinforms, function clarifies. Circ.Cardiovasc.Genet. 1: 79–80.

19. Perla-Kajan J, Jakubowski H. (2010) Paraoxonase 1 protects against protein N-homocysteinylation in humans. FASEB J 24: 931–6.

20. Draganov DI, Stetson PL, Watson CE, Billecke SS, La Du BN. (2000) Rabbit serum paraoxonase 3 (PON3) is a high density lipoprotein-associated lactonase and protects low density lipoprotein against oxidation. J Biol Chem 275: 33435–42.

21. Ng CJ, Wadleigh DJ, Gangopadhyay A, Hama S, Grijalva VR, et al. (2001) Paraoxonase-2 is a ubiquitously expressed protein with antioxidant properties and is capable of preventing cell-mediated oxidative modification of low density lipoprotein. J Biol Chem 276: 44444–9.

22. Pauer GJ, Sturgill GM, Peachey NS, Hagstrom SA. (2010) Protective effect of paraoxonase 1 gene variant Gln192Arg in age-related macular degeneration. Am J Ophthalmol. 149: 513–22.

23. Jin M, Yaung J, Kannan R, He S, Ryan SJ, et al. (2005) Hepatocyte growth factor protects RPE cells from apoptosis induced by glutathione depletion. Invest Ophthalmol.Vis.Sci. 46: 4311–9.

24. Charlton-Menys V, Liu Y, Durrington PN. (2006) Semiautomated method for determination of serum paraoxonase activity using paraoxon as substrate. Clin.Chem 52: 453–7.

25. Ekinci D, Senturk M, Beydemir S, Kufrevioglu OI, Supuran CT. (2010) An alternative purification method for human serum paraoxonase 1 and its interactions with sulfonamides. Chem Biol Drug Des 76: 552–8.

26. Thyagarajan B, Jacobs DR Jr, Carr JJ, Alozie O, Steffes MW, et al. (2008) Factors associated with paraoxonase genotypes and activity in a diverse, young, healthy population: the Coronary Artery Risk Development in Young Adults (CARDIA) study. Clin.Chem 54: 738–46.

27. Cabana VG, Reardon CA, Feng N, Neath S, Lukens J, et al. (2003) Serum paraoxonase: effect of the apolipoprotein composition of HDL and the acute phase response. J Lipid Res. 44: 780–92.

28. Angayarkanni N, Barathi S, Seethalakshmi T, et al. (2008) Serum PON1 arylesterase activity in relation to hyperhomocysteinaemia and oxidative stress in young adult central retinal venous occlusion patients. Eye (Lond) 22: 969–74.

29. Koubaa N, Hammami S, Nakbi A, Ben Hamda K, Mahjoub S, et al. (2008) Relationship between thiolactonase activity and hyperhomocysteinemia according to MTHFR gene polymorphism in Tunisian Behcet's disease patients. Clin.Chem Lab Med 46: 187–92.

30. Cheng X, Klaassen CD. (2012) Hormonal and chemical regulation of paraoxonases in mice. J Pharmacol Exp Ther. 342: 688–95.

31. Yuan J, Devarajan A, Moya-Castro R, Zhang M, Evans S, et al. (2010) Putative innate immunity of antiatherogenic paraoxanase-2 via STAT5 signal transduction in HIV-1 infection of hematopoietic TF-1 cells and in SCID-hu mice. J Stem Cells 5: 43–8.

32. Fuhrman B, Gantman A, Khateeb J, Volkova N, Horke S J, et al. (2009) Urokinase activates macrophage PON2 gene transcription via the PI3K/ROS/MEK/SREBP-2 signalling cascade mediated by the PDGFR-beta. Cardiovasc.Res. 84: 145–54.

33. Bayram B, Ozcelik B, Grimm S, Roeder T, Schrader C, et al. (2012) A diet rich in olive oil phenolics reduces oxidative stress in the heart of SAMP8 mice by induction of Nrf2-dependent gene expression. Rejuvenation.Res.15: 71–81.

34. S.P.. Deakin, R. W . James. (2008) Transcriptional Regulation of the Paraoxonase Genes. The Paraoxonases: Their Role in Disease Development and Xenobiotic Metabolism Proteins And Cell Regulation Volume 6, pp 241–250, 241–250. Springer. Available at http://link.springer.com/book/10.1007%2F978-1-4020-6561-3. Accessed 7June 2014.

35. Shiner M, Fuhrman B, Aviram M. (2007) Macrophage paraoxonase 2 (PON2) expression is up-regulated by pomegranate juice phenolic anti-oxidants via PPAR gamma and AP-1 pathway activation. Atherosclerosis;195: 313–21.

36. Boyes WK, Tandon P, Barone S Jr, Padilla S. (1994) Effects of organophosphates on the visual system of rats. J Appl.Toxicol. 14: 135–43.

37. Dementi B. (1994) Ocular effects of organophosphates: a historical perspective of Saku disease. J Appl.Toxicol. 14: 119–29.

38. Tandon P, Padilla S, Barone S Jr, Pope CN, Tilson HA. (1994) Fenthion produces a persistent decrease in muscarinic receptor function in the adult rat retina. Toxicol.Appl.Pharmacol 25: 271–80.

39. Misra UK, Nag D, Misra NK, Murti CR. (1982) Macular degeneration associated with chronic pesticide exposure. Lancet 1: 288.

40. Verma RS, Srivastava N. (2001) Chlorpyrifos induced alterations in levels of thiobarbituric acid reactive substances and glutathione in rat brain. Indian J Exp Biol 39: 174–7.

41. Chiapella G, Flores-Martín J, Ridano ME, Reyna L, Magnarelli de Potas G, et al. (2013) The organophosphate chlorpyrifos disturbs redox balance and triggers antioxidant defense mechanisms in JEG-3 cells. Placenta 34: 792–8.

42. Draganov DI, Teiber JF, Speelman A, Osawa Y, Sunahara R, et al. (2005) Human paraoxonases (PON1, PON2, and PON3) are lactonases with overlapping and distinct substrate specificities. J Lipid Res. 46: 1239–47.

43. Devarajan A, Grijalva VR, Bourquard N, Meriwether D, Imaizumi S, et al. (2012) Macrophage paraoxonase 2 regulates calcium homeostasis and cell survival under endoplasmic reticulum stress conditions and is sufficient to prevent the development of aggravated atherosclerosis in paraoxonase 2 deficiency/apoE-/- mice on a Western diet. Mol.Genet.Metab 107: 416–27.

44. Shamir R, Hartman C, Karry R, Pavlotzky E, Eliakim R, et al. (2005) Paraoxonases (PONs) 1, 2, and 3 are expressed in human and mouse gastrointestinal tract and in Caco-2 cell line: selective secretion of PON1 and PON2. Free Radic.Biol Med 39: 336–44.

45. Kim KY, Rhim T, Choi I, Kim SS. (2001) N-acetylcysteine induces cell cycle arrest in hepatic stellate cells through its reducing activity. J Biol Chem 276: 40591–8.

46. Martinelli N, Consoli L, Girelli D, Grison E, Corrocher R, et al. (2013) Paraoxonases: ancient substrate hunters and their evolving role in ischemic heart disease. Adv.Clin.Chem 59: 65–100.

47. Arii K, Suehiro T, Ikeda Y, Kumon Y, Inoue M, et al. (2010) Role of protein kinase C in pitavastatin-induced human paraoxonase I expression in Huh7 cells. Metabolism 59: 1287–93.

48. Osaki F, Ikeda Y, Suehiro T, Ota K, Tsuzura S, et al. (2004) Roles of Sp1 and protein kinase C in regulation of human serum paraoxonase 1 (PON1) gene transcription in HepG2 cells. Atherosclerosis 176: 279–87.

49. Danell RM, Glish GL. (2000) A new approach for effecting surface-induced dissociation in an ion cyclotron resonance mass spectrometer: a modeling study. J Am Soc Mass Spectrom 11: 1107–17.

Does Occupational Exposure to Solvents and Pesticides in Association with Glutathione S-Transferase A1, M1, P1, and T1 Polymorphisms Increase the Risk of Bladder Cancer? The Belgrade Case-Control Study

Marija G. Matic[1,5]9, Vesna M. Coric[1,5]9, Ana R. Savic-Radojevic[1,5], Petar V. Bulat[2,5], Marija S. Pljesa-Ercegovac[1,5], Dejan P. Dragicevic[3,5], Tatjana I. Djukic[1,5], Tatjana P. Simic[1,5], Tatjana D. Pekmezovic[4,5]*

1 Institute of Medical and Clinical Biochemistry, Faculty of Medicine, University of Belgrade, Belgrade, Serbia, **2** Institute of Occupational Health, Belgrade, Serbia, **3** Clinic of Urology, Clinical Center of Serbia, Belgrade, Serbia, **4** Institute of Epidemiology, Faculty of Medicine, University of Belgrade, Belgrade, Serbia, **5** Faculty of Medicine, University of Belgrade, Belgrade, Serbia

Abstract

Objective: We investigated the role of the glutathione S-transferase A1, M1, P1 and T1 gene polymorphisms and potential effect modification by occupational exposure to different chemicals in Serbian bladder cancer male patients.

Patients and Methods: A hospital-based case-control study of bladder cancer in men comprised 143 histologically confirmed cases and 114 age-matched male controls. Deletion polymorphism of glutathione S-transferase M1 and T1 was identified by polymerase chain reaction method. Single nucleotide polymorphism of glutathione S-transferase A1 and P1 was identified by restriction fragment length polymorphism method. As a measure of effect size, odds ratio (OR) with corresponding 95% confidence interval (95%CI) was calculated.

Results: The glutathione S-transferase A1, T1 and P1 genotypes did not contribute independently toward the risk of bladder cancer, while the glutathione S-transferase M1-null genotype was overrepresented among cases (OR = 2.1, 95% CI = 1.1–4.2, p = 0.032). The most pronounced effect regarding occupational exposure to solvents and glutathione S-transferase genotype on bladder cancer risk was observed for the low activity glutathione S-transferase A1 genotype (OR = 9.2, 95% CI = 2.4–34.7, p = 0.001). The glutathione S-transferase M1-null genotype also enhanced the risk of bladder cancer among subjects exposed to solvents (OR = 6,5, 95% CI = 2.1–19.7, p = 0.001). The risk of bladder cancer development was 5.3–fold elevated among glutathione S-transferase T1-active patients exposed to solvents in comparison with glutathione S-transferase T1-active unexposed patients (95% CI = 1.9–15.1, p = 0.002). Moreover, men with glutathione S-transferase T1-active genotype exposed to pesticides exhibited 4.5 times higher risk in comparison with unexposed glutathione S-transferase T1-active subjects (95% CI = 0.9–22.5, p = 0.067).

Conclusion: Null or low-activity genotypes of the glutathione S-transferase A1, T1, and P1 did not contribute independently towards the risk of bladder cancer in males. However, in association with occupational exposure, low activity glutathione S-transferase A1 and glutathione S-transferase M1-null as well as glutathione S-transferase T1-active genotypes increase individual susceptibility to bladder cancer.

Editor: Keitaro Matsuo, Kyushu University Faculty of Medical Science, Japan

Funding: This work was supported by the Ministry of Education and Science of the Republic of Serbia (Grants number: 175052 and 175087. The funders had no role in study design, data collection and analysis, decision to publish, or preparation of the manuscript.

Competing Interests: The authors have declared that no competing interests exist.

* E-mail: pekmezovic@sezampro.rs

9 These authors equally contributed to this work.

Introduction

Bladder cancer is the second most common malignancy of the urinary tract and has the second highest mortality rate among urological neoplasms [1]. It affected 73,510 patients and lead to 14,880 deaths in 2012 worldwide [2]. Demographic characteristics associated with the greatest risk for bladder cancer include male gender, white race and the increasing age [3]. It is generally estimated that the male:female incidence ratio is 3.8:1.0 [3]. The most frequent pathohistological type of bladder cancer is urothelial carcinoma, also called transitional cell carcinoma (TCC), accounting for approximately 90% of all bladder cancers [3]. It has been known that uroepithelial cells are most vulnerable to metabolic end products of different compounds, including carcinogens. This malignancy is characterized by multifactorial etiology, involving both genetic and environmental factors.

The well established risk factors for bladder cancer include cigarette smoking (50% cases in men, 30% cases in women), but also exposure to occupational agents [3]. Occupational exposures account for 5 to 25% of all bladder cancer cases. [4]. Over 40 occupations have been associated with an elevated risk of bladder cancer in epidemiologic studies, but the evidence is compelling for only a few. Those established at risk industries include the manufacturing of products such as synthetic dyes and paints, cables, textiles, leather works, and aluminum and the petrochemical, coal tar, and rubber industries [5,6]. A number of specific occupations have also been identified to be associated with increased risk of bladder cancer. These include, but are not limited to, cooks and kitchen workers, electricians, hairdressers, leather workers, machinists, petroleum workers, rubber workers, coalminers, truckers, and vehicle mechanics, as summarized by Schulte et al. [7] in 1987, as well as coke oven workers, roofers, dry cleaners, chimney sweeps, and painters, as addressed by others in more recent literature [5,8–10].

Despite the fact that occupations associated with bladder cancer have been well established, the question still arises why individuals with seemingly equal exposure to occupational carcinogens develop bladder cancer in an unpredictable manner. This is probably attributed to genetic polymorphisms of the genes coding for the xenobiotic metabolizing enzymes, particularly glutathione S-transferase (GST). GSTs catalyze the conjugation of glutathione on electrophilic substrates and are an important line of defense in the protection of cellular components against reactive species. The most well characterized GST classes have been named alpha (GSTA), mu (GSTM), pi (GSTP) and theta (GSTT). Appreciable GST activities are seen in bladder epithelium [11]. GST enzymes that belong to various classes have different, but sometimes overlapping, substrate specificities. Several types of allelic variations have been identified within GST classes, with that in the GSTM1, GSTT1 and GSTP1 genes receiving the most attention in genetic epidemiological studies [12]. Individuals homozygous for the GSTM1*0 and GSTT1*0 alleles (frequently referred to as GSTM1-null and GSTT1-null genotypes), which comprise for 50% and 11–18% of white population, respectively [13,14], exhibit loss of GSTM1 and GSTT1 enzymatic activity. Single-nucleotide polymorphism (SNP) leading to amino acid substitution from isoleucine (Ile) to valine (Val) changes catalytic activity of the GSTP1 enzyme [15]. In healthy Caucasians, the frequencies of the genotype variants of GSTP Ile/Ile, -Ile/Val and -Val/Val are 51.5, 39.4, and 9.1%, respectively [15]. The role of GSTA1 polymorphism has emerged relatively recently in genetic epidemiological studies. It is represented by three, apparently linked, single nucleotide polymorphisms (SNPs): -567TOG, -69COT, -52GOA [16]. These substitutions result in differential expression with lower transcriptional activation of variant GSTA1*B (-567G, -69T, -52A) than common GSTA1*A allele (-567T, -69C,-52G) [16]. The relative frequencies of GSTA1-AA, AB and BB genotype in Caucasians are 38%, 48% and 14%, respectively [16].

Many of the well known occupational agents, such as polycyclic aromatic hydrocarbons, aromatic amines, halogenated hydrocarbons, associated with bladder cancer risk are substrates for GST. Although this reaction generally results in detoxification, in selected cases GST-mediated conjugation may lead to a more toxic or mutagenic metabolite. Still the data on association between GST gene variants and risk of occupational bladder cancer are scarce. We hypothesized that GST gene variants coding for enzymes involved in biotransformation of specific occupational agents may influence the risk of occupational bladder cancer.

Therefore, in this case-control study we investigated the role of the polymorphisms GSTA1, GSTM1, GSTP1 and GSTT1 gene and potential of effect modification by occupational exposure to different chemicals in Serbian male TCC patients.

Methods and Materials

Ethics Statement

This study was approved by the Ethical Committee of Faculty of Medicine, University of Belgrade and conducted according to the principles expressed in the Declaration of Helsinki. All the participants provided written informed consent.

Study subjects

A hospital-based case-control study of urinary bladder cancer in men was carried out between September 2007 and January 2010. A total of 143 histologically confirmed incident urinary bladder carcinoma male cases were recruited from the Clinics of Urology and Nephrology, Clinical centre of Serbia, Belgrade. This is the national reference center for urology and nephrology and the majority of bladder cancer patients from Serbia are diagnosed and treated at this clinic. The control group consisted of 114 male subjects which were recruited from individuals with nephrolithiasis admitted to the same hospital during the same period of time and had no history of any malignant disease. Urinary bladder carcinoma patients and corresponding controls did not differ with respect to mean age (Table 1).

After the informed consent was obtained, each subject was interviewed by well-trained interviewers using a standard questionnaire to collect information including demographic characteristics, history of cigarette smoking and occupational exposure. Response rate was 92% and the most frequent reason for no participation was personal.

In our study, smokers were defined as persons who reported every day smoking during a minimum of 60-day period prior to completing the questionnaire. Participants were asked about the number of cigarettes smoked per day and duration of smoking. The amount of pack-years was calculated using the following formula: pack-years = (cigarettes/day÷20)×(smoked years).

The life-time occupational history listed all jobs (including official jobs and jobs done outside normal working hours) lasting more than six months and consisted of the job title, the industry or type of business, employment dates and duration, company name and location, tasks as well as the exposure to at least one of the categories of agents under study, solvents and pesticides. In order to analyze occupational exposure occupational reports of patients were evaluated by experienced specialist in occupational medicine (author, P.B.). The exposure categories were defined as no exposure and exposure. Based on the evaluation patients were exposed to the following organic solvents: tetrachloroethylene, toluene, xylene, ethyl acetate, acetone, petrol ether and ethanol, as well as pesticides: organophosphate, carbamates, aminophosphonic analogues, chloroacetanilides, derivative of benzoic acid. All exposure data referred to a time period prior to the diagnosis of bladder cancer for the cases, and a corresponding period for the controls.

DNA extraction and genotyping

Genomic DNA was isolated from whole blood using the QIAGEN QIAmp (Qiagen, Inc., Chatsworth, CA, USA) 96-spin blood protocol according to the manufacturer's instructions. Blood was collected when patients were admitted to the clinic.

GSTM1 genotyping was performed by multiplex PCR method [17]. Primers used were GSTM1 forward: 5'-GAACTCCCT-

Table 1. Selected characteristics of male patients with bladder cancer and controls.

Characteristic	Cases	Controls	OR (95%CI)	P
	n (%)	n (%)		
Group				
Male	143	114		
Age (years)	63.6±10.7	61.1±9.9		N.S.
Smoking habits				
Never smokers	25 (18)	37 (34)	1.0 (reference group)	
Current smokers	112 (82)	72 (66)	2.3 (1.3–4.1)	0.005
No of pack-years of smoking	46.4±28.1	41.9±30.3	1.3 (0.7–2.5)	0.357
Occupational exposure				
No	77 (54)	80 (70)	1.0 (reference group)	
Yes	66 (46)	34 (30)	3.2 (1.6–6.6)[a]	0.001
Organic solvents	48 (34)	22 (19)	3.4 (1.5–7.3)[a]	0.002
Pesticides	15 (10)	9(8)	3.5 (0.9–12.9)[a]	0.058
Other chemicals	3 (2)	3 (3)	2.6 (0.4–17.7)[a]	0.323

N.S. not significant, *OR-* odds ratio, *CI*-confidence interval,
[a] *OR* adjusted by age and pack-years.

GAAAAGCTAAAGC-3′ and *GSTM1* reverse: 5′-GTTGGG-CTCAAATATACGGTGG-3′. Exon 7 of the *CYP1A1* gene was co-amplified and used as an internal control using the following primers: *CYP1A1* forward: 5′-GAACTGCCACTT CAGCTG-TCT-3; and *CYP1A1* reverse: 5′-CAGCTGCATTTG GAAGTG-CTC-3′. The presence of the *GSTM1-active* genotype was detected by the band at 215 bp, since the assay does not distinguish heterozygous or homozygous wild-type genotypes. Internal positive control (*CYP1A1*) PCR product corresponded to 312 bp.

GSTT1 genotyping was performed by multiplex PCR method [17]. Primers used were *GSTT1*-forward: 5′-TTCCTTACT-GGTCCTCACATCTC-3′ and *GSTT1*-reverse: 5′-TCACGG-GATCATGGCCAGCA-3′. Exon 7 of CYP1A1 genes were co-amplified and used as an internal control. The assay does not distinguish between heterozygous or homozygous wild-type genotypes; therefore, the presence of 480 bp bands was indicative for the *GSTT1-active* genotype. Internal positive control (*CYP1A1*) PCR product corresponded to 312 bp.

GSTP1 Ile105Val polymorphism was analyzed using the polymerase chain reaction–restriction fragment length polymorphism (PCR–RFLP) [18]. Primers used were: *GSTP1 Ile105Val* forward: 5′-ACCCCAGGGCTCTATGGGAA-3′ and *GSTP1 Ile105Val* reverse: 5′-TGAGGGCACAAGAAGCCCCT-3′. The amplification 176 bp products (20 μl) were digested by 10 U of restriction endonuclease Alw26Iat 37°C over night. The presence of restriction site resulting in two fragments (91 and 85 bp) indicated mutant allele (*Val/Val*), while if *Ile/Val* polymorphism incurred, it resulted in one more fragment of 176 bp.

GSTA1 C-69T polymorphism was determined by polymerase chain reaction–restriction fragment length polymorphism (PCR–RFLP) according to Coles et al [16]. The primers used were *GSTA1 C-69T* forward:5′-TGTTGATTGTTTGCCTGAAATT-3′ and *GSTA1 C-69T* reverse: 5′-GTTAAACGCTGT-CACCCGTCCT-3′. The amplification 481 bp products (20 μl) were digested by 10 U of restriction endonuclease Ear1 at 37°C over night. The presence of restriction site resulting in two fragments (385 and 96 bp) indicated mutant allele (*B/B*) and if *A/*

B polymorphism incurred, it resulted in one more fragment of 481 bp.

All genotyping was performed by laboratory personnel blinded to case-control status, and blinded quality control samples were inserted to validate genotyping identification procedures; concordance for blinded samples was 100%.

Statistical analysis

The distribution of the *GSTA1* and *GSTP1* polymorphisms for the case and control populations was tested for the Hardy–Weinberg equilibrium by χ^2 test. As a measure of effect size, odds ratio (OR) with corresponding 95% confidence interval (95%CI) was used to describe the strength of association between the genotypes and bladder cancer modified by occupational exposure. Unconditional logistic regression analysis is applied. Bearing in mind that age and smoking are well established risk factors for bladder cancer, we adjusted OR by these variables as potential confounders. Interactions between GST polymorphisms and occupational exposure were included in the logistic regression models and also adjusted by potential confounding variables. The probability level of ≤0.05 was considered statistically significant. For statistical analysis the SPSS 17.0 statistical software package (SPSS Inc, Chicago, IL, USA.) was used.

Results

Table 1 shows selected characteristics of male patients with bladder cancer and their controls. The smoking prevalence among cases was higher (82%) than the prevalence found in controls (66%) with the smokers being at 2.3-fold higher risk for TCC than non-smokers (95% CI = 1.3–4.1,p = 0.005). Furthermore, occupationally exposed men had 3.2 times higher risk for TCC than those unexposed (95% CI = 1.6–6.6, p = 0.001). We observed the significantly higher risk in those men occupationally exposed to organic solvents (OR = 3.4, 95% CI = 1.5–7.3, p = 0.002).

Genotyping was conducted for all recruited patients (Table 2). The *GSTA1* and *GSTP1* genotype frequencies were in Hardy-Weinberg equilibrium both for cases and controls (p>0.05). The observed genotype frequencies in controls were not significantly different from frequencies previously described among Caucasians. However, the frequency of *GSTT1-null* genotype in control group (28%) was higher than values reported among Caucasians (18.1%). As shown in Table 2, the frequencies of *GST null/low-activity* genotypes were higher in cases compared to controls with the exception of the *GSTT1-null* genotype. Although *GST A1, T1* and *P1* genotypes did not contribute independently toward the risk of TCC, the *GSTM1-null* genotype was overrepresented among cases (56%) compared to *M1-active* genotype with an adjusted OR of 2.1 reaching a statistical significance (95% CI = 1.1–4.2, p = 0.032).

Combined effects of *GSTA1, GSTM1, GSTP1* and *GSTT1* polymorphisms and occupational exposure on bladder cancer risk in male patients are shown in Table 3. When both cases and controls were dichotomized according to both genotype and occupational exposure, exposed subgroup was at TCC risk regardless of *GST* genotype. We found that occupationally exposed individuals with *GSTT1-active* genotype exhibited 4.3-fold increased risk compared to the unexposed *T1-active* subjects (95% CI = 1.7–10.6, p = 0.002). However, only for the *GSTP1* gene is there evidence of a gene–occupational exposure interaction (p = 0.017).

In order to test whether GST-occupational exposure interaction is modified by the specific type of exposure, cases and controls were further stratified into exposed to solvents and exposed to pesticides. Combined effect of occupational exposure to solvents and *GST* genotype on bladder cancer risk in male patients is shown on Table 4. The results of gene-occupational exposure to solvents interaction analyses indicated a significant effect between occupational exposure to solvents and all common *GST* polymorphisms tested. The most pronounced effect regarding occupational exposure to solvents and *GST* genotype on bladder cancer risk was observed for the *GSTA1* genotype, since men exposed to solvents with *GSTA1-low activity* genotype had 9 times higher risk of

bladder cancer than *GSTA1-active* unexposed men (95% CI = 2.4–34.7, p = 0.001). Similarly to that observed for *GSTA1-low activity*, the *GSTM1-null* genotype enhanced the risk of TCC among subjects exposed to solvents compared to the unexposed *GSTM1-active* individuals (OR = 6.5, 95% CI = 2.1–19.7, p = 0.001). These results point to the importance of antioxidant GSTA1 and GSTM1 activity protection against free radicals produced during solvent metabolism. The risk of TCC development was 5.3–fold elevated among *GSTT1-active* patients exposed to solvents in comparison with *GSTT1-active* unexposed patients (95% CI = 1.9–15.1, p = 0.002). Significant association was also found for *GSTP1 Ile/Ile* individuals who had 3.3 higher TCC risk compared to the unexposed *Ile/Ile* individuals (95% CI = 1.0–10.8, p = 0.047). However, only for GSTP1 statistically significant interaction between genotype and occupational exposure to solvents was found (p = 0.044)

Combined effect of occupational exposure to pesticides and *GST* genotype on bladder cancer risk in male patients is shown on Table 5. Men with *GSTT1-active* genotype exposed to pesticides exhibited 4.5 times higher risk in comparison with unexposed *GSTT1-active* subjects (95% CI = 0.9-22.5, p = 0.067).

Discussion

Our results showed that occupationally exposed men had 3 times higher risk for TCC. This result confirms the occupational exposure as a TCC risk factor [4]. Furthermore, the analysis of gene-occupational exposure interaction indicated a significant effect between occupationally exposed men and GSTP1 polymorphism. GSTP1 seems to play a role of particular importance in the detoxification of inhaled toxicants in occupationally exposed individuals since it is the most abundant GST isoform in the lung [19]. The mutated GSTP1 seems to be less effective in detoxification than the wild genotype [20]. Thus, Heuser et al. [18] showed that the mutated genotype (Ile/Val or Val/Val) was

Table 2. *GSTA1, GSTM1, GSTT1* and *GSTP1* genotypes in relation to bladder cancer risk in male patients.

GST genotype	Cases	Controls	OR (95%CI)	p
	n (%)	n (%)		
GSTA1				
AA	45 (31)	41 (36)	1.0 (reference group)	
AB	81 (57)	54 (47)	1.9 (0.9–4.2)	0.094
BB	17 (12)	19 (17)	1.1 (0.4–2.9)	0.875
AB+BB	98 (69)	73 (64)	1.7 (0.8–3.5)	0.171
GSTM1				
active[a]	63 (44)	58 (51)	1.0 (reference group)	
null[b]	80 (56)	56 (49)	2.1 (1.1–4.2)	0.032
GSTT1				
active[a]	101 (74)	82 (72)	1.0 (reference group)	
null[b]	36 (26)	32 (28)	1.0 (0.5–2.2)	0.999
GSTP1				
Ile/Ile	62 (43)	49 (43)	1.0 (reference group)	
Ile/Val	65 (46)	48 (42)	0.92 (0.5–1.9)	0.918
Val/Val	16 (11)	17 (15)	0.6 (0.2–1.9)	0.401
Ile/Val+Val/Val	81 (57)	65 (47)	0.9 (0.4–1.7)	0.876

[a]Active (present) if at least one active allele present.
[b]Inactive (null) if no active alleles present. *OR*- odds ratio adjusted for age and pack-years. *CI*- confidence interval.

Table 3. Combined effect of occupational exposure and *GST* genotype on bladder cancer risk in male male patients.

GST/exposure	Cases	Controls	OR (95%CI)	p
	n (%)	n (%)		
GSTA1				
AA/unexposed	21 (15%)	32 (28%)	1.0 (reference group)	
AB+BB/unexposed	56 (39%)	48 (42%)	2.4 (0.8–7.3)	0.121
AA/exposed	24 (17%)	9 (8%)	6.2 (1.4–27.1)	0.015
AB+BB/exposed	42 (29%)	25 (22%)	6.4 (2.0–20.2)	0.002
P interaction between genotype and occupational exposure = 0.104				
GSTM1				
active[a]/unexposed	35 (24%)	44 (39%)	1.0 (reference group)	
null[b]/unexposed	42 (29%)	36 (32%)	3.3 (1.2–9.4)	0.023
active/exposed	28 (20%)	14 (12%)	5.4 (1.9–15.8)	0.002
null/exposed	38 (27%)	20 (17%)	6.0 (2.2–16.5)	0.001
P interaction between genotype and occupational exposure = 0.601				
GSTT1				
active[a]/unexposed	54 (40%)	57 (50%)	1.0 (reference group)	
null[b]/unexposed	22 (16%)	23 (20%)	1.3 (0.5–3.9)	0.577
active/exposed	47 (34%)	25 (22%)	4.3 (1.7–10.6)	0.002
null/exposed	14 (10%)	9 (8%)	2.6 (0.8–8.9)	0.124
P interaction between genotype and occupational exposure = 0.770				
GSTP1				
Ile/Ile/unexposed	31 (22%)	32 (28%)	1.0 (reference group)	
Ile/Val+Val/Val/unexposed	46 (32%)	48 (42%)	0.8 (0.3–2.1)	0.605
Ile/Ile/exposed	31 (22%)	17 (15%)	2.8 (1.0–7.9)	0.049
Ile/Val+Val/Val/exposed	35 (24%)	17 (15%)	2.8 (1.0–8.0)	0.049
P interaction between genotype and occupational exposure = 0.017				

[a]Active (present) if at least one active allele present.
[b]Inactive (null) if no active alleles present. *OR*- odds ratio adjusted for age and pack-years. *CI*- confidence interval.

associated with greater DNA damage in Brazilian footwear workers than the wild (Ile/Ile) genotype [21]. These studies point to an interaction between the exposure and GSTP1 genotype. In our study, the most significant TCC risk was found for solvents. Epidemiologic evidence on the relationship between solvents and various cancers, such as gastrointestinal cancers, lung cancer and lymphohematopoietic malignancies, is well established [22]. Among compounds that have carcinogenic role halogenic aliphatic solvents have been mostly described. There are few reports about relationship between urinary bladder risk and solvents. Previous case-control studies reported significantly increased risks (between 3.1 and 8.8 times) among workers in the dyestuffs industry [23,24]. Several other investigators have reported elevated risks for spray painters [25,26], who have been reported to be exposed to many known or suspected carcinogens, including solvents. On the other hand, Lohi and others [27] found that among Finnish workers exposure to solvents was positively associated with the incidence of bladder cancer in women, but not in men.

It is important to note that risk imposed by occupational hazards was modified by GST polymorphism. We observed that individuals occupationally exposed to solvents with at least one *low activity GSTA1 allele* had the highest risk (about 9 times), while *GSTM1-null* carriers had 6.5 times higher bladder cancer risk when compared to unexposed *GSTA1 AA* and *GSTM1-active*

persons, respectively. This result was expected since in several malignant diseases, such as colorectal, prostate and hepatocellular cancer, *GSTA1*B allele* with lower transcriptional activity was associated with increased risk. GSTA1 protein belongs to the most promiscuous GSTs that acts upon a broad range of substrates which bind to its active site [28]. Our findings that *low-activity GSTA1* and *GSTM1-null* genotype increase susceptibility to bladder cancer in occupationally exposed men can be explained by the role of GST enzymes in detoxification and in antioxidant defense. Namely, GSTA1 and GSTM1 possess strong peroxidase activity and are key components in cellular defense against free radicals [29]. It may be speculated that free radicals are produced during solvent metabolism [30]. Regarding potential place of solvent detoxification, it is important to note that uroepithelial cells do not express GSTA1, while their GSTM1 protein level is also relatively low [31]. On the other hand, liver cells abundantly express GSTA1 and GSTM1 and thus participate in GSTA1 and GSTM1 mediated conjugation of different metabolites with glutathione, thereby enhancing their excretion in urine [32]. Taken together, these data suggest that liver, by its GSTs conjugating and peroxidase activity plays a key role in protection against bladder carcinogens present in halogenated solvents. On the other hand, *GSTT1-active* individuals occupationally exposed to solvents exhibited 5 times higher risk of TCC in comparison with *GSTT1-active* unexposed subjects. These results are biologically

Table 4. Combined effect of occupational exposure to solvents and *GST* genotype on bladder cancer risk in male patients.

GST/exposure	Cases	Controls	OR (95%CI)	p
	n (%)	n (%)		
GSTA1				
AA/unexposed	21 (1%)	32 (32%)	1.0 (reference group)	
AB+BB/unexposed	56 (46%)	48 (49%)	2.4 (0.8–7.3)	0.121
AA/solvents	14 (11%)	6 (6%)	5.9 (1.0–33.1)	0.046
AB+BB/solvents	31 (25%)	13 (13%)	9.2 (2.4–34.7)	0.001
P interaction between genotype and occupational exposure to solvents = 0.228				
GSTM1				
active[a]/unexposed	35 (28%)	44 (43%)	1.0 (reference group)	
null[b]/unexposed	42 (34%)	36 (35%)	3.3 (1.2–9.4)	0.023
active/solvents	21 (17%)	10 (10%)	4.7 (1.6–13.8)	0.006
null/solvents	27 (22%)	12 (12%)	6.5 (2.1–19.7)	0.001
P interaction between genotype and occupational exposure to solvents = 0.896				
GSTT1				
active[a]/unexposed	54 (46%)	57 (56%)	1.0 (reference group)	
null[b]/unexposed	22 (18%)	23 (22%)	1.3 (0.5–3.9)	0.577
active/solvents	34 (29%)	15 (15%)	5.3 (1.9–15.1)	0.002
null/solvents	8 (7%)	7 (7%)	1.7 (0.4–7.3)	0.470
P interaction between genotype and occupational exposure to solvents = 0.224				
GSTP1				
Ile/Ile/unexposed	31 (25%)	32 (31%)	1.0 (reference group)	
Ile/Val+Val/Val/unexposed	46 (37%)	48 (47%)	0.8 (0.3–2.1)	0.605
Ile/Ile/solvents	22 (18%)	9 (9%)	3.3 (1.0–10.8)	0.047
Ile/Val+Val/Val/solvents	26 (21%)	13 (13%)	2.6 (0.9–7.9)	0.089
P interaction between genotype and total occupational exposure to solvents = 0.044				

[a]Active (present) if at least one active allele present.
[b]Inactive (null) if no active alleles present. *OR*- odds ratio adjusted for age and pack years; *CI*- confidence interval.

plausible since GST-mediated conjugation with halogenated substrates may lead to a more toxic or mutagenic metabolites. Namely, substrates with ≥2 halogenes are activated because the conjugated product is unstable, leading to reactions with nucleophiles, particularly DNA and proteins [33]. The human polymorphic GSTT1 catalyze conjugation of halomethanes, dihalomethanes, ethylene oxide and a number of other industrial compounds. Our results confirm the assumption of Avima M Ruder et al. [34] that humans with fully functional *GST* genes produce enzymes that metabolize some solvents to cytotoxic metabolites; while those with less functional or nonfunctioning genes have little or no enzyme and apparently do not produce cytotoxic metabolites from solvent exposure. Until now, the association between *GST* polymorphism and occupationally related cancers has been studied mostly in renal cell carcinoma. Results of these studies showed that *GSTT1-active* genotype enhanced the risk of renal cell carcinoma among subjects exposed to solvents. Our results on higher bladder carcinoma risk in *GSTT1-active* individuals occupationaly exposed to solvents are in accordance with previously published results in renal cell carcinoma [35,36]. Regarding the potential mechanism of solvent metabolism by GST, it is generally assumed that the main site is liver, followed by a mandatory transfer of conjugates to the kidney. However, the initial bioactivation step of halogenated solvents, can take place in the kidney itself [37]. Uroepithelium is also capable of

metabolizing some procarcinogens to inactive or genotoxic metabolites, and is, therefore, not exposed only to preformed reactive metabolites in the urine [38]. As the renal parenchyma and uroepithelium are exposed to the same broad range of potentially genotoxic compounds, the potential genotoxicity of carcinogens also depends on the biotransformation capacity of these tissues. As a result of GST polymorphism, great interindividual differences in GST isoenzyme profiles exist, in both renal parenchyma and uroepithelial cells [37].

Although it has been postulated that exposure to pesticides and/or fertilizers might be responsible for higher urinary bladder risk, the evidence is still conflicting. Some studies have shown that TCC risk was significantly elevated among men in the landscape and horticultural services industry, as well as in gardeners, and lawn care service employees [39,40], while others did not [41]. Some suggestions of a possible relation between GST status and early markers of genotoxic effects in humans exposed to pesticides are available. An increased frequency of micronuclei in cultured peripheral lymphocytes has been found among pesticide exposed greenhouse workers with the *GSTM1-active* genotype [42]. Significantly higher levels of sister hromatid exchanges were also found among *GSTT1-active* individuals exposed to pesticides when compared to *GSTT1-null* workers similarly exposed [43]. Until now only one study investigated association between *GST* polymorphism and occupational exposure to pesticide with respect

Table 5. Combined effect of occupational exposure to pesticides and *GST* genotype on bladder cancer risk in male patients.

GST/exposure	Cases n (%)	Controls n (%)	OR (95%CI)	p
GSTA1				
AA/unexposed	21 (22%)	32 (36%)	1.0 (reference group)	
AB+BB/unexposed	56 (60%)	48 (54%)	2.4 (0.8–7.3)	0.121
AA/pesticides	8 (9%)	3 (3%)	4.2 (0.5–36.0)	0.190
AB+BB/pesticides	8 (9%)	6 (7%)	2.0 (0.5–7.9)	0.239
P interaction between genotype and occupational exposure to pesticides = 0.957				
GSTM1				
active[a]/unexposed	35 (37%)	44 (49%)	1.0 (reference group)	
null[b]/unexposed	42 (45%)	36 (41%)	3.3 (1.2–9.4)	0.023
active/pesticides	7 (8%)	3 (3%)	2.9 (0.7–12.2)	0.138
null/pesticides	9 (10%)	6 (7%)	1.9 (0.5–6.7)	0.264
P interaction between genotype and occupational exposure to pesticides = 0.125				
GSTT1				
active[a]/unexposed	54 (59%)	57 (64%)	1.0 (reference group)	
null[b]/unexposed	22 (24%)	23 (25%)	1.3 (0.5–3.9)	0.577
active/pesticides	11 (12%)	7 (8%)	4.5 (0.9–22.5)	0.067
null/pesticides	5 (5%)	2 (3%)	2.6 (0.4–20.6)	0.264
P interaction between genotype and occupational exposure to pesticides = 0.508				
GSTP1				
Ile/Ile/unexposed	31 (33%)	32 (36%)	1.0 (reference group)	
Ile/Val+Val/Val/unexposed	46 (49%)	48 (53%)	0.8 (0.3–2.1)	0.605
Ile/Ile/pesticides	9 (10%)	6 (7%)	2.9 (0.6–13.6)	0.181
Ile/Val+Val/Val/pesticides	7 (8%)	3 (4%)	2.4 (0.5–10.1)	0.231
P interaction between genotype and occupational exposure to pesticides = 0.320				

[a]Active (present) if at least one active allele present.
[b]Inactive (null) if no active alleles present. *OR*- odds ratio adjusted for age and pack years. *CI*- confidence interval.

to risk of carcinoma of urinary tract. Namely, Karami and others reported that renal cell carcinoma risk associated with pesticide exposure was highest among individuals with *active GSTM1/T1* genotypes [44]. Although we did not observe significant effect between exposure to pesticides and GST polymorphisms we found borderline significance for *GSTT1-active* genotype. One of the reasons for non-significant association between *GSTT1-active* genotype may be the relatively small number of pesticide exposed participants in both case and control groups. Nevertheless, it is well known that pesticides produced from halogenated alkanes, alkenes undergo bioactivation in the liver and kidney after conjugation to glutathione by GSTT1 [41]. Therefore, an active GSTT1 enzyme will be required to conjugate substrates and form more reactive intermediates that directly damage tissues. Conversely, the deleted variant of *GSTT1-genotype* will form an inactive enzyme and therefore metabolism of halogenated compounds will occur through oxidation, without formation of reactive intermediates [44].

The principal limitations of this study are the relatively small sample size which limiting the precision of the odds ratios, hospital-based control group and qualitative evaluation of occupational exposure. Concerning the actual sample size (143 cases and 114 controls), the statistical power is 66%. Furthermore, it is well known that relatively small numbers of both study participants and *GST* polymorphisms studied might be sources of

potential biases which may influence the study findings. However, we tested effects of four *GST* polymorphisms and occupational exposure on bladder cancer risk and therefore significantly decreased chance for publication bias. Additionally, we cannot entirely rule out the possibility that some of our results could be caused by confounding, although we included only men and adjusted all results by age and smoking status. Further studies with larger samples and more rigorous designs are needed to investigate the gene effects and the potential effect modification by environmental factors.

Conclusions

GSTM1-null genotype increased the risk of bladder cancer in males. Null or low-activity genotypes of the *GSTA1*, *GSTT1*, and *GSTP1* did not contribute independently towards the risk of bladder cancer in males. However, in association with occupational exposure, both *low activity GSTA1* and *GSTM1-null* genotype increase individual susceptibility to bladder cancer suggesting the protective role of these detoxification and antioxidant enzymes in metabolism of occupational hazards, specifically organic solvents. On the other hand, the presence of *GSTT1-active* genotype in occupationally exposed subjects, resulting in GSTT1 protein expression and GSTT1 mediated bioactivation, increases the risk of bladder cancer.

Acknowledgments

The authors would like to thank technician Miss Sanja Zivotic for collecting data and support in manuscript preparation as well as Professor Goran Trajkovic, for final statistical consultancy.

Author Contributions

Conceived and designed the experiments: MGM VMC TPS TDP. Performed the experiments: MGM VMC TID. Analyzed the data: MGM VMC ARSR MSPE TID TPS TDP. Contributed reagents/materials/ analysis tools: PVB DPD. Wrote the paper: MGM VMC ARSR MSPE TPS TDP.

References

1. Kim JJ (2012) Recent advances in treatment of advanced urothelial carcinoma. Curr Urol Rep 13:147–52.
2. Siegel R, Naishadham D, Jemal A (2012) Cancer statistics. CA Cancer J Clin 62:10–29.
3. American Cancer Society (2012) Bladder Cancer 2012. American Cancer Society, Atlanta, USA.
4. Olfert SM, Felknor SA, Delclos GL (2006) An updated review of the literature: risk factors for bladder cancer with focus on occupational exposures. South Med J 99:1256–63.
5. Clapp RW, Howe G, Lefevre MJ (2005) Environmental an Occupational Causes of Cancer, A Review of Recent Scientific Literature. Lowell Center for Sustainable Production. Lowe Mass, USA
6. International Agency for Research on Cancer (2010) Painting, fire- fighting, and shiftwork. IARC Monographs on the Evaluation of Carcinogenic Risks to Humans. pp. 43–394.
7. Schulte PA, Ringen K, Hemstreet GP, Ward E (1987) Occupational cancer of the urinary tract. Occup Med 2: 85–107.
8. International Agency for Research on Cancer (2010) Some aromatic amines, organic dyes, and related exposures. IARC Mono-graphs on the Evaluation of Carcinogenic Risks to Humans. pp. 1–692.
9. International Agency for Research on Cancer (2010) Some non-heterocyclic polycyclic aromatic hydrocarbons and some related compounds. IARC Monographs on the Evaluation of Carcinogenic Risks to Humans. pp. 754–759.
10. Pukkala E, Martinsen JI, Lynge E, Gunnarsdottir HK, Sparén P, et al. (2009) Occupation and cancer follow-up of 15 million people in five Nordic countries. Acta Oncologica 48: 646–790.
11. Simic T, Mimic-Oka J, Savic-Radojevic A, Opacic M, Pljesa M, et al. (2005) Glutathione S-transferase T1-1 activity upregulated in transitional cell carcinoma of urinary bladder. Urology 65: 1035–40.
12. Di Pietro G, Magno LA, Rios-Santos F (2010) Glutathione S-transferases: an overview in cancer research. Expert Opin Drug Metab Toxicol 6: 153–70.
13. Eaton DL, Bammler TK (1999) Concise review of the glutathione S-transferases and their significance to toxicology. Toxicol Sci 49: 156–64.
14. Landi S (2000) Mammalian class θ GST and differential susceptibility to carcinogens: a review. Mutat Res 463: 247–83.
15. Watson MA, Stewart RK, Smith GB Massey TE, Bell DA (1998) Human glutathione S-transferase P1 polymorphisms: relationship to lung tissue enzyme activity and population frequency distribution. Carcinogenesis 19: 275–80.
16. Coles FB, Kadlubar FF (2005) Human alpha class glutathione S-transferases: genetic polymorphism, expression, and susceptibility to disease. In: Helmut S, Lester P, editors. Glutathione Transferases and Gamma-Glutamyl Transpepti-dases, Methods Enzymology. London: Elsevier Academic Press. pp 9–42.
17. Abdel-Rahman SZ, El-Zein RA, Anwar WA, Au WW (1996) A multiplex PCR procedure for polymorphic analysis of GSTM1 and GSTT1 genes in population studies. Cancer Lett 107: 229–33.
18. Harries LW, Stubbins MJ, Forman D, Howard GC, Wolf CR (1997) Identification of genetic polymorphisms at the glutathione S-transferase Pi locus and association with susceptibility to bladder, testicular and prostate cancer. Carcinogenesis 18: 641–44.
19. Hirvonen A (2005) Gene–environment interaction and biological monitoring of occupational exposures. Toxicol Appl Pharmacol 207: 329–335.
20. Miller DP, Asomaning K, Liu G, Wain JC, Lynch TJ, et al. (2006) An association between glutathione S-transferase P1 gene polymorphism and younger age at onset of lung carcinoma. Cancer 107: 1570–7.
21. Heuser VD, Erdtmann B, Kvitko K, Rohr P, da Silva J (2007) Evaluation of genetic damage in Brazilian footwear-workers: Biomarkers of exposure, effect, and susceptibility. Toxicology 232: 235–47.
22. Lynge E, Anttila A, Hemminki A (1997) Organic solvents and cancer. Cancer Causes Control 8: 406–19.
23. Risch HA, Burch JD, Miller AB, Hill GB, Steele R, et al. (1988) Occupational factors and the incidence of cancer of the bladder in Canada. Br Industr Med 45: 361–367.
24. Bonassi S, Merlo F, Pearce N, Puntoni R (1989) Bladder cancer an occupational exposure to polycyc aromatic hydrocarbons. Int Cancer 44: 648–651.
25. La Vecchia C, Negri E, D'Avanzo B, Franceschi S (1990) Occupation and the risk of bladder cancer. Int J Epidemiol 19: 264–8.
26. Cordier S, Clavel J, Limasset JC, Boccon-Gibod L, Le Moual N, et al. (1993) Occupational risks of bladder cancer in France: A multicentre case-control study. Int J Epidemiol 22: 402–11.
27. Lohi J, Kyyrönen P, Kauppinen T, Kujala V, Pukkala E (2008) Occupational exposure to solvents and gasoline and risk of cancers in the urinary tract among Finnish workers. Am J Ind Med 51: 668–72.
28. Honaker MT, Acchione M, Zhang W, Mannervik B, Atkins WM (2013) Enzymatic detoxication, conformational selection, and the role of molten globule active sites. J Biol Chem 288: 18599–611.
29. Hayes JD, Strange RC (2000) Glutathione S-transferase polymorphisms and their biological consequences. Pharmacology 61: 154–66.
30. Weber LW, Boll M, Stampfl A (2003) Hepatotoxicity and mechanism of action of haloalkanes: carbon tetrachloride as a toxicological model. Crit Rev Toxicol 33: 105–36.
31. Savic-Radojevic A, Mimic-Oka J, Pljesa-Ercegovac M, Opacic M, Dragicevic D, et al. (2007) Glutathione S-transferase-P1 expression correlates with increased antioxidant capacity in transitional cell carcinoma of the urinary bladder. Eur Urol 52: 470–7.
32. Rossi AM, Guarnieri C, Rovesti S, Gobba F, Ghittori S, et al. (1999) Genetic polymorphisms influence variability in benzene metabolism in humans. Pharmacogenetics 9: 445–51.
33. Guengerich P (2005) Activation of alkyl halides by glutathione transferases. In: Helmut S, Lester P, editors. Glutathione Transferases and Gamma-Glutamyl Transpeptidases, Methods Enzymology. London: Elsevier Academic Press.pp 9–42.
34. Ruder AM, Yiin JH, Waters MA, Carreón T, Hein MJ, et al. (2013) The Upper Midwest Health Study: gliomas and occupational exposure to chlorinated solvents. Occup Environ Med 70: 73–80.
35. Buzio L, De Palma G, Mozzoni P, Tondel M, Buzio C, et al. (2003) Glutathione S-transferases M1-1 and T1-1 as risk modifiers for renal cell cancer associated with occupational exposure to chemicals. Occup Environ Med 60: 789–93.
36. Moore LE, Boffetta P, Karami S, Brennan P, Stewart PS, et al. (2010) Occupational trichloroethylene exposure and renal carcinoma risk: evidence of genetic susceptibility by reductive metabolism gene variants. Cancer Res 70: 6527–36.
37. Simic T, Savic-Radojevic A, Pljesa-Ercegovac M, Matic M, Mimic-Oka J (2009) Glutathione S-transferases in kidney and urinary bladder tumors. Nat Rev Urol 6: 281–9.
38. Thier R, Golka K, Brüning T, Ko Y, Bolt HM (2002) Genetic susceptibility to environmental toxicants: the interface between human and experimental studies in the development of new toxicological concepts. Toxicol Lett 127: 321–7.
39. Band PR, Le ND, MacArthur AC, Fang R, Gallagher RP (2005) Identification of occupational cancer risks in British Columbia: a population-based case-control study of 1129 cases of bladder cancer. J Occup Environ Med 47: 854–8.
40. Zahm SH (1997) Mortality study of pesticide applicators and other employees of a lawn care service company. J Occup Environ Med 39: 1055–67.
41. Viel F-F, Challier B (1995) Bladder cancer among French farmers: does exposure to pesticides in vineyards play a part? Occup Environ Med 52: 587–92.
42. Falck GC, Hirvonen A, Scarpato R, Saarikoski ST, Migliore L, et al. (1999) Micronuclei in blood lymphocytes and genetic polymorphism for GSTM1, GSTT1 and NAT2 in pesticide-exposed greenhouse workers. Mutat Res 441: 225–37.
43. Scarpato R, Migliore L, Hirvonen A, Falck G, Norppa H (1996) Cytogenetic monitoring of occupational exposure to pesticides: characterization of GSTM1, GSTT1, and NAT2 genotypes. Environ Mol Mutagen 27: 263–9.
44. Karami S, Boffetta P, Rothman N, Hung RJ, Stewart T, et al. (2008) Renal cell carcinoma, occupational pesticide exposure and modification by glutathione S transferase polymorphisms. Carcinogenesis 29: 1567–71.

Selection and Evaluation of Potential Reference Genes for Gene Expression Analysis in the Brown Planthopper, *Nilaparvata lugens* (Hemiptera: Delphacidae) Using Reverse-Transcription Quantitative PCR

Miao Yuan[1,9]**, Yanhui Lu**[1,9]**, Xun Zhu**[1]**, Hu Wan**[1]**, Muhammad Shakeel**[1]**, Sha Zhan**[1]**, Byung-Rae Jin**[2]**, Jianhong Li**[1]*****

1 Laboratory of Pesticide, College of Plant Science & Technology, Huazhong Agricultural University, Wuhan, China, **2** Laboratory of Insect Molecular Biology and Biotechnology, Department of Applied Biology, College of Natural Resources and Life Science, Dong-A University, Busan, Korea

Abstract

The brown planthopper (BPH), *Nilaparvata lugens* (Hemiptera, Delphacidae), is one of the most important rice pests. Abundant genetic studies on BPH have been conducted using reverse-transcription quantitative real-time PCR (qRT-PCR). Using qRT-PCR, the expression levels of target genes are calculated on the basis of endogenous controls. These genes need to be appropriately selected by experimentally assessing whether they are stably expressed under different conditions. However, such studies on potential reference genes in *N. lugens* are lacking. In this paper, we presented a systematic exploration of eight candidate reference genes in *N. lugens*, namely, actin 1 (ACT), muscle actin (MACT), ribosomal protein S11 (RPS11), ribosomal protein S15e (RPS15), alpha 2-tubulin (TUB), elongation factor 1 delta (EF), 18S ribosomal RNA (18S), and arginine kinase (AK) and used four alternative methods (BestKeeper, geNorm, NormFinder, and the delta Ct method) to evaluate the suitability of these genes as endogenous controls. We examined their expression levels among different experimental factors (developmental stage, body part, geographic population, temperature variation, pesticide exposure, diet change, and starvation) following the MIQE (Minimum Information for publication of Quantitative real time PCR Experiments) guidelines. Based on the results of RefFinder, which integrates four currently available major software programs to compare and rank the tested candidate reference genes, RPS15, RPS11, and TUB were found to be the most suitable reference genes in different developmental stages, body parts, and geographic populations, respectively. RPS15 was the most suitable gene under different temperature and diet conditions, while RPS11 was the most suitable gene under different pesticide exposure and starvation conditions. This work sheds light on establishing a standardized qRT-PCR procedure in *N. lugens*, and serves as a starting point for screening for reference genes for expression studies of related insects.

Editor: Xiao-Wei Wang, Zhejiang University, China

Funding: This research was supported by China Hubei Province Science & Technology Department (No. 2009BFA011). The funders had no role in study design, data collection and analysis, decision to publish, or preparation of the manuscript.

Competing Interests: The authors have declared that no competing interests exist.

* E-mail: jianhl@mail.hzau.edu.cn

⑨ These authors contributed equally to this work.

Introduction

The brown planthopper (BPH), *Nilaparvata lugens* (*N. lugens*), is the most devastating rice pest in extensive areas throughout Asia [1]. The BPH ingests nutrients specifically from the phloem of rice plants with its stylet, causing the entire plant to become yellow and dry rapidly, a phenomenon referred to as hopperburn [2]. In addition, BPH is a vector of viruses that cause diseases in rice, such as *Rice ragged stunt virus* (RRSV) and *Rice grassy stunt virus* (RGSV) [3]. In recent years, *N. lugens* outbreaks have occurred more frequently in the Yangtze River Delta areas and in the South of China [4,5]. Because of its long-distance migration, quick adaptation to resistant rice varieties and development of high

resistance to pesticides, *N. lugens* infestations are difficult to control [6].

Quantitative real-time reverse-transcription polymerase chain reaction (qRT-PCR) is the most sensitive and accurate method to measure variations in mRNA expression levels of a single gene in different experimental and clinical conditions [7,8]. At present, RNA interference (RNAi) is an effective tool to control important insect pests via gene silencing [9,10,11,12,13]. Interestingly, several studies have shown that injection or ingestion of dsRNAs in *N. lugens* can reduce the transcript levels of target genes [14,15,16]. On the other hand, the sequencing of *N.lugens* genome has been recently included in the 5000 insect genome initiative (http://arthropodgenomes.org/wiki/i5K), somehow reflecting the economic importance of this pest. Meanwhile, enormous progress has been made by means of the sequencing of *N. lugens* ESTs from various tissues [17], transcriptome analysis [18], and pyrosequencing the midgut transcriptome [19]. These data provided comprehensive gene expression information at the transcriptional level that could facilitate our understanding of the molecular mechanisms underlying various physiological aspects including development, wing dimorphism and sex difference in BPH. For precise and reliable gene expression results, normalization of quantitative real-time PCR data is required against a control gene, which is typically a gene that shows highly uniform expression in living organisms during various phases of development under different environmental or experimental conditions [20]. Quantitative assays frequently use housekeeping genes such as β-actin, glyceraldehyde-3-phosphate dehydrogenase (GAPDH), tubulin, and 18S ribosomal RNA (rRNA) because they are necessary for survival and are synthesized in all nucleated cell types. It is often considered that there are only a few fluctuations in the transcription of these genes compared to others [21,22,23]. However, numerous studies show that the expression levels of these housekeeping genes also vary in different situations [24,25].

Although qRT-PCR is a highly reliable method for measuring gene transcript levels, if the reference genes are not selected properly, it will result in inaccurate calculation of the normalization factor and consequently obscure actual biological differences among samples. Therefore, it is necessary to validate the expression stability of control genes under specific experimental conditions before using them for normalization. Reference genes in qRT-PCR studies on BPH have often been selected based on consensus and experience in other species rather than empirical evidence in support of their efficacy [1,14,15,16]. There is therefore a definite need to analyze the expression of these genes in different body parts in different populations, under different experimental conditions, and at different stages of development. This study examined the stability of eight reference genes, actin 1 (ACT), muscle actin (MACT), ribosomal protein S11 (RPS11), ribosomal protein S15e (RPS15), alpha 2-tubulin (TUB), elongation factor 1 delta (EF), 18S ribosomal RNA (18S), and arginine kinase (AK), in *N. lugens* in terms of different factors (developmental stage, body part, geographic population, temperature variation, pesticide treatment, diet change, and starvation).

Materials and Methods

Insects

Unless stated, the laboratory population of *N. lugens* was originally collected from Changsha, Hunan, People's Republic of China in 2009 and artificially maintained in our lab since. The laboratory strain and other populations used in this experiment are from different fields which no specific permissions were required, because these fields are the experimental plots of Huazhong Agricultural University, Wuhan, Hubei, China. The insects were reared on rice (Shanyou 63) in a thermostatic chamber. The chamber was maintained at 80% relative humidity, 25°C±2°C temperature and a 14:10 h light:dark cycle.

Treatments

(1) Developmental stage: For each treatment group, 6 samples each of about 50 one-day-old eggs, 50 1st instar nymphs, 30 2nd instar nymphs, 20 3rd instar nymphs, 20 4th instar nymphs, 20 5th instar nymphs, 20 adult females, and 20 adult males of *N. lugens* were collected.

(2) Body part: A dissection needle and a tweezer (Dumont, World Precision Instruments, USA) were used to obtain head, thorax, and abdomen from virgin adult males and females

from the *N. lugens* laboratory population. Besides, virgin adult males and females were collected as whole-body samples. For each treatment group, 6 samples of 20 insects each were collected.

(3) Geographic population: One geographic population was originally collected from Changsha, Hunan, China, which was maintained with no exposure to insecticides. The other population was generously provided by Dr. Manqun Wang (Huazhong Agricultural University), which was originally collected from Wuhan, Hubei, China. These two places are approximately 310 kilometers apart. Both these populations have been maintained for more than 3 years in our laboratory. Third instar nymphs and adults were collected. For each treatment group, 6 samples of 20 insects each were collected.

(4) Temperature-induced stress: Third instar nymphs were divided into 10 groups and then each group was exposed for 5 min to each temperature: extremely low temperatures (4°C, 8°C, and 12°C), low temperatures (16°C and 20°C), average temperatures (24°C and 28°C), and high temperatures (32°C, 36°C and 40°C). For each treatment group, 6 samples of 20 insects each were collected. There was no mortality in response to the temperature treatment.

(5) Pesticide-induced stress: The stability of candidate reference genes was tested in 3rd instar nymphs subjected to 6 different pesticide treatments: compound pesticide (abamectin 3.6 mg/L+nitenpyram 0.2 mg/L), nitenpyram (0.4 mg/L), pymetrozine (42.08 mg/L), buprofezin (1.19 mg/L), isoprocarb (34.91 mg/L), and chlorpyrifos (52.27 mg/L). The concentration of pesticide was LC_{50} and opted by the results of bioassay (Table S1). The testing pesticide solutions were made using water containing 0.1% w/v Triton X-100 (Beijing Solarbio Science and Technology Co. Ltd., China). The roots of the rice seedlings were tightly packaged by the absorbent cotton. The seedlings were completely dipped in the testing solutions for 5 s and then air dried for 10–15 min depending on the ambient relative humidity (http://www.irac-online.org/content/uploads/2009/09/Method_005_v3_june09.pdf). Third instar nymphs were collected from the laboratory population and then transferred into the transparent plastic tube which contained the testing seedlings. Water containing 0.1% w/v Triton X-100 was used as a separate control group for each pesticide treatment. Because of the different mechanism of action of the testing pesticide, the living insects were collected after 4, 4, 7, 5, 3 and 3 days for compound pesticide, nitenpyram, pymetrozine, buprofezin, isoprocarb, and chlorpyrifos treatments, respectively [26,27,28]. For each treatment group, 6 samples of 50 insects each were collected.

(6) Diet-induced stress: Our third treatment condition involved the stability of reference gene expression in *N. lugens* challenged with different diets: artificial diet [29], Taichung Native 1 rice (TN1), Minghui 63 rice (MH63), transgenic rice Huahui 1 (HH1), Shanyou 63 rice (SY63), and transgenic rice Bt Shanyou 63 rice (BTSY63). The seeds of TN1, MH63, HH1, SY63, and BTSY63 were generously provided by Dr. Yongjun Lin (Huazhong Agricultural University). Newly hatched nymphs were collected and then reared on different diets. From each diet group, 3rd instar nymphs and adults were collected. For each treatment group, 6 replications of 20 insects each were collected.

(7) Starvation-induced stress: Third instar nymphs and adults were collected in separate glass cylinders (15.0 cm in length and 2.5 cm in diameter) covered by Parafilm M (Bemis, USA)

Table 1. Function, primer sequence and amplicon characteristics of the candidate reference genes used in this study.

Gene symbol	Gene name	(putative) Function	Gene ID	Primer sequences [5′→3′]	L (bp)[a]	E (%)[b]	R²[c]
ACT	actin 1	Involved in cell motility, structure and integrity	ABY48093.1	*For* 5′ TGCGTGACATCAAGGAGAAG 3′ *Rev* 5′ GTACCACCGGACAGGACAGT 3′	283	96.7	0.997
MACT	muscle actin	Involved in cell motility, structure and integrity	ADB92676.1	*For* 5′ CTTGGCTGGTCGTGACTTGACCGA 3′ *Rev* 5′ ACTTCTCCAGGGAGGTGGAGGCG 3′	179	101.7	0.997
RPS11	ribosomal protein S11	Structural constituent of ribosome	ACN79505.1	*For* 5′ CCGATCGTGTGGCGTTGAAGGG 3′ *Rev* 5′ ATGGCCGACATTCTTCCAGGTCC 3′	159	93.5	0.997
RPS15	ribosomal protein S15	Structural constituent of ribosome	ACN79501.1	*For* 5′ TAAAAATGGCAGACGAAGAGCCCAA 3′ *Rev* 5′ TTCCACGGTTGAAACGTCTGCG 3′	150	101.5	0.999
TUB	α-tubulin	Cytoskeleton structural protein	ACN79512.1	*For* 5′ ACTCGTTCGGAGGAGGCACC 3′ *Rev* 5′ GTTCCAGGGTGGTGTGGGTGGT 3′	174	101.7	0.995
EF	elongation factor 1 delta	Structural constituent of ribosome	DQ445523.1	*For* 5′ GAAGTAGCTCTGGCACAGGA 3′ *Rev* 5′ TTGACGAGCCTTTGCTACCT 3′	150	103.9	0.996
18S	18S ribosomal RNA	Cytosolic small ribosomal subunit	JN662398.1	*For* 5′ GTAACCCGCTGAACCTCC 3′ *Rev* 5′ GTCCGAAGACCTCACTAAATCA 3′	170	107.2	0.990
AK	arginine kinase	Key enzyme for cellular energy metabolism	AAT77152.1	*For* 5′ ACCACAACGACAACAAGACCTTCC 3′ *Rev* 5′ TGGGACAGAAAGTCAGGAATCCCA 3′	186	98.3	0.998

[a]Length of the amplicon.
[b]Real-time qPCR efficiency (calculated by the standard curve method).
[c]Reproducibility of the real-time qPCR reaction.

with no food in a thermostatic chamber; they were kept there for two days. We used a satiation group (3rd instar nymphs and adults fed on SY63) as the control group. For each treatment group, 6 samples of 50 insects each were collected. The mortality rate was approximately 30%.

Total RNA Extraction and cDNA Synthesis

All collected insects were preserved in a clean micro-centrifuge tube (1.5 ml) and stored at −80°C after freezing in liquid nitrogen. Six total RNA samples were prepared for each developmental and treatment group. Subsequently, total RNA was extracted using a SV Total RNA Isolation System (Promega, USA). According to the manufacturer's protocol, total RNA was incubated for 15 min at 20–25°C after adding 5 μl DNase I enzyme (Promega, USA). The quality and quantity of RNA were assessed with a UV-1800 spectrophotometer (SHIMADZU, Japan). Only samples with a 260/280 ratio of 1.9 to 2.1, which indicates no protein contamination, and a 260/230 ratio of 2.0 to 2.4, which indicates no guanidine thiocyanate contamination were considered. Total RNA concentration ranged from 447 to 1071 ng/μl according to spectrophotometric determination. The $A_{260}:A_{280}$ values of the isolated total RNA ranged from 1.914 to 1.966, indicating the high purity of the total RNA. The integrity of total RNA was confirmed by 1% agarose gel electrophoresis. CDNA was produced using the PrimeScript 1st Strand cDNA Synthesis Kit (TAKARA, Japan) in a total volume of 20 μl, with 4 μl 5×PrimeScript Buffer, 1 μg of total RNA, 1 μl oligo dT primer, 1 μl PrimeScript RTase (200 U/μl), and 0.5 μl RNase Inhibitor (40 U/μl). Following the manufacturer's protocol, the 20 ul mixture was incubated for 60 min at 42°C. No-template and no-reverse-transcription controls were included for each reverse-transcription run for the control treatment. CDNA was stored at −20°C for later use.

Primer Design

The sequences of all candidate reference genes were downloaded from GenBank (http://www.ncbi.nlm.nih.gov/genbank/) and UNKA (BPH) EST BLAST database (http://bphest.dna.affrc.go.jp/). The PCR primer sequences used for quantification of the expression of the genes encoding ACT, MACT, RPS11, RPS15, TUB, EF, 18S, and AK are shown in Table 1. The secondary structure of the template was analyzed with UNAFold using the DNA folding form of the mfold web server (http://mfold.rna.albany.edu/?q = mfold/DNA-Folding-Form) [30] with the following settings: melting temperature, 60°C; DNA sequence, linear; Na^+ concentration, 50 mM; Mg^{2+} concentration, 3 mM. The other parameters were set by default. The primers were designed on the NCBI-Primer-BLAST website (http://www.ncbi.nlm.nih.gov/tools/primer-blast/index.cgi?LINK_LOC = BlastHome). The settings in NCBI-Primer-BLAST were as follows: primer melting temperature, 57–63°C; primer GC content, 40–60%; and PCR product size, 150–300 base pairs. The excluded regions were determined using mfold, and the other parameters were set by default. Four primer pairs were designed for each gene. The length of PCR products was assessed using gel electrophoresis, and the identity of the PCR products was confirmed by sequence analysis. Only primers which could not amplify non-specific products and dimmers were employed. A 10-fold dilution series of cDNA from the whole body of adults was employed as a standard curve, and the reverse-transcription qPCR efficiency was determined for each gene and each treatment, using the linear regression model [31]. The corresponding qRT-PCR efficiencies (E) were calculated according to the equation: $E = (10^{[-1/slope]} - 1) \times 100$ [32]. After detecting the efficiencies of the chosen primers, the primers which displayed a coefficient of correlation greater than 0.99 and efficiencies between 95% and 108% were selected for the next qRT-PCR (Table 1).

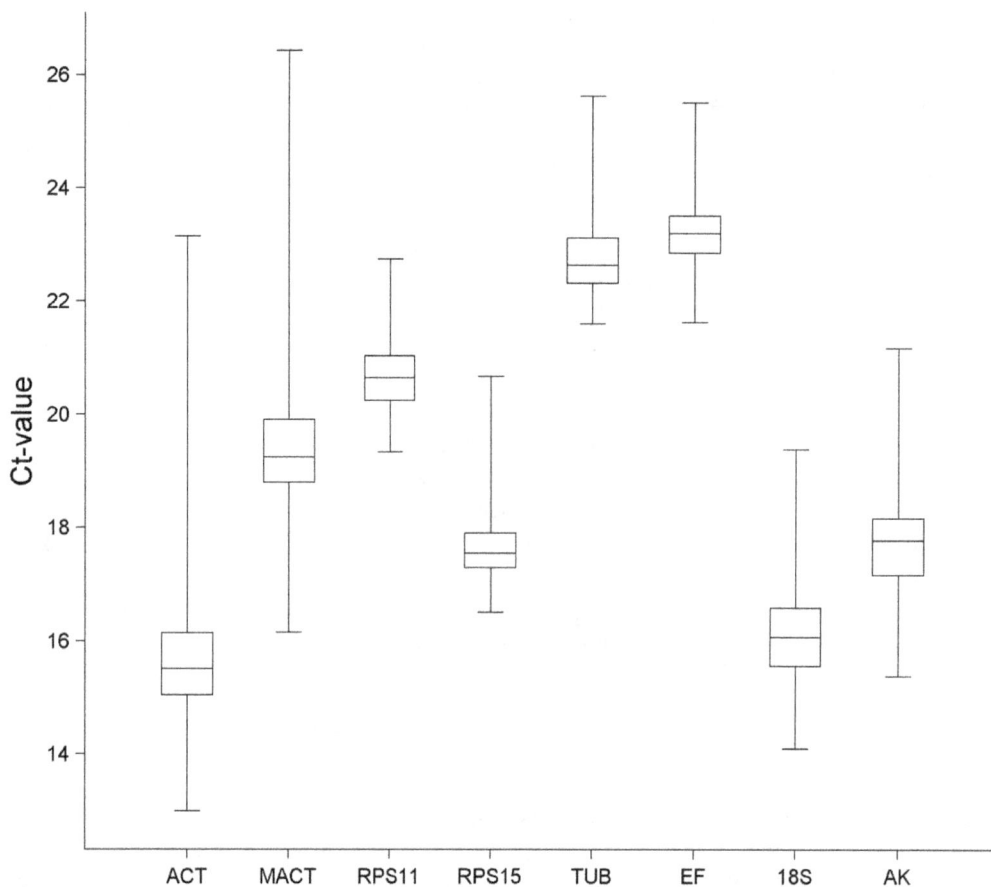

Figure 1. Expression levels of candidate reference genes. The expression level of candidate *N. lugens* reference genes in the total samples is shown in terms of the cycle threshold number (Ct-value). The data are expressed as whisker box plots; the box represents the 25th–75th percentiles, the median is indicated by a bar across the box, the whiskers on each box represent the minimum and maximum values.

Reverse-transcription qPCR Assays

Triplicate 1st-strand DNA aliquots for each treatment served as templates for qRT-PCR using SsoFast™ EvaGreen® Supermix (Bio-Rad) on a Bio-Rad iQ2 Optical System (Bio-Rad). Amplification reactions were performed in a 20 μl volume with 1 μl of cDNA and 100 nM of each primer, in iQ™ 96-well PCR plates (Bio-Rad) covered with Microseal "B" adhesive seals (Bio-Rad). Thermal cycling conditions were as follows: initial denaturation temperature, 95°C for 30 s, followed by 40 cycles at 95°C for 5 s and 60°C for 10 s. After the reaction, a melting curve analysis from 65°C to 95°C was applied to ensure consistency and specificity of the amplified product.

Data Mining and Selection of Reference Genes

Expression levels were determined as the number of cycles needed for the amplification to reach a fixed threshold in the exponential phase of the PCR reaction [33]. The number of cycles is referred to as the threshold cycle (Ct) value. The threshold was set at 500 for all genes. Four freely available software tools, BestKeeper [34], geNorm version3.5 [35], NormFinder version 0.953 [36], and the delta Ct method [37] were used to evaluate gene expression stability. The Excel based tool Bestkeeper, uses raw data (Ct values) and PCR efficiency (E) to determine the best-suited standards and combines them into an index by the coefficient of determination and the P value [34]. Quantities transformed to a linear scale (the highest relative quantity for each

gene was set to 1) were used as input data for geNorm and NormFinder. geNorm algorithm first calculates an expression stability value (M) for each gene and then compares the pairwise variation (V) of this gene with the others. Reference genes are ranked according to their expression stability by a repeated process of stepwise exclusion of the least stably expressed genes. The geNorm program also indicates the minimum number of reference genes for accurate normalization by the pairwise variation value. The value of Vn/n+1 under 0.15 means that no additional genes are required for normalization [35]. NormFinder provides a stability value for each gene which is a direct measure for the estimated expression variation enabling the user to evaluate the systematic error introduced when using the gene for normalization [36]. The delta Ct method compares relative expression of pairs of genes within each sample to confidently identify useful housekeeping genes [37]. A user-friendly web-based comprehensive tool, RefFinder (http://www.leonxie.com/referencegene. php?type = reference) was used, integrating four currently available major software programs to compare and ranking the tested candidate reference genes. Based on the rankings from each program, RefFinder assigns an appropriate weight to an individual gene and calculates the geometric mean of their weights for the overall final ranking. According to the results of RefFinder, candidate genes with the lower ranking were considered to be most stably expressed under tested experimental conditions, and thus could be selected as ideal reference genes.

Results

Expression Profiles of Candidate Reference Genes

In order to evaluate gene expression levels of all studied housekeeping genes within the whole sample set of *N.lugens*, mRNA expressions for every gene were measured. Gene expression levels showed a broad range of variance between Ct-value 12.99 (ACT) and 26.43 (MACT) (Figure 1). Out of eight studied genes, ACT (mean Ct-value 15.71) and 18S (mean Ct-value 16.16) were expressed at the highest levels; TUB (mean Ct-value 22.79) and EF (mean Ct-value 23.25) at the lowest levels. The lowest expression variability within all samples was observed for the gene RPS11 (mean Ct-value±SD, 20.65±0.58) and RPS15 (17.74±0.69). ACT (15.71±1.36) and MACT (19.37±1.39) showed the most variable expression within the sample set.

Analysis of Gene Expression Stability

(1) Developmental stage: The stability ranking generated by the Delta Ct method was largely similar with the results obtained from BestKeeper and NormFinder. However, the most stable genes ranking by geNorm analysis were different to the results generated by the other three methods. All four programs identified ACT and MACT as the least stable genes, and RPS11, RPS15, and EF as the most stable genes except geNorm (Table 2). According to the results of RefFinder, the stability ranking from the most stable to the least stable in the developmental stages was RPS15, RPS11, TUB, EF, 18S, AK, ACT, and MACT (Table S2). As can be noticed, TUB was the most stable gene across different nymphal stages and across different sexes (Table S3). With geNorm, the V value of 0.154 obtained for the RPS15-RPS11 pair was near the proposed cut-off value of 0.15. Moreover, the inclusion of additional reference genes did not lower the V value below the proposed 0.15 cut-off value until the fourth gene was added (Figure 2). According to geNorm, four reference genes (RPS15, TUB, 18S, and EF) should be required for a suitable normalization in the different developmental stages.

(2) Body part: All four programs, except BestKeeper, identified RPS11, RPS15, and 18S as the most stable genes (Table 2). According to the results of RefFinder, the stability ranking from the most stable to the least stable gene in different body parts was RPS11, TUB, RPS15, 18S, ACT, MACT, EF, and AK (Table S2). RPS11 was the most stable gene across the different body parts of female and male adults (Table S4). TUB was the most stable gene between males and females in the head, thorax, and whole body (Table S5). However, TUB displayed high instability between males and females in the abdomen (Table S5). GeNorm analysis revealed that the pairwise variation values were all above the cut-off value and decreased with the added reference genes (Figure 2). These results indicated that normalization with three stable reference genes (RPS11, 18S, and RPS15) was required (as suggested by the geNorm manual).

(3) Population: The stability ranking generated by the Delta Ct method was largely similar with the results obtained by NormFinder. All four programs, except geNorm, identified TUB as the most stable gene (Table 2). According to the results of RefFinder, the stability ranking from the most stable to the least stable gene in the two different populations was TUB, RPS11, EF, RPS15, AK, ACT, 18S, and MACT (Table S2). EF and TUB showed high expression stability in the nymphs and adults of these two populations, respectively.

Interestingly, RPS15 showed high instability in the adults of both different populations, and was ranked one of the least stable genes in the 3rd instar nymphs of two different populations (Table S6). GeNorm analysis revealed that all the pairwise variation values were below the proposed 0.15 cut-off, except for V2/3 (Figure 2). According to geNorm, three reference genes (RPS11, EF, and RPS15) should be required for a suitable normalization in these two different geographic populations.

(4) Temperature: All four programs identified RPS15 and TUB as the most stable genes, and identified ACT as the least stable gene (Table 2). From the results of RefFinder, the stability ranking from the most stable to the least stable gene in the temperature-stressed samples was RPS15, TUB, EF, RPS11, AK, MACT, 18S, and ACT (Table S2). Under extremely low temperature stress, AK was ranked one of the most stable genes, while it was ranked one of the least stable genes under low temperature stress (Table S7). TUB was the most stable gene at average temperatures (Table S7). MACT, which was ranked one of the least stable genes under extremely low temperature, low temperature, and average temperature, showed high expression stability under high-temperature stress (Table S7). ACT was ranked as the least stable gene in all temperature conditions (Table S7). GeNorm analysis revealed that all the pairwise variation values were below the proposed 0.15 cut-off (Figure 2). According to geNorm, three reference genes (RPS15, TUB, and EF) should be required for a suitable normalization in the different temperature treatment samples.

(5) Pesticide treatment: The stability ranking generated by the Delta Ct method was same as the results obtained from NormFinder and geNorm. The stability ranking generated by BestKeeper was largely similar with the one obtained by the other three methods. All four programs identified RPS11 and EF as the most stable genes (Table 2). According to RefFinder, the stability ranking from the most stable to the least stable in the pesticide-stressed samples was RPS11, EF, TUB, RPS15, 18S, AK, MACT, and ACT (Table S2). As can be noticed, RPS11 was also the most stable gene in all pesticide-treated samples (Table S2), compound-pesticide-treated samples, buprofezin-treated samples, and isoprocarb-treated samples (Table S8). EF and TUB were the most stable genes in the nitenpyram-treated samples and chlorpyrifos-treated samples (Table S8), respectively. MACT, which was ranked one of the least stable genes in other pesticide treatments, showed the highest stability in pymetrozine-treated samples (Table S8). GeNorm analysis revealed that all the pairwise variation values were below the proposed 0.15 cut-off value (Figure 2). According to geNorm, three reference genes (RPS11, EF, and TUB) should be required for a suitable normalization in the pesticide-stressed samples.

(6) Diet: All four programs identified RPS15 as the most stable gene, and identified ACT and MACT as the least stable genes (Table 2). According to RefFinder, the stability ranking from the most stable to the least stable in the different diets treatments was RPS15, TUB, RPS11, EF, AK, 18S, ACT, and MACT (Table S2). RPS15 was the most stable gene in *N. lugens* reared on artificial diet, TN1, HH1 and SY63, and was ranked second in the *N. lugens* reared on MH63 (Table S9). However, RPS15 was the least stable gene in *N. lugens* reared on BTSY63 (Table S9). The results also showed that RPS15 and RPS11 were the most stable genes in *N. lugens* reared on non-genetically modified rice and genetically modified rice,

respectively (Table S10). In *N. lugens* nymphs reared on non-genetically modified rice, TUB was the most stable gene (Table S10), while in *N. lugens* adults reared on non-genetically modified rice, RPS15 was still the most stable gene (Table S10). RPS15 and 18s were the most stable genes in the *N. lugens* nymphs and adults reared on genetically modified rice, respectively (Table S10). With geNorm, the V value of 0.176 obtained by the RPS15 and TUB pair was near the proposed 0.15 cut-off value. Moreover, the inclusion of additional reference genes did not lower the V value below the proposed 0.15 cut-off until the 4th gene was added (Figure 2). According to geNorm, four reference genes (RPS15, TUB, EF and RPS11) should be required for a suitable normalization in the different diets treatments.

(7) Starvation: The gene stability of the starvation group compared to a satiation group (SY63) was analyzed. All four programs identified ACT and MACT as the least stable genes, and identified RPS11 as the most stable gene except BestKeeper (Table 2). According to RefFinder, the stability ranking from the most stable to the least stable in the starvation treatments was RPS11, TUB, RPS15, AK, 18S, EF, ACT, and MACT (Table S2). RPS11 was the most stable gene both in starved nymphs and starved adults (Table S11). GeNorm analysis revealed that all the pairwise variation values were below the proposed 0.15 cut-off (Figure 2). According to geNorm, three reference genes (RPS11, AK, and EF) should be required for a suitable normalization in the starvation treatments.

Ranking of *N. lugens* Reference Genes Over all Treatments

All four programs identified ACT and MACT as the least stable genes, and RPS11 and RPS15 as the most stable genes except geNorm (Table 2). According to RefFinder, the stability ranking from the most stable to the least stable across the different developmental stages, body parts, populations, and stressors was RPS11, RPS15, EF, TUB, AK, 18S, ACT, and MACT (Table S2).

Discussion

This work analyzed the expression stability of eight candidate reference genes in *N. lugens* across different treatments and developmental stages using qRT-PCR. A major result of this study is that 18S showed unacceptable variation in response to certain treatments. Previously, 18S ribosomal RNA has been considered as an ideal reference gene due to its apparent relatively invariable rRNA expression levels with respect to other genes [38]. 18S rRNA was found to be one of the most suitable housekeepers in the different developmental stages of *Lucilia cuprina* [39], in different organs of *Rhodnius prolixus* under diverse conditions [40,41], and in the planthopper *Delphacodes kuscheli* infected by the plant fijivirus *Mal de Río Cuarto virus* (MRCV) [42]. However, in our study, 18S ranked as one of the least stable genes in the total samples and almost in all experimental conditions indicating that 18S was not suitable as a reference gene for *N. lugens* under our experimental conditions (Tables S2, S3, S4, S5, S6, S7, S8, S9, S10, S11). This result is in line with the earlier studies indicating that 18S rRNA is not stable enough in *Bactrocera dorsalis* under specified experimental conditions [43]. The transcription by a separate RNA polymerase is proposed to be a reason why rRNA could not be considered as a suitable reference gene [44]. On the other hand, one of the major limitations of using the 18S gene as a normalizer in qRT-PCR is that an imbalance of rRNA and mRNA fractions can occur between samples [38]. Our study suggests that 18S rRNA could not be used for correcting sample-to-sample variation of mRNA quantity in *N. lugens*.

Like 18S rRNA, actin is another commonly used reference gene which encodes a major component of the protein scaffold that supports the cell and determines its shape, and is expressed at moderately abundant levels in most cell types. Actin has been highly ranked as a suitable reference gene in studies of gene expression in *Apis mellifera* [45], *Schistocera gregaria* [46], *Drosophila melanogaster* [47], *Plutella xylostella* [48], and *Chilo suppressalis* [48]. Actin gene has as well been selected as reference gene in gene expression studies in *N. lugens* [12,13,14]. However, compared with the other candidate genes examined here, the expression levels of ACT and MACT were highly variable across the different treatments (Tables S2, S3, S4, S5, S6, S7, S8, S9, S10, S11). ACT and MACT, which participate in many important cellular processes including muscle contraction, cell motility, cell division and cytokinesis, ranked one of the least stable genes in the total

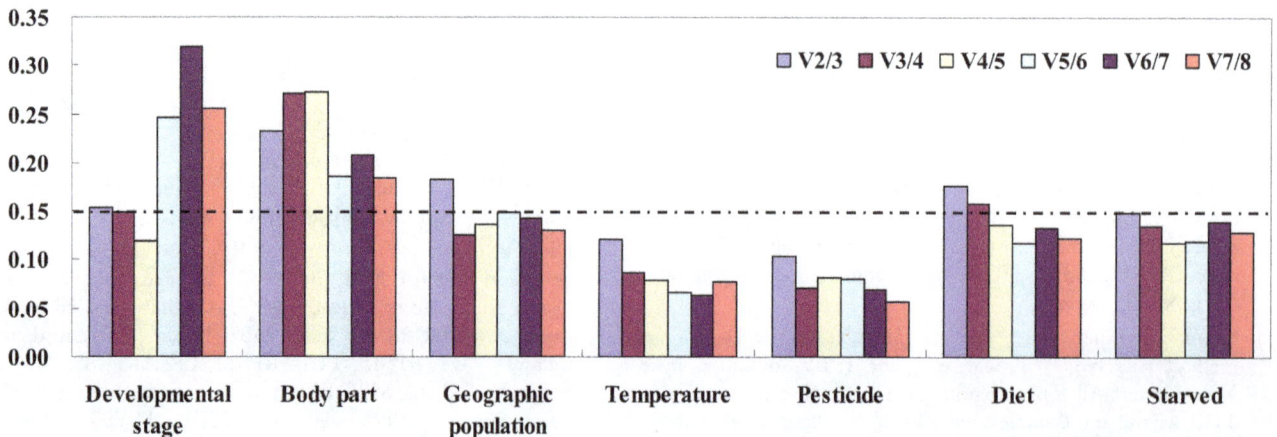

Figure 2. Determination of the optimal number of reference genes for accurate normalization calculated by geNorm. The value of Vn/Vn+1 indicates the pairwise variation (Y axis) between two sequential normalization factors and determines the optimal number of reference genes required for accurate normalization. A value below 0.15 indicates that an additional reference gene will not significantly improve normalization.

Table 2. Ranking order of the candidate reference genes of *N. lugens* in different experimental conditions.

Experimental conditions	Rank	Delta Ct Gene name	Standard deviation	BestKeeper Gene name	Standard deviation	NormFinder Gene name	Stability value	geNorm Gene name	Stability value
Different	1	RPS11	1.190	RPS11	0.380	RPS11	0.407	RPS15/TUB	0.425
developmental	2	RPS15	1.204	RPS15	0.520	RPS15	0.705		
stages	3	EF	1.274	EF	0.541	EF	0.827	18S	0.480
	4	TUB	1.355	18S	0.557	AK	0.876	EF	0.566
	5	18S	1.401	TUB	0.605	TUB	1.069	RPS11	0.614
	6	AK	1.532	AK	0.816	18S	1.144	AK	0.915
	7	ACT	2.047	MACT	1.539	ACT	1.864	ACT	1.309
	8	MACT	2.148	ACT	1.582	MACT	2.004	MACT	1.519
Different body parts	1	RPS11	1.096	RPS15	0.465	RPS11	0.203	RPS11/18S	0.620
	2	RPS15	1.210	TUB	0.501	18S	0.628		
	3	18S	1.212	RPS11	0.557	RPS15	0.741	RPS15	0.717
	4	ACT	1.427	AK	0.928	TUB	1.093	TUB	0.935
	5	TUB	1.455	EF	0.953	ACT	1.100	EF	1.149
	6	MACT	1.458	18S	0.963	MACT	1.152	ACT	1.193
	7	EF	1.610	ACT	1.001	AK	1.411	MACT	1.294
	8	AK	1.703	MACT	1.013	EF	1.421	AK	1.396
Different geographic	1	TUB	0.708	TUB	0.590	TUB	0.145	RPS11/EF	0.212
populations	2	RPS11	0.728	EF	0.637	RPS11	0.362		
	3	RPS15	0.774	RPS15	0.637	RPS15	0.412	RPS15	0.440
	4	EF	0.785	RPS11	0.706	EF	0.506	TUB	0.501
	5	AK	0.922	ACT	0.756	AK	0.709	AK	0.594
	6	ACT	0.936	AK	0.794	ACT	0.750	ACT	0.707
	7	MACT	1.122	MACT	0.824	18S	1.016	MACT	0.803
	8	18S	1.156	18S	0.980	MACT	1.017	18S	0.891
Temperature-stress	1	RPS15	0.433	RPS15	0.204	RPS15	0.221	RPS15/TUB	0.287
treatments	2	TUB	0.450	TUB	0.235	TUB	0.265		
	3	EF	0.478	RPS11	0.277	EF	0.305	EF	0.356
	4	RPS11	0.500	AK	0.282	MACT	0.342	AK	0.379
	5	AK	0.501	MACT	0.325	AK	0.345	RPS11	0.408
	6	MACT	0.505	18S	0.345	RPS11	0.351	MACT	0.429
	7	18S	0.544	ACT	0.357	18S	0.414	18S	0.454
	8	ACT	0.688	EF	0.547	ACT	0.608	ACT	0.512
Pesticide-stress	1	RPS11	0.435	EF	0.245	RPS11	0.253	RPS11/EF	0.277
treatments	2	EF	0.435	RPS11	0.248	EF	0.257		
	3	TUB	0.439	TUB	0.267	TUB	0.271	TUB	0.318
	4	RPS15	0.445	RPS11	0.296	RPS15	0.277	RPS15	0.328
	5	18S	0.518	MACT	0.465	18S	0.391	18S	0.379
	6	AK	0.544	AK	0.473	AK	0.430	AK	0.430
	7	MACT	0.557	ACT	0.539	MACT	0.443	MACT	0.469
	8	ACT	0.557	18S	0.583	ACT	0.443	ACT	0.491
Different diet	1	RPS15	0.730	RPS15	0.490	RPS15	0.362	RPS15/TUB	0.421
treatments	2	TUB	0.792	RPS11	0.527	TUB	0.485		
	3	RPS11	0.850	EF	0.565	RPS11	0.559	EF	0.513
	4	EF	0.851	AK	0.584	AK	0.578	RPS11	0.603
	5	AK	0.872	TUB	0.603	EF	0.626	18S	0.670
	6	18S	0.906	18S	0.639	18S	0.666	AK	0.723
	7	ACT	0.989	ACT	0.658	ACT	0.778	ACT	0.814

Table 2. Cont.

Experimental conditions	Rank	Delta Ct		BestKeeper		NormFinder		geNorm	
		Gene name	Standard deviation	Gene name	Standard deviation	Gene name	Stability value	Gene name	Stability value
	8	MACT	1.106	MACT	0.812	MACT	0.957	MACT	0.887
Starvation-stress	1	RPS11	0.680	TUB	0.247	RPS11	0.282	RPS11/AK	0.372
treatments	2	TUB	0.720	RPS15	0.283	TUB	0.304		
	3	RPS15	0.778	RPS11	0.379	18S	0.480	EF	0.446
	4	18S	0.804	18S	0.506	RPS15	0.506	RPS15	0.521
	5	AK	0.826	AK	0.585	AK	0.624	TUB	0.573
	6	EF	0.896	EF	0.595	EF	0.767	18S	0.645
	7	ACT	0.952	ACT	0.621	ACT	0.785	ACT	0.759
	8	MACT	1.102	MACT	0.736	MACT	1.009	MACT	0.845
All above conditions	1	RPS11	0.946	RPS11	0.463	RPS11	0.370	RPS15/EF	0.488
	2	RPS15	1.011	RPS15	0.504	RPS15	0.655		
	3	TUB	1.037	TUB	0.524	TUB	0.671	TUB	0.611
	4	EF	1.107	EF	0.549	AK	0.806	RPS11	0.666
	5	AK	1.174	AK	0.672	EF	0.832	18S	0.788
	6	18S	1.203	18S	0.694	18S	0.900	AK	0.914
	7	ACT	1.354	ACT	0.842	ACT	1.146	ACT	1.077
	8	MACT	1.372	MACT	0.869	MACT	1.175	MACT	1.151

The expression stability was also measured using the Delta Ct method, BestKeeper, NormFinder, and geNorm and ranked from the most stable to the least stable.

samples and under almost all experimental conditions. And not surprisingly, its transcript level varies among developmental stages and different cell types, since it has functions in various cellular processes. In *N. lugens*, ACT and MACT should not be used as reference genes under certain treatments.

Our results also demonstrated that the best-suited reference genes can be different in response to diverse factors (Table S2). Reference genes need to be appropriately selected under different experimental conditions. However, the expression of several reference genes from *N. lugens* were comparatively stable across selected experimental conditions. Ranking of the genes differed somewhat for geNorm, NormFinder, BestKeeper, and the delta Ct method probably because the programs have different algorithms and different sensitivities toward co-regulated reference genes. In spite of the slight discrepancies, all the programs identified both RPS11 and RPS15 as the same ideal reference genes for most of the experimental conditions assessed here (Table S2). Ribosomal proteins compose the ribosomal subunits involved in the cellular process of translation in conjunction with rRNA. RPS11 and RPS15 encode the component of the 40S ribosomal subunit which is the small subunit of eukaryotic 80S ribosomes [49]. Considering the function of ribosomal proteins, it is not surprising that their transcription level varies among different cell types and developmental stages in the brown planthopper. Our result is in line with the earlier studies on ribosomal protein genes in *A. mellifera* [45], *S. gregaria* [46], *Tribolium castaneum* [50,51], *D. melanogaster* [47], *B. mori* [48], *C. suppressalis* [48], and *Bemisia tabaci* [52].

Arginine kinase, which is the only phosphagen kinase in two major invertebrate groups, namely arthropods and mollusks, was one of the most stable genes in *Bombus terrestris* [53]. In our study, AK was also the most stable gene in BPH under extremely low temperature stress (Table S7), and the second most stable gene in nymphs (Table S3). Elongation factor which plays an important role in translation by catalyzing the GTP-dependent binding of aminoacyl-tRNA to the acceptor site of the ribosome exhibited the second most stable expression in the BPH under pesticide-stress (Table S2). EF was found to be the most stable genes for the labial gland and fat body of *Bombus lucorum* [53] and for reliable normalization of qRT-PCR assays studying density-dependent behavioral change in *Chortoicetes terminifera* [54]. However, arginin kinase and elongation factor didn't show acceptable stable expression in most treatments (Table S2). Even for housekeeping genes, whose products are indispensable for every living cell and are relatively stably expressed, there are tissue-specific differences based upon extra demands in the required rate at which new housekeeping proteins need to be produced to maintain cell function [55].

Multiple reference genes are increasingly used to analyze gene expression under various experimental conditions, because one reference gene is usually insufficient to normalize the expression results of target genes [56]. After measuring the expression of 20 candidate reference genes and 7 target genes in 15 *Drosophila* head cDNA samples using qRT-PCR, 20 reference genes exhibited sample-specific variation in their expression stability and the most stable normalizing factor variation across samples did not exhibit a continuous decrease with pairwise inclusion of more reference genes; these results suggest that either too few or too many reference genes may detriment the robustness of data normalization [57]. When several reference genes are used simultaneously in a given experiment, the probability of biased normalization decreases. GeNorm determines the pairwise variations (V) in normalization factors (the geometric mean of multiple reference genes) using n or n +1 reference genes. Our results showed that the best-suited reference genes were different across different experimental conditions (Figure 2). This implies that the expression stability of putative control genes needs to be verified before each qRT-PCR experiment.

Conclusion

To our knowledge this is the first study to evaluate candidate reference genes for gene expression analyses in *N. lugens*. Most importantly, we identified reference genes which should be used for accurate elucidation of the expression profiles of functional genes. We concluded that RPS15, RPS11, and TUB were the most suitable reference genes for the analysis of developmental stage, body part, and geographic population, respectively (Table S2). And that RPS15, RPS11, RPS15, and RPS11 were the most suitable reference genes under temperature, pesticide, diet, and starvation stress, respectively (Table S2). This work emphasizes the importance of establishing a standardized reverse-transcription quantitative PCR procedure following the MIQE guidelines in *N. lugens*, and serves as a resource for screening reference genes for expression studies in other insects.

Supporting Information

Table S1 Insecticides toxicity to 3rd instar *N. lugens* larvae.

Table S2 Expression stability of the candidate reference genes in the total samples. The average expression stability of the reference genes was measured using the Geomean method of RefFinder (http://www.leonxie.com/referencegene. php?type = reference). A lower rank indicates more stable expression.

Table S3 Expression stability of the candidate reference genes across different nymphal stages and across different sexes. The average expression stability of the reference gene was measured using the Geomean method of RefFinder (http://www.leonxie.com/referencegene. php?type = reference). A lower rank indicates more stable expression.

Table S4 Expression stability of the candidate reference genes different body parts of female and male adults. The average expression stability of the reference gene was measured using the Geomean method of RefFinder (http://www. leonxie.com/referencegene.php?type = reference). A lower rank indicates more stable expression.

Table S5 Expression stability of the candidate reference genes across males and females in the heads, thoraxes, abdomens, and whole bodies. The average expression stability of the reference gene was measured using the Geomean method of RefFinder (http://www.leonxie.com/ referencegene.php?type = reference). A lower rank indicates more stable expression.

Table S6 Expression stability of the candidate reference genes across two different *N. lugens* geographic

populations. The average expression stability of the reference gene was measured using the Geomean method of RefFinder (http://www.leonxie.com/referencegene.php?type = reference). A lower rank indicates more stable expression.

Table S7 Expression stability of the candidate reference genes across different temperatures. The average expression stability of the reference gene was measured using the Geomean method of RefFinder (http://www.leonxie.com/ referencegene.php?type = reference). A lower rank indicates more stable expression.

Table S8 Expression stability of the candidate reference genes under different pesticide stresses. The average expression stability of the reference gene was measured using the Geomean method of RefFinder (http://www.leonxie.com/ referencegene.php?type = reference). A lower rank indicates more stable expression.

Table S9 Expression stability of the candidate reference genes of *N. lugens* fed on different diets. The average expression stability of the reference gene was measured using the Geomean method of RefFinder (http://www.leonxie.com/ referencegene.php?type = reference). A lower rank indicates more stable expression.

Table S10 Expression stability of the candidate reference genes of *N. lugens* fed on non-genetically modified rice and genetically modified rice. The average expression stability of the reference gene was measured using the Geomean method of RefFinder (http://www.leonxie.com/referencegene. php?type = reference). A lower rank indicates more stable expression.

Table S11 Expression stability of the candidate reference genes of straved *N. lugens*. The average expression stability of the reference gene was measured using the Geomean method of RefFinder (http://www.leonxie.com/referencegene. php?type = reference). A lower rank indicates more stable expression.

Acknowledgments

Special thanks go to Dr. Mariana del Vas (Instituto de Biotecnología, CICVyA, Instituto Nacional de Tecnología Agropecuaria (IB-INTA), Argentina) for comments on an earlier draft, to Prof. Manqun Wang (Huazhong Agricultural University, China) for supplying the insects, and to Prof. Yongjun Lin (Huazhong Agricultural University, China) for supplying the rice seeds of TN1, HH1, MH63, SY63, and BTSY63.

Author Contributions

Conceived and designed the experiments: MY XZ YL JL. Performed the experiments: MY. Analyzed the data: MY YL. Contributed reagents/materials/analysis tools: SZ BJ HW MS. Wrote the paper: MY.

References

1. Dong XL, Zhai YF, Zhang JQ, Sun ZX, Chen J, et al. (2012) Fork head transcription factor is required for ovarian mature in the brown planthopper, *Nilaparvata lugens* (stål). BMC Mol Biol 12: 53.

2. Otake A (1978) Population characteristics of the brown planthopper, *Nilaparvata lugens* (Hemiptera: Delphacidae), with special reference to differences in Japan and the tropics. J Appl Ecol 15: 385–394.

3. Hibino H (1996) Biology and epidemiology of rice viruses. Annu Rev Phytopathol 34: 249–274.

4. Wang YH, Wang MH (2007) Factors affecting the outbreak and management tactics of brown planthopper, *Nilaparvata lugens* (Stål) in China in recent years (in Chinese). Pestic Sci Admin 29: 49–54.

5. Wang Y, Chen J, Zhu YC, Ma C, Huang Y, et al. (2008) Susceptibility to neonicotinoids and risk of resistance development in the brown planthopper, *Nilaparvata lugens* (Stål) (Homoptera: Delphacidae). Pest Manag Sci 64: 1278–1284.

6. Zhang Y, Fan HW, Huang HJ, Xue J, Wu WJ, et al. (2012) Chitin synthase 1 gene and its two alternative splicing variants from two sap-sucking insects, *Nilaparvata lugens* and *Laodelphgax striatellus* (Hemiptera: Delphacidae). Insect Biochem Mol Bio 42(9): 637–646.

7. Gibson UEM, Heid CA, Williams PM (1996) A novel method for real-time quantitative RT-PCR. Genome Methods 6: 995–1001.

8. Heid CA, Stevens J, Livak KJ, Williams PM (1996) Real-time quantitative PCR. Genome Methods 6: 986–994.

9. Baum AJ, Bogaer T, Clinton W, Heck GR, Feldmann P, et al. (2007) Control of coleopteran insect pests through RNA interference. Nat Biotechnol 25: 1322–1326.

10. Mao YB, Cai WJ, Wang JW, Hong GJ, Tao XY, et al. (2007) Silencing a cotton bollworm P450 monooxygenase gene by plant-mediated RNAi impairs larval tolerance of gossypol. Nat Biotechnol 25: 1307–1313.

11. Price DRG, Gatehouse JA (2008) RNAi-mediated crop protection against insects. Cell 26 (7): 393–400.

12. Whyard D, Singh AD, Wong S (2009) Ingested double-stranded RNAs can act as species-specific insecticides. Insect Biochem Mol 39 (11): 824–832.

13. Huvenne H, Smagghe G. (2010) Mechanisms of dsRNA uptake in insects and potential of RNAi for pest control: A review. J Insect Physiol 56 (3): 227–235.

14. Chen J, Zhang D, Yao Q, Zhang J, Dong X, et al. (2010) Feeding-based RNA interference of a trehalose phosphate synthase gene in the brown planthopper, *Nilaparvata lugens*. Insect Mol Biol 19 (6): 777–786.

15. Liu SH, Ding ZP, Zhang CW, Yang BJ, Liu ZW (2010) Gene knockdown by intro-thoracic injection of double-stranded RNA in the brown planthopper, *Nilaparvata lugens*. Insect Biochem Mol 40 (9): 666–671.

16. Zha WJ, Peng XX, Chen RZ, Du B, Zhu LL, et al. (2011) Knockdown of midgut genes by dsRNA-transgenic plant-mediated RNA interference in the Hemipteran Insect *Nilaparvata lugens*. PLoS ONE 6 (5): e20504.

17. Hiroaki N, Sawako K, Yoko K, Kageaki M, Qaing Z, et al. (2008) Annotated ESTs from various tissues of the brown planthopper *Nilaparvata lugens*: A genomic resource for studying agricultural pests. BMC Genomics 9: 117.

18. Xue J, Bao YY, Li BL, Cheng YB, Peng ZY, et al. (2010) Transcriptome analysis of the brown planthopper *Nilaparvata lugens*. PLoS ONE 5(12): e14233.

19. Peng X, Zhang W, He R, Lu T, Zhu L, et al. (2011) Pyrosequencing the midgut transcriptome of the brown planthopper *Nilaparvata lugens*. Insect Mol Biol 20(6): 745–762.

20. Jain M, Nijhawan A, Tyagi AK, Khurana JP (2006) Validation of housekeeping genes as internal control for studying gene expression in rice by quantitative real-time PCR. Biochem Biophys Res Co 345: 646–651.

21. Thellin O, Zorzi W, Lakaye B, Borman BD, Coumans B, et al. (1999) Housekeeping genes as internal standards: use and limits. J Biotechnol 75: 291–295.

22. Schmittgen TD, Zakrajsek BA (2000) Effect of experimental treatment on housekeeping gene expression: validation by real-time, quantitative RT-PCR. J Biochem Biophys Methods 46: 69–81.

23. Provenzano M, Mocellin S (2007) Complementary techniques: validation of gene expression data by quantitative real-time PCR. Eurekah Bioscience 2 (6): 510–513.

24. Selvey S, Thompson EW, Matthaei K, Lea RA, Irving MG, et al. (2001) β-Actin–an unsuitable internal control for RT-PCR. Mol Cell Probe 15: 307–311.

25. Radonić A, Thulke S, Mackay IM, Landt O, Siegert W, et al. (2004) Guideline to reference gene selection for quantitative real-time PCR. Biochem Biophys Res Co 313: 856–862.

26. Yanhua W, Jin C, Zhu YC, Chongyong M, Yue H, et al. (2008) Susceptibility to neonicotinoids and risk of resistance development in the brown planthopper, *Nilaparvata lugens* (Stål) (Homoptera: Delphacidae). Pest Manag Sci 64: 1278–1284.

27. Liu F, Li H, Qiu J, Zhang Y, Huang L, et al. (2010) Monitoring of resistance to several insecticides in brown planthopper (*Nilaparvata lugens*) in Huizhou. Chinese Bulletin of Entomology 47(5): 991–993.

28. Ling Y, Huang F, Long L, Zhong Y, Yin W, et al. (2011) Studies on the pesticide resistant of *Nilaparvata lugens* (Stål) in China and Vietnam. Chinese Journal of Applied Entomology 48(5): 1374–1380.

29. Fu Q, Zhang ZT, Hu C, Lai FX, Sun ZX (2001) A chemically defined diet enables continuous rearing of the brown planthopper, *Nilaparvata lugens* (Stål) (Homoptera: Delphacidae). Appl Entomol Zool 36 (1): 111–116.

30. Markham NR, Zuker M (2005) DINAMelt web server for nucleic acid melting prediction. Nucleic Acids Res 33: W577–581.

31. Pfaffl MW (2001) A new mathematical model for relative quantification in real-time RT-PCR. Nucleic Acids Res 29: 9.

32. Radonic A, Thulke S, Mackay I, Landt O, Siegert W, et al. (2004) Guideline to reference gene selection for quantitative real-time PCR. Biochem Bioph Res Co 313: 856–862.

33. Walker NJ (2002) A technique whose time has come. Science 296: 557–559.

34. Pfaffl MW, Tichopad A, Prgomet C, Neuvians TP (2004) Determination of stable housekeeping genes, differentially regulated target genes and sample integrity: BestKeeper—Excel-based tool using pairwise correlations. Biotechnology Letters 26: 509–515.

35. Vandesompele J, De Preter K, Pattyn F, Poppe B, Van Roy N, et al. (2002) Accurate normalization of real-time quantitative RT-PCR data by geometric averaging of multiple internal control genes. Genome Biol 3: RESEARCH0034.

36. Andersen CL, Ledet-Jensen J, Ørntoft T (2004) Normalization of real-time quantitative RT-PCR data. a model-based variance estimation approach to identify genes suited for normalization, applied to bladder and colon cancer data sets. Cancer Res 64: 5245–5250.

37. Nicholas Silver SB, Jiang J, Thein SL (2006) Selection of housekeeping genes for gene expression studies in human reticulocytes using realtime PCR. BMC Molecular Biology 7: 33.

38. Bustin SA (2000) Absolute quantification of mRNA using real-time reverse-transcription polymerase chain reaction assays. J Mol Endocrinol 25: 169–193.

39. Bagnall NH, Kotze AC (2010) Evaluation of reference genes for real-time PCR quantification of gene expression in the Australian sheep blowfly, *Lucilia cuprina*. Med Vet Entomol 24: 176–181.

40. Majerowicz D, Alves-Bezerra M, Logullo R, Fonseca-de-Souza AL, Meyer-Fernandes JR, et al. (2011) Looking for reference genes for real-time quantitative PCR experiments in *Rhodnius prolixus* (Hemiptera: Reduviidae). Insect Mol Biol 20(6): 713–722.

41. Paim RM, Pereira MH, Ponzio RD, Rodrigues JO, Guarneri AA, et al. (2012) Validation of reference genes for expression analysis in the salivary gland and the intestine of *Rhodnius prolixus* (Hemiptera, Reduviidae) under different experimental conditions by quantitative real-time PCR. BMC Research Notes 5: 128.

42. Maroniche GA, Sagadín M, Mongelli VC, Truol GAM, del Vas M (2011) Reference gene selection for gene expression studies using RT-qPCR in virus-infected planthoppers. Virology J 8: 308–315.

43. Shen GM, Jiang HB, Wang XN, Wang JJ (2010) Evaluation of endogenous references for gene expression profiling in different tissues of the oriental fruit fly, *Bactrocera dorsalis* (Diptera: Tephritidae). BMC Mol Biol 11: 76.

44. Tricarico C, Pinzani P, Bianchi S, Paglierani M, Distante V, et al. (2002) Quantitative real-time reverse transcription polymerase chain reaction: normalization to rRNA or single housekeeping genes is inappropriate for human tissue biopsies. Anal Biochem 309: 293–300.

45. Scharlaken B, Graaf DC, Goossens K, Brunain M, Peelman LJ, et al. (2008) Reference gene selection for insect expression studies using quantitative real-time PCR: the head of the honeybee, *Apis mellifera*, after a bacterial challenge. J Insect Sci 8: 33.

46. Hiel MBV, Wielendaele PV, Temmerman L, Soest SV, Vuerinckx K, et al. (2009) Identification and validation of housekeeping genes in brains of the desert locust *Schistocerca gregaria* under different developmental conditions. BMC Mol Biol 10: 56.

47. Ponton F, Chapuis MP, Pernice M, Sword GA, Simpson SJ (2011) Evaluation of potential reference genes for reverse-transcription-qPCR studies of physiological responses in *Drosophila melanogaster*. J Insect Physiol 57: 840–850.

48. Teng XL, Zhang Z, He GL, Yang LW, Li F (2012) Validation of reference genes for quantitative expression analysis by real-time RT-PCR in four lepidopteran insects. J Insect Sci 12: 60.

49. Campbell MG, Karbstein K (2011) Protein-Protein Interactions within Late Pre-40S Ribosomes. PLoS ONE 6(1): e16194.

50. Lord JC, Hartzer K, Toutges M, Oppert B (2010) Evaluation of quantitative PCR reference genes for gene expression studies in *Tribolium castaneum* after fungal challenge. J Microbiol Meth 80: 219–221.

51. Toutges MJ, Hartzer K, Lord J, Oppert B (2010) Evaluation of reference genes for quantitative polymerase chain reaction across life cycle stages and tissue types of *Tribolium castaneum*. J Agric Food Chem 58: 8948–8951.

52. Li R, Xie W, Wang S, Wu Q, Yang N, et al. (2013) Reference Gene Selection for qRT-PCR Analysis in the Sweetpotato Whitefly, *Bemisia tabaci* (Hemiptera: Aleyrodidae). PLoS ONE 8(1): e53006.

53. Horňáková D, Matoušková P, Kindl J, Valterová I, Pichová I (2010) Selection of reference genes for real-time polymerase chain reaction analysis in tissues from *Bombus terrestris* and *Bombus lucorum* of different ages. Anal Biochem 397: 118–120.

54. Chapuis MP, Donya TE, Dodgson T, Blodin L, Ponton F, et al. (2011) Assessment and validation of a suite of reverse-transcription-quantitative PCR reference genes for analyses of density-dependent behavioral plasticity in the Australian plague locust. BMC Mol Biol 12: 7.

55. Thorrez L, Van Deun K, Tranchevent L-C, Van Lommel L, Engelen K, et al. (2008) Using Ribosomal Protein Genes as Reference: A Tale of Caution. PLoS ONE 3(3): e1854.

56. Kylee J, Veazey, Michael C (2011) Golding selection of stable reference genes for quantitative RT-PCR comparisons of mouse embryonic and extra-embryonic stem cells. PLoS ONE 6: 27592.

57. Lin DJ, Salvaterra PM (2011) Robust RT-qPCR data normalization: validation and selection of internal reference genes during post-experimental data analysis. PLoS ONE 6(3): e17762.

PERMISSIONS

LIST OF CONTRIBUTORS

Martin Lechenet, Sandrine Petit and Nicolas M. Munier-Jolain
Institut National de la Recherche Agronomique, Unité Mixte de Recherche 1347 Agroécologie, Dijon, Côte d'Or, France

Vincent Bretagnolle
Centre d'Etudes Biologiques de Chizé – Centre National de Recherche Scientifique, Beauvoir sur Niort, Deux-Sévres, France

Christian Bockstaller
Institut National de la Recherche Agronomique, Unité de Recherche 1121 Agronomie et Environnement, Colmar, Haut-Rhin, France
Université de Lorraine, Vandoeuvre-lés-Nancy, Meurthe-et-Moselle, France

François Boissinot
Chambre d'Agriculture des Pays de la Loire, Angers, Maine-et-Loire, France

Marie-Sophie Petit
Chambre Régionale d'Agriculture de Bourgogne, Quetigny, Côte d'Or, France

Lewis H. Ziska
Crop Systems and Global Change Laboratory, United States Department of Agriculture, Agricultural Research Service, Beltsville, Maryland, United States of America

Yiran Liang, Peng Wang, Donghui Liu, Zhigang Shen, Hui Liu and Zhiqiang Zhou
Department of Applied Chemistry, China Agricultural University, Beijing, PR China

Zhixin Jia
Institute of Materia Medica, Chinese Academy of Medical Sciences and Peking Union Medical College, Beijing, PR China

Marco Katzenberger, Helder Duarte and Miguel Tejedo
Department of Evolutionary Ecology, Doñana Biological Station - Spanish Council for Scientific Research, Sevilla, Spain

John Hammond
Department of Biology, University of New Mexico, Albuquerque, New Mexico, United States of America

Cecilia Calabuig
Department of Animal Sciences, Federal Rural University of the Semiarid Region, Mossoró , Rio Grande do Norte, Brazil

Rick A. Relyea
Department of Biological Sciences, University of Pittsburgh, Pittsburgh, Pennsylvania, United States of America

Sébastien Marcombe
Center for Vector Biology, Rutgers University, New Brunswick, New Jersey, United States of America

Pasteur Institute, Vientiane, Laos Ary Farajollahi
Center for Vector Biology, Rutgers University, New Brunswick, New Jersey, United States of America
Mercer County Mosquito Control, West Trenton, New Jersey, United States of America

Sean P. Healy
Monmouth County Mosquito Extermination Commission, Eatontown, New Jersey, United States of America
Department of Entomology, Louisiana State University Agricultural Center, Baton Rouge, Louisiana, United States of America

Gary G. Clark
Mosquito and Fly Research Unit, Agriculture Research Service, United States Department of Agriculture, Gainesville, Florida, United States of America

Dina M. Fonseca
Center for Vector Biology, Rutgers University, New Brunswick, New Jersey, United States of America

Yu Yu, Feng-Chiao Su and Stuart A. Batterman
Environmental Health Sciences, University of Michigan, Ann Arbor, Michigan, United States of America

Brian C. Callaghan, Stephen A. Goutman and Eva L. Feldman
Department of Neurology, University of Michigan, Ann Arbor, Michigan, United States of America

Yi Xie, Binbin Wang, Fanchi Li, Lie Ma, Min Ni, Weide Shen and Bing Li
School of Basic Medicine and Biological Sciences, Soochow University, Suzhou, Jiangsu, P.R. China
National Engineering Laboratory for Modern Silk, Soochow University, Suzhou, Jiangsu, P.R. China

Fashui Hong
School of Basic Medicine and Biological Sciences, Soochow University, Suzhou, Jiangsu, P.R. China

Ken Tan
Key Laboratory of Tropical Forest Ecology, Xishuangbanna Tropical Botanical Garden, Chinese Academy of Science, Kunming, Yunnan Province, China
Eastern Bee Research Institute, Yunnan Agricultural University, Heilongtan, Kunming, Yunnan Province, China

Weiwen Chen, Shihao Dong, Xiwen Liu and Yuchong Wang
Eastern Bee Research Institute, Yunnan Agricultural University, Heilongtan, Kunming, Yunnan Province, China

James C. Nieh
Division of Biological Sciences, Section of Ecology, Behavior, and Evolution, University of California San Diego, La Jolla, California, United States of America

Oriol Vall and Oscar Garcia-Algar
Unitat de Recerca Infáncia i Entorn (URIE), Institut Hospital del Mar d'Investigacions Médiques (IMIM), Barcelona, Spain
Red de Salud Materno-Infantil y del Desarrollo (SAMID), Instituto Carlos III, Madrid, Spain
Departament de Pediatria, Obstetricia, Ginecologia i Medicina Preventiva, Universitat Autónoma de Barcelona, Barcelona, Spain

Mario Gomez-Culebras and Ernesto Rodriguez-Carrasco
Departamento de Cirugía Pediátrica, Hospital de la Candelaria, Universidad de Tenerife, Santa Cruz de Tenerife, Spain

Carme Puig and Xavier Joya
Unitat de Recerca Infáncia i Entorn (URIE), Institut Hospital del Mar d'Investigacions Médiques (IMIM), Barcelona, Spain
Red de Salud Materno-Infantil y del Desarrollo (SAMID), Instituto Carlos III, Madrid, Spain

Arelis Gomez Baltazar and Lizzeth Canchucaja
Unitat de Recerca Infáncia i Entorn (URIE), Institut Hospital del Mar d'Investigacions Médiques (IMIM), Barcelona, Spain
Departament de Pediatria, Obstetricia, Ginecologia i Medicina Preventiva, Universitat Autónoma de Barcelona, Barcelona, Spain

Meng-Xiao Lu, Xian-Jin Liu and Xiang-Yang Yu
Pesticide Biology and Ecology Research Center, Nanjing, Jiangsu, China

Key Laboratory of Food Safety Monitoring and Management of Ministry of Agriculture, Nanjing, Jiangsu, China

Wayne W. Jiang
Department of Entomology, Michigan State University, East Lansing, Michigan, United States of America

Jia-Lei Wang
Pesticide Biology and Ecology Research Center, Nanjing, Jiangsu, China

Qiu Jian
Institute for the Control of Agrochemicals, Ministry of Agriculture, Beijing, China

Yan Shen
Key Laboratory of Food Safety Monitoring and Management of Ministry of Agriculture, Nanjing, Jiangsu, China

Chunrong Xiong
Department of Pathogen Biology, Nanjing Medical University, Nanjing, Jiangsu, China
Jiangsu Province Key Laboratory of Modern Pathogen Biology, Nanjing, Jiangsu, China
Jiangsu Institute of Parasitic Diseases, Wuxi, Jiangsu, China

Fujin Fang, Lin Chen, Dan Zhou, Bo Shen, Lei Ma, Yan Sun, Donghui Zhang, Changliang Zhu
Department of Pathogen Biology, Nanjing Medical University, Nanjing, Jiangsu, China
Jiangsu Province Key Laboratory of Modern Pathogen Biology, Nanjing, Jiangsu, China

Qinggui Yang
Department of Pathogen Biology, Nanjing Medical University, Nanjing, Jiangsu, China
Jiangsu Province Key Laboratory of Modern Pathogen Biology, Nanjing, Jiangsu, China
National Key Laboratory of Vector Biology, Jiangsu Entry-Exit Inspection and Quarantine Bureau, Nanjing, Jiangsu, China

Ji He
Department of Pathogen Biology, Nanjing Medical University, Nanjing, Jiangsu, China
Jiangsu Province Key Laboratory of Modern Pathogen Biology, Nanjing, Jiangsu, China
National Key Laboratory of Surveillance and Detection for Medical Vectors, Xiamen Entry–Exit Inspection and Quarantine Bureau, Xiamen, Fujian, China

Wanyi Zhu, Christopher A. Mullin and James L. Frazier
Department of Entomology, Center for Pollinator Research, The Pennsylvania State University, University Park, Pennsylvania, United States of America

Daniel R. Schmehl
Honey Bee Research and Extension Laboratory, Department of Entomology and Nematology, University of Florida, Gainesville, Florida, United States of America

Armel Djénontin
Faculté des Sciences et Techniques/MIVEGEC (IRD 224-CNRS 5290-UM1-UM2), Université d'Abomey Calavi/Centre de Recherche Entomologique de Cotonou (CREC), Cotonou, Bénin

Cédric Pennetier, Barnabas Zogo, Koffi Bhonna Soukou and Marina Ole-Sangba
MIVEGEC (IRD 224-CNRS 5290-UM1-UM2), Centre de Recherche Entomologique de Cotonou (CREC), Cotonou, Bénin

Martin Akogbéto
Faculté des Sciences et Techniques/Centre de Recherche Entomologique de Cotonou (CREC), Université d'Abomey Calavi/Centre de Recherche Entomologique de Cotonou (CREC), Cotonou, Bénin

Fabrice Chandre
MIVEGEC (IRD 224-CNRS 5290-UM1-UM2), Laboratoire de lutte contre les Insectes Nuisibles (LIN), Montpellier, France

Rajpal Yadav
Department of Control of Neglected Tropical Diseases, World Health Organization, Geneva, Switzerland

Vincent Corbel
MIVEGEC (IRD 224-CNRS 5290-UM1-UM2)/ Department of Entomology, Kasetsart University, Ladyaow Chatuchak Bangkok, Thailand

Ying Zhang1, Song Han, Duohong Liang, Xinzhu Shi and Fengzhi Wang
Department of Epidemiology, Public Health School, Shenyang Medical College, Shenyang, China

Wei Liu, Li Zhang, Lixin Chen and Yingzi Gu
Department of Obstetrics, Central Hospital Affiliated to Shenyang Medical College, Shenyang, China

Ying Tian
School of Medicine, Shanghai JiaoTong University, Shanghai, China

Jason M. Hill and Glenn E. Stauffer
Pennsylvania Cooperative Fish and Wildlife Research Unit, Pennsylvania State University, University Park, Pennsylvania, United States of America

J. Franklin Egan
USDA-ARS Pasture Systems and Watershed Management Research Unit, University Park, Pennsylvania, United States of America

Duane R. Diefenbach
U.S. Geological Survey, Pennsylvania Cooperative Fish and Wildlife Research Unit, Pennsylvania State University, University Park, Pennsylvania, United States of America

Flora Mayhoub
Laboratoire PériTox, Unité mixte Université – INERIS (EA 4285-UMI 01), Université de Picardie Jules Verne, Amiens, France
Faculty of Medicine, Tishreen University, Latakia, Syria

Thierry Berton
Laboratoire PériTox, Unité mixte Université – INERIS (EA 4285-UMI 01), Université de Picardie Jules Verne, Amiens, France
Unité NOVA, Institut National de l'Environnement Industriel et des Risques, Verneuil en Halatte, France

Véronique Bach, Erwan Stéphan-Blanchard and Karen Chardon
Laboratoire PériTox, Unité mixte Université – INERIS (EA 4285-UMI 01), Université de Picardie Jules Verne, Amiens, France

Karine Tack
Unité NOVA, Institut National de l'Environnement Industriel et des Risques, Verneuil en Halatte, France

Caroline Deguines
Laboratoire PériTox, Unité mixte Université – INERIS (EA 4285-UMI 01), Université de Picardie Jules Verne, Amiens, France
Médecine Néonatale, Pôle Femme-Couple-Enfant, Centre Hospitalier Universitaire d'Amiens, Amiens, France

Adeline Floch- Barneaud
Laboratoire PériTox, Unité mixte Université – INERIS (EA 4285-UMI 01), Université de Picardie Jules Verne, Amiens, France
Unité ISAE, Institut National de l'Environnement Industriel et des Risques, Verneuil en Halatte, France

Sophie Desmots
Laboratoire PériTox, Unité mixte Université – INERIS (EA 4285-UMI 01), Université de Picardie Jules Verne, Amiens, France
Unité TOXI, Institut National de l'Environnement Industriel et des Risques, Verneuil en Halatte, France

Carlos Guerrero-Bosagna
Center for Reproductive Biology, School of Biological Sciences, Washington State University, Pullman, Washington, United States of America
Department of Physics, Biology and Chemistry, Linkö ping University, Linkö ping, Sweden

Shelby Weeks and Michael K. Skinner
Center for Reproductive Biology, School of Biological Sciences, Washington State University, Pullman, Washington, United States of America

Jagan Mohan Jasna, Kannadasan Anandbabu, Subramaniam Rajesh Bharathi, Narayanasamy Angayarkanni
R.S Mehta Jain Department of Biochemistry and Cell Biology, KBIRVO Block, Vision Research Foundation, Sankara Nethralaya, Chennai, India

Marija G. Matic, Vesna M. Coric, Ana R. Savic-Radojevic, Marija S. Pljesa-Ercegovac, Tatjana I. Djukic and Tatjana P. Simic
Institute of Medical and Clinical Biochemistry, Faculty of Medicine, University of Belgrade, Belgrade, Serbia
Faculty of Medicine, University of Belgrade, Belgrade, Serbia

Petar V. Bulat
Institute of Occupational Health, Belgrade, Serbia
Faculty of Medicine, University of Belgrade, Belgrade, Serbia

Dejan P. Dragicevic
Clinic of Urology, Clinical Center of Serbia, Belgrade, Serbia
Faculty of Medicine, University of Belgrade, Belgrade, Serbia

Tatjana D. Pekmezovic
Institute of Epidemiology, Faculty of Medicine, University of Belgrade, Belgrade, Serbia
Faculty of Medicine, University of Belgrade, Belgrade, Serbia

Miao Yuan, Yanhui Lu, Xun Zhu, Hu Wan, Muhammad Shakeel, Sha Zha and Jianhong Li
Laboratory of Pesticide, College of Plant Science & Technology, Huazhong Agricultural University, Wuhan, China

Byung-Rae Jin
Laboratory of Insect Molecular Biology and Biotechnology, Department of Applied Biology, College of Natural Resources and Life Science, Dong-A University, Busan, Korea

Index

A

Aedes Albopictus, 38-39, 41-47

Amyotrophic Lateral Sclerosis (als), 62-63, 65, 67, 69

Anopheline, 120-121, 125-126

Antidepressants, 93

Arable Farming, 1-3, 5, 7, 9-10

Asparagus Lettuce, 96-103

B

Bee Avoidance, 81, 83, 85, 87

Birth Outcomes, 92, 145-147, 149, 151, 153, 155, 157

Bombyx Mori, 71, 78-80

Brain Ultrastructure Evaluation, 72

Brassica Chinensis, 96-103

Breast Milk Analysis, 89, 91, 93, 95

Brown Planthopper (bph), 191

C

Canary Islands, 89-95

Chemicals And Reagents, 19

Chikungunya, 38, 46

Chlorpyrifos, 32, 96-103, 109-110, 112-117, 119, 127, 134-136, 173-175, 177, 179-182, 192

Chromosomic Alteration, 92

Cognitive Impairments, 81

Common Pesticides, 109, 111, 113, 115, 117-119

Composite Indices, 48-49, 51-53, 55, 57, 59, 61

Computation Of Sustainability Indicators, 9

Crop Diversity, 3, 8

Cropping Systems, 1-10

Culicine Mosquitoes, 120-121, 125

Cultivated Soybean, 11, 13, 15, 17

D

Determination Of Neurochemicals, 78

Dichlorodiphenyltrichloroethane (ddt), 38, 89, 95, 162

Digital Gene Expression Profile (dge), 71-72

Dissipation, 96-97, 99-103

E

Eastern Black Nightshade, 16

Economic Sustainability, 3

Enantioselective Metabolism, 19, 21, 23, 25

Energy Efficiency, 2-8, 10, 49, 60

Enhanced Pesticide Use, 11, 13, 15, 17

Environmental Risk Factors, 62-63, 65, 67-70

Environmental Stressors, 27, 29, 31, 33, 35-37

Environmental Sustainability, 1, 3, 5, 7, 9, 49

Enzymatic Phenotyping Of Ache1, 42-43

Epigenetic Transgenerational, 159, 161, 163, 165, 167-172

Epithelium, 69, 173, 175, 177, 179, 181-182, 184

F

Field Efficacy, 120-121, 125

Formulation Solvent, 109, 111, 113, 115, 117, 119

G

Geographic Population, 191-192, 199

Glutathione S-transferase, 183, 185, 187, 189-190

H

Habitat Availability, 137-141, 143

Histopathological Evaluation, 72, 78

Hive Environment, 109, 111, 113, 115, 117-119

Honey Bee Larvae, 109-111, 113-119

I

Imidacloprid Alters, 81, 83, 85, 87

Imidacloprid Concentrations, 82-86

Important Economic Insects, 71

Insecticide Acute Toxicity, 137-143

Insecticide Resistance, 38-39, 41, 43-47, 125

Intravenous Administration, 19-20, 23-24

L

Larval Bioassays, 39, 42, 44

Latitudinal Gradient, 11, 14

M

Maximum Residue Limits (mrls), 96, 100, 102

Melipona Quadrifasciata Anthidioides, 82, 87

Metabolites Measurements, 128

Molecular Mechanisms, 71, 73, 75, 77, 79, 160, 164-165

Morphology Of The Tadpoles, 29, 31

Mosquito Larvicide, 120-121, 125

Mosquito Strains, 39, 46

Multiple Correspondence Analysis (mca), 48, 50

N

Nanoparticles, 71, 73, 75, 77, 79-80

Nectar Collection Experiment, 82, 84

Neonatal Behavioral Neurological Assessment (nbna), 127, 129

Nerve Toxicity, 71-75, 77, 79

Neurobehavioral Development, 127-129, 131, 133, 135-136

Neurotransmitter Contents, 72, 75

Nilaparvata Lugens, 191, 199-200

O

Occupational Exposure, 62-64, 66-67, 69, 130, 145-149, 151, 153-157, 183-190

Organic Cropping Systems, 2-6

Organochlorine, 26, 32, 40, 42, 89, 94-95, 135

Organochlorine Compounds (ocs), 89

Organophosphate, 32, 38-40, 44-47, 79, 102, 108-110, 117-118, 127, 129, 131, 133, 135-136, 158, 173, 177, 179, 182, 184

Organophosphate Pesticides, 108, 127, 129, 131, 133, 135-136, 177, 179

Oxidative Stress, 65, 71, 73-74, 76-80, 132, 173, 175, 177, 179-182

P

Paraoxonase Enzyme, 173, 175, 177, 179, 181

Parental Exposure, 145, 147, 149, 151, 153, 155, 157

Pesticide Reduction, 1, 3, 5, 7, 9

Phoxim-exposed Brain, 71, 73, 75, 77, 79

Polymorphisms, 183-186, 189-190

Postnatal Exposure, 89, 91, 93, 95

Pro-environmental Behaviours, 48-49, 51-59, 61

Q

Quizalofop-ethyl, 19, 21, 23, 25-26

R

Resistance Measurement, 78

Reverse-transcription Quantitative Pcr, 191

S

Socio-demographic Variables, 50-52, 54, 57-58, 89

Sperm Epimutations, 159-160, 162, 164-165, 170-172

Swimming With Predators, 27, 29, 31, 33, 35, 37

T

Thermal Performance, 27-32, 35-37

Thermal Physiology, 27-29, 31-33, 35-37

Titanium Dioxide, 71, 73, 75, 77, 79

Trypsin-catalyzed Deltamethrin, 104-105, 107

V

Vectobac Gr, 120-121, 124-125

W

Weightlifting, 63, 66

www.ingramcontent.com/pod-product-compliance
Lightning Source LLC
Chambersburg PA
CBHW080654200326
41458CB00013B/4855